Hydrogen Safety

While hydrogen is of vital and growing importance in many industrial sectors, this volatile substance poses unique challenges, including easy leakage, low ignition energy, a wide range of combustible fuel–air mixtures, buoyancy, and its ability to embrittle metals that are required to ensure safe operation. Updated to include the latest advances in the decade since original publication, *Hydrogen Safety, Second Edition* highlights physiological, physical, and chemical hazards associated with hydrogen production, storage, distribution, and usage systems. Focused on providing a balanced view of hydrogen safety – one that integrates principles from physical sciences, engineering, management, and social sciences – this book is organized to address questions associated with the hazards of hydrogen and the ensuing risks associated with its industrial and public use. This book:

- Addresses issues of inherently safer design, safety management systems, and safety culture.
- Features updated case studies of significant accidents involving hydrogen, along with their detailed analysis and lessons learnt, and potential accident scenarios under certain conditions.
- Details current research trends and perspectives on materials-based hydrogen storage solutions, hydrogen use in vehicles, and hydrogen in construction materials.
- Describes Process Safety Management as applied to the process industries, in conjunction with the components of the US Department of Energy Safety Plant Elements for hydrogen safety, and covers activities of the European Commission (EC) Network of Excellence for Hydrogen Safety (HySafe).
- Includes updated codes for gaseous and liquefied hydrogen and the NFPA 2 Hydrogen Technologies Code.
- Concludes with research and legal requirements.

Offering a holistic view of hydrogen safety, from properties to safety systems, this book helps readers in chemical, industrial, safety, and related engineering subjects ensure a safe application and environment.

Fotis Rigas is a Professor of Chemical Engineering at the National Technical University of Athens in Greece and has been a Visiting Professor at the National Autonomous University of Mexico. His current research and academic activities are in the areas of process safety, chemical engineering kinetics, and bioremediation of contaminated sites. He has published or presented over 180 papers in the fields of his activities and is a reviewer of papers in 54 international scientific journals. He is within the top 2% worldwide of scientists in their main subfield discipline according to the latest Stanford University bibliometrics study (2023).

Hydrogen Safety

Production, Transport, Storage, Use, and the Environment

Second Edition

Fotis Rigas

CRC Press
Taylor & Francis Group
Boca Raton London New York

CRC Press is an imprint of the
Taylor & Francis Group, an **informa** business

Designed cover image: © Shutterstock, Scharfsinn

Second edition published 2025
by CRC Press
2385 NW Executive Center Drive, Suite 320, Boca Raton FL 33431

and by CRC Press
4 Park Square, Milton Park, Abingdon, Oxon, OX14 4RN

CRC Press is an imprint of Taylor & Francis Group, LLC

© 2025 Fotis Rigas

First edition published by CRC Press 2013

ISBN: 978-1-032-31723-6 (hbk)
ISBN: 978-1-032-32134-9 (pbk)
ISBN: 978-1-003-31300-7 (ebk)

DOI: 10.1201/9781003313007

Typeset in Times
by Newgen Publishing UK

Contents

Series Preface

Green Chemistry and Chemical Engineering
A Book Series by CRC Press/Taylor & Francis

Towards the late 20th century, the development of various environmentally friendly processes, techniques, and methodologies saw significant growth in the scientific community. The main driving force for such growth was due to the rising awareness of sustainable development, more stringent environmental regulation, and increasing costs of raw materials and waste treatment. After several decades of development, we now broadly term these environmentally friendly processes, techniques, and methodologies as *green/clean technologies.*

In the 21st century, the global community has experienced many extreme weather events such as prolonged drought, extreme heat, tornadoes, and wildfires. The scientific community believes that these extreme weather events are closely linked to climate change, and they are expected to increase in frequency and intensity in the future. Following the Paris Agreement and Glasgow Climate Pact, there is now an international commitment to limit the rise of global temperature to well below 2°C by the end of this century and to pursue efforts to limit temperature increase to 1.5°C above pre-industrial levels. Hence, it is believed that green/clean technologies will have a much bolder role to play in combating global climate change in the coming years.

It is also worth mentioning that the United Nations Sustainable Development Goals (UNSDG) that were launched in 2015 define 17 important goals to transform the world by 2030. It is believed that some of these goals may be addressed with the development of green/clean technologies. These include:

- Goal 6 – Clean Water and Sanitation: Ensure access to water and sanitation for all
- Goal 7 – Affordable and Clean Energy: Ensure access to affordable, reliable, sustainable, and modern energy
- Goal 12 – Responsible Consumption and Production: Ensure sustainable consumption and production patterns
- Goal 13 – Climate Action: Take urgent action to combat climate change and its impacts

The *Green Chemistry and Chemical Engineering* book series by CRC Press/Taylor & Francis focuses on the subset of green technologies dedicated to addressing the "2E" agenda, i.e., *environment* and *energy*. It involves the development of materials (e.g., catalysts, nanomaterials), methodologies (e.g., process optimization, footprint reduction, artificial intelligence), and processes (e.g., waste treatment) that will bring forth solutions to address pressing problems such as:

- Greenhouse gas management and reduction
- Sustainable water production
- Wastewater treatment and recycling
- Circular economy and waste reduction
- Renewable energy
- Sustainable use of energy resources

I am hopeful that this *Green Chemistry and Chemical Engineering* series will serve as a de facto source of reference materials and practical guides for academics and industrial practitioners looking to advance the discipline and aims of green chemistry and chemical engineering.

Dominic C. Y. Foo
Centre for Green Technologies
University of Nottingham, Malaysia

Preface

This second edition of the book *Hydrogen Safety* on the safe handling and use of hydrogen is intended, as much as its first edition, to provoke thought and dialogue as well as to provide practical guidance. The necessity for the new edition stemmed from the good reception by the scientific community of the first edition in 2013, the numerous research studies and published papers during the last decade, and also from the fact that the first edition was translated last year into Chinese by Professor Huixing Meng at the School of Mechatronical Engineering (Beijing Institute of Technology, Beijing, China).

The author Fotis Rigas hopes that readers will find the information presented in the book to be helpful in their endeavors to bring the envisaged "hydrogen economy" into reality. But he also hopes that important areas of safety research, often viewed as being nontechnological, will be welcomed into the discussion of hydrogen safety along with the more familiar technological topics.

The book is organized to first address questions associated with the hazards of hydrogen and the ensuing risks in its use within industry and by the public. What are the properties of hydrogen that can render it a hazardous substance? How have these hazards historically resulted in undesired incidents? How might these hazards arise in the storage of hydrogen and with its use in vehicular transportation?

The issues of inherently safer design follow and, in accordance with the previous comment on so-called nontechnological topics, safety management systems and safety culture. The European Commission (EC) Network of Excellence for Hydrogen Safety, HySafe, is singled out for separate coverage, as are various case studies associated with hydrogen and constructional materials. In the "Epilogue", a brief look at future research requirements and current legal requirements for hydrogen safety are presented.

Furthermore, new sections addressing the environmental concerns about climate change raised by the anticipated highly increased production and use of hydrogen due to the forthcoming hydrogen economy have been added in the second edition.

The approach, then, has been to attempt a balanced view of hydrogen safety. Such a perspective comes not only from the author's areas of expertise, but also from a collective belief that the safety of any material or activity is most effectively addressed through a combination of the physical sciences and engineering principles, together with the management and social sciences as well as the environmental issues. Non-hydrogen-related industries have undergone painful experiences in this regard; there should be no need for industrial applications involving hydrogen to experience the same challenges.

Author Fotis Rigas would like to express his gratitude to Allison Shatkin, Senior Publisher in Materials Science and Chemical Engineering of CRC Press/Taylor & Francis, for entrusting him for a second time with the task of writing this book.

1 Introduction

1.1 GENERAL CONSIDERATIONS

In examining the history of hydrogen usage, it is perhaps informative to turn to the development of another energy resource, petroleum. Edwin Drake drilled the first oil well in Pennsylvania in 1859 to use extracted oil as a substitute for whale oil, the main lighting source and feedstock for consumer and chemical products at that time. That resource was hazardous to obtain and dwindling because of heavy exploitation, as is currently the case with oil. Petroleum presented many advantages over whale oil and solved a great many ecological and resource security problems associated with the old resource. Moreover, the use of petroleum as an energy source was instrumental in addressing the enormous problems caused by the employment of horses for transportation in big cities that created the so-called "Great Horse-Manure Crisis of 1894". It was estimated that in 50 years after that date, every street in London would be buried under nine feet of manure! Thus, the problem of excessive horse manure thrown down in the streets was dealt with by the utilization of cars in big cities like London and New York. Nevertheless, after a century and a half of use, petroleum has created new problems related to environmental pollution and energy security [1]. To address these problems, efforts have been dedicated to the gradual substitution of all ordinary fossil fuels (oil, coal, and natural gas) with new energy vectors, primarily because of the exhaustion of natural resources and the simultaneous rapid increase of energy demands worldwide.

Hydrogen is considered one of the most promising fuels for generalized use in the future, mainly because it is an energy-efficient, low-polluting, and renewable fuel. Hydrogen is versatile and clean, and with environmental benefits in mind, the production of hydrogen from renewable energy sources such as biomass, wind, solar, and nuclear sources is under consideration [2, 3]. Yet discussion of potential problems that widespread use of hydrogen might bring, as with whale oil and petroleum before it, is lacking. The identification and control of potential adverse consequences is an ethical requirement whenever any new technology is introduced, and certainly in the case of hydrogen, with such world-influencing potential [1].

DOI: 10.1201/9781003313007-1

The strategic areas of research under investigation and financed by research organizations worldwide with regard to hydrogen are the following:

- Clean hydrogen production from existing and novel processes.
- Storage, including hybrid storage systems.
- Basic materials, including those for electrolyzers, fuel cells, and storage systems.
- Safety and regulatory issues required for the preparation of regulations and safety standards at a global level.
- Societal issues, including public awareness and preparations for the transition to a hydrogen energy economy.

Hydrogen could be an effective energy carrier if it can be rendered economically competitive and connected with an infrastructure providing a safe and environmentally acceptable energy system throughout the whole production, distribution, and end use chain. Global hydrogen demand is projected to increase from 70 million tonnes in 2019 to 120 million tonnes by 2024 [4]. Hydrogen has been, for more than a century, produced and used with a high safety record for commercial and industrial purposes, such as refinery and chemical processes and rocket propulsion, or as a radiolytically produced byproduct in nuclear boiling water reactors. Industry has significant experience in the safe handling of hazardous materials in chemical plants, where only well-trained personnel come into contact with hydrogen. Nevertheless, the widespread use of hydrogen as an energy carrier will result in its use by laypersons, necessitating different safety regulations and technologies, which are now under development [5, 6].

One of the major issues affecting the acceptance of hydrogen for public use is the safety of hydrogen installations (production and storage units), as well as its applications (i.e., as vehicle fuel or for home use). The hazards associated with the use of hydrogen can be characterized as physiological (frostbite and asphyxiation), physical (component failures and embrittlement), and chemical (burning or explosion), the primary hazard being inadvertently producing a flammable or explosive mixture with air [7, 8]. As far as European countries are concerned, hazardous chemical installations come under the so-called Seveso III Directive (2012/18/EU) for the control of major-accident hazards involving dangerous substances. Hydrogen is included in this directive, indeed with a stricter minimal quantity for the directive's implementation than any other ordinary fuel [9, 10].

In recent works, theoretical [11] and computational [12] safety comparisons between hydrogen and other fuels do not allow a clear conclusion to be made of which is the safest. Indeed, in the past, there were circumstances in which hydrogen applications gave rise to severe accidents with significant economic and societal cost, affirming the need for augmented safety measures wherever hydrogen is handled [13]. Undoubtedly, the need for safety measures should be pointed out when loss prevention and public safety are concerned. This supposes the knowledge of potential hazards plus the determination of risk zones around installations handling hydrogen.

As noted by Guy [14] in his article *The hydrogen economy*, hydrogen is largely produced as a synthesis gas for use in chemical production (e.g., ammonia and methanol) or recovered as a byproduct for use in oil refineries. He further comments that while the safe handling of hydrogen by industry (especially industrial gas companies) is well understood, use of hydrogen in the public realm can be problematic. This book demonstrates that the need for safer production, storage, distribution, and use of hydrogen in all application sectors must indeed be well understood and acted upon if the envisaged hydrogen economy is to materialize and endure.

1.2 ADVANTAGES AND DISADVANTAGES OF HYDROGEN

For hydrogen to be used as a fuel and an energy carrier in the new "hydrogen economy" era, its advantages and disadvantages should be taken into account first [15]:

Advantages of hydrogen

- *Burns cleanly, releasing only water and energy.*
- *Stores more energy per unit weight than most other fuels.*
- *Can be produced from low-carbon sources.*
- *Can be used as a fuel, as an energy carrier, for energy storage, or as a chemical feedstock.*
- *Can be used to decarbonize "hard to abate" sectors.*
- *Offers wider benefits for energy security, industrial strategy, and air quality.*

Disadvantages of hydrogen

- *Almost all production nowadays is from high-carbon sources.*
- *Currently expensive production and uncertain cost reductions.*
- *Bulky and expensive to transport and store.*
- *Low production efficiency and large overall energy production requirements.*
- *Complex supply and value chains needed for its use.*
- *New safety standards and societal acceptance needed.*

Yet, beyond the significant advantages, it should be emphasized that hydrogen comes with safety hazards, high infrastructure costs, and similar issues as fossil fuel supply and distribution, especially when considering carbon capture, use, and storage. Although there is significant consensus that hydrogen is necessary to avoid risky climate change by reaching net-zero emissions, there is a lack of agreement on how much hydrogen will be needed to achieve this target and in which sectors we will use it. In any case, hydrogen is an "energy carrier" or "energy vector" like electricity. However, in contrast to electricity, the energy stored in hydrogen is captured in a stable chemical compound rather than a more short-lived electrical charge. In this way, its energy is easier to store, transport, and convert into other fuels or chemical feedstocks [15].

1.3 PRODUCTION METHODS OF HYDROGEN AND ITS COLORFUL WORLD

According to Dawood et al. [16], hydrogen is a versatile fuel that can be produced using all types of energy sources (coal, oil, natural gas, biomass, renewables, and nuclear) via a variety of technologies, such as reforming, gasification, electrolysis, pyrolysis, water-splitting, photolysis, biolysis, and thermolysis. Thus, the *production of hydrogen* by most other technologies involves substantial use of fossil fuels and CO_2 emissions. Around 2.5–5 metric tons of carbon released as CO_2 per ton of hydrogen is currently produced by conventional means.

Hence, hydrogen's production can be carbon-free, only if genuinely carbon-free renewable energy sources are used for production. Although the use of *renewable* hydrogen instead of fossil fuels is the best way to prevent global warming, and despite all intensive efforts to render renewable hydrogen cost-effective, none of the available technologies were till 2019 economically viable [17, 18]. However, the situation has greatly changed since the Ukraine war in 2022.

As regards green hydrogen, according to Bethoux [19], *free hydrogen found on Earth* is also remarkable. Although it has until recently been assumed that it is impossible for hydrogen to exist as naturally occurring molecular hydrogen, several hydrogen emissions sources have been discovered. The first occurrences have been spotted on the seabed, yet it is not economically feasible to exploit them. Recently, other emanations have been identified in some places, such as Iceland, Sicily, California, and Russia. The origin of these sources has not yet been clarified. However, satellites are now actively searching for such hydrogen-emitting sources. To this end, the European Space Agency (ESA) initiated the sen4H2 project in 2019 to detect naturally occurring hydrogen sources on Earth.

Green hydrogen through water electrolysis production is emerging as one of the main and best alternatives to replace the use of fossil fuels and thus mitigate environmental pollution and its consequences for the planet. A sizable expansion of electrolytic hydrogen is expected in the coming years: in 2050, 22% of total worldwide hydrogen production will come from this route [20].

A water electrolyzer is an electrochemical device that converts electric and thermal energy into chemical energy stored in a fuel (hydrogen) [21]. In the water electrolysis process, water dissociates into hydrogen and oxygen under the influence of direct current [22].

Electrolyzers are expected to compete with fossil-based hydrogen in 2030 when electricity generated by renewable energy sources will be possibly cheaper, and industry in the European Union (EU) is planned to reach 40 GW capacities [23]. For this purpose, the principles of the electrolysis process are established, as well as the different ways to carry it out, among which are: alkaline electrolysis (AE); proton exchange membrane (PEM) electrolysis, and high-temperature electrolysis (HTE).

The associated *hazards and risks of electrolysis* have been investigated, and the Dow Fire and Explosion Index (F&EI) has been calculated for the three electrolysis methods, obtaining similar results for each. In addition, the Canadian Society for Chemical Engineering (CSChE) Process Safety Management (PSM) standard and the main international standards currently apply to electrolytic hydrogen production systems, such as ISO 31000:2018, ISO 15916:2015, and ISO 22734:2019. The Dow F&EI calculation indicates that the electrolysis processes mentioned have a similar risk of fire and explosion. But the electrolysis process through PEM presents the highest value, due to the process in which the highest pressures have to be reached. However, all processes can be executed safely and efficiently by applying the standards and preventive measures. It is established that the CSChE PSM standard applies to any process involving hydrogen and, therefore, it also applies to the electrolytic process. Moreover, this standard fits international standards, such as ISO 31000:2018, ISO 15916:2015, and ISO 22734:2019, which should apply in the electrolysis process.

Considering a hydrogen strategy for a climate-neutral approach, the EU defines *renewable or clean hydrogen* as hydrogen produced through water electrolysis and with the required electricity originating from renewable sources. The full *life-cycle greenhouse gas (GHG) emissions* of renewable hydrogen production are near zero. Low-carbon hydrogen includes fossil-based hydrogen with carbon capture and electricity-based hydrogen with significantly decreased full life-cycle GHG emissions compared to current hydrogen production methods. Today, neither renewable hydrogen (*green hydrogen*) nor low-carbon hydrogen, particularly fossil-based hydrogen with carbon capture (*blue hydrogen*), are cost-competitive against fossil-based hydrogen (*grey hydrogen*) [24].

In a critical review, Vostakola et al. [25] investigated the recent progress in the *integrated green-hydrogen production processes*, examining various aspects of the most promising thermochemical water-splitting cycles, with a particular focus on their capabilities to produce green hydrogen with high performance, redox pair stability, and the technological maturity and readiness necessary for commercial use. According to these authors, the water-splitting method is a potential technology for effectively converting renewable thermal energy sources into green hydrogen. This technique is principally based on recirculating an active material, capable of experiencing multiple reduction–oxidation (redox) steps through an integrated cycle to convert water into hydrogen and oxygen.

There is also so-called *turquoise hydrogen*, made by the thermal cracking of natural gas into hydrogen and solid carbon (molten metal pyrolysis). This new colored hydrogen is beneficial in two ways: it does not require carbon capture and sequestration (CCS), and instead of CO_2, it produces "natural graphite", a critical raw material of specific technological interest for the EU. Big corporations such as Russia's Gazprom and Germany's BASF are funding significant amounts of relevant research, but this technology is still in its infancy [26]. Diab et al. [27] conducted a study on the life-cycle assessment of hydrogen produced by the electrolysis of methane via thermal plasma. They found that using renewable natural gas as a feedstock leads to a negative hydrogen carbon intensity, which is the lowest compared to grey, blue, and even green hydrogen, thus making turquoise hydrogen a game-changer for the energy transition. The study has not included economic parameters to highlight the financial benefits of this method compared to other low-carbon methods like water electrolysis, as well as to assess the benefits of using this process with additional units to produce ammonia and compare it to conventional ammonia processes.

In addition to these, there are also other color codes to identify hydrogen. *Black hydrogen* refers to the gasification of bituminous coal, whereas *brown hydrogen* is produced from the gasification of lignite coal. However, both are very polluting processes, producing CO_2 and CO as byproducts, thus polluting the atmosphere. *Purple hydrogen* is made through chemo-thermal electrolysis splitting of water using nuclear power. *Pink hydrogen* is generated through water electrolysis using electricity from a nuclear power plant. *Red hydrogen* is produced through the high-temperature catalytic splitting of water using nuclear power as an energy source. Finally, *white hydrogen* refers to a naturally occurring form of hydrogen found in the Earth's crust [4, 28–31].

Surprisingly, it has been estimated by Howarth and Jacobson [32, 33] that the GHG footprint of blue hydrogen is more than 20% greater than burning natural gas or coal for heat and some 60% greater than burning diesel oil for heat. In their sensitivity analysis, GHG emissions from blue hydrogen are still greater than from simply burning natural gas and are only 18%–25% less than for grey hydrogen. In contrast, other authors [34, 35] find that blue hydrogen can result in significantly lower CO_2 emissions than the production through natural gas, as long as hydrogen production processes and CO_2 capture technologies are applied that ensure a high CO_2 capture rate, preferably above 90%, and a low-emission natural gas supply chain.

Considering any delay to taking measures to confront the climate crisis makes these worrying to a greater extent, and action should be taken more urgently. Although *blue hydrogen* looks like an interim solution, it is an enormous waste of time and money and is a distraction in our efforts for a substantial and permanent confrontation of global issues related to the increase of carbon dioxide in the atmosphere. The truth is that only sustainable clean energy sources should be used, such as green hydrogen, if we want to obtain our goal of a maximum of 1.5°C increase in temperature by 2050 [36].

To address the question of *how green the national strategies are worldwide*, Cheng and Lee [37] proposed a classification of regulation severity for green hydrogen, investigating the suggested strategies and their related policies, and assessing their severity for decarbonization. Their classification incorporated four parameters: fossil fuel punitive actions, hydrogen certificates, innovation activation, and a coal abandonment plan. Then, the authors classified the hydrogen strategies into the following groups: zero regulatory stringencies, scale first and clean later, and green hydrogen now. They found that most current strategies are of the group "scale first and clean later", with one or more regulatory measures running.

An illustrative hydrogen color spectrum as regards the technologies and the feedstock used to produce hydrogen and their impact on the environment expressed as a GHG footprint is shown in Figure 1.1 [38, 39].

As regards the *radiolytic yield of hydrogen*, the method is not currently considered promising as a commercially acceptable hydrogen production method due to the low energy efficiency of hydrogen production. On the contrary, radiolysis is considered a serious *safety problem* for water reactors since it may result in hydrogen explosions in nuclear reactors, thus jeopardizing the safety of reactors. Nevertheless, nuclear-generated hydrogen is more promising than other sources in a future world energy economy. Still, technical uncertainties exist in nuclear hydrogen processes that need to be addressed through vigorous research and development efforts [40, 41].

Considering *recent innovative hydrogen production methods*, comparative assessments show that geothermal and biomass sources have the highest sustainability performance, followed by hydroelectricity and solar. Coal has the lowest sustainability ranking as a source of hydrogen, followed by nuclear and natural gas. When hydrogen production systems are compared, thermochemical systems seem

	Terminology	Technology		Feedstock/ Electricity source	GHG footprint*
PRODUCTION VIA ELECTRICITY	Green			Wind | Solar | Hydro Geothermal | Tidal	0 $kgCO_2$/kgH_2
	Purple/Pink	Electrolysis		Nuclear	
	Yellow			Mixed-origin grid energy	**17.5** $kgCO_2$/kgH_2 Gas-fired **39.5** $kgCO_2$/kgH_2 Coal-fired
PRODUCTION VIA FOSSIL FUELS	Blue	Natural gas reforming + CCUS Gasification + CCUS		Natural gas | coal	**1** $kgCO_2$/kgH_2 with CCUS (90% capture rate)
	Turquoise	Pyrolysis	Solid carbon (by-product)	Natural gas	**4** $kgCO_2$/kgH_2 with CCUS (56% capture rate)
	Grey	Natural gas reforming			**9** $kgCO_2$/kgH_2 without CCUS
	Brown	Gasification		Brown coal (lignite)	**2** $kgCO_2$/kgH_2 with CCUS (90% capture rate)
	Black			Black coal	**20** $kgCO_2$/kgH_2 without CCUS

* GHG footprint given as a general guide but it is accepted that each category can be higher in some cases. Does not include CO_2 emissions linked to the transmission and distribution of hydrogen to the end users. CCUS captures only the feedstock-related CO_2

Source: IEA 2019 | The future of Hydrogen - Seizing today's opportunities

FIGURE 1.1 An illustrative hydrogen color spectrum. Courtesy of Global Energy Infrastructure (www.globalenergyinfrastructure.com).

Source: [38, 39].

to be the most favorable with the highest rankings, followed by photo-fermentation and artificial photosynthesis [42, 4]. Exotic methods of hydrogen production have been proposed, for instance hydrogen generation from the photocatalytic treatment of wastewater containing pharmaceuticals and personal care products by oxygen-doped crystalline carbon nitride [43].

Another alternative for hydrogen production is the *photocatalytic dehydrogenation of formic acid* which has been proposed recently by Navlani-Garcia et al. [44] instead of the thermal catalysts usually applied in this reaction. This process is considered more environmentally friendly and green as regards sustainability, yet considerable effort is still needed to optimize the method and the efficiency of the photocatalysts. These authors have focused on catalytic systems based on TiO_2, CdS, and C_3N_4.

The *thermochemical cycles* of hydrogen production are classified according to their operating temperatures: low-temperature cycles (<1100°C) and high-temperature cycles (>1100°C). Concerning the low-temperature processes, the copper chlorine cycle is more efficient and less costly for hydrogen production. The zinc oxide and ferrite cycles appear to be more suitable for large-scale high-temperature methods. Nevertheless, some issues, such as energy storage capacity, durability, and cost-effectiveness, should first be addressed before these technologies are scaled up into commercial plants for hydrogen production. Large-scale hydrogen production through nuclear plants is currently being studied, particularly through the sulfur–iodine cycle (S–I cycle) and the secondary systems interconnected to it [45]. The S–I cycle is a thermochemical water-splitting process that was first developed by the General

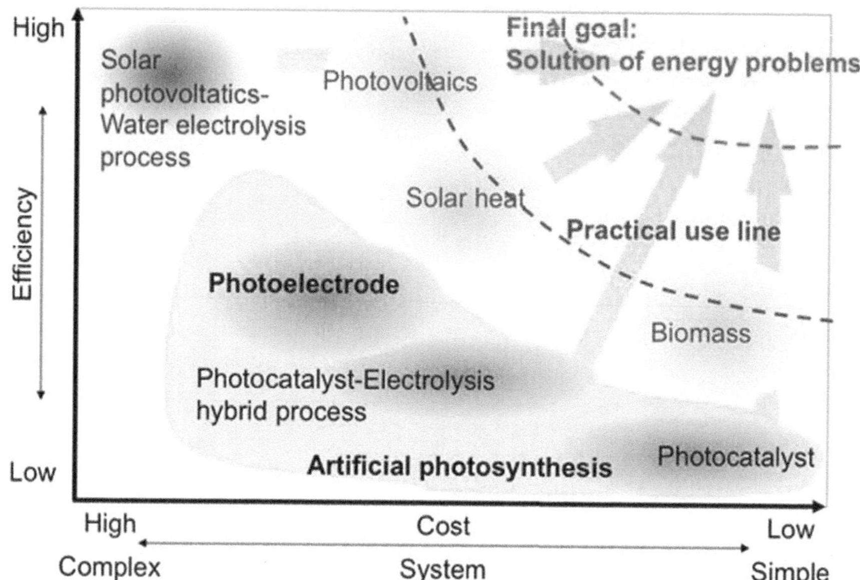

FIGURE 1.2 Comparison of various solar energy conversion technologies. Courtesy of AIST. (www.aist.go.jp/aist_e/list/latest_research/2012/20120511/20120511.html).

Source: [47].

Atomics corporation in the mid-1970s. It has been extensively studied and developed in the past decade, especially by the French Commissariat à l'Energie Atomique, the Institute of Nuclear and New Energy Technology in China, the Korea Institute of Energy Research in South Korea, the Japan Atomic Energy Agency, etc. [25].

A hybrid system to produce *hydrogen from seawater* has also been proposed. The system consists of parabolic trough collectors, a reverse osmosis desalination system, and a thermochemical water decomposition unit with a heat dissipation system using cooling towers [46].

A simplified graphical comparison of various *solar energy conversion technologies* to produce hydrogen is shown in Figure 1.2 as regards the efficiency versus cost and complexity of the hydrogen production system [47].

1.4 PRODUCTION COST ISSUES OF HYDROGEN

A comparative assessment study of *blue hydrogen* for natural gas-producing regions found that hydrogen costs are 1.22, 1.23, 2.12, 1.69, 2.36, 1.66, and 2.55 USD/kg H_2 for steam methane reforming (SMR), autothermal reforming (ATR), and natural gas decomposition (NGD), SMR-52%, SMR-85%, ATR with carbon capture and sequestration (ATR-CCS), and NGD with CCS (NGD-CCS), respectively. Blue hydrogen from ATR has the lowest life-cycle GHG emissions of 3.91 kg CO_2 eq/kg H_2, followed by blue hydrogen from NGD (4.54 kg CO_2 eq/kg H_2), SMR-85% (6.66 kg CO_2 eq/kg H_2), and SMR-52% (8.20 kg CO_2 eq/kg H_2) [48].

Nevertheless, and despite all challenges, *low-carbon hydrogen production can become competitive* within the next decade. The main obstacle to low-carbon hydrogen is the cost gap with hydrogen from fossil fuels. For the time being, the production of hydrogen from fossil fuels is the cheapest method with the levelized cost of hydrogen production from natural gas ranging from USD 0.5 to USD 1.7 per kg. Carbon capture, usage, and storage (CCUS) is a technology that captures the CO_2 emitted by various activities and uses it effectively or stores it safely. Applying CCUS technologies to reduce the CO_2 emissions resulting from hydrogen production increases the levelized cost of production to around USD 1 to USD 2 per kg. Moreover, the use of renewable electricity to produce hydrogen further increases costs from USD 3 to USD 8 per kg [49, 50].

A combination of *solar and wind energy* was used by Armijo and Philibert [51] to model the production of *hydrogen and ammonia from water* in their techno-economic analysis. The case studies presented concerned Chile and Argentina, and the estimated production costs were around 2 USD/kg for green hydrogen and below 0.5 USD/kg for green ammonia, which is quite close to the methods employing production based on fossil fuels.

In another study, Fasihi et al. [52] estimated that by 2030, the *cost of producing ammonia* based on hybrid photovoltaic and wind power would be 370–450 €/tonne in most places, whereas this could be as low as 285–350 €/tonne NH_3 at the best sites by 2050.

Considering the *competition among clean energies* and fossil fuels, the roles of wind, solar, tidal current, geothermal, algal biodiesel, biohydrogen, hydrogen fuel cell, biodiesel, and bioethanol power sources have been investigated by Dincer [8]. The results showed that biohydrogen, hydrogen fuel cells, biofuel, and wind technologies will perform well without external support. It is estimated that biohydrogen and hydrogen fuel cells will lead all clean energies, even under competition with other energy sources. More stringent carbon constraints accompanying reductions in the cost of hydrogen from electrolysis will increase the possibility that natural gas- and coal-based hydrogen production assets could become stranded [53].

Comparing *hydrogen versus electricity*, in terms of energy and economic expenditure of these technologies it is preferable to use electricity in the case of short-term energy storage (less than 50–110 h), whereas in long-term energy storage, the hydrogen system is more efficient. Thus, we may anticipate that the hydrogen economy and electricity economy will coexist in the future energy industry, and each energy carrier will find its specific applications [54]. Hydrogen produced through fossil fuels emits GHGs, even when CCS techniques are applied. Conversely, electrolytically produced hydrogen does not emit GHG but is more expensive. It is anticipated that shortly water electrolysis for hydrogen production based on renewable energy could become cheaper than fossil fuel with CCS technology [53].

Rosenow and Eyre [55] state that because *electricity* cannot serve all end uses, other energy carriers will be needed in a net-zero economy. It is for this reason that in recent years hydrogen has emerged as a topic of considerable interest on the agenda. The informed analysis focuses on energy uses that will be difficult to electrify, in particular industrial processes in which chemical reduction is needed as well as heating (notably primary steel-making) and long-range transportation where battery weight might be prohibitive (e.g. in aviation, shipping, and heavy road freight). In addition, hydrogen may have a role in energy storage, especially where storage times exceed a few hours and therefore battery storage is uneconomical. Hydrogen-powered light vehicles also focus on limited changes to refueling infrastructure. However, most analyses confirm that electrification is a better option in most low-temperature heating and light vehicle options and that broader claims are part of a hype cycle that has now peaked [36]. The fundamental disadvantage of hydrogen is cost. *Hydrogen is not a resource but an energy carrier that needs to be produced.* Because of the transformation costs and energy losses, hydrogen is more expensive than the fossil fuels or electricity from which it is manufactured. The implication is that hydrogen is unlikely to be the preferred zero-carbon fuel where electricity is an option. In sectors in which hydrogen is likely to be used, the high unit cost will provide an additional stimulus for efficiency. For example, it will strengthen the case for efficient fuel cells in heavy freight and long-distance transport. In general, the economic case for energy efficiency in a world with widespread hydrogen use improves rather than diminishes [55].

In addition to the "hydrogen economy", the term *"methanol economy"* has also been introduced by Bockris [56], suggesting methanol as a zero-carbon fuel. This idea is based on the production of methanol from green hydrogen and CO_2 taken from the atmosphere, which when burnt would give back the same amount of carbon to the atmosphere, thus being a zero-carbon fuel. Furthermore, since methanol is a liquid, it would act as a "liquid hydrogen", a name accredited to the Nobelist G.A. Olah.

1.5   STORAGE AND TRANSPORTATION ISSUES OF HYDROGEN

Hydrogen can be *stored and transported* in many ways: as a pressurized gas in huge containers, in pipelines (facing hydrogen embrittlement and escape through steel walls), as a liquid (involving high costs and use of about 35% of the total energy content to liquefy the hydrogen to −253°C), adsorbed in porous media, and in the form of *chemical carriers* with an obvious advantage from the safety point of view. These include liquid organic carriers (such as methylcyclohexane) and liquid inorganic carriers, such as ammonia (NH_3).

Surplus hydrogen can be stored in underground formations such as subsurface aquifers, salt caverns, and/or exhausted hydrocarbon reservoirs in so-called Underground Seasonal Hydrogen Storage (seasonal UHS). Hydrogen in seasonal UHS is stored safely for several months and may be returned for electrification when it is needed. Underground storage of natural gas has been practiced successfully for many decades [4].

Tarkowski and Uliasz-Misiak [57] reviewed the issues concerning the large-scale *underground storage* of hydrogen. The authors stressed that underground storage of large quantities of hydrogen from surplus renewable energy production is of interest to government institutions interested in the construction of hydrogen storage sites, geological services, large renewable energy source electricity producers, and chemical and petrochemical plants. It also offers the possibility of long-term, safe storage of this gas at relatively low costs. Yet the prospect of quick implementation of underground hydrogen storage (UHS) technology on an industrial scale is associated with the definition and overcoming of numerous barriers and obstacles that stand in its way today. Based on the recent literature, they found that the following have been identified as significant barriers to the implementation of UHS: geological and reservoir constraints, technical and safety limitations, legal barriers, conflicts of interest, and social acceptance of UHS.

Zivar et al. [58] emphasized in a review that *underground storage* is a proven way to store an enormous amount of energy (electricity) after converting it into hydrogen as it has higher energy content per unit mass than other gases such as methane and natural gas. The paper reviews the technical aspects and feasibility of the underground storage of hydrogen in depleted hydrocarbon reservoirs, aquifers, and artificial underground cavities (caverns). The mechanisms of UHS, as well as numerous phenomena such as hydrodynamics, geochemical, physiochemical, bio-chemical, or microbial reactions, have been discussed. Modeling studies have also been incorporated into the literature to assess the feasibility of the process that is reviewed in the paper. Finally, technical challenges along with proper remedial techniques and economic viability have been briefly discussed.

Liu et al. [59] performed a feasibility evaluation study of large-scale *underground hydrogen storage* in bedded salt rocks in China, taking the Jiangsu province of China as an example. Large-scale UHS in salt caverns is proposed to realize peak shaving for wind power. A bedded salt rock, the Jintan salt mine in Jiangsu, is selected as a potential site for UHS. To this end, the regional geological conditions of the Jintan salt mine were analyzed. It was found that this mine has good stratigraphic trapping and meets the site-selection requirements for UHS. Then, a numerical simulation model was established to analyze the stability and applicability of the proposed UHS caverns. The tightness of the UHS salt caverns was also investigated. Finally, it was estimated that salt caverns for 36.9 TWh-scale UHS in Jiangsu can be mainly provided by salt mine enterprises, but more attention should be paid to the usability of caverns during extraction.

To determine if a power-to-gas pilot-scale plant would be possible in Oregon, USA, a feasibility study was conducted by Taie et al. [60] that assessed the technical, political, economic, environmental, safety, and policy aspects of a potential project using *UHS* for seasonal energy shifting in the region. There appears to be the political motivation and the technical feasibility to install a demonstration-scale power-to-gas plant in the region to assess the technical and economic performance of the system when exposed to real-world boundary conditions. However, preliminary economic analyses show the system will be challenged by low-capacity factor operation, and that a renewable gas incentive would need to be charged to willing customers who support the decarbonization mission.

The bulk transport of compressed green hydrogen by *submarine pipelines* and compressed and liquefied transport by ship was studied by d'Amore-Domenech et al. [61]. The authors analyzed two methods of liquid hydrogen (LH_2) transport by ship, two methods of compressed hydrogen (cH_2) transport by ship, and two methods of hydrogen transport by pipeline. The results indicated that for a distance of 100 km, the best option is the pipeline. Aimed at 2500 km and 100 kt/y, the best alternative is cH_2 shipping without storage facilities at the ports. For distances of around 2500 km and cargos of 1 Mt/y, and distances of 5000 km and cargos of 100 kt/y, the best options are cH_2 or LH_2 shipping. For all other scenarios, the best option is LH_2 shipping.

Current natural gas transmission pipelines can transport hydrogen–methane mixtures with slight modifications to their construction standards. Nevertheless, since a hydrogen–methane blend has different properties compared with pure methane or hydrogen, the addition of hydrogen could pose some safety hazards depending on the percentage of hydrogen added (e.g., fugitive emissions are greater with hydrogen) as well as the reactivity of the gas mixture [62–65]. More than 46 pre-industrial pilot projects are currently running worldwide, investigating the impact of hydrogen addition on existing natural gas infrastructure and appliances [66].

The lower heating value and density of fuel gas are significant tools in evaluating the proper *interchangeability of gases*. In addition, the Wobbe index is a parameter of great importance in the natural gas industry since, as a first approximation, it allows the establishment of interchangeability limits between fuel gases. However, this parameter considers only the energy provided by the injector and does not take into account other parameters involved in the combustion process, such as incomplete combustion, lifting flame, blowout flame, flashback flame, and yellow tip formation. Finally, evaluation of the interchangeability of gases is both a theoretical and practical process since the composition of natural gas changes depending on the country, even in different regions of the same country. Therefore, an experimental evaluation for each specific context is necessary [67].

The estimated gross length of global *hydrogen pipelines* is about 5000 km, whereas the length of natural gas pipelines is 3 million km. Nevertheless, a decision of the German Supreme Court allowed the use of natural gas pipelines for hydrogen. Thus, this conversion could contribute to the competitiveness of the hydrogen-based energy supply systems in Europe [68]. Furthermore, the most relevant synergies between liquified natural gas (LNG) and low-carbon hydrogen appear to be commercial. The LNG industry has decades of experience developing specialized infrastructure and supply chains, with associated high investment risks and high capital requirements useful in the development of LNG-based low-carbon hydrogen supply chains [69].

1.6 THE HYDROGEN ECONOMY AND GLOBAL WARMING ISSUES

The effectiveness in mitigating *environmental emissions* has recently been assessed by *life-cycle analyses* throughout the supply chain of proposed hydrogen energy carriers such as LH_2, methylcyclohexane (MCH), and ammonia (NH_3) that are promising energy vectors in the clean energy systems currently being developed. Life-cycle analyses have shown that H_2 liquefaction, MCH dehydrogenation, and NH_3 production are CO_2 hotspots for the target supply chains, and the use of renewable electricity in processes for the H_2 supply chain and the application of CCS to NH_3 production processes should be effective at establishing low-carbon supply chains [70, 4].

A study has been performed by Ampah et al. [46] investigating the evolutionary trends and critical enablers of hydrogen production technologies using econometric tools to determine the key drivers behind the field's development to combine patent life-cycle analysis with econometric analysis. This study adds to the ongoing discussion about hydrogen production from the standpoint of technological trends since the former reviews are distinguished by a lack of current trends, restrictions on individual hydrogen production sources, a limited application area, or limited reviewed stages beyond the production stage (such as storage). Thus, this study closes these gaps by identifying the technological *life-cycle of both fossil fuel and renewable hydrogen production technologies.*

Aiming at the 2°C scenario, the first step to address climate change could be accomplished if the world takes crucial decisions to lower energy-related carbon dioxide emissions by 45% relative to the 2010 level by 2030 and net zero by midcentury. Moreover, given that the population will grow by more than 2 billion people by 2030, considerable reductions in other GHGs should be accomplished [15]. Although it seems that the current efforts are not yet enough to limit the global temperature rise to 2°C, this goal still seems technically attainable [71–72].

Nevertheless, in the 1.5°C *scenario* of climate change, *water electrolysis* to produce hydrogen as an environmentally friendly fuel would necessitate massive amounts of additional renewable electricity generation. In this scenario, approximately 31,320 TWh of electricity would be required to power electrolyzers, which is more than is currently produced globally from all sources combined [4].

The International Energy Agency (IEA) prepared the report *"The Future of Hydrogen"* [73] for the G20 Osaka Summit in 2019, presenting and analyzing the existing hydrogen technologies and their potential to participate in the energy system transformation. Based on this report, the IEA composed the "Global Hydrogen Review 2021" [49] in which, considering the climate crisis, they concluded that more decisive actions should be taken.

Even though hydrogen seems to be adopted as a clean fuel and research is accelerated, this is not enough to reach the desired *net-zero emissions by 2050.* "Net Zero by 2050" [74] is a roadmap aimed at reaching net-zero carbon emissions by 2050 that would limit the global temperature rise to 1.5°C. The study determines hundreds of actions to be taken by all energy-related sectors to change the global economy from a fossil fuel economy into a renewable (mainly solar and wind) energy economy. To this end, huge amounts of investment, innovation, proficient design and implementation, technological development, infrastructure construction, international cooperation, and accomplishment in many related areas are needed. In other words, we have to hurry up because, simply: *"We are running out of time".*

Despite all these uncertainties, considerable successes have been obtained, with the cost of automotive *fuel cells* brought down by 70% from 2008 to 2021 due to significant technological achievements and the increased sales of fuel cell electric vehicles. Concerning electrolyzers, Europe is heading the electrolyzer technology, possessing 40% of the global installed capacity, and is anticipated to remain the largest market because of pioneering research and development [49].

Hydrogen is a crucial factor in the energy adjustment process that is required to limit global warming to 2°C above pre-industrial levels and to try to restrict the temperature increase to 1.5°C, aiming at the same time to significantly reduce the impacts of climate change as was affirmed by the 26th United Nations Climate

Change Conference in Glasgow in November 2021 (short name *COP26*) [75]. The achievements of COP26 at a glance are:

- *Mitigation: ensured global net zero and future strengthening of mitigation measures. The countries agreed to bring new commitments next year, a new climate program on mitigation measures, and they completed the Paris Rulebook. They also promised commitments to abandon coal power and deforestation programs, minimize methane emissions and accelerate electric vehicle programs.*
- *Adaptation & Loss and Damage: enhanced efforts to face climate impacts. Among the participating countries, 80 of them were already covered by adaptation plans to increase preparedness for climate risks, whereas 45 countries promised to submit plans by next year.*
- *Finance: mobilized billions and trillions. Developed countries have made efforts to deliver the USD100 billion climate finance target and promised to reach it by 2023 at the latest. Thirty-four countries and five finance institutions promised to stop financial support for undiminished fossil fuel plans by next year. Financial institutions and central banks intend to canalize trillions towards global net zero.*
- *Collaboration: worked together to deliver. The Glasgow Breakthroughs will enhance collaboration between governments, businesses, and society to fulfill climate goals faster, while collaborative partners in energy, electric vehicles, shipping, and commodities would accept commitments.*

At the 27th United Nations *Climate Change Conference (COP27)*, held in Sharm el-Sheikh, Egypt, 6–20 November 2022, many countries joined forces to work toward achieving the world's combined climate goals as agreed upon under the Paris Agreement. The participating countries decided upon, amongst other things, [76]:

- *Reaffirming the outcomes of all previous Conferences of the Parties.*
- *Reaffirming the critical role of multilateralism based on United Nations values and principles, including in the context of the implementation of the Convention and the Paris Agreement.*
- *Noting the importance of transition to sustainable lifestyles and sustainable patterns of consumption and production for efforts to address climate change.*
- *Acknowledging that climate change is a common concern of humankind.*
- *Noting the importance of ensuring the integrity of all ecosystems.*
- *Emphasizing that enhanced effective climate action should be implemented in a manner that is just and inclusive.*
- *Recognizing the fundamental priority of safeguarding food security and ending hunger, and the particular vulnerabilities of food production systems to the adverse impacts of climate change.*
- *Recognizing the critical role of protecting, conserving and restoring water systems and water-related ecosystems.*

- *Underlining the urgent need to address, in a comprehensive and synergetic manner, the interlinked global crises of climate change and biodiversity loss.*
- *Acknowledging that the impacts of climate change exacerbate the global energy and food crises, and vice versa, particularly in developing countries.*
- *Stressing that the increasingly complex and challenging global geopolitical situation and its impact on the energy, food, and economic situations should not be used as a pretext for backtracking, backsliding, or de-prioritizing climate action.*

They also took specific decisions on science and urgency, energy, mitigation of global warming, early warning and systematic observation, financing, technology transfer and deployment, ocean and climate change, forests and agriculture, finalizing on enhancing implementation and the actions needed to be taken by non-Party stakeholders.

The 28th United Nations Climate Change Conference (COP28) was held in Dubai, United Arab Emirates, and was the biggest of its kind [77]. Some 85,000 participated in the conference from 30 November to 13 December 2023, including more than 150 heads of state and government.

COP28 was particularly important as it signaled the first "global stock take" of the global efforts to address climate change under the Paris Agreement. Considering that the progress since Paris was not sufficient in all fields of climate action, the countries decided to speed up efforts across all areas by 2030. This includes a call on governments to hasten the transition from fossil fuels to renewables including wind and solar power.

Comparing the global warming potential (GWP) of the currently applied hydrogen production methods, the fossil-based steam reforming method has a higher GWP (11.9567 per kg of hydrogen) than hydrogen mercury cells, hydrogen diaphragm cells, and membrane cells with GWPs of 1.0510, 0.9090, and 0.8872 per kg of hydrogen, respectively. The GWP value is sufficiently low for hydrogen production from electrolysis (non-fossil-based) (0.0325 per kg of hydrogen), as compared to hydrogen production from alternative energy sources [78].

Using hydrogen as an energy carrier can effectively reduce air pollution and GHG emissions associated with fossil fuels. Nevertheless, its use on a large scale, with eventually large leaks into the atmosphere, might have some adverse environmental effects, such as increasing the lifetime of methane, worsening climate effects, and causing some depletion of the ozone layer. Till now, there have only been a few studies on this issue, and the uncertainties within them are considerable, but the existing studies show that the related *global atmospheric impacts* are likely to be small [79–81].

Rhee et al. [82] investigated the significant role of soils in the cycle of hydrogen in the atmosphere. They observed that the removal of molecular hydrogen from the atmosphere is controlled by the absorption in soils. Nevertheless, estimates of the extent of this process on a global scale are highly doubtful. They estimated that 82% of the annual turnover of tropospheric H_2 is due to soil uptake. This figure equals 88 ($\pm$11) Tg a^{-1}, of which the northern hemisphere's share is 62 ($\pm$10) Tg a^{-1}. Their estimations further showed that tropospheric H_2 has a lifetime of only 1.4 ($\pm$0.2) years.

1.7 OVERALL SAFETY AND THE GEOPOLITICAL ROLE OF HYDROGEN

To estimate the *overall safety of hydrogen* throughout its existence, that is its *life-cycle safety*, assessments, precautions, and measures should be taken starting from the production of hydrogen (either from carbonaceous raw materials or from water electrolysis) through storage, supply, and distribution after conversion to synthetic fuels or transported by truck, ship, or pipeline, and finally to its end use for transport, industry, electricity generation, fertilizer production, or heating purposes. All this life-cycle of hydrogen is vividly sketched in Tom Prater's isometric infographic in Figure 1.3 [15].

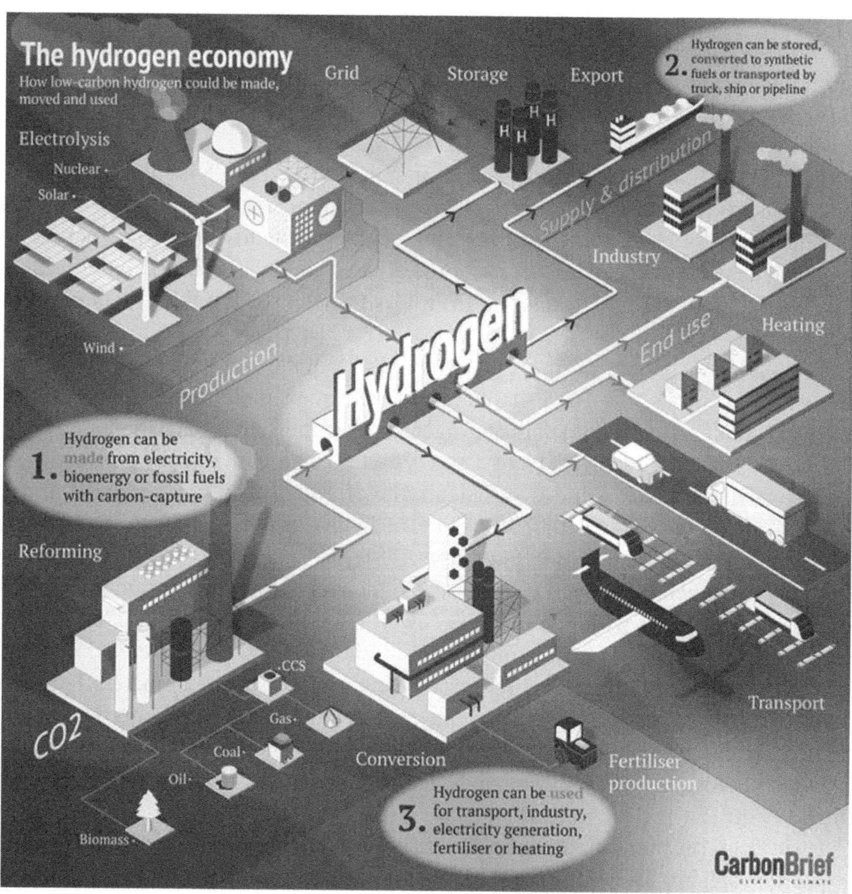

FIGURE 1.3 An isometric infographic for the hydrogen economy from production to end use. Graphic by Tom Prater. Courtesy of Carbon Free (www.carbonbrief.org/in-depth-qa-does-the-world-need-hydrogen-to-solve-climate-change).

Source: [15].

Hydrogen could play a new *geopolitical role* since some countries may take advantage of the *energy–geopolitical issues*, such as security of supply/demand and diversification. Thus, geopolitics will increasingly take into account technological dominance, along with resource availability. The current oil and natural gas major producers may try to position themselves as secure and reliable hydrogen exporters, to gain a significant geopolitical role, as well as increase their revenues. The development of low-carbon hydrogen pathways, alongside other renewable energy technologies aiming at fighting climate change, should be supported and envisioned from a global perspective. National strategies will have a limited effect on the climate crisis without a broader focus on the global picture since they risk widening the gap across countries and worsening existing inequalities. In our divided world, reaching the challenging targets required to limit climate change will be ineffective and discriminatory [83–84].

Concerning the EU, competitive and secure supplies of renewable hydrogen are a key policy priority aiming to accelerate the worldwide transition to decarbonization [85]. Yet no major hydrogen pipeline networks exist in the EU today, and no liquefied hydrogen ships are in use, which could have a significant impact on supply and demand. The EU, prioritizing its *energy independence*, develops internal, self-sufficient renewable hydrogen markets. In a way, hydrogen demand could be met with internal production but at higher costs. In addition, the EU prioritizes cost reduction with regional imports from neighboring countries that would hopefully decrease supply costs, but also reproduce energy dependencies of the past. Even though the geopolitical center of gravity would shift from East to South, this would not considerably improve the EU's strategic autonomy on energy. Furthermore, the EU prioritizes energy security and cost optimization by combining *long-distance imports* from other gas producers with regional imports and internal production.

From the security of supply perspective, *more global and less regionalized hydrogen trade* flows would reconfigure the geopolitical balance between suppliers and consumers by offering the possibility to switch providers. This, combined with the creation of strategic reserves, would probably end the dependence on dominant suppliers [86].

In a bibliometric analysis of green-hydrogen research [87], the outcomes of *policy decisions* in the United States, Europe, India, and China will profoundly impact green-hydrogen production and storage over the next five years. If these policies are implemented, these countries will account for two-thirds of this growth. Asia will account for the most significant part and become the second-largest producer globally.

Even *small countries* like Greece are currently moving forward rapidly on the hydrogen economy. Thus, Greece is to become the new hub for hydrogen in Europe as it achieved great success in July 2022 when two projects within the framework of the so-called IPCEI (significant projects of common European interest) were approved by the European Commission in the first phase. The massive imports of hydrogen from the Middle East and North Africa are to be directed to Europe via Greece. True to the motto issued by the EU after the Ukraine war, the EU will import 10 million tons of renewable hydrogen produced annually until 2030. From today's perspective, this share will even increase [88].

The first major hydrogen projects that meet the demands of industrial production have already been launched in *Saudi Arabia*. In the northwest of the country, a region the size of Belgium has been declared an *exclusive renewable and hydrogen zone,* called NEOM. Significant production volumes are expected here as of 2024, and large-scale production of renewable hydrogen will begin in 2026. In addition, there is also the use of the last large gas field, which is still not explored, from 2024. The Saudi government has announced that this gas field is not to be used to export gas but hydrogen.

Hydrogen as a gas is best transported via *pipelines*. It is the cheapest and safest way to send large volumes over relatively long distances. On the other hand, in certain countries like Greece, possessing an enormous merchant *transportation fleet,* would be a successful alternative, especially considering the time needed until the pipelines are completed. Moreover, the possibility of transporting ammonia by ships as a hydrogen substitute instead of gaseous or liquified hydrogen would be a more economical option, especially during the transitional phase of construction of a reliable pipeline system. Thus, it would be the right time for Greece and other similar countries with significant merchant fleets to adapt quickly to this option and to build the appropriate ships through suitable conversions. Additionally, the Eastern Mediterranean Basin is well known for its rich offshore salt caverns under the sea. It is a perfect storage facility for all the hydrogen produced in the Middle East–North African region. It would also serve perfectly as a strategic hydrogen reserve for the whole EU [88].

Ficco et al. [89] presented and discussed a project to *recover an abandoned industrial site* located in central Italy for the production, distribution, and consumption on a local scale of green hydrogen. The project aims to promote hydrogen use in the industrial and public transport sectors and to support the research, technology development, and innovation relating to the hydrogen supply chain. The cost–benefit analysis results have shown that a very significant initial investment is required and that it would be sustainable only with a relevant public grant. On the other hand, in the long term (i.e., up to 2070), the project becomes sustainable since the benefits significantly outweigh the costs.

Concerning *energy security,* the American economic and social theorist Jeremy Rifkin in his 2002 book *The Hydrogen Economy* [90] envisaged a world in which the widespread rollout of hydrogen will dramatically increase global security:

> *The road to global security lies in lessening our dependence on Middle East oil and making sure that all people on Earth have access to the energy they need to sustain life.* ***The hydrogen economy is a promissory note for a safer world.***

Yet Rifkin is not the only one optimistic about the anticipated benefits of hydrogen's geopolitical prevalence in the energy market. Professor Thijs Van de Graaf of Ghent University, in a paper with his colleagues titled *The new oil?* [91], claims that:

> *Over time, cross-border maritime trade in **hydrogen has the potential to fundamentally redraw the geography of global energy trade**, create a new class of energy exporters, and reshape geopolitical relations and alliances between countries.*

1.8 ARE WE REALLY MOVING INTO THE "HYDROGEN ECONOMY"?

The term *"hydrogen economy"* was first proposed in 1970 by *Professor John Bockris* [56], who envisaged a hydrogen-powered world stemming from solar and/or nuclear energy.

On the question of whether hydrogen could help tackle climate change, *Dr. Saehoon Kim*, head of the fuel cell department at Hyundai, told an energy webinar in the UK in July 2020 [15]:

> *In the past, our technology and industry was all about collecting oil, delivering oil, and using oil. And now, in the future, it will be collecting sunshine, delivering sunshine, and using sunshine – and what will make that possible is hydrogen.*

Japan, especially, has been investigating hydrogen as an energy source since the 1970s, and in its 2017 hydrogen strategy announced programs to create the first "hydrogen-based society".

China, the world's largest producer and user of hydrogen, has already been developing hydrogen fuel cells for around 20 years. Former science minister Wan Gong, who pioneered the nation's electric car strategy, has said it "should look into establishing a hydrogen society". In the meantime, the USA has held back research and development in hydrogen fuel cars, even though it has previously been a leader in hydrogen technology [15].

Regarding the *controversy between electric vehicles (EV) and fuel cell-powered vehicles (FCEV)*, although both are electricity-powered vehicles, there appear to be well-defined pros and cons. Overall, both hydrogen and electric cars have their own safety issues, and it is necessary to take them into account before making a decision [92, 93].

Thus, considering the *lifetime carbon footprint*, there is no winner, as the comparison depends on the materials' sources. Concerning FCEVs, the only byproduct of converting hydrogen into electricity is water, so there are no hazardous gas emissions. Yet hydrogen sourcing is a crucial factor here. For instance, the grey hydrogen produced by steam reforming is produced using fossil fuels, which can therefore significantly increase the overall amount of carbon emissions from a vehicle. On the other hand, green hydrogen derived from renewable energy has a very low footprint on the environment. On the other hand, most EVs are powered by lithium-ion batteries, which do not emit CO_2 when used to power a vehicle. However, lithium-ion batteries require many toxic materials, the sourcing of which can significantly increase their lifetime carbon footprint.

With regard to *efficiency*, battery-powered EVs are 80% energy-efficient, while FCEVs are less than 40%. This means that for every 100 W of energy produced, nearly 80 W will be used to power the vehicle when using an EV but less than 40 W will when using an FCEV.

Concerning existing infrastructure and the predominance of battery-powered EVs as compared with FCEVs, the former option has a distinct advantage over the latter considering *refueling/recharging and availability*. Yet FCEVs have much faster *charging times* and increased *ranges*.

Regarding production and *disposal*, green-hydrogen FCEVs have a strong advantage over battery-powered EVs since the sourcing of lithium batteries may cause profound environmental damage with tons of carbon emissions.

With hydrogen fuel, a specialized commodity for the general public, the small network of retail stations naturally charges high *prices*. Thus, currently, hydrogen fuel costs twice as much as electricity to power a car.

Nevertheless, hydrogen fuel cells have some disparagers, with Tesla's straightforward chief executive *Elon Musk* naming them "fool cells" and rejecting the technology as "mind-bogglingly stupid".

The hydrogen FCEV industry suffers from the well-known "chicken-and-egg" problem, since not enough cars are produced to lower their prices, and this results in limited refueling stations that, in turn, limit the demand for hydrogen FCEVs.

Bill Gates believes that cheap *green hydrogen* would be a massive breakthrough in clean energy (Gates Notes) [94] and states the following:

When most people picture greenhouse gas emissions, they think about cars and electricity. That's because they turn keys, press buttons, and flip switches every day. The good news is, we already have ways to decarbonize these types of emissions (solar, wind, and nuclear power and lithium-ion batteries). The bad news is, they add up to only about one third of the total.

*The other two-thirds – almost 35 billion tons – are much harder for most people to see. For example, we all use products made of cement, plastic, and steel, but most of us don't manufacture them or load them onto cargo ships. **To zero out emissions on these products, we need new technologies. Enter clean hydrogen. It has so many potential uses that some people refer to it as the Swiss Army knife of decarbonization**.* (Figure 1.4)

FIGURE 1.4   A Swiss Army knife. (From Google Images.)

Moreover, concerning *energy security*, Bill Gates adds [94]:

> *So, **the potential of clean hydrogen is tantalizing, and its necessity is becoming clearer every day.** Take Russia's war in Ukraine, which has made hydrogen not just a climate change issue but also an energy security issue. The EU has already announced its intention to produce and import 20 million tons of green hydrogen by 2030, enough to reduce its dependence on Russian natural gas imports by at least a third.*

1.9 ORGANIZATION OF THE CHAPTERS

Following this Introduction (Chapter 1), the remainder of the book is organized in the following manner. Chapter 2 provides an overview of hydrogen accidents over the years. This is followed in Chapter 3 by a discussion of the various properties of hydrogen associated with hazards in the gas and liquid phases. The manifestation of these properties as physiological, physical, and chemical hazards is discussed in Chapter 4. Additional attention is given in Chapter 4 of this second edition of the book to hydrogen control and risk mitigation and especially to environmental concerns from the anticipated extended utilization of hydrogen. The hazards and ensuing risks associated with hydrogen storage facilities and hydrogen use as a transportation fuel are addressed in Chapters 5 and 6, respectively. In Chapter 5, special attention is given to ammonia as a hydrogen carrier and storage material. In Chapter 6, more information has been added now on ground transportation (vehicles and locomotives), maritime transportation, space exploration, and hydrogen-powered aviation.

Chapter 7 deals with the application of the principles of inherently safer design [14, 95] to the field of hydrogen safety. Enhancements in the safer handling and use of hydrogen by means of an appropriate safety management system are then covered in Chapter 8.

Chapter 9 describes a unique effort aimed at facilitating the safe introduction of hydrogen as an energy carrier and removing safety-related obstacles – the European Commission Network of Excellence, known as HySafe [96]. This is followed in Chapter 10 by an exposition of the value of studying lessons learned from case histories in making safety improvements in all areas, specifically here with respect to hydrogen usage. In addition, Chapter 10 has been enriched with recent references concerning the safety challenges stemming from the probable use of hydrogen as a fuel in nuclear fusion reactors.

Chapter 11 deals with the important issue of the effects of hydrogen on materials of construction; the subject of Chapter 12 is future needs in the area of hydrogen safety. Finally, Chapter 13 provides an overview of various guidelines and procedural approaches to hydrogen safety, and Chapter 14 gives concluding remarks and summarizes the old as well as the up-to-date knowledge on crucial hydrogen safety challenges as a result of the considerable progress achieved over the last decade.

The material has been arranged in a sequence that is intended to integrate the traditional view of hydrogen safety – hazardous properties and industrial risks in storage and use – with the equally important, yet often overlooked, safety aspects associated with the management and social sciences (e.g., safety management systems and safety culture). It is therefore hoped that this book will be helpful to practitioners and researchers in a variety of fields and with a range of backgrounds and experience.

REFERENCES

1. Cherry, R.S., A hydrogen utopia? *International Journal of Hydrogen Energy*, 29, 125, 2004.
2. Akansu, S.O., Dulger, Z., Kahraman, N., and Veziroglu, T.N., Internal combustion engines fueled by natural gas: Hydrogen mixtures, *International Journal of Hydrogen Energy*, 29, 1527, 2004.
3. Rigas, F. and Sklavounos, S., Evaluation of hazards associated with hydrogen storage facilities, *International Journal of Hydrogen Energy*, 30, 1501, 2005.
4. Osman, A.I., Mehta, N., Elgarahy, A.M., Hefny, M., Al-Hinai, A., Al-Muhtaseb, A.H., and Rooney, D.W., Hydrogen production, storage, utilisation and environmental impacts: A review, *Environmental Chemistry Letters*, 20, 153–188, 2022.
5. Momirlan, M. and Veziroglu, T.N., Current status of hydrogen energy, *Renewable and Sustainable Energy Reviews*, 6, 141, 2002.
6. EUR 22002, *Introducing Hydrogen as an Energy Carrier*, European Commission, Directorate-General for Research Sustainable Energy Systems, Luxembourg: Office for Official Publications of the European Communities, 2006.
7. Schulte, I., Hart, D., and van der Vorst, R., Issues affecting the acceptance of hydrogen fuel, *International Journal of Hydrogen Energy*, 29, 677, 2004.
8. Dincer, I., Technical, environmental and exergetic aspects of hydrogen energy systems, *International Journal of Hydrogen Energy*, 27, 265, 2002.
9. EEC, European Economic Community, Seveso III, Directive 2012/18/EU of the European Parliament and of the Council, *On the Control of Major-Accident Hazards Involving Dangerous Substances*, amending and subsequently repealing, Council Directive 96/82/EC, Brussels, 4 July 2012.
10. Kirchsteiger, C., Availability of community level information on industrial risks in the EU. *Process Safety and Environmental Protection*, 78, 81, 2000.
11. IChemE, Institute of Chemical Engineers, *Accident Database*, Loughborough, UK, 1997. www.icheme.org/knowledge-networks/safety-centre/resources/accident-data/
12. Taylor, J.R., *Risk Analysis for Process Plants, Pipelines and Transport*, Chapman & Hall, London, 1994, 102.
13. CCPS, Center for Chemical Process Safety, *Guidelines for Hazard Evaluation Procedures*, American Institute of Chemical Engineers, New York, 1992, 69.
14. Guy, K.W.A., The hydrogen economy, *Process Safety and Environmental Protection*, 78 (4), 324–327, 2000.
15. Evans, S. and Gabbatiss, J., *In-Depth Q&A: Does the World Need Hydrogen to Solve Climate Change? Carbon Brief*, 2021. www.carbonbrief.org/in-depth-qa-does-the-world-need-hydrogen-to-solve-climate-change
16. Dawood, F., Anda, M., and Shafiullah, G.M., Hydrogen production for energy: An overview, *International Journal of Hydrogen Energy*, 45, 3847–3869, 2020.
17. Abbasi, T. and Abbasi, S.A., Renewable' hydrogen: Prospects and challenges, *Renewable and Sustainable Energy Reviews*, 15, 3034, 2011.
18. Sampson, J., Renewable hydrogen, made cost competitive, *GasWorld*, 2019. www.gasworld.com/feature/renewable-hydrogen-made-cost-competitive/2087542.article/
19. Bethoux, O., Hydrogen fuel cell road vehicles and their infrastructure: An option towards an environmentally friendly energy transition, *Energies*, 13, 6132, 2020.

20. Nicita, A., Maggio, G., Andaloro, A.P.F., and Squadrito, G., Green hydrogen as feedstock: Financial analysis of a photovoltaic-powered electrolysis plant. *International Journal of Hydrogen Energy*, 45, 11395–11408, 2020.

21. Domínguez, R., Calderón, E., and Bustos, J., Process safety in electrolytic green hydrogen production, *Production Management and Process Control*, 36, 185–195, 2022.

22. Chi, J. and Yu, H., Water electrolysis based on renewable energy for hydrogen production, *Chinese Journal of Catalysis*, 39, 390–394, 2018.

23. Kovac, A., Paranos, M., and Marcius, D., Hydrogen in energy transition: A review, *International Journal of Hydrogen Energy*, 46, 10016–10035, 2021.

24. European Commission, *A Hydrogen Strategy for a Climate-Neutral EUROPE*, Brussels, 2020. chrome-extension://efaidnbmnnnibpcajpcglclefindmkaj/https://energy.ec.europa.eu/system/files/2020-07/hydrogen_strategy_0.pdf.

25. Vostakola, M.F., Salamatinia, B., and Horri, B.A., A review on recent progress in the integrated green hydrogen production processes, *Energies*, 15, 1209, 2022. https://doi.org/10.3390/en15031209

26. van Renssen, S., The hydrogen solution? *Nature Climate Change*, 10,799–801, 2020.

27. Diab, J., Fulcheri, L., Hessel, V., Rohani, V., and Frenklach, M., Why turquoise hydrogen will be a game changer for the energy transition, *International Journal of Hydrogen Energy*, 47, 25831–25848, 2022.

28. H2 Bulletin, Hydrogen Colours Codes. www.h2bulletin.com/knowledge/hydrogen-colours-codes/

29. von Döllen, A., Hwang, Y.S., and Schlüter, S., The future is colorful—An analysis of the CO2 Bow wave and why green hydrogen cannot do it alone, *Energies*, 14, 5720, 2021.

30. Hermesmann, M. and Müller, T.E., Green, turquoise, blue, or grey? Environmentally friendly hydrogen production in transforming energy systems, *Progress in Energy and Combustion Science*, 90, 100996, 2022.

31. Boretti, A., White is the color of hydrogen from concentrated solar energy and thermochemical water splitting cycles, *International Journal of Hydrogen Energy*, 46, 20790–20791, 2021.

32. Howarth R.W. and Jacobson M.Z., How green is blue hydrogen? *Energy Science Engineering*, 00:1–12, 2021. https://doi.org/10.1002/ese3.956

33. Howarth, R.W. and Jacobson, M.Z., Reply to comment on "How green is blue hydrogen?", *Energy Science Engineering*, 1–6, 2022. https://doi:10.1002/ese3.1154

34. Romano M.C., Antonini C., Bardow A., et al. Comment on "How green is blue hydrogen?" *Energy Science Engineering*, 1–11, 2022. doi:10.1002/ese3.1126.

35. Pettersen J., Steeneveldt R., Grainger D., Scott T., Holst L.-M., and Hamborg E.S., Blue hydrogen must be done properly, *Energy Science Engineering*, 1–17, 2022. doi:10.1002/ese3.1232.

36. Rosenow, J. and Lowes, R., Will blue hydrogen lock us into fossil fuels forever?, *One Earth*, November 2021. doi:10.1016/j.oneear.2021.10.018.

37. Cheng, W. and Lee, S., How green are the national hydrogen strategies? *Sustainability*, 14, 1930, 2022. https://doi.org/10.3390/su14031930

38. Hydrogen-data telling a story, Global Energy Infrastructure. https://globalenergyinfrastructure.com/articles/2021/03-march/hydrogen-data-telling-a-story/

39. Togni, L., Fakhoury, R., Deen, A., Stekli, J., Schäfer, F., N. Babenhauserheide, N., Zahn, M., Steinbach, C. (Project Team), National Hydrogen Strategies, World Energy Council, Working Paper, September 2021. chrome-extension://efaidnbmnnnibpcajpcglclefindmkaj/https://www.worldenergy.org/assets/downloads/Working_Paper_-_National_Hydrogen_Strategies_-_September_2021.pdf

40. IAEA Nuclear Energy Series, *Hydrogen Production Using Nuclear Energy*, No. NP-T-4.2, 2013.

41. Ramshesh, W., Safety aspect concerning radiolytic gas generation in reactors, *Science Progress*, 84 (1), 69–85, 2001.

42. Dincer I. and Acar C., Innovation in hydrogen production, *International Journal of Hydrogen Energy*, 42, 14843, 2017.

43. Wang, Y., Zhou, J., Wang, F., Xie, Y., Liu, S., Ao, Z., and Li, C., Hydrogen generation from photocatalytic treatment of wastewater containing pharmaceuticals and personal care products by Oxygen-doped crystalline carbon nitride, *Separation and Purification Technology*, 296, 121425, 2022. https://doi.org/10.1016/j.seppur.2022.121425

44. Navlani-Garcia, M., Salinas-Torres, D., Mori, K., Kuwahara, Y., and Yamashita, H., Photocatalytic approaches for hydrogen production via formic acid decomposition, *Topics in Current Chemistry*, 377, 27, 2019.

45. Rodriguez, S.B. et al., *Development of Design and Simulation Model and Safety Study of Large-Scale Hydrogen Production Using Nuclear Power*, Sandia Report SAND2007-6218, 2007.

46. Ampah, J.D. et al., Investigating the evolutionary trends and key enablers of hydrogen production technologies: A patent-life cycle and econometric analysis, *International Journal of Hydrogen Energy*, article in press. https://doi.org/10.1016/j.ijhydene.2022.07.258.

47. AIST, National Institute of Advanced Industrial Science and Technology, Achieving the World's Highest Efficiency in Hydrogen Production by Water Electrolysis Using an Oxide Photoelectrode. www.aist.go.jp/aist_e/list/latest_research/2012/20120511/20120511.htm

48. Oni, A.O., Anaya, K., Giwa, T., Di Lullo, G., and Kumar A., Comparative assessment of blue hydrogen from steam methane reforming, autothermal reforming, and natural gas decomposition technologies for natural gas-producing regions, *Energy Conversion and Management*, 254, 115245, 2022. https://doi.org/10.1016/j.enconman.2022.115245

49. IEA International Energy Agency, Energy Technology Policy (ETP) Division of the Directorate of Sustainability, Technology and Outlooks (STO) *Global Hydrogen Review*, 2023. chrome-extension://efaidnbmnnnibpcajpcglclefindmkaj/https://iea.blob.core.windows.net/assets/ecdfc3bb-d212-4a4c-9ff7-6ce5b1e19cef/GlobalHydrogenReview2023.pdf

50. Ahshan, R., Potential and economic analysis of solar-to-hydrogen production in the sultanate of Oman, *Sustainability*, 13, 9516, 2021.

51. Armijo, J. and Philibert, C., Flexible production of green hydrogen and ammonia from variable solar and wind energy: Case study of Chile and Argentina, *International Journal of Hydrogen Energy*, 45, 1541–1558, 2020.

52. Fasihi, M., Weiss, R., Savolainen, and J., Breyer, C., Global potential of green ammonia based on hybrid PV-wind power plants, *Applied Energy*, 294, 116170, 2021.

53. Longden, Th., Beck, F.J., Jotzo, F., Andrews, R., and Prasad, M., 'Clean' hydrogen? – Comparing the emissions and costs of fossil fuel versus renewable electricity based hydrogen, *Applied Energy*, 306, 118145, 2022. https://doi.org/10.1016/j.apenergy.2021.118145

54. Marchenko, O.V. and Solomin, S.V., The future energy: Hydrogen versus electricity, *International Journal of Hydrogen Energy*, 40, 3801, 2015.

55. Rosenow, J. and Eyre, N., Reinventing energy efficiency for net zero, *Energy Research & Social Science*, Post-print version of article, 2022.

56. Bockris, J., The hydrogen economy: Its history, *International Journal of Hydrogen Energy*, 38, 2579–2588, 2013.

57. Tarkowski, R. and Uliasz-Misiak, B., Towards underground hydrogen storage: A review of barriers, *Renewable and Sustainable Energy Reviews*, 162, 112451, 2022.
58. Zivar, D., Kumar, S., and Foroozesh, J., Underground hydrogen storage: A comprehensive review, *International Journal of Hydrogen Energy*, 46, 23436–23462, 2021.
59. Liu, W., Zhang, Z., Chen, J., Jiang, D., Wu., F., Fan, J., and Li, Y., Feasibility evaluation of large-scale underground hydrogen storage in bedded salt rocks of China: A case study in Jiangsu province, *Energy*, 198, 117348, 2020.
60. Taie, Z., Villaverde, G., Morris, J.S., Lavrich, Z., Chittum, A., White, K., and Hagen, C., Hydrogen for heat: Using underground hydrogen storage for seasonal energy shifting in northern climates, *International Journal of Hydrogen Energy*, 46, 3365–3378, 2021.
61. d'Amore-Domenech, R., Meca, V.L., Pollet, B.G., and Leo, T.J., On the bulk transport of green hydrogen at sea: Comparison between submarine pipeline and compressed and liquefied transport by ship, *Energy*, 267, 126621, 2023.
62. Messaoudani, Z.L., Rigas, F., Hamid, M.D.B., and Hassan Ch. R. Ch., Hazards, safety and knowledge gaps on hydrogen transmission via natural gas grid: A critical review, *International Journal of Hydrogen Energy*, 41, 17511–17525, 2016.
63. Dell'Isola, M., Ficco, G., L. Moretti, Perna, A., Candelaresi, D., and Spazzafumo, G., Impact of hydrogen injection on thermophysical properties and measurement reliability in natural gas networks, 76° Italian National Congress ATI, *E3S Web of Conferences,* 312, 01004, 2021.
64. Kuczynski, S., Laciak, M., Olijnyk, A., Szurjej, A., and Wlodek, T., Thermodynamic and technical issues of hydrogen and methane-hydrogen mixtures pipeline transmission, *Energies*, 12, 569, 2019.
65. Ferrario, A.M., Cigolotti, V., Ruz, A.M., Gallardo, F., Garcia, J, and Monteleone, G., Role of hydrogen in low-carbon energy future, in *Technologies for Integrated Energy Systems and Networks,* 1st ed., Edited by Giorgio Graditi and Marialaura Di Somma, Wiley GmbH, 2022. https://doi.org/10.1002/9783527833634.ch4
66. Kopsacheilis, I., Ventikos, N.P., and Drivas, D., Hydrogen enriched natural gas: a critical review on safety and risk analysis on pilot projects, in *Pipeline Technology Conference*, Berlin, 2022. www.pipeline-conference.com/abstracts/hydrogen-enriched-natural-gas-critical-review-safety-and-risk-analysis-pilot-projects
67. Escalante, E.S.R. and Carvalho Jr., J.A., Interchangeability analysis from natural gas to natural gas/hydrogen mixtures: The Wobbe index as a first approximation, in *Rio Oil & Gas Expo and Conference*, 2022. doi: https://doi.org/10.48072/2525-7579.rog.2022.
68. Sadik-Zada, E.R., Political economy of green hydrogen rollout: A global perspective, *Sustainability*, 13, 13464, 2021.
69. Al-Kuwari, O., Schonfisch, M., The emerging hydrogen economy and its impact on LNG, *International Journal of Hydrogen Energy*, 47, 2080, 2022.
70. Ozawa, A., Kudoh, Y., Kitagawa, N., and Muramatsu, R., Life cycle CO_2 emissions from power generation using hydrogen energy carriers, *International Journal of Hydrogen Energy*, 44, 11219, 2019.
71. Birol, F., Cozzi, L., and Gül, T., *Redrawing The Energy-Climate Map*, World Energy Outlook Special Report. International Energy Agency, Paris, 2013.
72. Fischedick, M., Marzinkowski, J., Winzer, P., and Weigel, M., Techno-economic evaluation of innovative steel production technologies. *Journal of Cleaner Production,* 84, 563–580, 2014. https://doi:10.1016/j.jclepro.2014.05.063
73. IEA, International Energy Agency, *The Future of Hydrogen: Seizing today's opportunities*, Fatih Birol Executive Director, Report prepared by the IEA for the G20, Japan, 2019. www.iea.org/reports/the-future-of-hydrogen

74. IEA, International Energy Agency, *Net Zero by 2050: A Roadmap for the Global Energy Sector*, Fatih Birol Executive Director, 2021b. at: www.iea.org/reports/net-zero-by-2050

75. COP26, The glasgow climate pact, in *Glasgow 26th World Climate Conference*, November 2021. https://ukcop26.org/the-glasgow-climate-pact/

76. COP27, Sharm el-Sheikh Implementation Plan, in *27th World Climate Conference Decision -/CP.27*, 20 November 2022.

77. COP28, in First global stocktake, Proposal by the President, *UN Climate Change Conference in Dubai, the United Arab Emirates*, 30 November – 12 December 2023. https://unfccc.int/cop28

78. Sharma, S. Agarwal, S., and Jain, A., Significance of hydrogen as economic and environmentally friendly fuel, *Energies*, 14, 7389, 2021.

79. Tromp, T.K., Shia, R.L., Allen, M., Eiler, J.M., and Yung, Y.L., Potential environmental impact of a hydrogen economy on the stratosphere, *Science*, 300, 1740–1742, 2003.

80. Ball, P., Hydrogen fuel could widen ozone hole, *Nature*, 2003. https://doi.org/10.1038/news030609-14

81. Derwent, R.G., *Department for Business, Energy and Industrial Strategy, Hydrogen for Heating: Atmospheric Impacts*, BEIS Research Paper Number: no. 21, 2018.

82. Rhee, T.S., The overwhelming role of soils in the global atmospheric hydrogen cycle, Atmospheric *Chemistry and Physics Discussion*, 5, 11215–11248, 2005.

83. Noussan, M., Raimondi, P.P., Scita, R., and Hafner, M., The role of green and blue hydrogen in the energy transition—A technological and geopolitical perspective, *Sustainability*, 13, 298, 2021.

84. Vincent, D., *Défis technologiques et économiques de la gouvernance mondiale de l'hydrogène pour atteindre les objectifs de l'Accord de Paris*, Mémoire de Master, Faculté d'Economie de Grenoble, Université Grenoble Alpes, 2021.

85. Espegren, K., Damman, S., Pisciella, P., and Graabak, I., The role of hydrogen in the transition from a petroleum economy to a low-carbon society, *International Journal of Hydrogen Energy*, 46, 23125–23138, 2021.

86. Nuñez-Jimenez, A. and De Blasio, N., *The Future of Renewable Hydrogen in the European Union, Hydrogen in the European Union, Market and Geopolitical Implications*, Report, Belfer Center for Science and International Affairs, Harvard Kennedy School, March 2022. https://nrs.harvard.edu/URN-3:HUL.INSTREPOS:37372476

87. Raman, R., Nair, V.K., Prakash, V., Patwardhan, A., and Nedungadi, P., Green-hydrogen research: What have we achieved, and where are we going? Bibliometrics analysis, *Energy Reports*, 8, 9242–9260, 2022.

88. Hydrogen Central, Greece is Becoming the New Hub for Hydrogen in Europe – Hydrogen Europe, August 2022. https://hydrogen-central.com/greece-becoming-new-hub-hydrogen-europe-hydrogen-europe/

89. Ficco, G., Arpino, F., Dell'Isola, M., Grimaldi, M., Lisi, S., Development of a hydrogen valley for exploitation of green hydrogen in Central Italy, *Energies*, 15, 8072, 2022.

90. Rifkin, J., *The Hydrogen Economy: The Creation of the Worldwide Energy Web and the Redistribution of Power on Earth*, TarcherPerigee, 2002.

91. De Graaf, Th., Overland, I., Scholten, D., Westphal, K., The new oil? The geopolitics and international governance of hydrogen, *Energy Research & Social Science*, 70, 101667, 2020.

92. Fletcher, C., *Hydrogen vs. Electric Cars: Comparing Innovative Sustainability*, Earth Org, August 26, 2022. https://earth.org/hydrogen-vs-electric-cars/

93. Voelcker, J., Hydrogen Fuel-Cell Vehicles: Everything You Need to Know, Car and Driver, September 26, 2022. www.caranddriver.com/features/a41103863/hydrogen-cars-fcev/
94. Gates, Bill, GatesNotes. www.gatesnotes.com/Energy/Clean-Hydrogen
95. Kletz, T., and Amyotte, P., *Process Plants: A Handbook for Inherently Safer Design*, CRC Press/Taylor & Francis Group, Boca Raton, FL, 2010.
96. Jordan, T., et al., Achievements of the EC network of excellence HySafe, *International Journal of Hydrogen Energy*, 36 (3), 2656–2665, 2011.

2 Historical Survey of Hydrogen Accidents

2.1 GENERAL CONSIDERATIONS

A historical survey reveals that there have been quite a few severe accidents in the industrial and transport sectors because of hydrogen use. The determinant causes for a major event may be classified into the following categories [1]:

- Mechanical or material failure.
- Corrosion attack.
- Overpressurization.
- Enhanced embrittlement of storage tanks at low temperatures.
- Boiling liquid expanding vapor explosion.
- Rupture due to impact by shock waves and missiles from adjacent explosions.
- Human error.

Well-known accident databases, such as those of the United Nations Environment Programme and the Organisation for Economic Co-operation and Development, MHIDAS (Major Hazardous Incident Data Service), and BARPI (Bureau d'Analyse des Risques et Pollutions Industriels), have been used to collect and classify past accidents related to hydrogen applications. The results are shown in Table 2.1 [2–8].

Although this is not a complete list of all accidents involving hydrogen, Table 2.1 gives a clear idea of the potential hazards stemming from careless use, storage, or transmission of hydrogen. This is why such databases should be continuously updated. These characteristic examples prove that, in practice, hydrogen is liable to cause major chemical accidents posing a considerable risk not only for on-site but also for off-site damage.

Typical examples of such accidents and their consequences for people and property are quoted in Section 2.2 and may be found in detail in the relevant literature and databases [7–15].

DOI: 10.1201/9781003313007-2

TABLE 2.1
Summary of Major Accidents Related to Hydrogen

Year/Date	Location	Activity/Origin of Accident	Death	Injury	Evacuated
2019/06.10	Kjørbo, Norway	Explosion/hydrogen fueling station	-	3	NA
2019/06.01	Santa Clara, USA	Explosion/hydrogen production and storage company	NA	NA	NA
2019/05.23	Gangwon, South Korea	Explosion/energy research center	2	6	Nearby businesses and homes.
2014/NA	Yokkaichi City, Japan	Explosion/metal and chemical company	5	NA	NA
2014/NA	Pristina, Kosovo	Explosion/coal-fired power plant	3	NA	NA
2011/03.11	Fukushima Daiichi, Japan	Explosion/nuclear power plant	-	16	20-km radius
2002/10.15	Brousseval, France	Explosion/iron foundry	-	14	-
2001/05.01	Oklahoma, USA	Fire/transport (trailer)	1	1	15
2000/09.03	Gonfreville-l'Orcher, Le Havre, France	Explosion/chemical plant	-	12	-
2000/02.10	Tomakomai, Hokkaido, Japan	Fire/refinery	-	-	-
1999/05.07	Panipat, India	Fire/refinery	5	-	-
1999/04.08	Hillsborough, Tampa, Florida, USA	Fire and explosion/power station	3	50	38
1998/06.08	Auzouer-en-Touraine, France	Explosion and fire/fine chemicals	-	1	200
1995/10.31	Bolbec, France	Explosion/ manufacture of pharmaceutical products	-	5	NA
1993/NA	Krasnaya Tura, Russia	Cloud explosion/pipeline	NA	4	NA
1992/04.22	Jarrie, Isère, France	Fire/hydrogen peroxide plant	1	2	-
1992/01.18	Pennsylvania, USA	Fire/NA	1	3	-
1992/10.16	Sodegaura, Japan	Explosion/refinery	10	7	-
1992/08.08	Wilmington, Delaware, USA	Explosion/refinery	-	16	-
1992/08.28	Hong Kong	Explosion/power plant	2	19	-
1991/05.16	Kakuda, Miyagi, Japan	Explosion/space rocket facilities	-	-	-
1991/02.14	Daesan, South Korea	Explosion/petrochemicals	-	2	-

(continued)

TABLE 2.1 (Continued)
Summary of Major Accidents Related to Hydrogen

Year/Date	Location	Activity/Origin of Accident	Death	Injury	Evacuated
1991/10	Hanau, Frankfurt, Germany	Explosion/optical fiber production	NA	NA	NA
1990/07.25	Birmingham, UK	Fire and gas cloud	-	>60	70,050
1990/NA	Prague, then Czechoslovakia	Explosion and fire	15	26	NA
1990/04.29	Ottmarsheim, France	Fire	NA	NA	NA
1989/10.23	Pasadena, Texas, USA	Explosion/chemical plant	23	314	NA
1989/08.09	Saint-Fons, France	Deflagration/manufacture of chemicals	1	2	NA
1989/02.23	Le Grand-Quevilly, France	Explosion/manufacture of fertilizers	2	-	NA
1989	Boussens, Haute-Garonne, France	A hydrogen leak ignited into a jet flame	4	NA	NA
1989	Richmond, Virginia, USA	Explosion/gas refinery	NA	NA	NA
1988/06.15	Genoa, Italy	Explosion	3	2	15,000
1986/04.26	Chernobyl, Ukraine, then Soviet Union	Explosion/nuclear power plant	28 (4000–90,000 cancer related)	NA	30-km radius
1986/01.28	Kennedy Space Center, Florida, USA	*Challenger* explosion/space center	7	119	NA
1985/NA	Porsgrunn, Norway	Explosion/ammonia production plant	2	NA	NA
1979/03	Three Mile Island, USA	Severe nuclear core meltdown accident	NA	NA	NA
1977/NA	Gujarat, India	Explosion/chemical plant	5	35	NA
1975/04.05	Ilford, Essex, UK	Explosion/chemical plant	1	-	-
1972/NA	Dodewaard, Netherlands	Explosion/hydrogen processing facility	4	40	NA
1937/05.06	Lakehurst, New Jersey, USA (Figure 2.1)	*Hindenburg* explosion/airship	36	NA	NA

Note: NA, not available; -, none.

2.2 SIGNIFICANT DISASTERS

2.2.1 THE *HINDENBURG* DISASTER

The largest aircraft ever built were the Zeppelins LZ-129 *Hindenburg* and her sister ship LZ-130 *Graf Zeppelin II*. They were constructed by Luftschiffbau Zeppelin in 1935. The *Hindenburg,* named after the president of Germany Paul von Hindenburg and constructed with an aluminum framework, was 245 m long, 41 m in diameter, and contained 211,890 m^3 of gas in 16 cells, with a useful lift of 112,000 kg, powered by four 820 kW engines giving it a maximum speed of 135 km/h. It was covered with cotton fabric impregnated with cellulose acetate butyrate resin containing iron oxide and aluminum powder. The *Hindenburg* made its first flight in March 1936 and was filled with highly flammable hydrogen as the buoyancy gas, instead of noncombustible helium, due to a United States military embargo on helium. Aiming at the prevention of any possible fire on the aircraft from hydrogen leaks, the German engineers used various safety measures, among which was the special resin coating of the fabric to render it electrically conductive to avoid sparks from static electricity [16].

The disaster occurred at Lakehurst, New Jersey, on May 6, 1937 (Figure 2.1) and was for many years under investigation to identify the reasons that caused the ignition of hydrogen. It is remarkable that until that moment Zeppelins had an impressive safety record, with no passenger injured on any of these airships, and they were widely seen as symbols of German and Nazi superiority. Moreover, the *Graf Zeppelin* had flown safely for more than 1.6 million km (1 million miles), including the first complete circumnavigation of the globe by airship. Yet in this accident, which terminated the use of airships as passenger carriers, the ignition of hydrogen proceeded rapidly to

FIGURE 2.1 *Hindenburg* burning.

Source: From Google Images.

fire toward the tail section of the craft, completely destroying it. The fire was almost simultaneously succeeded by an explosion that quickly engulfed the 240-ton craft, causing it to crash to the ground. The accident resulted in 35 fatalities (13 passengers and 22 crew) out of the 97 people onboard (36 passengers and 61 crew), as well as one more fatality on the ground. Many assumptions have been formulated to determine the origin of ignition. The attention given to the accident destroyed the belief that one could rely on giant airships and ended the airship era [17].

The hypotheses posed for the starting point of the accident include sabotage, static electricity, lightning, and engine failure. The first hypothesis stated by Hugo Eckener, former head of the Zeppelin Company, assumed sabotage due to warning letters he had received before the accident, but after some investigation he adopted the static spark cause. Systematic analysis of the accident revealed that the cover shell and the paint of the airship were flammable and ignitable by static electricity sparks. The cellulose acetate butyrate used as the outer shell of the airship has been proven sufficiently flammable by itself, but its flammability was significantly enhanced by the additives used, iron oxide and aluminum powder, which are sometimes used as components of solid rocket propellants or thermite compositions [17].

In addition, the airship's outer shell was separated from the aluminum frame by nonconductive ramie cords, thus permitting static electricity accumulation on the shell. Prevailing atmospheric conditions at the time the accident occurred could generate considerable electrostatic discharge activity on the airship. Furthermore, the mooring lines were wet (i.e., conductive) and connected to the aluminum frame. When the wet mooring lines touched the ground, they grounded the aluminum frame, resulting in the formation of an electrical discharge from the electrically charged shell to the grounded frame [16].

Although the airship was completely destroyed when the hydrogen caught fire, the triggering cause was the use of hazardous materials for the outer shell and bad design with regard to permitting static electricity development on it. Since no death was directly related to the hydrogen burning, the airship would probably have been destroyed even if helium had been used instead of hydrogen due to the shell burning and the loss of buoyancy. Nevertheless, taking into account that hydrogen is extremely flammable, nowadays helium is used exclusively instead of hydrogen in buoyancy-driven airships, thus resulting in inherently safer systems (as discussed in Chapter 7).

2.2.2 HYDROGEN RELEASE DURING MAINTENANCE WORK

At about 1:00–1:05 pm on October 23, 1989, at Pasadena, Texas, a massive and destructive hydrogen vapor cloud explosion occurred in a polyethylene plant (Phillips 66 Company), causing loss of life to 23 people, injuries to 314 others (185 Phillips 66 and 129 contract employees), and extensive damage to the whole plant. All those killed were within 75 m of the point where the gas was initially released. The explosion strength was estimated to be equivalent to 2.4 tons of the explosive TNT. The accident resulted from a release of about 40,000 kg of process gas containing ethylene, isobutene, hexane, and hydrogen during maintenance on a reaction loop line. This massive gas cloud was ignited and consumed within less than two minutes. The initial explosion threw debris as far away as six miles and registered between 3 and 4 on the Richter scale on Rice University seismographs. There were also many secondary explosions.

In addition to the deaths and injuries, estimates of the property damage at the company and the lost income due to the disruption of business are USD715.5 million and USD700 million, respectively. Phillips 66 Company also agreed to pay a USD4 million fine and to establish process safety management procedures at four of its facilities [7, 18].

According to the investigation of this incident by Silas and Cox of the Phillips 66 Company, serious design and human errors triggered the accident [19]:

Examination of the evidence after the accident indicates that the lockout device had been removed and the air hoses had been reconnected to the valve operator on the Demco® valve of the No. 4 leg. The valve was open, and the settling leg was open to the atmosphere at the bottom of the leg where a wedge spool leading to the product take off valve should have been connected. Block valves to the air lines for the Demco and the piping leading to them had been damaged as a result of the explosion and moved, making their position meaningless. The evidence indicates the release occurred through this No. 4 open Demco valve and settling leg.

The US Occupational Safety and Health Administration (OSHA) detailed numerous defects in the management of the installation [20], which in brief were the following:

- Lack of a hazard assessment study (as discussed in Chapter 8).
- Inadequate safety distances between the control room and the reactor that did not allow emergency shutdown actions and safe evacuation of personnel during the initial vapor release.
- Lack of an effective permit system for the control of maintenance activities (as discussed in Chapter 8).
- Erroneous design of building ventilation intakes, so that in the case of gas release from the process unit, the escaping gas could be trapped in adjacent buildings.

The findings of OSHA's investigation of the Phillips 66 disaster involve deficiencies in what we now refer to as process safety management, emergency planning and response, building or facility egress and escape routes, and employee training. Recommended layout criteria and separation distances were made available and proposed by OSHA to the company [21].

2.2.3 PRESSURIZED HYDROGEN TANK RUPTURE

In 1991 at Hanau, Frankfurt, Germany, a tank containing 100 m^3 of hydrogen pressurized at 45 bar burst without apparent reason while it was stored outdoors in an industrial plant. The shock wave and the missiles of the tank shell caused heavy damage to the other units of the plant [22]. Investigations showed that welding in the metallic shell suffered from extensive cracks from the inner side to the outside. Likely, the corners along the welding caused concentration of stresses, so that the first cracks appeared in those parts. Under the influence of hydrogen, cracks grew much faster than normal, leading to material failure and subsequent mechanical explosion. As a result, all comparable tanks in the country were checked, manufacturing rules were revised to provide upper limits for corners, and new test methods for detecting cracks in early stages during operation were established. Undoubtedly, this accident contributed to improvements in hydrogen tank safety.

2.2.4 THE *CHALLENGER* DISASTER

The first launch of the space shuttle *Challenger* was on April 12, 1981, after seven years of development (1972–1979). Its three main engines from Rocketdyne operated with liquid hydrogen and liquid oxygen as fuel and oxidizer, respectively. The external tank from Martin Marietta contained 616 tons of liquid oxygen (1991 L) and 102 tons of liquid hydrogen (14,500 L). The solid rocket boosters (SRBs) from Thiokol each contained 503 tons of fuel (16% atomized aluminum powder as fuel, 69.83% ammonium perchlorate as oxidizer, 0.17% iron powder as catalyst, 12% polybutadiene–acrylic acid–acrylonitrile terpolymer (PBAN) as binder, and 2% epoxy curing agent).

On January 28, 1986, at 11:38 am Eastern Standard Time, the *Challenger* space shuttle left Pad 39B at the Kennedy Space Center in Florida for Mission 51-L. This was the tenth flight of the orbiter *Challenger*. Seventy-three seconds later, the space shuttle was completely destroyed owing to an explosion of the hydrogen tank (Figure 2.2). All seven crew members (six astronauts and one civilian) were killed.

FIGURE 2.2 Space shuttle *Challenger* at liftoff, its smoke plume after in-flight breakup, and icicles that draped the Kennedy Space Center on that day. (From https://images.nasa.gov/.)

The cause of the explosion was determined to be an O-ring failure in the right SRB. The unusually cold weather (see Figure 2.2), beyond the tolerances for which the rubber seals were approved, most likely caused the O-ring failure. The temperature at ground level at Pad 39B was 36°F (2.2°C); this was 15°F (8.3°C) colder than any other previous launch by NASA [16, 23].

The record-low temperatures of the launch reduced the elasticity of the rubber O-rings, reducing their ability to seal the joints. As a result, the broken seals caused a breach into the joint 73 seconds after liftoff. This allowed pressurized hot gas from within the SRB to leak and burn through the wall to the adjacent external hydrogen fuel tank, which led to the separation of the right-hand SRB's aft attachment and its crash into the fuel tank, which exploded [23].

2.3 INCIDENT REPORTING

Reporting of incidents related to hydrogen and analysis of the principal causes are useful for sharing lessons learned with the private and public sectors. For this purpose, the H_2 Incidents Database has been created by the Pacific Northwest National Laboratory with funding from the US Department of Energy (available at https://h2tools.org/lessons). In this database, incidents and near-misses are reported without including the names of the companies and other details; this confidentiality encourages reporting of events. The incidents are classified according to contributing factors, equipment, damage and injuries, probable causes, and contributing factors.

The most important records from this database in percentages of incidents by mid-year 2022 are shown in Figures 2.3 to 2.6. Thus, the percentage of hydrogen incidents in various contributing factors as reported in the H_2 Incidents Database for a total of 435 incidents is depicted in Figure 2.3. This figure shows that human error and situational awareness are by far the most frequent causes nowadays, followed by change in procedures, equipment, or materials, but this is expected to change in the years to come as there is a movement from the intense hydrogen research of today to the more widespread utilization of hydrogen.

The percentage of types of equipment involved in these incidents is shown in Figure 2.4 for a total of 353 incidents. As is the case for technological incidents in general, the simplest types of equipment (e.g., piping and valves) are most frequently involved in an incident because less attention is usually paid to them.

It is clear from Figure 2.5, in which percentages of damage and injuries during hydrogen incidents are shown, that for a total of 288 incidents only a small proportion resulted in loss of human life (5.2%). This is because special mitigation measures are usually taken, given knowledge of the severity of such incidents.

With regard to probable causes of hydrogen incidents, Figure 2.6 verifies for 347 incidents a situation that is well known in the process industries. Indeed, although equipment failure is by far the most common cause of incidents, in fact human error is hidden inside almost every other cause, including equipment failure, design flaws, inadequate maintenance, and failure to follow standard operating procedures.

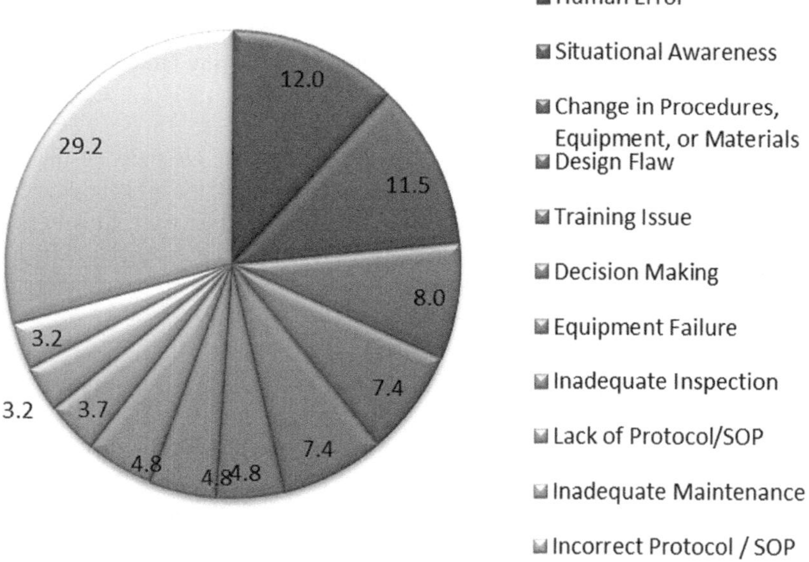

FIGURE 2.3 Percentage of incidents related to hydrogen in contributing factors as reported in the H_2 Incidents Database.

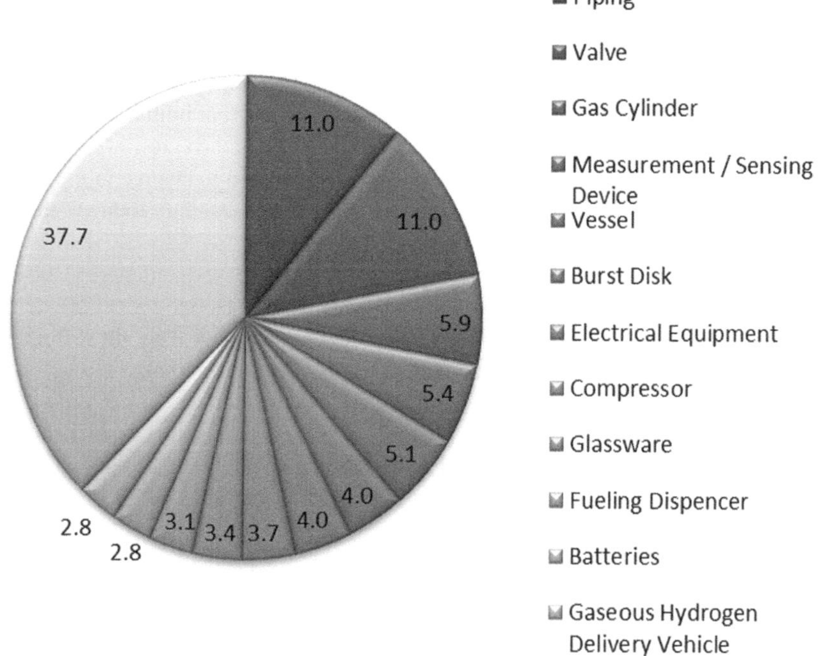

FIGURE 2.4 Percentage of different types of equipment involved in the incidents reported in the H_2 Incidents Database.

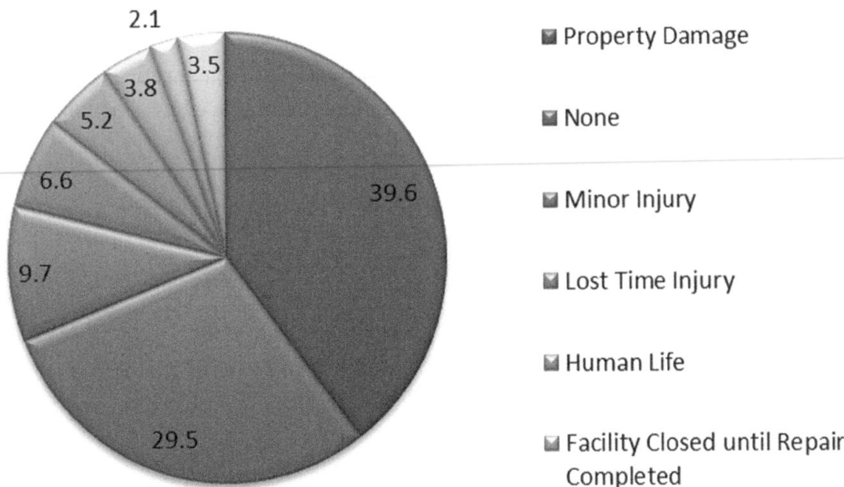

FIGURE 2.5 Percentage of types of damage and injuries from hydrogen incidents as reported in the H$_2$ Incidents Database.

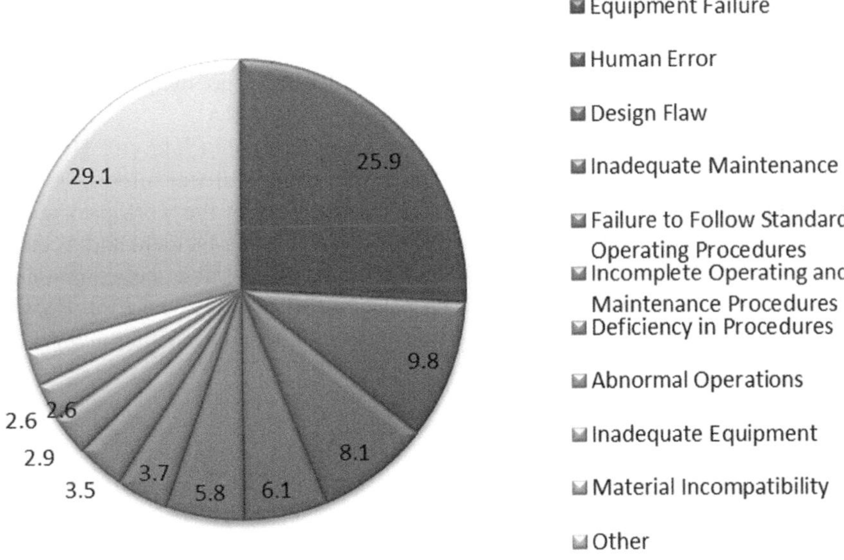

FIGURE 2.6 Percentage of probable causes for hydrogen incidents as reported in the H$_2$ Incidents Database.

The lessons learned from this analysis are the following:

- The most frequent contributing factors in hydrogen incidents are human error and situational awareness, followed by a change in procedures, equipment, or materials.
- Most incidents occur in the simplest equipment, such as piping and valves, to which one does not usually pay enough attention.
- Most incidents result in property damage, but only a few in injuries, probably due to the limited quantities of hydrogen still used and the mitigation measures taken.
- The most probable cause is equipment failure, while human error is hidden inside almost every other cause, including equipment failure, design flaws, inadequate maintenance, and failure to follow standard operating procedures.

The most recent incidents due to hydrogen that were recorded in 2022, according to the H_2 Incidents Database, are shown in Table 2.2. This cause and consequences analysis gives an idea of the broad range of factors that can lead to an accident when using, storing, or transporting hydrogen. Figure 2.7 shows one example from the database.

In addition to the H_2 Incidents Database, many other remarkable efforts have been developed or are under way, aimed at collecting and offering valuable information on past accidents to assist in the composition of new safety regulations, codes of practice, and standards, and in the prevention of similar accidents by supplying useful data for both qualitative and quantitative risk assessment (QRA).

Among them, the Network of Excellence project HySafe (Safety of Hydrogen as an Energy Carrier) greatly contributes to the successful transition of Europe to sustainable development based on hydrogen use. One of the work packages of HySafe that is under construction is the European Hydrogen Incident and Accident Database (HIAD 2.0) [25], dedicated to offering a collaborative and communicative database in the form of an open web-based information system [24]. Because of its significance, the HySafe project is presented in a separate chapter in this book (Chapter 9).

TABLE 2.2
Most Recent Accidents due to Hydrogen That Were Reported to the H_2 Incidents Database (Intended for Public Use)

Incident Type and Description	Damage and Injuries	Cause and Ignition	Lessons Learned/Suggestions/ Mitigation Measures
Hydrogen blast led to deaths at a US specialty silicones plant (see Figure 2.8). *Incident Date:* May 3, 2019	*Severity:* Incident *Injuries:* The explosion killed four workers and seriously injured a fifth.	*Leak*: Yes *Ignition*: The hydrogen explosion took place in an area where the company made a silicon hydride emulsion. The blast was felt 30 km from the facility. The force from the explosion was felt up to 20 miles away	AB Specialty Silicones commissioned a private independent study that determined the root cause to be "human error that resulted in the mistaken addition of an erroneous ingredient". The Chemical Safety Board (CSB) released a report detailing the events that led up to the massive explosion and fire and proposed prevention and mitigation measures. Among others the facility lacked functioning detectors for flammable gases.
A hydrogen storage tank at a new-energy research center exploded, destroying the center's 5100 m² building and damaging the windows and structures of neighboring buildings. *Incident Date*: May 23, 2019	*Severity*: Incident *Injuries*: Two people died and six others were injured.	*Leak*: Yes This reservoir was part of a research project combining the production of electricity by photovoltaic panels and a water electrolysis process.	The Korean government is still investigating the accident.

(continued)

TABLE 2.2 (Continued)
Most Recent Accidents due to Hydrogen That Were Reported to the H$_2$ Incidents Database (Intended for Public Use)

Incident Type and Description	Damage and Injuries	Cause and Ignition	Lessons Learned/Suggestions/Mitigation Measures
Pressure sensor diaphragm rupture on hydrogen compressor. *Incident Date*: Jan 15, 2019	*Severity*: Near-miss *Injuries*: No Facility closed until repair completed.	*Leak*: Yes *Ignition*: No The sensing diaphragm of a pressure transducer unexpectedly ruptured and released approximately 0.1 kg of hydrogen into the atmosphere from the compressor discharge line.	Maintain an internal process for verifying component wetted material compatibility for intended use as part of the procurement process for hydrogen system equipment.
An explosion followed by a fire occurred at a hydrogen production and storage company. Upon arriving, the firefighters saw several parked tankers that were burning and were likely to contain liquid hydrogen (tank containing 2–4 t of hydrogen) or gaseous hydrogen (cylinders totaling 200–500 kg of hydrogen). *Incident Date*: Jun 1, 2019	*Severity*: Incident *Injuries*: No	*Leak*: Yes *Ignition*: Yes	Nearby businesses and homes were evacuated. The fire was put out in just over one hour with no further explosions. No infrastructure outside the site was damaged. Firefighters used thermal imaging and took air samples to ensure that hydrogen no longer posed a threat. The hydrogen filling station was destroyed. The operator took its tanker fleet off the road to be inspected. The accident downed nine out of the 11 hydrogen fueling stations in the area, affecting 1000 owners of fuel-cell cars for several weeks. Site employees said that a few minutes before the explosion, hydrogen began leaking while it was being transferred to a distribution trailer.

Fueling station high pressure storage leak. *Incident Date:* Jun 10, 2019	*Severity*: Incident *Injuries*: No Facility closed until repair completed.	*Leak*: Yes *Ignition*: Yes A hydrogen leak originating from a tank within a high-pressure storage unit serving a hydrogen vehicle fueling station resulted in fire and explosion.	Implement rigorous assembly, verification, and documentation procedures for equipment.
An explosion followed by a fire occurred at a hydrogen fueling station. Hydrogen was produced on-site by an electrolyzer. Emergency services arrived seven minutes after the explosion. A 500 m cordon was set up. The motorway and nearby roads were closed to traffic. Firefighters brought the fire under control two and a half hours after the explosion. *Incident Date*: Jun 10, 2019	*Severity*: Incident *Injuries*: Yes The blast of the explosion caused the airbags of nearby vehicles to deploy, slightly injuring three people. The nation's hydrogen supply was interrupted. Makers of fuel-cell vehicles put deliveries of new vehicles on hold. All the operator's hydrogen fueling stations, whether featuring the same technology or not, were temporarily closed while an investigation was carried out in Europe, the USA, and South Korea.	*Leak*: Yes *Ignition*: Yes	After 17 days of investigating, the operator determined that the incident was caused by a leak on the high-pressure hydrogen storage unit (cylinders containing hydrogen compressed at 200 bar). The bolts on the ring between the coupling flange and one of the cylinders had not been adequately torqued. This allowed hydrogen to leak out and form a highly explosive mixture with air. The source of ignition has yet to be determined. The operator scheduled inspections at stations featuring the same assembly systems (four in Norway, three in Iceland, and three in Germany).

(continued)

TABLE 2.2 (Continued)
Most Recent Accidents due to Hydrogen That Were Reported to the H$_2$ Incidents Database (Intended for Public Use)

Incident Type and Description	Damage and Injuries	Cause and Ignition	Lessons Learned/Suggestions/Mitigation Measures
			It updated its assembly procedures and made improvements to the quality of its verifications (double checks). It looked into ways to improve detection of hydrogen leaks and avoid the presence of ignition sources at its sites (flat surfaces free of gravel, better ventilation).
Unauthorized field modified equipment; drain lines cause control cabinet explosion. *Incident Date:* Mar 22, 2018	*Severity:* Incident *Injuries:* Yes. The technician was forcefully hit by the flying metal panel holding the switch and sustained serious injuries requiring lengthy hospitalization and rehabilitation. Two were hospitalized. Two others were injured. Significant damage to the indoor facility also occurred.	*Leak:* Yes. *Ignition:* Yes. A sealed, unclassified electrical control enclosure, part of a listed and certified force-ventilated commercial hydrogen processing unit enclosure, exploded when the equipment manufacturer's technician pressed the machine stop switch to complete factory commissioning procedure.	Do not interconnect hydrogen drain trap lines with other drain lines. Thoroughly inspect equipment installation to manufacturer's installation instructions. Warn against unauthorized field modifications. Consult the manufacturer and listing agency before any field modifications are undertaken.

Partially spent ammonia borane (AB) reaction with water. *Incident Date*: Feb 07, 2014	*Severity*: Incident *Injuries*: No. Project delay.	*Leak*: Yes *Ignition*: Yes As part of preparing for material disposal, a small fire occurred within a fume hood as a researcher was combining several spent AB samples that had previously been stored uncovered in the back of the hood for over six months.	The procedure for disposal of spent or partially spent AB has been modified so that it does not include the use of water. Instead, the AB is removed from containers and transferred for disposal by rinsing with mineral oil, silicone oil, or other similar inert materials. It is then disposed of as a slurry.
Reacting AB exposure to air. *Incident Date*: Sep 19, 2013	*Severity*: Incident *Injuries*: No Laboratory equipment damage.	*Leak*: Yes *Ignition*: Yes Spontaneous ignition. While performing hydrogen gas release experimentation by thermally reacting a slurry of AB powder in silicone oil in a plug flow reactor, a discharge port on the test reactor became loose.	The researcher's failure to pull the fire alarm was an oversight of required facility practice. Hot, reacting AB produces hydrogen as well as other pyrophoric impurities. Reactions should be carried out in inert atmospheres or purged with inert gases. As a consequence of this incident, an updated procedure was put in place to check critical fittings before each subsequent test and to purge the apparatus with argon in the area surrounding the reactor/fittings.
Pipe ruptures during hydrogen regeneration of catalyst. *Incident Date*: Nov 04, 2009	*Severity*: Incident *Injuries*: Minor injury Substantial damage to residential homes adjacent to the facility.	*Leak*: Yes *Ignition*: Yes A distillate dewaxing unit at an oil refinery was undergoing hot hydrogen regeneration of the catalyst when an explosion occurred.	An investigative communication notes that "mechanical integrity programs at refineries repeatedly emphasize inspection strategies rather than the use of inherently safer design to control the damage mechanisms that ultimately cause major process safety incidents".

(*continued*)

TABLE 2.2 (Continued)
Most Recent Accidents due to Hydrogen That Were Reported to the H_2 Incidents Database (Intended for Public Use)

Incident Type and Description	Damage and Injuries	Cause and Ignition	Lessons Learned/Suggestions/ Mitigation Measures
Rupture of refinery heat exchanger. *Incident Date*: Apr 02, 2010	*Severity*: Incident *Injuries*: The rupture fatally injured seven employees working in the immediate vicinity of the heat exchanger at the time of the incident.	*Leak*: Yes *Ignition*: Yes A petroleum refinery experienced a catastrophic rupture at one bank of three heat exchangers in a catalytic reformer/naphtha hydrotreater unit because of high temperature hydrogen attack (HTHA).	Carbon steel Nelson curve methodology cannot be depended on to prevent HTHA equipment failures and cannot be reliably used to predict the occurrence of HTHA equipment damage. Given the difficulty of inspecting for HTHA because the damage might not be detected, inherently safer design is a better approach to prevent HTHA. Process hazards analysis (PHA) and damage mechanism hazard reviews (DMHRs) need to carefully consider all assumptions. Effective programs need to be in place to manage and provide oversight for hazardous nonroutine work.
Hydrogen fire from valve packing during maintenance shutdown at chemical manufacturing plant. *Incident Date*: Mar 26, 2009	*Severity*: Incident *Injuries*: No Property damage.	*Leak*: Yes *Ignition*: Yes Ignition source unknown. A chemical plant experienced a valve failure during a planned shutdown for maintenance that caused hydrogen to leak from a valve and catch fire.	An important aspect of the reliability of a valve is the condition of the stem seal which tends to deteriorate with time and wear. Valves used in hydrogen service should be packed with the correct valve packing material and periodically checked for leaks as part of a regular maintenance program.

Unexpected burst disk rupture during testing activity. *Incident Date*: Jul 25, 2013	*Severity:* Incident *Injuries:* No	*Leak*: Yes *Ignition*: No A partial pressure sensor for an automated gas environment system (AGES) was not functioning correctly for pure hydrogen flow. While personnel were troubleshooting the problem, a burst disk ruptured resulting in a leak of hydrogen gas and actuation of a flammable gas alarm.	Personnel were focused on the AGES system test and results, not the compatibility of the test equipment. The manual valve was needed to successfully test the system, however the fact that this particular valve could not accommodate the full cylinder pressure was overlooked.
Hydrogen cylinder leak at fueling station. *Incident Date*: Mar 13, 2012	*Severity*: Incident *Injuries*: No	*Leak:* Yes *Ignition:* No An alarm sounded at a recently inaugurated hydrogen fueling station in a major metropolitan area. One out of a total of 120 high-pressure hydrogen cylinders, located on the roof of the fueling station, failed in service. Gaseous hydrogen was leaking from a screw fitting of the cylinder, but the hydrogen was not ignited.	The hydrogen supplier installed a fire-resistant material board adjacent to the high-pressure hydrogen storage banks to prevent any potential jet flames from affecting adjacent high-pressure cylinders for several minutes. The hydrogen supplier installed a semi-automated sprinkler system to cool the high-pressure hydrogen storage banks to prevent any potential escaping hydrogen gas that might ignite in jet flames from affecting other hydrogen cylinders.

(continued)

TABLE 2.2 (Continued)
Most Recent Accidents due to Hydrogen That Were Reported to the H_2 Incidents Database (Intended for Public Use)

Incident Type and Description	Damage and Injuries	Cause and Ignition	Lessons Learned/Suggestions/ Mitigation Measures
Hydrogen gas explosion in municipal refuse incineration facility. The workers injected water to remove some clinker blockage, and the water reacted with incinerated aluminum ash to form hydrogen, which caused the explosion. *Incident Date:* Jul 06, 1995	*Severity:* Incident *Injuries:* Yes Three workers were seriously burned by high-temperature gas that spouted from the inspection door, and one of them died ten days later. Property damage.	*Leak:* Yes *Ignition:* Yes. *Probable causes:* Human error and deficiency in procedures. *Ignition source:* Either hot clinker or sparks from chisel.	Aluminum should have been separated from the refuse prior to feeding it to the incinerator.
Hydrogen explosion and fire in a styrene plant in a large petrochemical complex. The explosion followed the release of about 30 kg of 700 psig hydrogen gas from a burst flange into a compressor shed. *Incident Date:* Apr 20, 1984	*Severity:* Incident *Injuries:* Yes Two men were killed and two others injured. Property damage.	*Leak:* Yes *Ignition:* Yes The operators were bringing the plant online increasing the hydrogen circulation pressure. *Probable causes:* Equipment failure, inadequate equipment, and hazard not identified in advance. *Ignition source:* Unknown.	Hydrogen gas was released through a failed 19 inch diameter gasket and ignited under the roof of the compressor shed where it was partially confined. Areas where hydrogen may be released should not be confined with walls and roofs.

Fatal accident due to bacterial hydrogen production in storage tank. The accident took place at an onshore processing facility for slop water from offshore petroleum industry. *Incident Date*: May 27, 2008	*Severity:* Incident *Injuries*: Yes One operator was killed. Two operators were trying to remove the lid from a manhole on top of a 1600 m³ storage tank. To open the rusted bolts holding the lid in place, they used a cutting disk. One man was thrown over from the explosion and killed. Property damage.	*Leak:* Yes *Ignition:* Yes The hydrogen inside the tank from biological fermentation exploded because of sparks from the cutting disk. *Probable causes:* Human error, flammable mixture in confined area, and hazard not identified in advance. *Ignition source:* Sparks from angle grinder.	Avoid anaerobic and stagnant conditions in slop water by providing sufficient circulation and/or aeration. Use inert atmosphere or open-top floating roof tank to avoid space where an explosive mixture could form. Use nonsparking tools where explosive atmospheres may develop.
Pickup collision with tractor pulling hydrogen tube trailer (Figure 2.7). The collision caused damage to tubes, valves, piping, and fittings, resulting in hydrogen release and fire. The hydrogen ignited and burned the rear of the semitrailer. *Incident Date*: May 01, 2001	*Severity:* Incident *Injuries*: Yes The tractor-semitrailer driver was killed as a result of blunt force trauma. The driver of the pickup truck received non-life-threatening injuries. Damage, cleanup, and lost revenues were estimated at USD155,000.	*Leak*: Yes *Ignition*: Yes *Probable cause:* Vehicle collision. *Ignition source:* No ignition source defined.	Increase physical protection, shielding, and securing of transported hydrogen tubes.

(*continued*)

TABLE 2.2 (Continued)
Most Recent Accidents due to Hydrogen That Were Reported to the H_2 Incidents Database (Intended for Public Use)

Incident Type and Description	Damage and Injuries	Cause and Ignition	Lessons Learned/Suggestions/ Mitigation Measures
Refinery hydrocracker pipe rupture and release of explosive mixture with hydrogen. A refinery hydrocracker effluent pipe section ruptured and released a mixture of gases, including hydrogen, which instantly ignited on contact with the air, causing an explosion and a fire. *Incident Date*: Jan 21, 1997	*Severity:* Incident *Injuries*: Yes An operator who was checking a field temperature panel at the base of the reactor and trying to diagnose the high-temperature problem was killed. A total of 46 other plant personnel were injured. Property damage.	*Leak*: Yes *Ignition*: Yes Excessive high temperature, likely in excess of 760°C, initiated in one of the reactor beds spread to adjacent beds and raised the temperature and pressure of the effluent piping to the point where it failed. *Probable causes*: Failure to follow standard operating procedures, human error, and incorrect engineering hazard analysis calculation. *Ignition source:* No ignition source defined.	Management must provide an operating environment conducive for operators to follow emergency shutdown procedures when required. Some backup system of temperature indicators should be used so that the reactors can be operated safely in case of instrument malfunction. Process hazard analysis must be based on actual equipment and operating conditions.

Source: [11, 12, 25].

FIGURE 2.7 Scene of a hydrogen tube trailer accident. (From H$_2$ Incidents Database intended for public use https://h2tools.org/sites/default/files/imports/files//Attachment%25201-Photos.pdf.)

FIGURE 2.8 The factory remains of a hydrogen release explosion at AB Specialty Silicones on May 3, 2019, that killed four workers and seriously injured a fifth. Courtesy of US Chemical Safety and Hazard Investigation Board (CSB). Public domain.

Source: [26, 27].

REFERENCES

1. Federal Institute for Materials Research and Testing, *Hydrogen Safety*, German Hydrogen Association, Brussels, 2002, 18–21.
2. Rigas, F., and Sklavounos, S., Hydrogen safety, in *Hydrogen Fuel: Production, Transport and Storage,* R. Gupta (Ed.), CRC Press/Taylor & Francis, Boca Raton, FL, 2008, 537–538.
3. Rigas, F., and Sklavounos, S., Evaluation of hazards associated with hydrogen storage facilities, *International Journal of Hydrogen Energy*, 30, 1501–1510, 2005.
4. Institution of Chemical Engineers (IChemE), *Accident Database*, Loughborough, UK, 1997. www.icheme.org/knowledge-networks/safety-centre/resources/accident-data/
5. Rosyid, O.A., System-Analytic Safety Evaluation of the Hydrogen Cycle for Energetic Utilization, Dissertation, Otto-von-Guericke University, Magdeburg, Germany, 2006.
6. Khan, F.I., and Abbasi, S.A., Major accidents in process industries and an analysis of causes and consequences, *Journal of Loss Prevention in the Process Industries*, 12, 361–378, 1999.
7. Bethea, R.M., *Explosion and Fire at Pasadena, Texas,* American Institute of Chemical Engineers, New York, 1996.
8. Ministère chargé de l'environnement, France, *Explosion dans une unité de craquage d'une raffinerie*, No 19423, 2000.
9. Center for Chemical Process Safety, *Guidelines for Hazard Evaluation Procedures*, American Institute of Chemical Engineers, New York, 1992, 69.
10. Warren, P., *Hazardous Gases and Fumes*, Butterworth-Heinemann, Oxford, 1997, 96.
11. ARIA Database, Accidentology Involving Hydrogen, Ministry of Ecology, Energy, Sustainable Development and Town and Country Planning, France. www.aria.develo ppement-durable.gouv.fr/wp-content/files_mf/SY_hydrogen_GB_2009.pdf
12. FLASH ARIA, Feedback on technological accidents, Hydrogen and Transport: The Risks Should Not Be Underestimated, Ministry of Ecological Transition, France, April 2024.
13. Yanez, J., Kuznetsov, M., Souto-Iglesias, A., An analysis of the hydrogen explosion in the Fukushima-Daiichi accident, *International Journal of Hydrogen Energy*, 40, 8261–8280, 2015.
14. Sovacool, B.K., et al., Balancing safety with sustainability: Assessing the risk of accidents for modern low-carbon energy systems, *Journal of Cleaner Production*, 112, 3952–3965, 2016.
15. Pacific Northwest National Laboratory / Battelle / US Department of Energy, *Hydrogen Incident Examples*, PNNL-29731, 2020.
16. Zuettel, A., Borgschulte, A., and Schlapbach, L. (Eds.), *Hydrogen as a Future Energy Carrier*, Wiley-VCH Verlag, Berlin, Germany, 2008, 14–20.
17. Wikipedia, Hindenburg disaster, https://en.wikipedia.org/wiki/Hindenburg_disaster
18. Mahoney, D.G. (Ed.), *Large Property Damage Losses in the Hydrocarbon-Chemical Industries: A Thirty-Year Review*, 15th ed., M & M Protection Consultants, New York, NY, 1993.
19. Silas, C.J., and Cox, G.A., *Phillips 66 Company's Response to OSHA Citations*, Phillips Petroleum Co., Bartlesville, OK, 1990.
20. OSHA, *Occupational Safety and Health Administration: Phillips 66 Company Houston Chemical Complex Explosion and Fire: A Report to the President*, US Department of Labor, Washington, DC, 1990.
21. Lees, F.P., *Loss Prevention in the Process Industries, Vol. 1, Ch. 10 (Plant Siting and Layout)*, Butterworths, London, 1980, 211–230.

22. Hord, J., Is hydrogen a safe fuel? *International Journal of Hydrogen Energy*, 3, 157, 1978.
23. Wikipedia, Space Shuttle Solid Rocket Booster, https://en.wikipedia.org/wiki/Space_Shuttle_Solid_Rocket_Booster
24. Kirchsteiger, C., Vetere Arellano, A.L., and Funnemark, E., Towards establishing an international hydrogen incidents and accidents database (HIAD), *Journal of Loss Prevention in the Process Industries,* 20, 98, 2007.
25. HIAD 2.0 for users, European Hydrogen Safety Panel (EHSP), *Fuel Cell and Hydrogen Joint Undertaking (FCH 2 JU)*, Statistics, lessons learnt and recommendations from the analysis of the Hydrogen Incidents and Accidents Database, 2022.
26. Johnson, Jeff, *Hydrogen blast led to deaths at US silicones plant*, Chemical and Engineering News, December 19, 2019, https://cen.acs.org/safety/industrial-safety/Hydrogen-blast-led-deaths-US/97/web/2019/12
27. CSB, U.S. Chemical Safety and Hazard Investigation Board (CSB), Explosion and Fire at AB Specialty Silicones Facility on May 3, 2019, No. 2019-03-I-IL, Factual Update Published: December 2019.

3 Hydrogen Properties Associated with Hazards

3.1 GENERAL CONSIDERATIONS

The atomic number of hydrogen is 1, and its mean atomic weight is 1.00794. Although it constitutes around 75% of universal mass, it is scarce on Earth as free hydrogen. The principal isotope of hydrogen is protium (symbol ^{1}H), composed of one proton, one electron, and no neutrons. It is negatively charged in ionic compounds (H$^-$) and positively charged in organic compounds (H$^+$). Hydrogen reacts with most elements, mainly with oxygen to create water, and is a constituent of most organic compounds. Being the simplest atom, an analytic solution to the Schrödinger equation has been found for it, thus contributing to the development of quantum mechanics.

Hydrogen gas was first prepared by Robert Boyle in 1671, via the reaction of metals with strong acids. In 1766, Henry Cavendish discovered the hydrogen element. A name was given to this element by Antoine Lavoisier in 1783, from the Greek word ύδρο – hydro, meaning "water", and γενής – genes, meaning "creator". In standard conditions, hydrogen is a colorless, odorless, tasteless, flammable diatomic gas with the molecular formula H_2. It is noteworthy that François Isaac de Rivaz made the first internal combustion engine powered by a mixture of hydrogen and oxygen as early as 1806 [1].

With regard to the relative rotation of the nuclear spin of the individual atoms, the following forms of hydrogen are defined: o-H_2, *ortho*-hydrogen; p-H_2, *para*-hydrogen; e-H_2, equilibrium hydrogen; and n-H_2, normal hydrogen. The form e-H_2 corresponds to the equilibrium concentration at definite temperatures.

Normal hydrogen is a mixture of 75% *ortho*-hydrogen and 25% *para*-hydrogen in thermodynamic equilibrium at room temperature (293.15 K or 20°C). These two forms are distinguished by the relative rotation of the nuclear spin of the individual atoms in the molecule. In *ortho*-hydrogen the molecules spin in the same direction (symmetric with parallel nuclear spins, ↑↑), whereas in *para*-hydrogen they spin in the opposite direction (antisymmetric with antiparallel nuclear spins, ↑↓) [2].

During the process of the cooling of hydrogen from room temperature to its normal boiling point (NBP = 21.2 K) the equilibrium concentration of *ortho*-hydrogen drops from 75% at room temperature to 50% at 77 K and finally to 0.2% at NBP. The

DOI: 10.1201/9781003313007-3

noncatalyzed conversion rate is very slow, with the half-life of the conversion being greater than one year at 77 K. The conversion reaction from *ortho-* to *para*-hydrogen is exothermic, and the heat of conversion is temperature dependent [3]. The heat of conversion is 670 J/g for 75% o-H_2, which is much higher than the heat of evaporation, 447 J/g. As a result of this internal heating, the loss of hydrogen from a storage vessel is enormous and reaches 30% after only 48 h [4]. Thus, in LH_2 (liquefied hydrogen) production facilities, catalysts should be employed to accelerate this conversion. The transformation from *ortho-* to *para*-hydrogen can be catalyzed by a number of surface-active and paramagnetic materials. For instance, n-H_2 can be adsorbed on charcoal cooled with liquid hydrogen and desorbed in the equilibrium mixture (e-H_2). The conversion may take only a few minutes, if a highly active form of charcoal is used. Other suitable *ortho–para* catalysts are metals such as tungsten, nickel, or any paramagnetic oxides like chromium or gadolinium oxides. The nuclear spin is reversed without breaking the H–H bond [2, 4].

The physical properties of these two molecular forms differ slightly, but their chemical properties are practically identical, thus presenting the same chemical behavior with regard to chemical hazards [5].

Table 3.1 gives selected thermophysical, chemical, and combustion properties of hydrogen [2, 5, 6]. Some of the key overall properties of hydrogen that are relevant to its employment as an engine fuel are also listed in this table together with the corresponding values of methane, the other promising gaseous fuel for engine applications, and those of gasoline [7]. Where property values for gasoline could not be found, these are represented by iso-octane vapor [8] or the arithmetic average of normal heptane and octane [9].

3.2 HYDROGEN GAS PROPERTIES RELATED TO HAZARDS

Detection: In atmospheric conditions, hydrogen gas is colorless, odorless, and not detectable in any concentration by human senses. It is not toxic but can cause asphyxiation by diluting the oxygen in the air, with a limiting oxygen concentration equal to 19.5% by volume, under which the atmosphere is considered oxygen deficient. Furthermore, the inability to detect hydrogen renders it a latent fuel, ready to ignite.

The major hazards caused by an accidental release of hydrogen are related to the possibility of creating an explosive mixture with air. In this respect, hydrogen is more hazardous than other flammable gases like methane, propane, or gasoline vapors, especially in confined spaces, due to its wide flammability and detonability ranges and its very low ignition energy. Even though hydrogen possesses the highest buoyancy of all gases, and this property makes the risks of an unwanted release decrease rapidly to acceptable levels in outdoor release or the presence of adequate ventilation, the installation of effective systems for the detection of explosive atmospheres should always be taken into consideration as a safety measure. Choosing from the hydrogen sensors available on the market, the electrochemical, catalytic, and thermal conductivity sensors are mainly used in industry, the semiconductor-based sensors are usually used in research

TABLE 3.1

Selected Comparative Properties of Hydrogen, Methane, and Gasoline

Property	Value			
	Hydrogen	**Methane**	**Gasoline**	**References**
Molecular weight	2.016	16.043	~107	7,9,38,25
	2.016	16.043	-	1,39
	2.02	16.04	-	40,41
Normal melting point (K)	14.1	90.68	213	38,6
	13.99	90.694	-	1
	13.95	-	-	42
	14.01	-	-	39
Normal boiling point (K)	20.268	111.632	310–478	5,7,9,38,6,39
	20.271	111.6	-	1
	20.35	111.57	310–477	42,25
Critical temperature (K)	32.97–33.1	190	-	38,6
	33.1	190.56	-	1
	32.938	-	-	42
	33.2-33.25	190.65	-	40,41,39
Critical pressure (MPa or atm)	1.8	4.6	-	38,6
	1.269	4.599	-	1
	1.297	-	-	42,39
	1.315	4.55	-	40,41
Density of vapor at NBP	1.338	-	-	2,5,6
(kg/m^3)	1.331	1.82	-	42
	1.34	1.816	-	40,41
	1.438	-	-	43
	1.338	-	-	39
	1.312	1.819	-	25
Density of liquid at NBP	70.78	423.8	745 (at STP)	2,5,6
(kg/m^3)	67.76	-	-	42
	70.78	422.62	-	39,25
Density of gas at NTP (g/m^3)	82 (at 300 K)	717	5,110	8
	83.764	651.19	~4400	9
	84	650	-	44
	83.8	651	-	1,40,41
	83.45	668	-	25,39
Density of gas at STP (g/m^3)	84	650 657	4400	2
	89.87	(at 298.2 K)	-	38, 6
	89.88	717	-	1
	89.9	717	-	43,39
Heat of fusion at 14.1 K (kJ/kg)	58	58.59	-	38
	58.04	-	-	1
	58.09	-	-	42

TABLE 3.1 (Continued)
Selected Comparative Properties of Hydrogen, Methane, and Gasoline

Property	Value			References
	Hydrogen	Methane	Gasoline	
Heat of vaporization (kJ/kg)	445.6	509.9	250–400	2,39
	447	-	-	38
	445.6	-	-	25
	448.4	-	-	1
	446	-	-	42
	451.9	760	349 (at 15.6°C)	25
Heat of combustion (low) (kJ/g)	119.93	50.02	44.5	2,5,9,44,25
	119.7	46.72	44.79	8
	120	48	-	40,41
Heat of combustion (high) (kJ/g)	141.86	55.53	48	2,5,9,25
	141.8	55.3	48.29	2
	141.7	52.68	-	8
	141.87	55.54	-	1
	142	53	-	40,41
	143.05	-	-	39
Flammability limits in NTP air (vol%)	4–75	5.3–15	1–7.6	2,5,7,9
	4–75	5.3–15	1.2–6	8,39
	4–77	4.4–16.5	1.4–7.6	6,44
	4–75	4.4–17	-	1
	4–75	-	-	42
	4.1–74	5.3–15	-	40,41
Flammability limits in NTP oxygen (vol%)	4.1–94	-	-	5
Detonability limits in NTP air (vol%)	18.3–59	6.3–13.5	1.1–3.3	2,5,7,9
	13.5–70	-	-	45
	18–59	6.3–14	-	44
	18.2–58.9	5.7–14	-	40,41
	13–70	-	-	39
Detonability limits in NTP oxygen (vol%)	15–90	-	-	5
	-	8.25–55.8	-	46
Stoichiometric composition in air (vol%)	29.53	9.48	1.76	2,5,9
	29.5	9.5	-	44
Minimum ignition energy in air (mJ)	0.017	0.29	0.24	5,9
	0.02	0.29	0.24	2,44
	0.02	0.28	0.25	8
	0.14	-	0.024	7
	0.019	-	-	39
	0.017	0.25	-	47
Minimum ignition (detonation) energy in air (mJ)	$1.0 \cdot 10^7$	$2.3 \cdot 10^{11}$	-	47

(continued)

TABLE 3.1 (Continued)
Selected Comparative Properties of Hydrogen, Methane, and Gasoline

Property	Value			
	Hydrogen	Methane	Gasoline	References
Autoignition temperature (K)	858	813	501–744	2,5,9
	858	813	500–750	8
	833	813	501–744	44
	773	810	-	1
	673.15	-	-	42
	844	813	-	40
Adiabatic flame temperature in air (K)	2,318	2,148	~2,470	2,5,8,9,39
	—	2,190	—	8
Thermal energy radiated from flame to surroundings (%)	17–25	23–33	30–42	5,7,9
Burning velocity in NTP air (cm/s)	265–325	37–45	37–43	5,9
	346	43	-	44
	190	38	37-43	8
Burning velocity in STP air (cm/s)	346	45	176	2
Detonation velocity in NTP air (km/s)	1.48–2.15	1.39–1.64	1.4–1.7	5,9
Detonation velocity in STP air (km/s)	1.48–2.15	1.4–1.64	1.4–1.7	2
Energy of stoichometric mixture in NTP air (MJ/m^3)	3.58	3.58	3.91	48
Velocity of sound of vapor at NBP (m/s)	305	--	--	5
	355			39
Velocity of sound of liquid at NBP (m/s)	1273	--	--	5
	1093			39
Diffusion coefficient in NTP air (cm^2/s)	0.61	0.16	0.05	4,6,8
Diffusion coefficient in STP air (cm^2/s)	0.61	0.16	0.05	2
Buoyant velocity in NTP air (m/s)	1.2–9	0.8–6	Nonbuoyant	5,9
Limiting oxygen index (vol%)	5	12.1	11.6	2,9
Maximum experimental safe gap in NTP air (cm)	0.008	0.12	0.07	5,9
	0.008	-	0.96-1.02	25
Quenching gap in NTP air (cm)	0.064	0.203	0.2	5,8,9
Detonation induction distance in NTP air	L/D ~100	-	-	9

Note: STP (standard temperature and pressure): 273.15 K (0°C), 101.3 kPa (1 atm); NTP (normal temperature and pressure): 293.15 K (20°C), 101.3 kPa; NBP (normal boiling point): boiling point at 101.3 kPa; L/D: length to diameter ratio.

laboratories, and the MEMS (micro-electro-mechanic systems) are used in the aeronautic and space industries. Obviously, hydrogen detection devices themselves should not be a source of ignition and their response times should be sufficiently rapid [10].

Volumetric leakage: Since hydrogen is odorless, colorless, and tasteless, human senses are unable to detect a leak. By comparison, natural gas is also odorless, colorless, and tasteless, and this is the reason the industry adds a sulfur-containing odorant, such as a mercaptan, to make it detectable to people. In contrast, all known odorants are unsuitable for hydrogen detection because they contaminate fuel cells[1]. It is noteworthy that even minimal-scale hydrogen leaks can maintain stable flames at very low flow rates, as low as 4 μg/s. This flow rate limit is about an order of magnitude lower than the corresponding limits for methane and propane [11].

It is also remarkable that hydrogen gas leaking into external air may spontaneously ignite. Moreover, a hydrogen fire, although extremely hot, is almost impossible to see and thus can lead to accidental burns[1].

With the anticipated large-scale introduction of hydrogen as an energy carrier, its volumetric leakage from containers and pipelines is expected to be 1.3–2.8 times as large as gaseous methane leakage and approximately four times that of air under the same conditions. Thus comes the rule: "airproof is not hydrogen-proof". On the other hand, any released hydrogen has the potential to disperse rapidly by fast diffusion, turbulent convection, and buoyancy, thus considerably limiting its presence in the hazardous zone[2].

Ignition, autoignition, and spontaneous ignition phenomena: Apart from open flames, electrostatic sparks, and hot work, hot surfaces can also trigger the ignition of a flammable gaseous mixture. A hot surface can exist during normal operations, for instance from not properly lubricated bearings of pumps or motors, or may occur as a result of mechanical action, such as intense friction and tool impact during repair works. The term "hot surfaces" includes both hot spots and hotplate ignition [12, 13]. A smoldering layer of dust of a combustible material on a hot surface of an improperly cleaned machine can also result in the ignition of a combustible hydrogen–air mixture due to the extremely low minimum ignition energy of hydrogen in the air [14, 15]. In addition to friction sparking that may trigger a hydrogen–air ignition, thermite sparking produces relatively hotter sparks than friction sparking and it involves aluminum dust on rusted steel. The mixture can be ignited by mechanical energy like friction or a tool impact.

Large high-pressure hydrogen leaks would likely ignite spontaneously within fractions of a second, as shown by some jet release experiments [12, 16]. This information is important for risk estimation and risk reduction measures for such situations. The problems stemming from large-scale releases would be that there are no known precaution measures to minimize ignition sources, and there would be insufficient time for any action to be undertaken. Thus, there is a severe knowledge gap, and more work will be needed to better understand these phenomena. Numerous experimental and numerical studies can be found in the literature concerning the spontaneous ignition of high-pressure release of hydrogen in the air [17–24].

Electrostatic ignition: Since hydrogen is an electrical insulator in both gaseous and liquid phases, the flow or agitation of hydrogen gas or liquid may create static electricity charges, as happens with all non-conductive liquids or gases. For this reason, all hydrogen conveyors must be fully grounded. High-velocity hydrogen flow may also be responsible for the generation of static electrical charges in suspended particulates by so-called "triboelectricity". This is a type of electrification in which certain materials become electrically charged after being rubbed down with different materials and then separated [25]. As shown in Table 3.1, a stoichiometric mixture of hydrogen with air has a very low minimum ignition energy of 0.017 mJ. That means hydrogen is far more sensitive to ignition than most other gaseous or vaporized flammable materials. Therefore, the potential for electrostatic ignition is much greater. There are three main types of electrostatic discharges: spark, brush, and corona. Spark discharges are characterized by a single plasma channel between a high-potential conductor and an earthed conductor. The discharge lasts a very short time, and the discharge energy is calculable from the equation [26]

$$E = 1/2\ CV^2 \tag{3.1}$$

where E is energy in Joules, C is capacitance in farads, and V is voltage in volts. The discharge of static electricity accumulated on a human body via triboelectricity can be simulated as follows. The human body is considered as a capacitor of 100 to 400 pF charged to a voltage of 4000 to 50,000 V. Then, in touch with an object, this energy is discharged in less than a microsecond creating a spark with a maximum energy depending on the capacitance of the affected body [27].

Taking the values of 100 and 400 pF as the capacitance of a person, and the minimum ignition energy of 0.017 mJ for a stoichiometric mixture of hydrogen with air, then the voltage required to produce a spark of sufficient energy to ignite the atmosphere is calculated from Equation 3.1 as 290 to 580 V, which is much lower than the voltage created by triboelectricity on a human body.

Consequently, special antistatic measures should be taken everywhere a hydrogen leak may occur. The usual ways to discharge the buildup of static electricity on human bodies are bonding and earthing. Air humidifiers, ionizers, and antistatic safety boots with soles with good conductivity are suitable prevention measures. Yet, antistatic shoes should not be confused with insulating shoes, which provide just the opposite benefit, i.e. some protection against serious electric shocks from the mains voltage [27].

Brush discharges: These are typified by a discharge between a charged insulator and a conducting earthed point. They consist of many separate plasma channels, and they usually involve insulating plastics. A capacitance and hence an energy cannot be determined for the charged surface since this is a non-conductor. Gibson and Harper [28] introduced the term incendivity or equivalent energy which is when the brush discharge can ignite an atmosphere with the same minimum ignition energy as a spark capable of igniting this flammable atmosphere. Typical equivalent energies were found to be about 4 mJ for brush discharges from flat polyethylene sheets. Nevertheless, Ackroyd and Newton [29] found that certain plastics and thin plastic coatings on an earthed metal substrate (as is the case with lined pipework and vessels) had higher equivalent energies, and consequently, incendive brush discharges would not occur.

Corona discharges: They are silent, usually continuous discharges characterized by current but no plasma channel, which can ignite a hydrogen–air mixture. The usual practice for preventing the ignition of hydrogen, deliberately venting to the atmosphere, is possible by using a polished toroidal ring at the end of the vent aiming at the creation of a sufficiently enlarged tip radius. Nevertheless, corrosion and dirt deposits create small surface protrusions, which affect the polished finish and may result in corona discharges. Studies on hydrogen vents have shown that ignition was rare during fine weather, but it was more frequent during thunderstorms, sleet, falling snow, and cold frosty nights [26].

Buoyancy: As shown in Table 3.1, hydrogen gas is about 14 times lighter than air in normal conditions (normal temperature and pressure (NTP)), and this is why any leak quickly moves upward, thus reducing ignition hazards. Nevertheless, saturated vapor is heavier than air and will remain close to the ground until the temperature rises. Buoyant velocities range from 1.2 to 9 m/s in NTP air since they depend on the difference in air and vapor densities. Thus, the cold dense fuel vapors produced by LH_2 spills will initially remain close to the ground and then rise more slowly than standard temperature and pressure fuel gases [5, 30, 49].

Flame visibility: A hydrogen–air–oxygen flame is nearly invisible in daylight, irradiating mostly in the infrared (IR) and ultraviolet (UV) regions. Thus, any visibility of a hydrogen flame is caused by impurities such as moisture or particles in the air (Figure 3.1). Nevertheless, hydrogen fires are readily visible in the dark, and large hydrogen fires are detectable in daylight by the "heat ripples" and the thermal radiation to the skin [9]. At reduced pressures a pale blue or purple flame may be visible. Compared to hydrocarbon combustion, hydrogen flames radiate considerably less heat. Consequently, human physical perception of this heat is not possible until direct contact is made with a flame. An obvious hazard resulting from this property may be severe burns on persons exposed to hydrogen flames due to the ignition of hydrogen gas escaping from leaks. Therefore, a hydrogen fire may remain undetected and aggravate without a sign in areas where hydrogen can leak, accumulate, and finally form potentially combustible or explosive mixtures. As a result, the extinguishing of a hydrogen fire is not allowed unless the leak has stopped [31]. Hydrogen fire detectors can help to take instant actions in case of a leak, and they can be either fixed for continuous monitoring of remote operations or portable for field operations. Flame-visibility hydrogen fire detectors usually rely on UV and IR light detection detecting specific spectral radiation emitted during the combustion process by the various chemical species (ions, radicals, and molecules) that are either intermediates or final products of combustion. The hydroxyl radical (OH) and water are the main emitting chemical species found in hydrogen flames. These detection techniques assume that no interfering shields exist between the flame and the UV/IR detector [10].

FIGURE 3.1 The Space Shuttle main engine burnt hydrogen with oxygen, producing a nearly invisible flame at full thrust. From NASA, public domain. (From Wikimedia Commons, the free media repository https://commons.wikimedia.org/wiki/File:Shuttle_Main_Engine_Test_Firing_cropped_edited_and_reduced.jpg.).

Flame temperature: The flame temperature for 19.6 vol% hydrogen in air has been measured as 2318 K [2]. Nevertheless, the radiation emitted from a hydrogen flame is very low due to strong absorption by the ambient water vapor (emissivity $\varepsilon < 0.1$) unlike hydrocarbon flames ($\varepsilon \sim 1$) [32]. More information for deflagration and detonation temperatures and pressures are given in Chapter 4, Section 4.3.4.2, derived by the Gordon–McBride code [33, 34].

Burning velocity: Burning velocity in air is the subsonic velocity at which a flame of a flammable fuel–air mixture propagates. For hydrogen, this velocity ranges from 2.65 to 3.46 m/s, depending on pressure, temperature, and mixture composition. This high burning velocity of hydrogen, which is one order of magnitude higher than that of methane (maximum burning velocity in air at standard temperature and pressure (STP): 0.45 m/s), indicates its high explosive potential and the difficulty of confining or arresting hydrogen flames and explosions [2, 5].

Deflagration-to-detonation transition: Deflagration-to-detonation transition (DDT) is a sudden increase from the deflagration velocity of a flame, which is around the speed of sound for a near-stoichiometric hydrogen–air mixture, to the detonation velocity, which is about twice the maximum deflagration velocity [25]. The flame propagation velocity affects the severity of an explosion and, taking into account the quenching gap as well, is crucial for the design of flame arresters. Furthermore, a higher propagation velocity points to a higher drift for the combustible gas mixture to develop a DDT in tunnels or pipes of considerable length. Deflagration is an explosion front propagating with a flame velocity equal to subsonic or up to sonic velocity. In deflagration, the energy transfer from the reaction zone to the unreacted zone is achieved through well-known transport phenomena such as heat and mass transfer. Detonation is a more energetic explosion that propagates at a supersonic flame front velocity with the energy transferred from the reaction zone to the unreacted zone via a highly reactive shock wave [9].

Maximum experimental safe gap in NTP air: This is the maximum permissible clearance, between flat parallel steel surfaces, that prevents the propagation of dangerous flames or sparks through the gap. It is the largest gap size that does not permit ignition outside the test enclosure and is very important in the design and manufacture of explosion-proof equipment [9].

Quenching gap in NTP air: This is defined as the spark gap between two flat parallel plate electrodes at which ignition of combustible fuel–air mixtures is suppressed, i.e. smaller gaps have the effect of totally suppressing spark ignition and flame propagation [9].

Quenching distance in NTP air:

- Hydrogen flames are difficult to quench. Paradoxically, premixed hydrogen–air combustion can be intensified by applying heavy sprays of water. This is caused by induced turbulence and the ability of the mixture to burn around the droplets.
- Hydrogen has the lowest quenching distance compared to other flammable gases.
- Usually, quenching distance is reported as the minimum pipe diameter through which a premixed flame can pass. For example, the technical report of ISO/TR 15916:2004 states that the quenching gap in NTP air for hydrogen is 0.64 mm.
- The quenching distance decreases with the increase of pressure and temperature and depends on mixture composition [49].

Thermal energy radiation from flame: Exposure to hydrogen fires can result in significant damage from thermal radiation, which depends largely on the amount of water vapor in the atmosphere. In fact, atmospheric moisture absorbs the thermal energy radiated from a fire and can reduce it considerably. The intensity of radiation from a hydrogen flame at a specific distance depends on the amount of water vapor present in the atmosphere and is expressed by the equation [5]:

$$I = I_0 \cdot e^{-0.0046wr} \tag{3.2}$$

where I_0 is the initial intensity (energy/time·area), w is water vapor content (percent by weight), and r is distance (meters).

Limiting oxygen index: The limiting oxygen index is the minimum concentration of oxygen that will support flame propagation in a mixture of fuel vapors and air. For hydrogen, no flame propagation is observed under NTP conditions if the mixture contains less than 5 vol% oxygen [5]. This limiting oxygen index for hydrogen is much lower than those needed for methane and gasoline fumes, which demonstrates the increased risk of hydrogen leaks even in a partially inert atmosphere. The gases used to obtain inert gas dilution are nitrogen and carbon dioxide for release scenarios where the leak rate is small, i.e. where it will take minutes to build up any dangerous gas clouds. Hydrofluorocarbons (HFCs) are normally used in situations where the leak rate is large and protection will be needed in seconds rather than minutes. Yet HFC gases should only be applied in situations with a very low leak frequency but potentially severe consequences due to environmental concerns. Examples of application areas could be airplanes, submarines, and nuclear power plants where a passive autocatalytic recombiner (PAR) is normally used. Passive autocatalytic recombiners are devices that promote the recombination of hydrogen with oxygen available as a constituent in the air, forming water. Consequently, these devices provide a hydrogen sink and may serve to avoid, remove, or at least slow down the formation of flammable mixtures caused by the accidental increase of hydrogen in a closed area [10, 35, 36].

Considering the pre-ignition of a potentially flammable hydrogen mixture, inert gas dilution or another method to prevent its formation is necessary as a countermeasure in spaces where people are present, and it would not be wise to maintain an inert or partially inert atmosphere constantly. In this case, an alternative to prevent ignition would be to activate inert gas dilution when a leak is detected. In the case of a post-ignition situation where hydrogen has already ignited in the form of a jet fire, fine water-mist dilution to reduce flammability or sprinklers to improve mixing/dilution can be used, amongst other methods [10].

Joule–Thomson effect: When gases are expanded through a porous plug or a small aperture or nozzle from high to low pressure, they are usually cooled. However, the temperature of some real gases increases when they are expanded at a temperature and pressure beyond the temperature and pressure conditions that define their Joule–Thomson inversion curve. This maximum inversion temperature for hydrogen is 202 K at an absolute pressure of zero [2]. So, at temperatures and pressures greater than these, the temperature of hydrogen will increase upon expansion. With regard to safety, the increase of temperature as a result of the Joule–Thomson effect is not normally sufficient to ignite a hydrogen–air mixture. For instance, the temperature of hydrogen increases from 300 K to 346 K when it expands from a pressure of 100 MPa to 0.1 MPa. This increase in temperature is not sufficient to ignite hydrogen, whose autoignition temperature is 858 K at 1 atm and 620 K at low pressures [5, 49].

Hydrogen may be transported for long-distance delivery of hydrogen using existing natural gas pipelines after blending with natural gas (NG), but this has pros and cons. Thus, the added hydrogen can significantly change the Joule–Thomson coefficient of natural gas, which is a crucial factor for the liquefaction of natural gas and the formation of natural gas hydrate in pipelines. Specifically, hydrogen blending reduces the risk of gas hydrate blockage at the pipeline valves. On the other hand, regarding natural gas liquefaction, the temperature liquefaction reduction of the hydrogen-blended natural gas is smaller than that of natural gas without hydrogen under the same pressure drop. To obtain the same cooling effect, a higher pressure drop is required for the natural gas–hydrogen blend, thus increasing the cost [37].

Toxicity: Although hydrogen is not toxic, as an asphyxiant it can cause dizziness without warning, coma, and even death when breathed for sufficient time. Asphyxia or asphyxiation is a condition of insufficient supply of oxygen to the body that can cause hypoxia, which hinders the ample oxygenation of tissues and organs. Asphyxia is defined as the difficulty of a person getting enough oxygen through breathing for an extended time[1, 31].

Hydrogen embrittlement: Hydrogen dissolves in many metals and this dissolution is followed by leaking out. This may have adverse effects on the affected metals, such as hydrogen embrittlement leading to cracks and explosions due to loss of containment [1]. More details of this phenomenon are given in Chapter 4 (Section 4.2.1).

3.3 LIQUEFIED HYDROGEN PROPERTIES RELATED TO HAZARDS

All hazards accompanying hydrogen gas (GH_2) also exist with liquefied hydrogen (LH_2) due to its easy evaporation. Additional hazards should be taken into account when handling or storing liquid hydrogen because of that ease of evaporation.

Low boiling point: The boiling point of hydrogen at sea level pressure is 20.3 K. Any liquid hydrogen splashed on the skin or in the eyes can cause frostbite burns or hypothermia. Inhaling vapor or cold gas initially produces respiratory discomfort, and further breathing in can cause asphyxiation.

Ice formation: Vents and valves in storage vessels and dewars may be blocked by accumulation of ice formed from moisture in the air. Excessive pressure may then result in mechanical failure (container or component rupture) with jet release of hydrogen and potentially in a boiling liquid expanding vapor explosion (BLEVE) accompanied by high overpressures and a fireball with significant heat and radiation loads if ignited [10].

Rapid phase transition: Releases in water may result in rapid phase transition (RPT), a phenomenon due to the spontaneous and eventually explosive boiling of liquid hydrogen. It is caused by the favorable heat transfer conditions and a practically unlimited reservoir of heat available when large water quantities are involved, for instance in lakes and the sea. Emergency responses to liquid hydrogen spillages on the water should include a warning to boats in the area against approaching the gas cloud. In close proximity, even car traffic may have to be stopped or re-directed. Warning people in the area in some cases, especially downwind of the release, is also crucial, though the gas cloud is not likely to have a long duration due to its high buoyancy when warmed with the air blending, and so any benefit from the evacuation of people is not significant [10].

Continuous evaporation: The storage of hydrogen as a liquid in a vessel results in continuous evaporation, changing its state to gaseous hydrogen. Even with highly insulated containers, it is difficult to maintain such a low temperature to keep hydrogen in the liquid phase. As a result, the hydrogen will gradually leak away with a typical evaporation rate of 1% per day. To equalize pressure, hydrogen gas must be vented to a safe location or temporarily collected safely. Storage vessels should be kept under positive pressure to prevent air from entering and producing flammable mixtures. Liquefied hydrogen may be contaminated with air condensed and solidified from the atmosphere or with trace air accumulated during liquefaction of hydrogen. The quantity of solidified air can increase during repeated refilling or pressurization of storage vessels, producing an explosive mixture with hydrogen.

Pressure rise: Liquefied hydrogen confined, for instance, in a pipe between two valves will eventually warm to ambient temperature, resulting in a significant pressure rise. Standard storage system designs usually assume a heat leak equivalent to 0.5% of the liquid contents per day. Considering liquefied hydrogen as an ideal gas, the pressure resulting from a trapped volume of liquefied hydrogen at 1 atm vaporizing and being heated to 294 K is 85.8 MPa. However, the pressure is 172 MPa when hydrogen compressibility is considered [5].

High vapor density: The high density of the saturated vapor resulting immediately after release from a liquefied hydrogen storage vessel that is leaking causes the hydrogen cloud to move horizontally or downward for some time [49]. This was shown experimentally by the National Aeronautics and Space Administration (NASA) [5] in the Langley Research Center at the White Sands test facility in 1980 and simulated effectively later using a computational fluid dynamics (CFD) approach [13].

Electric charge buildup: Since the electrical resistivity of liquefied hydrogen is about 10^{19} ohm-cm at 25 V, the electric current-carrying capacity is small and more or less independent of the imposed voltage. Investigation has shown that electric charge buildup in flowing liquefied hydrogen is not a great concern [5].

REFERENCES

1. Hydrogen, Wikipedia, https://en.wikipedia.org/wiki/Hydrogen
2. Zuettel, A., Borgschulte, A., and Schlapbach, L. (Eds.), *Hydrogen as a Future Energy Carrier*, Wiley-VCH Verlag, Berlin, Germany, 2008, Chap. 4.
3. Sullivan, N.S., Zhou, D., and Edwards, C.M., Precise and efficient *in situ* ortho– para-hydrogen converter, *Cryogenics*, 30, 734, 1990.
4. Yucel, S., Theory of ortho-para conversion in hydrogen adsorbed on metal and paramagnetic surfaces at low temperatures, *Physics Review B*, 39, 3104, 1989.

5. ANSI, *Guide to Safety of Hydrogen and Hydrogen Systems,* American Institute of Aeronautics and Astronautics, Reston, VA, American National Standard ANSI/AIAA G-095-2004, Chap. 2. 2004.

6. Wolfram Alpha, Computational Knowledge Engine, http://wolframalpha.com

7. Adamson, K.A., and Pearson, P., Hydrogen and methanol: A comparison of safety, economics, efficiencies and emissions, *Journal of Power Sources,* 86, 548, 2000.

8. Karim, A., Hydrogen as a spark ignition engine fuel, *International Journal of Hydrogen Energy,* 28, 569, 2003.

9. Hord, J., Is hydrogen a safe fuel? *International Journal of Hydrogen Energy,* 3, 157, 1978.

10. HySafe, Hydrogen Safety Barriers and Safety Measures, Chapter V, Version 1.0, May 2006a. chrome-extension://efaidnbmnnnibpcajpcglclefindmkaj/http://www.hysafe. org/download/1038/BRHS_Chap5_Prevention%20version_0_9_0.pdf

11. Butler, M.S., Moran, C.W., Sunderland, P.B., and Axelbaum, R.L., Limits for hydrogen leaks that can support stable flames, *International Journal of Hydrogen Energy,* 34, 5174–5182, 2009.

12. Groethe, M., Merilo, E., Colton, J., Chiba, S., Sato, Y., and Iwabuchi H., Large-scale hydrogen deflagrations and detonations, *International Journal of Hydrogen Energy,* 32, 2125–2133, 2007.

13. Grune, J., Sempert, K., Kuznetsov, M., and Jordan, T., Experimental study of ignited unsteady hydrogen releases from a high pressure reservoir, *International Journal of Hydrogen Energy,* 39, 6176–6183, 2014.

14. El-Sayed, S.A., and Abdel-Latif, A.M., Smoldering combustion of dust layer on hot surface, *Journal of Loss Prevention in the Process Industries,* 13, 509–517, 2000.

15. Zanoni, M.A.B., Torero, J.L., and Gerhard, J.I., Delineating and explaining the limits of self-sustained smouldering combustion, *Combustion and Flame,* 201, 78–92, 2019.

16. Rudy, W., Dabkowski, A., and Teodorczyk, A., Experimental and numerical study on spontaneous ignition of hydrogen and hydrogen-methane jets in air, *International Journal of Hydrogen Energy,* 39, 20388–20395, 2014.

17. Asahara, M., Yokoyama, A., Hayashi, A.K., Yamada, E., and Tsuboi, N., Numerical simulation of auto-ignition induced by high-pressure hydrogen release with detailed reaction model: Fluid dynamic effect by diaphragm shape and boundary layer, *International Journal of Hydrogen Energy,* 39, 20378–20387, 2014.

18. Xu, B.P. and Wen, J.X., The effect of tube internal geometry on the propensity to spontaneous ignition in pressurized, hydrogen release, *International Journal of Hydrogen Energy,* 39, 20503–20508, 2014.

19. Kim, Y.R., Lee, H.J., Kim, S., and Jeung, I.S., A flow visualization study on self-ignition of high pressure hydrogen gas released into a tube, *Proceedings of the Combustion Institute,* 34, 2057–2064, 2013a.

20. Kitabayashi, N., Wada, Y., Mogi, T., Saburi, T., and Hayashi, A.K., Experimental study on high pressure hydrogen jets coming out of tubes of 0.1e4.2 m in length, *International Journal of Hydrogen Energy,* 38, 8100–8107, 2013.

21. Kim, S., Lee, H.J., Park, J.H., and Jeung, I.S., Effects of a wall on the self-ignition patterns and flame propagation of high-pressure hydrogen release through a tube, *Proceedings of the Combustion Institute,* 34, 2049–2056, 2013b.

22. Morii, Y., Terashima, H., Koshi, M., and Shimizu, T., Numerical study of the effect of obstacles on the spontaneous ignition of high-pressure hydrogen, *Journal of Loss Prevention in the Process Industries,* 34, 92–99, 2015.

23. Gong, L., Duan, Q., Jiang, L., Jin, K., and Sun, J., Experimental study on flow characteristics and spontaneous ignition produced by pressurized

hydrogen release through an Omega-shaped tube into atmosphere, *Fuel*, 184, 770–779, 2016.

24. Kim, Y.R., Lee, H.J., Kim, S., and Jeung, I.S., A flow visualization study on self-ignition of high pressure hydrogen gas released into a tube, *Proceedings of the Combustion Institute*, 34, 2057–2064, 2013c.

25. Molkov, V., Fundamentals of hydrogen safety engineering, Part I, 2012, www.bookb oon.com

26. Astbury, G.R. and Hawksworth, S.J., Spontaneous ignition of hydrogen leaks: A review of postulated mechanisms, *International Journal of Hydrogen Energy*, 32, 2178–2185, 2007.

27. Static electricity, Wikipedia, https://en.wikipedia.org/wiki/Static_electricity

28. Gibson, N. and Harper, D.J., Parameters for assessing electrostatic risk from non-conductors—A discussion, *Journal of Electrostatics*, 21, 27–36, 1988.

29. Ackroyd, G.P. and Newton, S.G. An investigation of the electrostatic ignition risks associated with a plastic coated metal, *Journal of Electrostatics,* 59, 143–51, 2003.

30. Rigas, F. and Sklavounos, S., Evaluation of hazards associated with hydrogen storage facilities, *International Journal of Hydrogen Energy*, 30, 1501–1510, 2005.

31. CAMEO Chemicals, Hydrogen Chemical Datasheet, 2021.

32. HySafe, Biennal Report hydrogen Safety, Chapter I: Hydrogen Fundamentals, version 1.0, October 2006c. chrome-extension://efaidnbmnnnibpcajpcglclefindmkaj/http://www.hysafe.org/download/1040/BRHS_Chap1_Fundamentals-version%201_0_1.pdf

33. Rigas, F. and Skalvounos, S., *Hydrogen Safety" in "Hydrogen Fuel: Production, Transport and Storage*, CRC Press, Taylor & Francis, Boca Raton, Florida, USA, 2008, Chap. 16.

34. Gordon, S. and McBride, B.J., *Computer Program for Calculation of Complex Chemical Equilibrium Compositions and Applications*, NASA Reference Publication 1311, 1994.

35. Lopez-Alonso, E., Papini, D., and Jimenez, G., Hydrogen distribution and Passive Autocatalytic Recombiner (PAR) mitigation in a PWR-KWU containment type, *Annals of Nuclear Energy*, 109, 600–611, 2017.

36. Liang, Z., Gardner, L., Clouthier, T., and Thomas, R., Experimental study of effect of ambient flow condition on the performance of a passive autocatalytic recombiner, *Nuclear Engineering and Design*, 301, 49–58, 2016.

37. Li, J., Su, Y., Yu, B., Wang, P., and Sun, D., Influences of hydrogen blending on the Joule–Thomson coefficient of natural gas, *ACS Omega*, 6, 16722–16735, 2021.

38. Kirk-Othmer Encyclopedia of Chemical Technology, *Fundamentals and Use of Hydrogen as a Fuel,* 3rd ed., Vol. 4, Wiley, New York, 1992.

39. HySafe, Hydrogen Safety Barriers and Safety Measures, Chapter I, Version 1.0, May 2006b. chrome-extension://efaidnbmnnnibpcajpcglclefindmkaj/http://www.hysafe.org/download/1038/BRHS_Chap5_Prevention%20version_0_9_0.pdf

40. Tabkhi, F., Azzaro-Pantel, C., Pibouleau, L., and Domenech, S., A mathematical framework for modelling and evaluating natural gas pipeline networks under hydrogen injection, *International Journal of Hydrogen Energy*, 33, 6222–6231, 2008.

41. Messaoudani, Z. L., Rigas, F., Hamid, M.D.B., and Hassan, C.R.C., Hazards, safety and knowledge gaps on hydrogen transmission via natural gas grid: A critical review, *International Journal of Hydrogen Energy*, 41, 17511–17525, 2016.

42. Rivkin, C., Burgess, R., and Buttner, W., *Hydrogen Technologies Safety Guide*, Technical Report NREL/TP-5400-60948, National Renewable Energy Laboratory, 15013 Denver West Parkway Golden, CO 80401, U.S. Department of Energy, 2015.

43. Air Liquide Gas Encyclopedia, https://encyclopedia.airliquide.com/

44. Farias, C.B.B. et al., Use of hydrogen as fuel: A trend of the 21st century, *Energies*, 15, 311, 2022.

45. Baker, W.E. and Tang, M.J., *Gas, Dust and Hybrid Explosions*, Elsevier, Amsterdam, 1991, 42.

46. Michels H.J., Munday, G., and Ubbelohde, A.R. Detonation limits in mixtures of oxygen and homologous hydrocarbons, Proceedings of the Royal Society of London. A. Mathematical and Physical Sciences, 319, 461–477, 1970.

47. Tretsiakova-McNally, S., *Dealing with hydrogen explosions*, HyResponse, Grant agreement No: 325348, 30 September 2016.

48. Rosyid, A. and Hauptmanns, U., System analysis: Safety assessment of hydrogen cycle for energetic utilization, in *Proceedings of the International Congress Hydrogen Energy and Exhibition*, Istanbul, 2005.

49. HyResponder, *Hydrogen properties relevant to safety*, Project for Fuel Cells and Hydrogen 2 Joint Undertaking (JU), European Union's Horizon 2020, grant agreement No 875089, https://hyresponder.eu/wp-content/uploads/2021/04/Lecture-2-slides.pdf

4 Hydrogen Hazards

The hazards associated with hydrogen can be physiological (frostbite and asphyxiation), physical (component failures and embrittlement), or chemical (burning or explosion), the primary hazard being inadvertently producing a flammable or explosive mixture with air. Safety can be obtained only when designers and operational personnel are aware of all hazards related to the handling and use of hydrogen. In general, hydrogen safety concerns are not more severe than those we are accustomed to with gasoline or natural gas, but they are simply different [1].

Based on safety, a comparison study of hydrogen with methane, natural gas, propane, and gasoline shows that, qualitatively, the overall risks of the four fuels are close. Nevertheless, gasoline is the most toxic. For small leaks, hydrogen has the highest ignition probability, and these three gaseous fuels have the highest risk of a burning jet or a potentially hazardous cloud formation [2].

Nevertheless, other authors claim that hydrogen represents a more significant hazard than methane and gasoline due to the broader flammability limits, lower ignition energy, and higher deflagration index [3].

Strangely, most hydrogen hazards stem from the fact that hydrogen gas is odorless, colorless, and tasteless, so leaks are not detected by human senses. This is why hydrogen sensors are often used in industry to successfully detect hydrogen leaks. By comparison, natural gas is also odorless, colorless, and tasteless, but in industry mercaptans are usually added as odorants to make it detectable by people. Unfortunately, all known odorants contaminate fuel cells (a popular application for hydrogen) and are not acceptable in food applications (hydrogenation of edible oils) [1].

Virtual sensors for hydrogen safety prediction have been proposed for hydrogen safety using artificial neural networks and neuro-fuzzy inference systems [4]. Thermochemical hydrogen facial gas sensors composed of platinum-decorated graphene sheets and a thermoelectric polymer nanocomposite have been fabricated [5]. Palladium film hydrogen sensors based on suspended micro-hotplates have also been fabricated to operate at elevated temperatures with low power consumption [6]. All these are innovative ways of avoiding expensive instrumentation for hydrogen monitoring.

DOI: 10.1201/9781003313007-4

4.1 PHYSIOLOGICAL HAZARDS

Harm to people can be expressed in terms of injury or death after exposure to flames, radiant heat fluxes, extremely low temperatures, or air blast waves. Asphyxiation hazard exists if the oxygen content drops below 18 vol% because of hydrogen accumulation in the air. Direct skin contact with cold gaseous or liquid hydrogen leads to numbness and a whitish coloring of the skin and finally to frostbite. The risk is higher than that with liquid nitrogen because of the lower temperature and the higher thermal conductivity of hydrogen. Thus, personnel present during leaks, fires, or explosions of hydrogen mixtures with air can incur several types of injury [1, 7–9].

4.1.1 ASPHYXIATION

Hydrogen is not toxic and does not pose any acute or long-term physiological hazards. The only side effect of inhalation of the gas is sleepiness and a high-pitched voice. However, asphyxiation may occur when entering a region where hydrogen (as with any other nontoxic gas) has displaced the air, lowering the oxygen concentration below 19.5 vol%. The stages of asphyxiation based on the oxygen concentration are shown in Table 4.1.

4.1.2 THERMAL BURNS

Thermal burns result from the radiant heat emitted by a hydrogen fire and absorbed by a person. Due to the absence of carbon and the presence of heat-absorbing water vapor created when hydrogen burns, a hydrogen fire has significantly less radiant heat compared to a hydrocarbon fire. Absorbed radiant heat is directly proportional to many factors, including exposure time, burning rate, heat of combustion, size of the burning surface, and atmospheric conditions (mainly wind and humidity). Hydrogen flames are nearly invisible in daylight, and this has been the reason for fatal thermal

TABLE 4.1
Stages of Asphyxiation in the Presence of Hydrogen in Air

Oxygen in Air (vol%)	Consequences
15–19	Decreased ability to perform tasks and early symptoms in persons with heart, lung, or circulatory problems may be induced
12–15	Deeper respiration, faster pulse, poor coordination
10–12	Giddiness, poor judgment, slightly blue lips
8–10	Nausea, vomiting, unconsciousness, ashen face, fainting, mental failure
6–8	Death in 8 min. 50% chance of recovery with treatment in 6 min, 100% recovery with treatment in 4 to 5 min
4	Coma in 40 s, convulsions, respiration ceases, death

Source: Data from References [1,7,8,9].

burns suffered by victims approaching an invisible jet fire. Hydrogen fires last only one-fifth to one-tenth of the time of hydrocarbon fires. Thus, the fire damage is less severe because of:

- High burning rate resulting from rapid mixing and high propagation velocity.
- High buoyant velocity.
- High rate of vapor generation of liquid hydrogen.

Although the maximum flame temperature of hydrogen is not very different from that of other fuels, the thermal energy radiated from the flame is only a fraction of that of, for instance, a natural gas flame.

The amount of damage that a burn can cause depends on its location, its depth, and how much body surface area it involves. Burns are classified on the basis of their depth in the victim's body:

- A *first-degree burn* is superficial and causes local inflammation of the skin, which is characterized by pain, redness, and a mild amount of swelling.
- A *second-degree burn* is deeper, and in addition to the pain, redness, and inflammation there is also blistering of the skin.
- A *third-degree burn* is deeper still, involving all layers of the skin, in effect killing that area of skin. Because the nerves and blood vessels are damaged, third-degree burns appear white and leathery and tend to be relatively painless.

Burns are not static and may mature. Over a few hours, a first-degree burn may involve deeper structures and become second degree. Think of a sunburn that blisters the next day. Similarly, second-degree burns may evolve into third-degree burns.

In general, thermal radiation flux exposure levels may have the consequences shown in Table 4.2 [1, 7–10].

It is interesting that the radiant heat effects on humans are a function of both the heat flux intensity and the duration of exposure. Thus, it is generally accepted that

---

TABLE 4.2
Radiant Heat Flux Harm Criteria for People

Thermal Radiation Intensity (kW/m^2)	Consequences in Humans
1.6	No harm for long exposures
4–5	Pain felt in 20 s; first-degree burns in 30 s
9.5	Immediate skin reactions; second-degree burns after 20 s
12.5–15	First-degree burns after 10 s; 1% lethality in 1 min
25	Significant injury in 10 s; 100% lethality in 1 min
35–37.5	1% lethality in 10 s

Source: Data from References [1,7,8,9].

harm from radiant heat has to be expressed in terms of a thermal dose unit, as given by the equation:

$$\text{Thermal Dose Unit} = I^{4/3}t \tag{4.1}$$

where I is the radiant heat flux in kW/m^2 and t is the exposure duration in seconds.

Thermal doses from ultraviolet or infrared radiation that result in first-, second-, and third-degree burns are shown in Table 4.3 [10]. As can be concluded from this table, which was based on experiments using animal skin or nuclear blast data, infrared radiation is more hazardous than that in the ultraviolet spectrum.

Based on these thermal dose levels, *dangerous dose* levels have been defined as the dose resulting in 1% of deaths in the exposed population. Furthermore, LD_{50} values (lethal dose to 50% of the exposed population) have been determined only for infrared radiation, but with a wide range of values. The Health and Safety Executive (HSE) of the United Kingdom proposes the value of 2000 (kW/m^2)$^{4/3}$ to be used in offshore and gas facilities [10].

Nevertheless, such point values are useful for application in quantitative risk assessment (QRA) consequence studies. Thus, the *probit function* approach appears more suitable in risk assessment of the probability of injury or fatality from a certain dose level. In probability and statistics, the probit (probability unit) function is the inverse cumulative distribution function, or quantile function, associated with the standard normal distribution. Using this function, the probability of a certain degree of harm to humans (e.g., fatality) or structures (e.g., destruction) can be obtained from the equation:

$$P(\text{fatality}) = 50 \left[\frac{1+(Y-5)}{|Y-5|} + erf \frac{|Y-5|}{\sqrt{2}} \right] \tag{4.2}$$

where Y is the value of the probit function calculated from Table 4.4, and $erf(x)$ is the error function.

TABLE 4.3
Radiation Burn Data Caused by Ultraviolet or Infrared Radiation

	Threshold Dose (kW/m^2)$^{4/3}$	
Burn Severity	**Ultraviolet**	**Infrared**
First degree	260–440	80–130
Second degree	670–1,100	240–730
Third degree	1220–3100	870–2640

Source: [10]. With permission from Elsevier.

TABLE 4.4
Thermal Dose Probit Functions for Humans

Probit	Probit Equation	Comment
First-degree burn (TNO) [6]	$Y = -39.83 + 3.0186 \ln [V]^a$	Based on Eisenberg model but accounts for infrared radiation
Second-degree burn (TNO) [6]	$Y = -43.14 + 3.0186 \ln [V]^a$	Based on Eisenberg model but accounts for infrared radiation
Fatality (Eisenberg) [7]	$Y = -38.48 + 2.56 \ln [V]^a$	Based on nuclear data from Hiroshima and Nagasaki (ultraviolet radiation)
Fatality (Tsao and Perry) [8]	$Y = -36.38 + 2.56 \ln [V]^a$	Eisenberg model modified to account for infrared (2.23 factor)
Fatality (TNO) [9]	$Y = -37.23 + 2.56 \ln [V]^a$	Tsao and Perry model modified to account for clothing (14%)
Fatality (Lees) [10]	$Y = -29.02 + 1.99 \ln [V]^b$	Accounts for clothing, based on porcine skin experiments using ultraviolet source to determine skin damage, uses burn mortality information

TNO: The Netherlands Organisation for Applied Scientific Research, with headquarters at The Hague, the Netherlands.

[a] $V = I^{4/3}t$ = thermal dose in $[(W/m^2)^{4/3} \, s]$.

[b] $V = F \times I^{4/3}t$ = thermal dose in $[(W/m^2)^{4/3} \, s]$, where $F = 0.5$ for normally clothed population and 1.0 when clothing ignition occurs.

Source: [10]. With permission from Elsevier.

The error function is a special function of sigmoid shape which occurs in probability, statistics, and partial differential equations, and is defined as:

$$\mathrm{erf}(x) = \frac{2}{\sqrt{\pi}} \int_0^x e^{-t^2} \, dt \tag{4.3}$$

The probit functions available to be used to determine the probability of first-degree or second-degree burns or a fatality as a function of radiation heat flux are shown in Table 4.4 [10–15].

Of the equations given in Table 4.4, the Tsao and Perry probit is expected to give conservative results for exposure to hydrogen fires. The Eisenberg probit may result in lower estimations of fatalities for hydrogen fires because it does not include the infrared spectrum. In addition, the Eisenberg probit is more suitable for hydrocarbon fires than the Tsao and Perry probit. As regards hydrogen, the best values for hydrogen fires lie between the results given by these two probit functions, but closer to the Eisenberg probit.

4.1.3 CRYOGENIC BURNS

Cryogenic burns (frostbite) may result from contact with extremely cold fluids or cold vessel surfaces. However, to keep hydrogen, as well as other cryogenic liquids, ultra-cold today, liquid hydrogen containers are double-walled, vacuum-jacketed, super-insulated containers that are designed to vent hydrogen safely in gaseous form if a breach of either the outer or inner wall is detected. The robust construction and redundant safety features dramatically reduce the likelihood of human contact.

Frostbite is damage to tissues from freezing, due to the formation of ice crystals inside cells, rupturing and destroying them in this way. Analogous to thermal burns, frostbites are classified on the basis of their depth:

- A *first-degree injury* occurs when only the surface skin is frozen, and then the injury is called *frostnip*, starting with itching and pain. Subsequently, the skin turns white and the area becomes numb. Frostnip generally does not lead to permanent damage. However, frostnip can lead to long-term sensitivity to heat and cold.
- A *second-degree injury* will occur when freezing continues. In this case, the skin may become frozen and hard, while the deep tissues are spared and remain soft and normal. This type of injury generally blisters 1–2 days after freezing. The blisters may become hard and blackened. However, they usually look worse than they actually are. Most of these injuries heal over 3–4 weeks, although the area may remain permanently sensitive to heat and cold.
- *Third- and fourth-degree injuries* occur when further freezing continues, leading to deep frostbite. The extremity is hard and feels woody, and use is lost temporarily or, in severe cases, permanently. The affected area appears deep purple or red, with blisters that are usually filled with blood. This type of severe frostbite may result in the loss of fingers and toes.

4.1.4 HYPOTHERMIA

Exposure to large liquefied hydrogen (LH_2) spills may result in hypothermia if proper precautions are not taken. Hypothermia occurs when body temperature drops below 35°C and becomes life threatening below 32.2°C. The major initial sign of hypothermia is a decrease in mental function that leads to impaired ability to make decisions, which may lead to further safety implications. Tiredness or lethargy, changes in speech, and disorientation are also typical. The affected persons will act as if they are drunk. The body gradually loses protective reflexes such as shivering, which is an important heat-generating defense. Other muscle functions also disappear so that the person cannot walk or stand. Eventually the person loses consciousness.

Recognizing hypothermia may be difficult for inexperienced persons, because the symptoms at first resemble other causes of change in mental and motor functions, such as diabetes, stroke, and alcohol or drug use. The most important thing is to be aware of the possibility and be prepared to intervene. Treatment involves slow heating of the body using blankets or other ways of increasing body warmth. Body temperature should increase by no more than a couple of degrees per hour.

TABLE 4.5
Damage to Humans from Overpressure Events

Overpressure (kPa)	Description of Damage
	Direct Effects on Humans
13.8	Threshold for eardrum rupture
34.5–48.3	50% probability of eardrum rupture
68.9–103.4	90% probability of eardrum rupture
82.7–103.4	Threshold for lung hemorrhage
137.9–172.4	50% probability of fatality from lung hemorrhage
206.8–241.3	90% probability of fatality from lung hemorrhage
48.3	Threshold of internal injuries by blast
482.6–1,379	Immediate blast fatalities
	Indirect Effects on Humans
10.3–20.0	People knocked down by pressure waves
13.8	Possible fatality by being projected against obstacles
55.2–110.3	People standing up will be thrown a distance
6.9–13.8	Threshold of skin lacerations by missiles
27.6–34.5	50% probability of fatality from missile wounds
48.3–68.9	100% probability of fatality from missile wounds

Source: Data from References [10] and [16].

4.1.5 OVERPRESSURE INJURY

The effects of blast waves on humans may be direct or indirect. The principal direct effect is the sudden increase of pressure that can cause damage to pressure-sensitive organs such as the lungs and ears. Indirect effects are caused by the impact on the human body from fragments, shrapnel, and debris generated by an explosion event, by collapsing structures, or by a violent shift of the body due to the impulse generated by an explosion and subsequent collision against a hard surface. Examples of the level of overpressure required to cause damage to humans are given in Table 4.5 [10, 16].

Nevertheless, as previously discussed for the effects of thermal radiation, blast waves from explosions will cause overpressure injury or death as a result of a combination of overpressure and duration of the effect. Again, when making assessments based on a QRA approach, one can find in the literature probit equations for many cases, including peak overpressure or impulse generated by an explosion as variables, and death from lung hemorrhage, head or whole body impact, or fragments of various sizes [10].

4.2 PHYSICAL HAZARDS

Among the most important physical hazards of hydrogen are those related either to its very small molecular size or to its storage being generally at low temperatures.

4.2.1 HYDROGEN EMBRITTLEMENT

Hydrogen embrittlement (also discussed in Chapter 11, Section 11.1) is the process by which a metal becomes brittle and develops small cracks due to exposure to hydrogen. It is a long-term effect and occurs from prolonged use of a hydrogen system. The mechanical properties of the metallic and nonmetallic materials of containment systems may then degrade and fail, resulting in spills and leaks, which will create hazards in the surroundings. Most of the damage caused in the case of leaks is usually incurred by ignition of the hydrogen following the rupture. Thus, all repairs and modifications to piping and equipment that handle hydrogen must be carefully engineered and tested.

The mechanisms of hydrogen embrittlement are not well understood, but certain factors are known to affect the rate of embrittlement, including hydrogen concentration, pressure, temperature of the environment, the purity, concentration, and exposure time of the hydrogen, the stress state, physical and mechanical properties, microstructure, surface conditions, and the nature of the crack front of the material (Figure 4.1) [17, 18].

Wang et al. investigated the hydrogen embrittlement fracture mechanism of AISI 430 ferritic stainless steel (FSS) under different hydrogen charging times. The microstructure and morphology of the tensile fracture before and after hydrogen charging, as characterized by scanning electron microscopy (SEM) and electron backscatter diffraction (EBSD), and the fracture mechanism of AISI 430 FSS were comprehensively described in comparison with the reported literature [19].

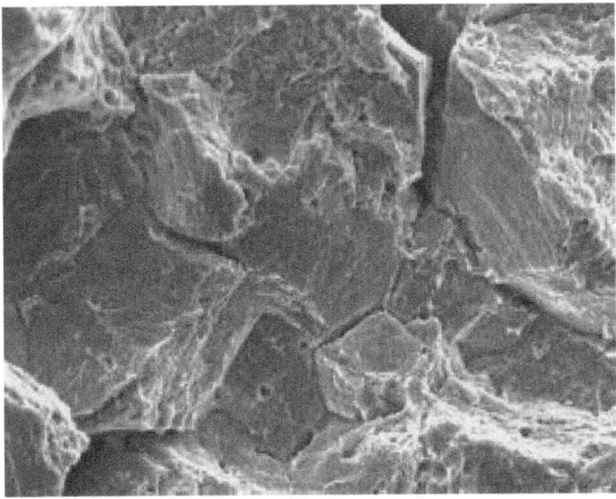

FIGURE 4.1 Scanning electron microscope image of fracture resulting from hydrogen embrittlement. This failure mode results from the absorption of hydrogen atoms into a component, which then accumulate at the grain boundaries, exerting an internal tensile stress. Magnification 2000×. (With permission from MAI Metallurgical Associates, Inc. http://metas soc.com/site/services/scanning-electron-microscopy/sem-eds-application-examples/).

4.2.1.1 Types of Embrittlement

Hydrogen embrittlement types include:

- *Environmental hydrogen embrittlement* of metals and alloys that may be plastically deformed in a gaseous hydrogen environment, resulting in increased surface cracks, loss of ductility, and decrease of fracture stress, with cracks starting at the surface.
- *Internal hydrogen embrittlement* caused by absorbed hydrogen, resulting in premature failure of certain metals, with cracks starting internally.
- *Hydrogen reaction embrittlement* caused by the chemical reaction of the absorbed hydrogen with one or more of the constituents of the metal to form a brittle metal hydride or methane with the carbon in steels. This phenomenon is favored by elevated temperatures.

4.2.1.2 Mechanical Properties' Deterioration

Considerable deterioration of mechanical properties of many metals and alloys in the presence of hydrogen has been observed. Studies [20, 21] have demonstrated that:

- The susceptibility of a metal (or alloy) due to hydrogen increases with the strength of the metal.
- Internal and environmental hydrogen embrittlement rates are maximum in the temperature range of 200 to 300 K (–73 to 27°C), whereas hydrogen reaction embrittlement occurs at temperatures above room temperature.
- The susceptibility of steel to hydrogen embrittlement increases with hydrogen purity.
- The susceptibility to embrittlement generally increases with the tensile stress.
- Embrittlement usually results in fatigue of the metal due to crack growth.

4.2.1.3 Main Factors of Hydrogen Embrittlement

Surface and surface films: The hydrogen compatibility of metastable austenitic stainless steels such as SS type 304 depends considerably on metal surface finish. The extent of surface cracking and ductility loss may be minimized by removing the layer produced by machining. It has been found that the oxides that form naturally on the metal surfaces restrict hydrogen absorption and control the degree of embrittlement due to their lower permeability than the base metals. Synthetic surface films used to reduce hydrogen absorption should be ductile at their operating temperature. Copper and gold, which remain ductile in a broad temperature range, are usually recommended [7, 22].

Effect of electrical discharge machining: Hydrogen embrittlement can be increased by electrical discharge machining commonly used for producing holes, notches, slots, or other cavities in metals. In this way, electrical discharges can introduce hydrogen into a machined component. Hydrogen is produced by the decomposition of the dielectric fluid (usually oil or kerosene) used when it is ionized by the electrical discharge.

Effect of trapping sites: Hydrogen may be trapped at sites within the structure of the metal, including dislocations, grain and phase boundaries, interstitial or vacancy clusters, voids or gas bubbles, oxygen or oxide inclusions, carbide particles, and other material defects. Trapping is most marked at low temperatures and is important at ambient temperatures where hydrogen embrittlement is also clearly visible.

4.2.1.4 Hydrogen Embrittlement Effects of Hydrogen/Methane Gas Mixtures in Pipelines

Considering that the most economical means of transporting large quantities of hydrogen over large distances is by using the existing natural gas pipeline network, the leakage diffusion and spread behavior of hydrogen-blended natural gas through urban distribution systems and the evolution of explosion incidents should be investigated.

The safety, reliable operation, and longevity of pipelines transmitting hydrogen–natural gas mixtures depend on the mechanical performance of construction materials. Nguyen et al. [23] studied experimentally the effect of hydrogen concentration in a methane/hydrogen gas mixture destined to be used in hydrogen transportation in pipelines. They found three types of pipeline steel, API X42, X65, and X70, were at the highest risk of hydrogen-induced crack initiation in methane–hydrogen mixtures, resulting in brittle fractures even at 1% hydrogen concentration. On the other hand, the X42 and X70 steels exhibited minor defects in their fracture mechanism in a hydrogen concentration of up to 30% gas mixture, with their ductility remaining unchanged. However, the hydrogen embrittlement indices of these three types of steel presented insignificant differences under a 10 MPa gas pressure.

To prevent pipeline hydrogen embrittlement, internal polymeric coating materials have been proposed, and a theoretical model of hydrogen permeation through coated steel has been developed by Lei et al. [24]. The authors tested 12 commercially available coatings based on crosslinked poly(vinyl alcohol) (PVA), poly(vinyl chloride) (PVC), and bisphenol-A diglycidyl ether (DGEBA)/polyetheramine (D-400) epoxy resins. The films made from two commercial epoxy resins had hydrogen permeabilities of 0.40 Barrer and 0.35 Barrer, respectively, which renders them suitable coating materials. A crosslinked PVA coating achieved a hydrogen permeability equal to 0.0084 Barrer, which indicates that this material is the most promising of all coatings tested. A model was also prepared to simulate an unsteady-state hydrogen diffusion through coated steel to assess the effect of a coating film on preventing hydrogen embrittlement. Application of this model indicated that with a 2 mm crosslinked PVA coating, hydrogen permeation inside the coating reduced the final hydrogen concentration on the steel surface by 44% compared to that without a coating.

The intended use of pipelines to transmit hydrogen has increased the necessity for higher strength steels possessing sufficient hydrogen embrittlement resistance. Ogawa and Iwata [25] investigated the fatigue crack growth (FCG) properties of pearlitic steel in a high-pressure, i.e., 90 MPa, hydrogen gas environment to assess its microstructure capability in future hydrogen-compatible structural materials. As-modified eutectoid steel with 1080 MPa tensile strength demonstrated a hydrogen-induced FCG resistance better than that of martensite iron with similar strength and sometimes better than pure ferritic iron's resistance. As-modified eutectoid steel superiority over martensite was increased when slow loading-rate conditions were applied, resulting in suppression of time-dependent cracking.

Zhou et al. [26] developed a three-dimensional computational fluid dynamics (3D CFD) simulation model to be applied in urban streets for the spread process and explosion characteristics of hydrogen-blended natural gas. The effect of hydrogen addition ratios and ambient wind speed on the consequences of explosion accidents was analyzed. The hazardous area, as affected by various environmental wind conditions, was estimated to investigate the most dangerous scenarios.

4.2.1.5 Hydrogen Embrittlement Control

In general, stainless steel is more resistant to hydrogen embrittlement than ordinary steels, and both pure aluminum and many aluminum alloys are even more resistant than stainless steel if the gas is dry. As also described at various points in Chapters 7, 10, and 11, all components of hydrogen fuel systems must be constructed of materials known to be compatible with hydrogen [17].

Oxide coatings, elimination of stress concentrations, additives to hydrogen, proper grain size, and careful alloy selection are among the measures to prevent embrittlement. In addition, the following measures [7, 22] can be used to effectively eliminate hydrogen embrittlement in metals:

- Use of aluminum as a structural material due to its low susceptibility to hydrogen.
- Design of components made of medium-strength steel (for gaseous hydrogen) and stainless steel (for liquid hydrogen) on the basis of increased thickness, surface finish, and proper welding techniques.
- In the absence of data, design of metal components assuming a substantial (up to five-fold) decrease in resistance to fatigue.
- Not using cast iron and hydride-forming metals and alloys as structural materials for hydrogen service.
- Taking into consideration that, in general, exposure temperatures below room temperature retard hydrogen reaction embrittlement.
- Taking into consideration that, in general, environmental and internal hydrogen embrittlement are increased in the temperature range of 200 to 300 K (–73 to 27°C).

4.2.2 Thermal Stability of Structural Materials

4.2.2.1 Low-Temperature Mechanical Properties

For safety reasons, selection of structural materials is based primarily on their mechanical properties, such as yield and tensile strength, ductility, and impact strength. The materials should address minimum design values of these properties over the entire temperature range of operation, taking into account non-operational conditions such as a hydrogen fire. The material properties should be stable without phase changes in the crystalline structure with time or repeated thermal cycling [7].

The main considerations of metal and alloy behavior at low-temperature conditions are the following:

- The transition from ductile to brittle behavior at low temperatures.
- Certain unconventional modes of plastic deformation encountered at very low temperatures.
- Mechanical and elastic property changes due to phase transformations in the crystalline structure at low temperatures.

Especially for the selection of a material for LH_2 service, the main thermal properties to be considered are low-temperature embrittlement and thermal contraction.

4.2.2.2　Low-Temperature Embrittlement

At lower temperatures, and especially at cryogenic temperatures, many materials change from ductile to brittle. This change may lead to failure of a hydrogen storage vessel or pipe and cause an accident. Such an accident occurred in Cleveland, Ohio in 1944 due to low-temperature embrittlement of a liquid natural gas (LNG) storage vessel made of 3.5% nickel steel with a capacity of 4248 m^3. The vessel ruptured and released 4163 m^3 of LNG, which ignited after having spread into nearby storm sewers. As a domino effect, a nearby storage vessel collapsed from the fire and spilled its contents, which also burned, with flames up to about 850 m in height. The accident resulted in 128 deaths, 200 to 400 injuries, and an estimated USD6,800,000 (at 1944 prices) in property damage [7].

The ductility of a material is commonly determined by the Charpy impact test. The results of the Charpy impact test at various temperatures for several materials are shown in Figure 4.2 [7].

The gradual loss of ductility of 9% nickel steel is clearly shown in this figure, as well as the stepwise embrittlement of 201 stainless steel at temperatures below 280 K (7°C) and C1020 carbon steel below 120 K (–153°C). This indicates that these materials should not be used for liquid hydrogen storage. The ductility of aluminum 2024-T4 does not change considerably at low temperatures, but its low strength should be taken into account in storage vessel design. Strangely, the Charpy impact strength of 304 stainless steel increases when the temperature decreases, indicating that this is a suitable material for the construction of LH_2 storage vessels and pipes.

The difference between the yield and tensile strengths of a material may also be used to measure its ductility. As a rule, a material is considered ductile when these two values differ considerably. Thus, in the case of 5986 aluminum, the values of yield and tensile strengths stand out as the temperature decreases in the cryogenic region, indicating that this is a suitable material for LH_2. However, these values are similar in the cryogenic region for AISI 430 stainless steel, indicating its unsuitability for use with LH_2.

4.2.2.3　Thermal Contraction

Different structural materials have different thermal contraction coefficients, meaning that accommodations should be made for their different dimensions at cryogenic

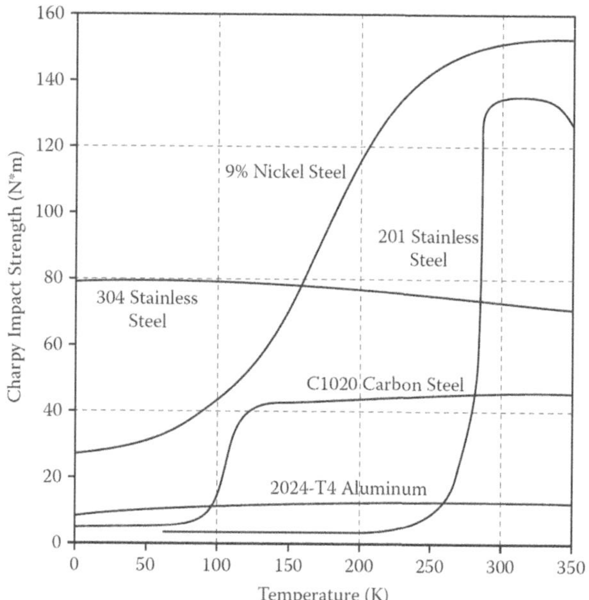

FIGURE 4.2   Charpy impact strength as a function of temperature for various materials.

Source: [7, 18]. With permission from the American Institute of Aeronautics and Astronautics.

temperatures. If not, problems associated with safety (e.g., leaks) may arise. In general, the contraction of most metals from room temperature (300 K) to a temperature close to the liquefaction temperature of hydrogen (20 K) is less than 1%, whereas the contraction for most common structural plastics is from 1 to 2.5% [7].

4.3   CHEMICAL HAZARDS

4.3.1   General Considerations and Accident Statistics

As shown in Table 3.1, major emphasis should be placed on containment, detection, and ventilation because the minimum energy of gaseous hydrogen (GH_2) for ignition in air at atmospheric pressure is about 0.02 mJ. Experience has shown that escaped hydrogen is very easily ignited.

For hydrogen (as fuel) to catch fire, it needs to be mixed with air (oxygen as oxidant) and the mixture must be within the flammability limits. In addition, and in accordance with the familiar fire triangle, an ignition source with sufficient energy must be present for hydrogen to burn. Leaks occur for every gas and especially for hydrogen, which even the best efforts cannot contain.

Rigas and Sklavounos [27] and Hansen [28] stress the lack of acknowledgment that LH_2 released into the air will initially show dense gas behavior rather than rising into the sky and disappearing like releases of compressed hydrogen gas. These studies aim to clarify the release and dispersion mechanisms of LH_2 and to explain that lower flammability limit (LFL) distances for LH_2 release can be significantly longer than for comparable LNG releases, not the opposite. Some further hazards of concern about using LH_2 for maritime applications are also discussed by Hansen [28], illustrating why hydrogen-fueled vessels may need to be designed to higher safety standards than current LNG-fueled vessels.

Ekoto et al. [29] experimentally investigated hydrogen release from fuel cell-powered forklifts and eventual ignition in enclosed areas. Release, dispersion, and combustion parameters were measured in a constructed test facility under scale. Mitigation measures such as active and passive ventilation and pressure relief panels were also investigated. Since the possible facility configurations and accident scenarios are numerous, it is impossible to design and experimentally perform and evaluate all probable combinations. Thus, systematic characterization of the experimental boundary conditions was performed so that the collected datasets could be used to validate specific computational models.

The hazards of a hydrogen jet flame created by a high-pressure hydrogen gas release were investigated by Mogi and Horiguchi [30]. In the experimental set-up, hydrogen jets were horizontally released through circular nozzles with diameters ranging from 0.1 to 4 mm. The release pressure varied from 0.01 to 40 MPa (gauge), and the blow-off limits were determined from the nozzle diameter and the release pressure. The flame sizes were measured, and experimentally validated equations were obtained for the length and width of the flame. It was found that the flame dimensions depended not only on the nozzle diameter but also, as expected, on the release pressure. When experimenting with slit nozzles, the flame length depended on the width of the shorter side of the slit nozzle. The radiation from the hydrogen flame could be estimated from the flow rate of the gas and the distance from the flame.

Safety measures include elimination of all likely sources of ignition. An investigation of industrial accidents has shown that 53% of them occurred because of leaks, off-gassing, and equipment rupture, as listed in Table 4.6 [31]. Purging or vent-exhaust

TABLE 4.6
Industrial Hydrogen Accidents

Category	Number of Accidents	Percentage of Total Accidents
Undetected leaks	32	22
Hydrogen–oxygen off-gassing explosions	25	17
Piping and pressure vessel ruptures	21	14
Inadequate inert gas purging	12	8
Vent and exhaust system incidents	10	7
Hydrogen–chlorine incidents	10	7
Others	35	25
Total	145	100

Source: Adapted from [31].

TABLE 4.7
Hydrogen Accidents in Ammonia Plants

Classification	Number	Percentage of Total Accidents
Gaskets	46	37
Equipment flanges	23	18
Piping flanges	16	13
Valve flanges	7	6
Valve packing	10	8
Oil leaks	24	19
Transfer header	9	7
Auxiliary boiler	8	6
Primary reformer	7	6
Cooling tower	3	2
Electrical	2	2
Miscellaneous	16	13
Total	125	100

Source: [32]. With permission.

TABLE 4.8
Hydrogen Accidents in the Aerospace Industry

Description	Accidents Involving Release of Hydrogen	Percentage of Total Accidents
Accidents involving release of liquid or gaseous hydrogen	87	81
Location of Hydrogen Release:		
To atmosphere	71[a]	66
To enclosures (piping, containers, etc.)	26[a]	24
Ignition of Hydrogen Releases:		
To atmosphere	44	41
To enclosures	24	22

[a] Hydrogen was released to both locations in ten accidents.

Source: [33]. With permission.

incidents account for 15%, and the remaining 32% were other types of incidents. In ammonia plants, most of the accidents are due to leaking of gaskets and valve packing, as shown in Table 4.7 [32]. The aerospace industry has paid a great toll, with 107 hydrogen incidents in 1974 alone, of which 87 involved releases of GH_2 or LH_2, as shown in Table 4.8 [33]. When accidents were not caused by equipment failure,

TABLE 4.9
HIAD 2.0 Events Classified by Sector

Sector	Number of Events by Sector	Percentage of Total Events
Chemical/petrochemical industry	259	62.1
Hydrogen transport and distribution	43	10.3
Nuclear power plants	23	5.5
Laboratory/R&D	15	3.6
Power generation	13	3.1
Hydrogen production	10	2.4
Aerospace	5	2.0
Entertainment	3	0.7
Hydrogen-powered vehicle	2	0.5
Stationary fuel cell	0	0
Other/unknown	44	10.6
Total	417	

Source: Adapted from [34].

they primarily occurred when procedures were not prescribed or when prescribed procedures were not followed. Ignition sources responsible for these incidents were electrical short circuits and sparks (25%), static charges (18%), welding or cutting torches, metal fracture, gas impingement, and the rupture of safety disks (3 to 6%), as accident reports have shown [7].

Table 4.9 shows the percentage of events that occurred in different industrial sectors from the analysis of the Hydrogen Incidents and Accidents Database (HIAD 2.0), as collected by the European Hydrogen Safety Panel (EHSP) [34]. Among the different sectors, the chemical and petrochemical industry has by far the largest share of the incidents, with 62.1%. It is followed by hydrogen transport and distribution with 10.3% and nuclear power plants with 5.5%. The other sectors under consideration account for only small shares.

Concerning statistics on causes and operational mode, Table 4.10 lists the number of events according to their causes. It is noteworthy that some events had multiple causes and hence are counted more than once here [34].

4.3.2 Flammability of Hydrogen

The flammability limits of mixtures of hydrogen with air, oxygen, or other oxidizers depend on the ignition energy, temperature, pressure, presence of diluents, and size and configuration of the equipment, facility, or apparatus. Such a mixture may be diluted with either of its constituents until its concentration shifts below the lower flammability limit (LFL) or above the upper flammability limit (UFL). The flammability limits of hydrogen mixtures with either air or oxygen widen for upward flame propagation and narrow for downward flame propagation.

TABLE 4.10
HIAD 2.0 Events Classified by Cause

Cause	Number of Events by Cause	Percentage of Total Events
System design error	126	79.7
Material/manufacturing error	127	80.4
Installation error	38	24.1
Job factors	98	62.0
Individual/human factors	94	59.5
Organization and management factors	158	100.0

The range for detonation limits for hydrogen–air mixtures is from 18.3 to 59% hydrogen.

Source: Adapted from [34].

Mixtures of LH_2 and liquid oxygen or solid oxygen as oxidizer are not hypergolic. In accidental fires of these mixtures during the mixing process the system caught fire because the required ignition energy is very small [7]. Nevertheless, LH_2 and liquid or solid oxygen can detonate when initiated by a shock wave.

The flammability limits of hydrogen in dry air for upward propagation in tubes at 101.3 kPa (1 atm) and ambient temperature range from 4.1% (LFL) to 74.8% (UFL). The flammability limits of hydrogen in oxygen for upward propagation in tubes at 101.3 kPa (1 atm) and ambient temperature are from 4.1% (LFL) to 94% (UFL). With a reduction in pressure below 101.3 kPa, the range of flammability limits narrows, as shown in Table 4.11 [7].

4.3.2.1 Hydrogen–Air Mixtures

The lowest pressure at which a low-energy ignition source ignites a hydrogen–air mixture is approximately 6.9 kPa (0.07 atm) at a hydrogen concentration between 20 and 30 vol% [7].

Hydrogen–air mixtures ignited by a 45 mJ spark ignition source at 311 K (38°C) environment temperature have an LFL equal to 4.5 vol% over the pressure range 34.5 to 101.3 kPa (0.34 to 1 atm). An increasingly higher LFL was required to obtain combustion below 34.5 kPa (0.34 atm). The lowest pressure at which a low-energy ignition source could inflame the mixture was 6.2 kPa (0.06 atm) at a hydrogen–air mixture of between 20 and 30 vol% hydrogen [35]. Yet, using a strong ignition source, the lowest pressure at which ignition can occur is 0.117 kPa (0.0012 atm).

For downward propagation, the LFL of hydrogen–air decreases from 9.0 to 6.3 vol% hydrogen and the UFL increases from 75 to 81.5 vol% hydrogen when the temperature is increased from 290 to 673 K (17 to 400°C) at a pressure of 101.3 kPa (1 atm).

Compared to methane–air mixtures, hydrogen–air mixtures are more hazardous with respect to accidental ignition due to the much wider flammability limits and the order of magnitude lower ignition energy, as shown in Figure 4.3 [36]. Its minimum

TABLE 4.11

Flammability Limits of Hydrogen–Air and Hydrogen–Oxygen Mixtures

| | Hydrogen Content (vol%) | | | | | |
| | Upward Propagation | | Downward Propagation | | Horizontal Propagation | |
Conditions	LFL	UFL	LFL	UFL	LFL	UFL
Hydrogen in air and oxygen at 101.3 kPa (1 atm)						
H_2 + air:						
Tubes	4.1	74.8	8.9	74.5	6.2	71.3
Spherical vessels	4.6	75.5	—	—	—	—
H_2 + oxygen	4.1	94.0	4.1	92.0	—	—
Hydrogen plus inert gas mixtures at 101.3 kPa (1 atm)						
H_2 + He + 21 vol% O_2	7.7	75.7	8.7	75.7	—	—
H_2 + CO_2 + 21 vol% O_2	5.3	69.8	13.1	69.8	—	—
H_2 + N_2 + 21 vol% O_2	4.2	74.6	9.0	74.6	—	—

Hydrogen in air at reduced pressure with a 45 mJ ignition source

| | 25 cm Tube | | 2 L Sphere | |
Pressure (kPa)	LFL	UFL	LFL	UFL
20	~4	~56	~5	~52
10	~10	~42	~11	~35
7	~15	~33	~16	~27
6	20–30		20–25 (at 6.5 kPa)	

Note: Dashes indicate no information available.

Source: [7, 18]. With permission.

ignition energy in the air (20 µJ) is more than ten times lower than that of propane or petrol. This energy can be provided easily by electrostatic discharges even from the human body, e.g. after stepping a short distance on an insulated plastic floor. In pure-oxygen environments (case of electrolyzers), the necessary energy is just 3 µJ. Inerting during critical phases (start-up/shutdown, including emergency stops), as well as compressor shutdown devices, reduces the risk of hydrogen–air mixtures (or hydrogen–pure air mixtures) [37].

4.3.2.2 Hydrogen–Oxygen Mixtures

The flammability limits for hydrogen–oxygen mixtures at 101.3 kPa (1 atm) range from 4 to 94 vol% of hydrogen for upward propagation in tubes. Reduced pressures increase the LFL [7]. The lowest pressure observed for ignition is 57 kPa (0.56 atm) at a hydrogen concentration of 50 vol% when a high-energy ignition source was used.

At elevated pressures, the LFL does not change with pressure up to 12.4 MPa (122 atm), while at 1.52 MPa (15 atm) the UFL is 95.7 vol% hydrogen [7]).

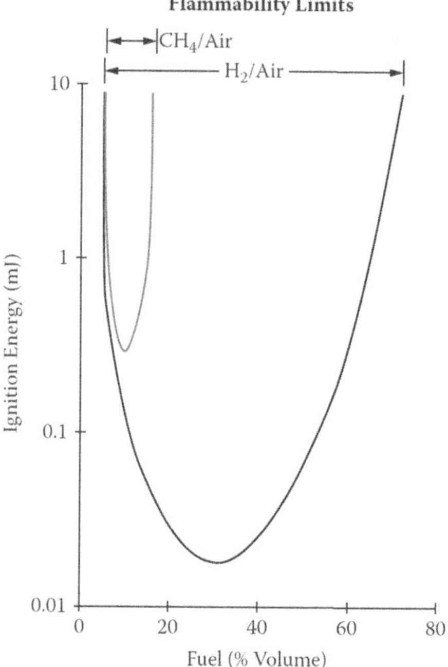

FIGURE 4.3   Minimum ignition energy of hydrogen–air and methane–air mixtures at a pressure of 101.3 kPa (1 atm) and a temperature of 298 K (25°C).

Source: [36]. With permission from Elsevier.

The LFL decreases from 9.6 to 9.1 vol% hydrogen and the UFL increases from 90 to 94 vol% hydrogen, when the temperature rises from 288 to 573 K (15 to 300°C).

4.3.2.3   Effects of Diluents

The flammability limits for hydrogen–oxygen–nitrogen mixtures are shown in Figure 4.4 [2]. Table 4.9 shows flammability limits for GH_2 and gaseous oxygen with equal concentrations of added inert gases (helium, carbon dioxide, and nitrogen). Table 4.12 shows the qualitative effects of helium, carbon dioxide, nitrogen, and argon diluents for various tube sizes [7]. Argon is the least effective diluent in reducing the flammable range for hydrogen in air.

The effects of helium, carbon dioxide, nitrogen, and water vapor on the flammability limits of hydrogen in air are shown in Figure 4.5 [7]. Measurements were performed at 298 K (25°C) and 101.3 kPa (1 atm), except for the water vapor measurements, which were performed at 422 K (149°C). Water was the most effective diluent in reducing the flammability range of hydrogen in air.

In a recent experimental study [38], the inhibiting effect of N_2 and CO_2 was investigated for hydrogen/air flames in a closed channel. Both N_2 and CO_2 have a remarkable inhibiting effect on hydrogen/air-premixed flame. Comparatively, CO_2

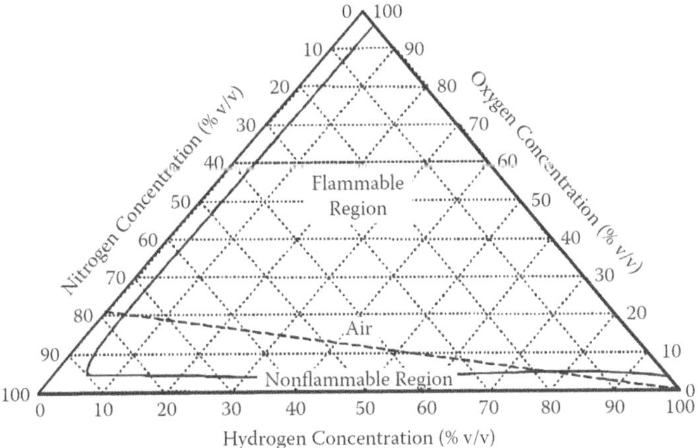

FIGURE 4.4 Flammability limits of hydrogen–oxygen–nitrogen mixtures at a pressure of 101.3 kPa (1 atm) and a temperature of 298 K (25°C).

Source: [7, 18]. With permission from the American Institute of Aeronautics and Astronautics.

TABLE 4.12
Effects of Equal Concentrations of Diluents on Flammable Range for Hydrogen in Air

Tube Diameter (cm)	Rating of Diluents at Reducing Flammable Range
Wide tubes	$CO_2 < N_2 < He < Ar$
2.2	$CO_2 < He < N_2 < Ar$
1.6	$He < CO_2 < N_2 < Ar$

Source: [7, 18]. With permission.

---

inhibition is more prominent due to its higher specific heat, higher collision efficiency, and more intense kinetic effects on the reactions of hydrogen combustion than N_2 [39].

4.3.2.4 Effects of Halocarbon Inhibitors

The effects of halocarbon inhibitors on the flammability limits of hydrogen–oxygen mixtures are shown in Figure 4.6 [40].

The effects of N_2, CH_3Br, and $CBrF_3$ on the extinguishment of hydrogen diffusion flames in air are compared in Table 4.13 [41]. The halocarbon inhibitors were more effective when added to the air stream; nitrogen was more effective when added to the fuel stream.

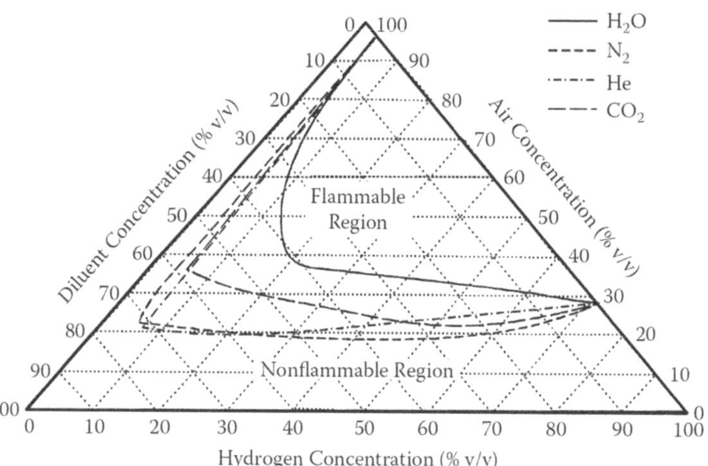

FIGURE 4.5   Effects of N_2, He, CO_2, and H_2O diluents on flammability limits of hydrogen in air at 101.3 kPa (1 atm). (The effects of N_2, He, and CO_2 are at 298 K [25°C] and H_2O is at 422 K [149°C].)

Source: [7, 18]. With permission from the American Institute of Aeronautics and Astronautics.

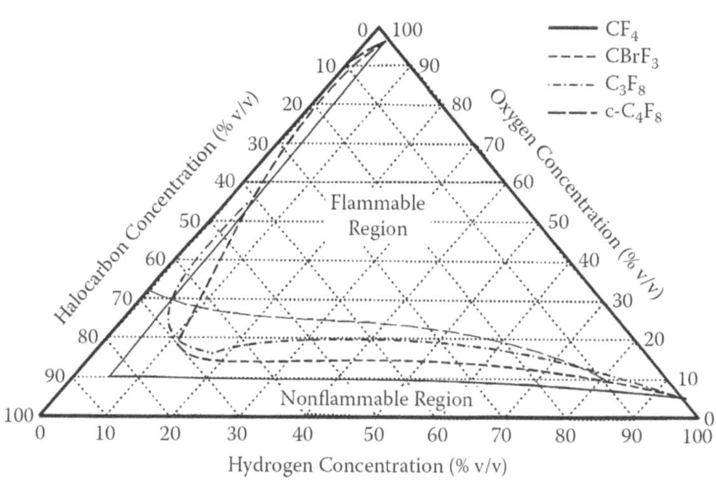

FIGURE 4.6   Effects of halocarbon inhibitors on flammability limits of hydrogen–oxygen mixtures at a pressure of 101.3 kPa (1 atm) and a temperature of 298 K (25°C).

Source: [7, 18]. With permission from the American Institute of Aeronautics and Astronautics.

TABLE 4.13
Inhibitor for Extinction of Hydrogen Diffusion Flames

Inhibitor	Concentration at Flame Extinction (vol%)
Added to Air:	
Nitrogen	94.1
CH_3Br	11.7
$CBrF_3$	17.7
Added to Fuel:	
Nitrogen	52.4
CH_3Br	58.1
$CBrF_3$	56.6

Source: [41]. With permission.

4.3.2.5 Autoignition Temperature

The autoignition temperature of hydrogen in air differs only slightly from that in oxygen. It is interesting to note that autoignition temperatures in both cases depend not only on GH_2 concentration and pressure but also on the surface treatment of containers. So, the autoignition temperatures found in the literature are very dependent on the system, and values should be applied only to similar systems. At 101.3 kPa (1 atm) the range of reported autoignition temperatures for stoichiometric hydrogen in air is from 773 to 850 K (500 to 577°C), while in stoichiometric concentration with oxygen it is from 773 to 833 K (500 to 560°C). At pressures from 20 to 50 kPa (0.20 to 0.49 atm), GH_2–air ignitions have occurred at 620 K (347°C).

In an experimental study on shock waves and spontaneous ignition produced by pressurized hydrogen release through a tube into the atmosphere, Duan et al. [42] found that the possibility of spontaneous ignition increases with increasing length of the downstream tube. And ignition was more likely to happen in the tube with a small diameter. They also confirmed that the diaphragm rupture process has an important influence on the formation of a shock wave, spontaneous ignition, and explosion outside the tube.

4.3.2.6 Quenching Gap in Air

Quenching distance is the passage gap dimension required to prevent the propagation of an open flame through a flammable fuel–air mixture that fills the passage. In a parallel-plate configuration, the quenching gap is defined as the spark gap between two flat electrodes at which ignition of combustible fuel–air mixtures is suppressed.

The quenching gap for hydrogen is 0.6 mm in air at normal temperature and pressure (NTP). Nevertheless, this value depends on the temperature, pressure, and composition of the combustible gas mixture, as well as the electrode configuration.

Faster burning gases generally have smaller quenching gaps. Consequently, flame arresters for faster burning gases require smaller apertures. The lowest quenching distance found in the literature for hydrogen is 0.076 mm [43]. There are three major considerations in determining the quenching distance for a gaseous fuel such as hydrogen: ignition energy, mixture composition, and pressure.

Quenching gap depends on the *ignition energy*. For instance, a low ignition energy of the order of 0.001 mJ corresponds to a small gap of the order of 0.01 cm. In contrast, a high ignition energy of 10 mJ corresponds to a larger gap of 1 cm [7].

Pressure and *composition* also affect the quenching distance, with this parameter increasing rapidly at very low pressures. The effect of *mixture ratio* is less well known. Nevertheless, quenching distance appears to be constant for a given pressure between the UFL and LFL. Since specific values for hydrogen–air mixtures are not available, the effect of pressure as a function of tube diameter for deflagration and detonation (to be defined in Section 4.3.4) of acetylene–air mixtures can be used as a guideline [7].

4.3.3 Ignition Sources

All ignition sources such as open flames, electrical equipment, or heating equipment should be eliminated or safely isolated in buildings or special rooms containing hydrogen systems. Operations should be conducted as if unforeseen ignition sources could occur.

Potential ignition sources for hydrogen systems are shown in Table 4.14 [7].

The ignition of GH_2–air mixtures usually results in deflagration, with explosion consequences significantly less severe than if detonation results. It is possible that in a confined or partially confined enclosure a deflagration can evolve into a detonation, a phenomenon known as deflagration-to-detonation transition (DDT). The geometry and flow conditions (turbulence) have a strong effect on DDT.

4.3.3.1 Sparks

Electrical sparks can be the result of electrical discharges between objects having different electrical potentials, such as breaking electrical circuits or discharges of static electricity.

Static electricity sparks can ignite hydrogen–air or hydrogen–oxygen mixtures. Static electricity is caused by many common articles, such as hair or fur when combed or stroked, or an operating conveyor belt. People can generate high-voltage charges of static electricity on themselves when walking on synthetic carpet or dry ground, moving while wearing synthetic clothing, sliding on automobile seats, or combing their hair. The flow of GH_2 or LH_2 in ducts or turbulence in containers can generate charges of static electricity, as with any other nonconductive liquid or gas. In addition, static charges may be induced during electrical storms [44–46].

Hard objects coming into shearing contact with each other can cause *friction sparks*, as in the cases of metal striking metal, metal striking stone, or stone striking stone. Friction sparks are particles of burning material, initially heated by the mechanical energy of friction and impact, that have been sheared off as a result of contact. Sparks from hand tools normally have low energy, while mechanical tools such as drills and pneumatic chisels can generate high-energy sparks.

TABLE 4.14
Potential Ignition Sources

Electrical Sources	Mechanical Sources	Thermal Sources	Chemical Sources
Static discharge	Mechanical impact	Open flame	Catalysts
Static electricity (e.g. two-phase flow)	Tensile rupture	Hot surface	Reactants
Static electricity (e.g. flow with solid particles included)	Friction and galling	Personnel smoking	
Electric arc	Mechanical vibration	Welding	
Lightning	Metal fracture	Exhaust from combustion engine	
Charge accumulation		Resonance ignition	
Electric charge generated by equipment operation		Explosive charge	
Electrical short circuits		High-velocity jet heating	
Electrical sparks		Shock wave from tank rupture	
Clothing (static electricity)		Fragment from bursting tank	

Note: This list should not be viewed as a complete list. Be alert for other possible ignition sources.

Source: [7, 18]. With permission.

Hard objects striking each other can also produce *impact sparks*. Impact sparks are usually produced by impact on a quartzitic rock, such as the sand in concrete, resulting in the throwing off of small particles of the impacted material.

Minimum spark energy for ignition is defined as the minimum spark energy required to ignite the most easily ignitable concentration of fuel in air and oxygen. The minimum spark energies of hydrogen in air are 0.017 mJ at 101.3 kPa (1 atm), 0.09 mJ at 5.1 kPa (0.05 atm), and 0.56 mJ at 2.03 kPa (0.02 atm). The minimum spark energy required for ignition of hydrogen in air is considerably less than that for methane (0.29 mJ) or gasoline (0.24 mJ). Nevertheless, the ignition energy for all three fuels is sufficiently low that ignition in air is rather certain in the presence of any weak ignition source, such as sparks, matches, hot surfaces, open flames, or even a weak spark caused by the discharge of static electricity from a human body.

4.3.3.2   Hot Objects and Flames

Objects at temperatures from 773 to 854 K (500 to 581°C) can ignite hydrogen–air or hydrogen–oxygen mixtures at atmospheric pressure. Substantially cooler objects, about 590 K (317°C), can also ignite these mixtures after prolonged contact at less than atmospheric pressure. Open flames can easily ignite hydrogen–air mixtures.

In cases where ignition sources are a required part of hydrogen use, provisions should be made to adequately contain any resulting deflagration or detonation. For instance, a combustor or engine should not be operated in atmospheres containing hydrogen without well-dispersed water sprays in its exhaust. Experience has shown that multiple bank sprays will partially suppress the detonation pressures and reduce the number and temperature of ignition sources in an exhaust system. Water sprays should not, however, be relied on as a means of avoiding detonations. Carbon dioxide may be used with the water spray to further reduce hazards.

4.3.4 Explosion Phenomena

4.3.4.1 Terms and Definitions

The general term *explosion* corresponds to a rapid energy release and pressure rise and can occur in both reactive and nonreactive systems. Pressurized gas vessel failures are typical examples of nonreactive explosions, whereas premixed gases of fuel and oxidizer permit a rapid energy release by chemical reactions in reactive systems. The term explosion is sometimes used interchangeably in the literature with any kind of violent pressure rise such as deflagration and detonation. Others limit the term explosion only to the bursting of an enclosure or a container due to the development of internal pressure from a deflagration [47].

Gas explosions do not necessarily require the transmission of a wave through the explosion source medium, although explosions in most cases involve some kind of wave such as a deflagration or detonation wave. For instance, in *volumetric explosions*, an explosive mixture contained in a vessel can be heated to a high enough temperature for fast reactions to occur [48].

Deflagration is the phenomenon in which the flame front moves through a flammable mixture in the form of a subsonic wave with respect to the unreacted medium. The terms deflagration, flash fire, combustion, flame, and burn are sometimes found in the literature being used interchangeably.

Detonation occurs when a flame front coupled to a shock wave propagates through a detonable mixture in the form of a supersonic wave with respect to the unreacted medium. The thermal energy of the reaction sustains the shock wave, and the shock wave compresses the unreacted material to sustain the reaction. In general, a detonation propagates two to three orders of magnitude faster than a deflagration and results in pressures at the detonation front 15 to 20 times higher than the initial pressure [48].

This explains why a detonation has much greater potential for causing personnel injury or equipment damage than a deflagration.

For gaseous mixtures exploding in the open, the term *unconfined vapor cloud explosion* (UVCE) is used, whereas when they explode in confined spaces the term *confined vapor cloud explosion* (CVCE) is used. Flame behavior and blast waves generated during unconfined hydrogen deflagrations have been experimentally studied using infrared photography. Infrared photography enabled expanding spherical flame behaviors to be measured and flame acceleration exponents to be evaluated [49–51].

A recent development has been the use of the term *vapor cloud explosion* (VCE) for either the confined or unconfined scenario. In these cases, either a detonation or a deflagration takes place depending on the factors mentioned earlier. In a very lean or

FIGURE 4.7 The BLEVE (boiling liquid expanding vapor explosion) is the worst possible outcome when a tank holding a pressure-liquefied gas, such as propane or hydrogen, fails due to fire contact or impact.

Source: Google images.

rich fuel mixture, the flame front travels in the cloud at a low velocity and insignificant pressure increase, a phenomenon known as a *flash fire*.

Another explosive phenomenon encountered in the storage of LH_2 under pressure is the *boiling liquid expanding vapor explosion* (BLEVE), which occurs from the sudden release of a large mass of pressurized liquid into the atmosphere. A typical primary cause is an external flame impinging on the shell of a vessel above the liquid level, thus weakening the shell and resulting in sudden rupture. Then the content of the vessel is released into the atmosphere and, if flammable, ignites forming a nearly spherical burning cloud, the so-called *fireball* (Figure 4.7). During burning, the combustion energy is emitted mostly in the form of radiant heat. The inner core of the cloud consists of almost pure fuel, whereas the outer layer, where the ignition first occurs, is a fuel–air mixture within the flammable limits. As the buoyancy forces of the hot gases begin to dominate, the burning cloud rises and becomes more spherical in shape [52, 53].

The mechanism of a BLEVE is based on the fact that in the absence of nucleation sites such as impurities, crystals, or ions in a liquid, its boiling point can be exceeded

TABLE 4.15

Superheat Limit States, Critical Properties, and Nucleation Rates for Hydrogen and Some Other Materials

| | Critical Properties | | Superheat Limit State | | Normal | Nucleation |
Substance	Temperature (K)	Pressure (bar)	Temperature (K)	Pressure (bar)[a]	Boiling Point (K)	Rate (nuclei/ cm³/s)
Hydrogen	32.98	12.93	27.7–29.5	5.54	20.3	10^{-2}
Methane	190.56	45.92	167.6	21.3	111.6	10^{5}
Ethane	305	49.9	269.2	21.7	184	10^{5}
Propane	370	43.6	326.4	18.4	231	10^{5}
Water	647	218	553.0	64.1	373	10^{21}

[a] Wolfram Alpha Computational Knowledge Engine was used to calculate these values.

Source: Data from References CCPS, [54]; Lide, [55].

without boiling, thus resulting in superheating of the liquid. Yet there is a limit above which the liquid cannot be further superheated at a given pressure, and when this limit is reached, microscopic vapor bubbles are formed spontaneously without nucleation sites. The very high nucleation rates accompanying this phenomenon result in the formation of a shock wave transmitted through the evaporating liquid. Superheat limit states, critical properties, and nucleation rates for hydrogen and some other cryogenic materials are shown for comparison in Table 4.15 [54, 55].

When using the van der Waals equation of state, the superheat limit temperature, T_{sl}, and critical temperature, T_c, have been correlated with the equation [56, 57]:

$$T_{sl} = 0.84\, T_c \tag{4.4}$$

or, when using the Redlich–Kwong equation of state, with the equation:

$$T_{sl} = 0.895\, T_c \tag{4.5}$$

It is obvious from Table 4.15 that the nucleation rate for hydrogen is millions of times lower than for other cryogenic gases and 10^{23} times lower than that of water. So, in this respect, occurrence of a shock wave is quite improbable during a sudden loss of containment of LH_2 compared to other liquefied flammable gases. Furthermore, vessels containing LH_2 are much safer than high-pressure steam boilers, which have been involved in many catastrophic explosions due to the BLEVE phenomenon (although with no possibility of a subsequent fireball of steam).

Based on a previous work of Ustolin et al. [58], Wingerden et al. [59] performed three experiments to determine the effect of a storage vessel filled with LH_2 on top of a fire. The vessels with a capacity of 1 m³ were equipped with double-walled vacuum insulation. The factors examined were the orientation of the vessel and the effect of the insulation material (perlite or multi-layer insulation). Only one of the storage vessels ruptured, resulting in a blast, a fireball, and fragments.

4.3.4.2 Detonation Limits

The maximum and minimum concentrations of a gas, vapor, mist, spray, or dust in the air or other gaseous oxidant for a stable detonation to occur are the so-called upper and lower *detonation limits*. These limits depend on the size and geometry of the surroundings as well as other factors. Therefore, detonation limits found in the literature should be used with caution. Detonation limits are sometimes confused with deflagration limits, and the term explosive limits is then used as a collective term [57].

As shown in Table 3.1, detonable concentrations are narrower than flammable concentrations for hydrogen as well as for all other gases. The worst case during an accidental release is reaching a near-stoichiometric concentration of a flammable gas and encountering an ignition source, with the end result being a detonation. The occurrence of such detonable concentrations is quite infrequent in an enclosed area and rather unlikely in the open air for most gases that have lower and upper detonation limits very close to each other. Unfortunately, with some gases, including hydrogen, this concentration span is quite broad and the detonability range is easily attained, leading to much more serious accidents than those arising from gases with narrow detonation ranges [60].

The higher molecular diffusion of GH_2 shown in Table 3.1 compared to other gases means that any release will quickly mix with the surrounding air, reach a hazardous range, and then come down to the safe region below the lower flammability limit. In an LH_2 leak, fast vaporization occurs first and molecular diffusion in the air follows. On the other hand, the positive buoyancy of GH_2 can move a release to either higher floors in a building, endangering these areas, or to safe heights in the open air.

Detonation limits for any fuel–oxidizer mixture depend, among other factors, on the nature and dimensions of confinement. The minimum dimensions for detonation of hydrogen–air mixtures at 101.3 kPa (1 atm) and 298 K (25°C) in different degrees of confinement are shown in Figure 4.8[61].

4.3.4.3 Ignition Energy

The *minimum ignition energy* for a detonation to occur depends on the hydrogen concentration in the detonable mixture. Thus, the minimum ignition energy is equivalent to about one gram of tetryl for hydrogen–air mixtures at the stoichiometric composition (possessing the highest sensitivity and leading to direct detonation), whereas this increases to several tens of grams of tetryl for very rich or very lean mixtures (close to the detonation limits) [7].

The ignition energy required for a stable detonation is large for lean or rich mixtures (i.e., far from the stoichiometric concentration). On the other hand, it is possible to produce overdriven detonations when a large ignition energy is used.

Stable detonation parameters such as detonation temperature and pressure can be calculated by the Chapman–Jouget approach using the Gordon–McBride computer code [62].

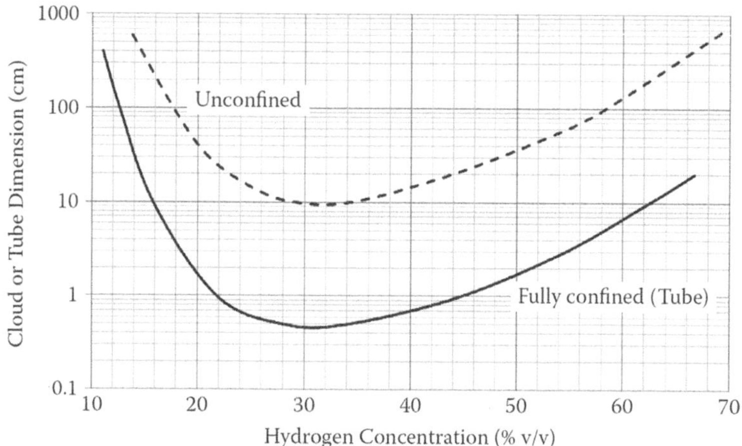

FIGURE 4.8   Minimum dimensions of GH_2–air mixtures for detonation at 101.3 kPa (1 atm) and 298 K (25°C).

Source: [7, 18]. With permission from the American Institute of Aeronautics and Astronautics.

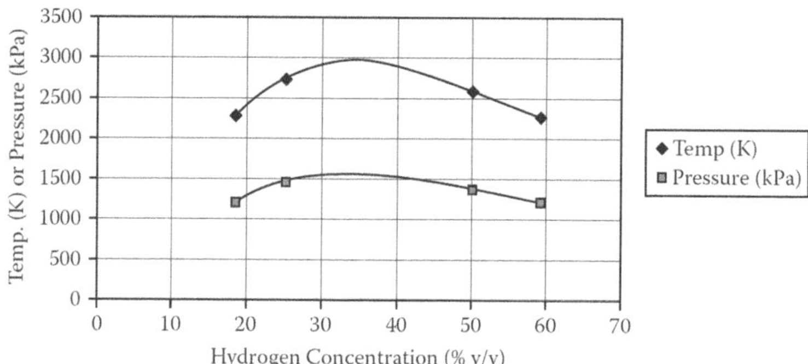

FIGURE 4.9   Detonation pressure and temperature of hydrogen–air mixtures starting from 101.3 kPa (1 atm) and 298 K (25°C). Chapman–Jouguet calculations using the Gordon–McBride code.

Source: [62].

The dependences of detonation temperature and pressure on hydrogen concentration for hydrogen–air mixtures and hydrogen–oxygen mixtures are shown in Figure 4.9 and Figure 4.10, respectively. Maximum values of approximately 3000 K and 1600 kPa for near-stoichiometric hydrogen–air mixtures are observed, whereas these maxima rise to about 3800 K and 2000 kPa for stoichiometric hydrogen–oxygen mixtures.

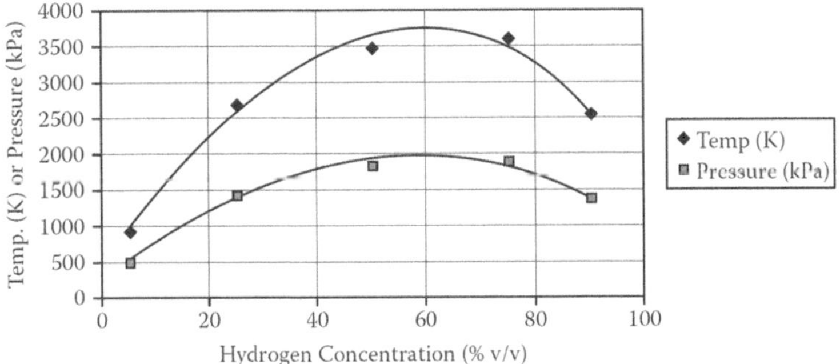

FIGURE 4.10 Detonation pressure and temperature of hydrogen–oxygen mixtures starting from 101.3 kPa (1 atm) and 298 K (25°C). Chapman–Jouguet calculations using the Gordon–McBride code.

Source: [62].

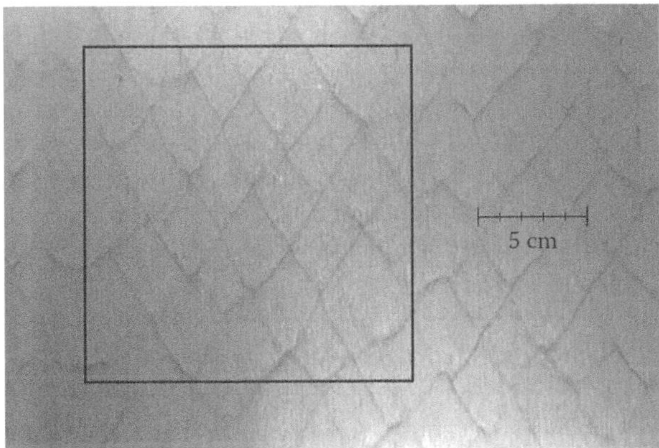

FIGURE 4.11 An actual soot foil photograph from foils aligned with the axis of the tube, which is used to record the detonation cell size with characteristic dimensions usually ranging from some millimeters to some centimeters. (Photo courtesy of Explosion Dynamics Laboratory, California Institute of Technology, https://shepherd.caltech.edu/EDL/PublicRe sources/CellImageProcessing/cellsize.html.)

4.3.4.4 Detonation Cell Size

A characteristic parameter related to detonation is the *detonation cell size* (λ). The wave front in a detonation is not planar but is composed of reaction cells, as verified by the pattern left by a detonation on a smoked plate (Figure 4.11).

A two-dimensional illustration of the actual structure is given in Figure 4.12 according to the Zel'dovich–von Neumann–Döring (ZND) model.

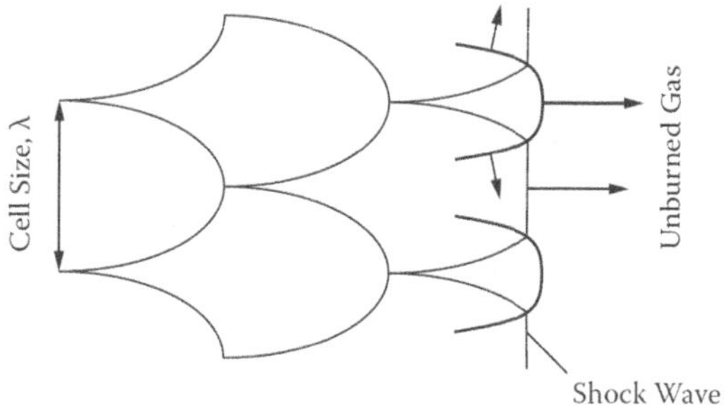

Pattern of an Actual Structure

FIGURE 4.12 ZND structure and pattern of an actual structure of a detonation front. The characteristic length scale of the cell pattern, the cell size, λ, is shown in the figure.

Source: www.gexcon.com/brochures/gas-explosion-handbook/

The cell size is valuable in the prediction of the onset of detonation and is related to key parameters with regard to hazardous situations [61, 63]. Cell lengths for stoichiometric hydrogen–air and hydrogen–oxygen mixtures at 101.3 kPa (1 atm) are 15.9 mm and 0.6 mm, respectively. Cell size decreases when pressure increases for hydrogen–air mixtures. The cell width of hydrogen–air detonations increases significantly with the concentration of diluents (e.g., carbon dioxide and water) [7].

The ratio of the length of a detonation cell (λ) to reaction zone width (δ) appears to depend on the mixture composition and initial conditions and varies in the range of about 10^2 (Figure 4.13).

4.3.4.5 Consequences of Hydrogen–Air Explosions

With regard to the consequences of an accidental release and subsequent explosion, the energy of explosion is often taken into consideration. Calculations such as one gram of stoichiometric hydrogen in NTP air is equivalent to 24 grams of TNT are often found in the literature, but they are misleading since they do not take into account the weight of the oxidant (air in this case). Including air in the calculations, it becomes clear that a stoichiometric hydrogen–air mixture produces only two-thirds of the explosion energy of TNT. This difference can be easily explained by the dilution effect of nitrogen in the air. In addition, it should be noticed that in an accidental release, mixing of hydrogen with air is not perfect – so only a small fraction of the theoretical energy will be produced.

Other factors that determine the catastrophic effects of an explosion are the initial density of the explosive (which is three orders of magnitude higher for TNT than for hydrogen–air mixtures) and the detonation velocity (which is three to five times higher for TNT than for hydrogen). So, the resulting pressure wave from a hydrogen explosion is considerably flatter (longer duration and lower maximum overpressure) than for TNT, and the destruction effects are caused mainly by impulse rather than overpressure.

Considering the effects of a deflagration, the parameters that are usually taken into account for designing deflagration suppression systems with the aid of a suitable

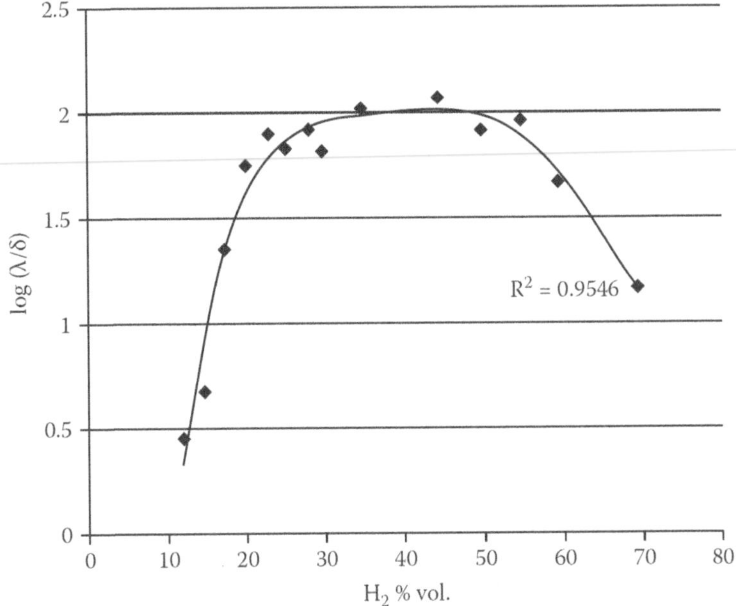

FIGURE 4.13 Dependence of the ratio of the experimental cell size (λ) to reaction zone width (δ) for hydrogen–air mixtures at 300 K. (Data from the Induced Chemical Reactions Laboratory, National Research Center, Kurchatov Institute, Russia.)

suppressant are maximum pressure, deflagration index, fundamental burning velocity, and effective burning velocity [64].

Maximum pressure, P_{max}, is the maximum pressure developed by a confined deflagration as determined by measurements over a wide range of concentrations.

Deflagration index, K (bar·m/s), defines the maximum rate of pressure rise with time, $(dP/dt)_{max}$, of a deflagration in a volume V, according to the equation:

$$K = \left(dP/dt\right)_{max} \times V^{1/3} \tag{4.6}$$

K_G is used for gases and K_{St} for dusts.

The rate of pressure rise is slower for larger volumes, thus allowing more time for explosion mitigation than for smaller volumes. Deflagration index measurements are undertaken in spheres of volumes of 20 L or 1 m³ (or greater) and the results scaled up to the process vessel's volume according to the cube-root scaling relationship given in equation (4.6).

Hazard classes (ST) have been determined for dusts based on the experimental K_{St} values as follows:

ST 0: K_{St} = 0 bar·m/s
ST 1: K_{St} = 1 to 200 bar·m/s

ST 2: K_{St} = 201 to 300 bar·m/s
ST 3: K_{St} > 300 bar·m/s

However, a similar classification system has not yet been developed for vapors and gases, and only limited results are given for gases. For comparison, typical P_{max} and deflagration indices for dusts and gases in larger test apparatus are given in Table 4.16 [64]. As shown in this table, hydrogen is classified as more hazardous compared to other gases, but it is much less hazardous than aluminum dust.

Lean hydrogen–air mixtures in vented vessels were experimentally tested to determine the effect of hydrogen concentration, ignition location, vent size, and obstacles on explosion overpressures [65]. Analysis of the experimental results revealed that the pressure maxima mostly depend on the interaction of the maximum flame area and the velocity of the explosion front.

Deflagration suppression systems for gas and vapor explosions are designed for either quiescent or turbulent mixture conditions. In fact, the effective turbulent burning velocity is typically at least four times the value of the laminar burning velocity of the same mixture.

4.3.4.6 Damage of Structures from Thermal Radiation and Blast Waves

Exposure of structures and equipment to radiant heat flux or direct flames may be destructive, as shown in Table 4.17. With regard to assessments based on a QRA approach, no probit functions are currently available in the literature relating to thermal effects on structures and equipment, as is the case for effects on humans. Thus, only simplified data can be used in this case, such as those shown in Table 4.18.

TABLE 4.16

Deflagration Maximum Pressures and Indices for Some Gases and Dusts – Larger Test Apparatus

Gas or Dust	P_{max} (bar)	K_G or K_{St} (bar·m/s)
Methane	8.4	58
Propane	8.3	103
Hydrogen	8.2	503
Coal dusts	8	70
Grain dusts	9	100
Polythene dusts	8	100
Cellulose dusts	8	140
PVC dusts	9	160
Flour	10	190
Stearate	7	225
Dyestuffs	11	340
Aluminum dusts	13	1000

Source: Adapted from [64].

TABLE 4.17
Damage to Structures and Equipment from Thermal Radiation

Thermal Radiation Intensity (kW/m²)	Type of Damage
4	Glass breakage (30 min exposure)
12.5–15	Piloted ignition of wood, melting of plastics (>30 min exposure)
18–20	Cable insulation degrades (>30 min exposure)
10 or 20	Ignition of fuel oil (120 or 40 s, respectively)
25–32	Unpiloted ignition of wood, steel deformation (>30 min exposure)
35–37.5	Process equipment and structural damage (including storage tanks) (>30 min exposure)
100	Steel structure collapse (>30 min exposure)

Source: [10]. With permission from Elsevier.

TABLE 4.18
Damage to Structures and Equipment from Overpressure Events

Overpressure (kPa)	Type of Damage
1	Threshold for glass breakage
15–20	Collapse of unreinforced concrete or cinderblock walls
20–30	Collapse of industrial steel frame structure
35–40	Displacement of pipe bridge; breakage of piping
70	Total destruction of buildings; heavy machinery damaged
50–100	Displacement of cylindrical storage tank; failure of pipes

Source: [10]. With permission from Elsevier.

Nevertheless, since the exposure duration needed for thermal radiation to be effective is rather long, the impact of thermal radiation from hydrogen on structures and equipment is not considered significant.

Some damage to structures and equipment from overpressure events is shown in Table 4.18. Probit functions relating peak overpressure and the impulse of shock waves with minor, major, and total damage or collapse of structures can be found in the literature to be used in QRAs [10].

4.3.4.7 Safety Distances from Hydrogen Storage Facilities

The safety distance may be defined as the acceptable minimum separation distance between a hazard source, such as a flammable gas leak, and the object to be protected (e.g., humans or buildings). This separation distance is a function of the quantity of hydrogen, and this is why we usually refer to a quantity–distance relationship. The consequences for the object and the safety distance depend on the specified threshold values of physically defined criteria such as the dose of thermal radiation, the toxic dose, or the peak overpressure of a blast wave.

Safety distances for GH_2 and LH_2 with regard to design and operation are regulated in the United States by OSHA (the Occupational Safety and Health Administration) as part of 29 CFR (Code of Federal Regulations). According to these regulations, the safety distance between a hydrogen installation and people or property is defined as 15.2 m for GH_2 in amounts more than 425 Nm^3. For LH_2 cryotanks containing more than 2.27 m^3, the safety distance from the same objects is at least 22.8 m [66]. In Europe, LH_2 storage tanks are covered by the recommendations given by EIGA (the European Industrial Gases Association) specifying minimum safety distances [67]. For instance, the specified minimum safety distance of LH_2 storage tanks from combustible liquids or solids, roads, railroads, overhead power lines, and technical buildings is 10 m. (For more details consult Chapter 13.)

4.3.4.8 Deflagration-to-Detonation Transition

As previously discussed, the oxidation (combustion) of hydrogen in air or other oxidant (e.g., pure oxygen) can be subsonic (deflagration) or supersonic (detonation) depending, among other factors, on the strength of the ignition source. A highly energetic source such as a shock wave can initiate a detonation, whereas a low-strength energy source such as a flame or a spark will typically result in a deflagration.

Nevertheless, there is a possibility that a deflagration will transition to a detonation after the flame has traveled for some time and distance by the following mechanism. Although a deflagration wave is subsonic, it can perturb the gas ahead of the flame front by compression waves generated due to the expansion of combustion products. A newly produced compression wave propagates at a higher velocity through the gas which is pre-compressed and adiabatically heated by the previous compression waves. Thus, a shock wave will be formed and strengthened when overtaken by successive compression waves and finally transformed to a shock wave. At the moment the strength of the precursor shock reaches a critical value, a sudden change of the propagation mechanism occurs from the diffusion and heat conduction-controlled subsonic propagation to adiabatic shock wave-driven propagation. This *deflagration-to-detonation transition* (DDT) is not a continuous process but rather a stepwise change [68, 69].

In general, the composition range in which a detonation can take place is narrower than that of a deflagration. The range for detonation limits found in Table 4.10 for hydrogen–air mixtures is from 18.3 to 59% hydrogen. However, with sufficiently strong ignition sources, these limits can be extended and, in addition, an overdriven detonation can be obtained; this results in a higher detonation velocity which gradually declines to a stable detonation at a thermodynamically determined lower velocity. Even in the absence of a strong initiator, a deflagration can be accelerated to detonation when other factors, such as sufficient degree of confinement and the presence of obstacles inducing turbulence in the flame front, limit energy losses and suitably mix the ingredients, thus triggering a detonation. The abrupt increase of overpressure and flame speed on the transition from deflagration to detonation for hydrogen–air mixtures has been modeled by Eichert, whose work was based on numerical calculations of one-dimensional tube combustion processes [69].

This phenomenon is highly influenced by obstacles creating more turbulence, thus causing intense mixing, enhanced by the much higher diffusion coefficient of hydrogen in the air compared to those of methane and gasoline fumes. In general, the time occurrence of DDT is not predictable, but the detonation induction distance in the air is of the order of 100 times the diameter of the pipe (or tunnel) in which the burning mixture propagates [70]. This distance depends on the combustible mixture constituents, the pressure, temperature, and concentration of the gaseous mixture, the enclosure geometry, and the strength of the ignition source [71]. In a stoichiometric hydrogen–air mixture in a tube, the experimentally observed distance from the ignition point to the location of DDT may be reduced to a length-to-diameter ratio of 15–40. The initiation of detonation during DDT is thought to happen in so-called "hot spots", which potentially could be located within the turbulent flame brush or ahead of it, e.g. in a focus of a strong shock reflection [70, 72]. The effect of diffusion time on the mechanism of deflagration to detonation transition in an inhomogeneous hydrogen–air mixture has also been investigated [73].

The deflagration of the stoichiometric cloud results in a pressure wave of only 0.01 MPa, which is below the level of eardrum injury. In contrast, the detonation of a stoichiometric hydrogen–air mixture could generate a blast with more than two orders of magnitude higher pressure of about 1.5 MPa, which is far above the fatal pressure of about 0.08–0.10 MPa. Consequently, safety measures to exclude the potential of a DDT are indispensable [70].

The minimum energy required to initiate a direct detonation of a hydrogen–air mixture is around $1 \cdot 10^7$ mJ, which is about nine orders of magnitude higher than the minimum ignition energy for deflagration (0.017 mJ). A crucial strategy to minimize the potential for flame acceleration or detonation is the avoidance of confinement and congestion where flammable hydrogen–air mixtures might form [72].

Since severe accidents have occurred in nuclear power plants due to the ignition of accidentally generated hydrogen, Zhang et al. [74] proposed an improved method for the prediction of an eventual hydrogen DDT. The method is based on reducing the oxygen concentration by using inerting diluents.

4.3.4.9 Accidental Hydrogen Generation in Nuclear Reactors

Hydrogen plays a major role in the development of water-cooled nuclear reactor accidents with severe impacts on public health, the economy, the nuclear reactor industry, agriculture, politics, and even on the roadmap of the energy demand worldwide. The most severe accidents in nuclear power plants in Chernobyl (1986), Three Mile Island (1979), and Fukushima Daiichi (2011) occurred due to a lack of control of the hydrogen's accidental production and its subsequent explosion (Figure 4.14).

Following the accidents in Chernobyl and Three Mile Island, numerous studies on hydrogen control have been carried out. Nevertheless, the results of these studies were not sufficient to predict the impact of hydrogen's accidental production in the similar severe accident in Fukushima [75, 76].

FIGURE 4.14 A photo from video footage taken on March 14, 2011, showing the hydrogen explosion at the Fukushima Daiichi nuclear power complex at the moment of the event and afterward. With permission. (Reuters/NTV photo via Reuters TV.)

The results of these studies were the basis for a deeper comprehension of the behavior of hydrogen production. The US Nuclear Regulatory Commission (USNRC) is:

…an independent agency created by Congress in 1974 to ensure the safe use of radioactive materials for beneficial civilian purposes while protecting people and the environment, and also regulating commercial nuclear power plants and other uses of nuclear materials.

Following its mission, USNRC published in 1983 a report on the role of hydrogen in normal and accidental situations in light water nuclear reactors (LWRs). With this goal, studies were conducted by USNRC on hydrogen deflagration, flammability limits, and the combustion characteristics of hydrogen–air–steam mixtures from 1984 to 1986. Furthermore, the evolution of various hydrogen explosions was studied from 1987 to 1998, and experimental results related to the operability of thermal igniters in a nuclear reactor were reported in 1989. At the same time, many countries, including Switzerland and France, were working on the behavior of hydrogen. Up to today, this is still one of the principal research activities and safety assessments of conventional LWRs worldwide [77].

4.3.4.9.1 Potential Hydrogen Sources

According to IAEA-TECDOC-1661 [78] potential hidden sources during the development of a severe accident in an LWR may come from:

- *In-vessel metal oxidation (Zr clads and grids and other metallic structures) or B_4C absorber material oxidation with steam or with water contained in the reactor pressure vessel lower plenum.*
- *Ex-vessel oxidation of metallic material (Zr, Cr, Fe...) during direct containment heating or into the water eventually contained in the cavity pit (short term event occurring at vessel lower head failure).*
- *Ex-vessel oxidation of metallic material (Zr, Cr, Fe...) during molten core concrete interaction (complete and rapid energetic oxidation of Zr and Cr during the first hour, then partial and slow oxidation of Fe up to the time of the complete basemat penetration by the corium).*

Corium is a molten material that is created in a nuclear reactor during a meltdown accident. It resembles volcanic lava and consists of nuclear fuel, fission products, control rods, structural materials from the destroyed parts of the reactor, reaction products with air, water, and steam, and, if the reactor vessel is ruptured, molten concrete from the floor of the reactor room [79].

Water radiolysis and the metal corrosion in the containment (mainly Al and Zr) are taken into account in hydrogen sources during a conventional design accident (e.g., loss of coolant accident). Yet these sources are considered negligible during the development of the investigation for a severe accident. In fact, in the event of a severe accident, several days are needed before the amount of hydrogen produced by these two processes renders the air flammable [78].

Concerning zircaloy, the mechanism by which hydrogen is produced in nuclear power plants is the following. At high temperatures, zircaloy, which is used as a metal cladding in nuclear reactors, can react with water to produce hydrogen. This exothermic reaction followed by a deflagration or detonation of the hydrogen–air mixture can produce chemical power to accelerate the progression of a severe accident. The zircaloy–water reaction, which turns into a runaway reaction at temperatures greater than about 1200°C, is as follows:

$$Zr + 2H_2O \rightarrow ZrO_2 + 2H_2 + \Delta H_{Zr}$$

where ΔH_{Zr} is the heat of the reaction per mole of zircaloy consumed, currently set at 616 MJ/kmol. This oxidation is assumed to take place at the Zr/ZrO_2 interface leading to an increase in the oxide layer thickness.

In severe accidents in nuclear power plants, substantial amounts of hydrogen can be generated, for instance, from the chemical reaction between zirconium cladding and hot water vapor, as well as from core–concrete interactions after a lower head failure of the vessel. Hydrogen thus generated may find its way into the compartments in the containment building with the potential to threaten the containment integrity by overpressurization from deflagrations or detonations. Moreover, even local hydrogen burning, which is not a threat to the global containment integrity, may also be a hazard to the survivability of safety-related equipment.

Hence, regulations for hydrogen control based on local hydrogen concentrations have been established in countries using nuclear reactors. They specify that the average hydrogen concentration in the containment to be controlled is not to exceed 10% during and following an accident that releases an equivalent amount of hydrogen as would be generated from a 100% fuel clad metal–water interaction, or that the post-accident atmosphere must not support hydrogen combustion. Therefore, equipment with hydrogen control units such as igniters or catalytic recombiners has been considered [80].

4.4 HYDROGEN CONTROL AND RISK MITIGATION

Severe damage to the containment that may lead to the release of some radioactive hydrogen can be avoided by specific mitigation measures. The main countermeasures are dilution of hydrogen, inertization of the atmosphere (dilution or removal of the oxygen), and removal of the hydrogen by burning or recombination [78, 81].

Thus, hydrogen risk mitigation measures that are often engaged are:

- Pre-inertization with nitrogen in most of the boiling water reactor (BWR) containments.
- Passive autocatalytic recombiners in large dry pressurized water reactor (PWR) containments.
- Igniters for PWR ice condenser and BWR containments.

4.4.1 Igniters and Recombiners

The igniters or the catalytic recombiners reduce the containment hydrogen concentration in these cases by intentionally burning the hydrogen generated to avoid severe consequences from hydrogen–air explosions. Continuous burning may prevent hydrogen accumulation and not result in a significant pressure spike in the containment building. Once pressure and temperature spikes are over, no more short-term challenges to containment integrity remain. On the basis of a best-estimate assessment, it has been shown that it is beneficial to utilize hydrogen igniters rather than do nothing with respect to the expected value of hydrogen concentration in the containment building during an accident [82].

4.4.2 Passive Autocatalytic Recombiners

Passive autocatalytic recombiners (PARs) are devices that consume amounts of oxygen as a reactant with hydrogen inside nuclear energy installations and result in an increased ratio of inert gas nitrogen to oxygen in the air. In addition, the product of oxygen consumption is water mist, which also reduces explosion hazards, and this is why they have been adopted by nuclear energy industries all over the world. PARs can effectively remove carbon monoxide from the air at the same time as they remove hydrogen [83].

Catalytic recombiners use platinum- and/or palladium-based catalysts to recombine (oxidize) hydrogen with atmospheric oxygen. They may carry out this chemical reaction at lower temperatures, with a higher hydrogen/oxygen concentration ratio, and even under steam-inerted conditions and over longer times than would be the case with gas combustion. PARs are available in two different designs, varying in the shape of the reacting surfaces (flat plate or pellets in cartridges) and the layout of the channel box. The functioning principles of these PARs are almost identical, but they vary in size and shape [81].

4.4.3　Active Systems

As well as PARs, active systems such as glow plug igniters have been developed and are already installed in most nuclear reactors in many countries following the guidelines of the Committee on the Safety of Nuclear Installations of the Organisation for Economic Co-operation and Development [81]. Igniter technology is a method of preventing catastrophic fires by oxidizing hydrogen near the limits of flammability. Igniters can treat higher hydrogen flow rates than PARs, but they have to be suitably placed and they need external power. Thus, a reliable power supply and proper igniter location are crucial for the operation of the igniters to reduce hydrogen concentration effectively. The proper application of igniters needs a prior study of gas flow scenarios to select the optimum locations. Both igniters and PARs can be used in combination for better hydrogen control [81].

4.4.4　Stochastic and Computational Assessments for Improving Hydrogen Safety

Network theory has been used to investigate hydrogen logistics incidents, thus providing insight into the causes of hydrogen logistics incidents by analyzing the interactions between factors and effects. It can also reveal the roles of each factor and its effect in the occurrence of a hydrogen logistics incident. This approach may be useful for enhancing safety in hydrogen logistics as well as developing associated precautions.

In a study, 100 hydrogen logistics incidents were collected, and their related incident chains were formulated [84]. After analyzing the incident chains and the interdependencies among factors and effects, eight significant factors and five effects were identified. The results revealed that each factor plays its role in the occurrence of an effect. Yet the occurrence of a hydrogen logistics incident cannot be significantly reduced by only considering the factors, since both factors and effects must be taken into account altogether. In this way, probabilistic estimations revealing the occurrence of significant factors and effects can be effective for controlling and reducing the occurrence of hydrogen logistics incidents, as well as developing associated

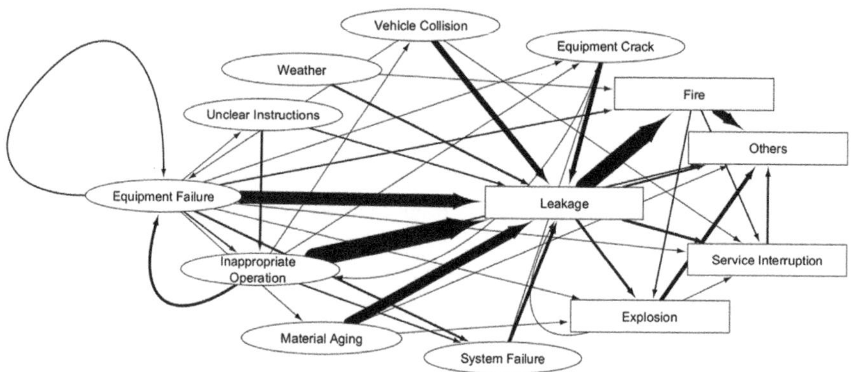

FIGURE 4.15 Network diagram showing interdependencies among factors and effects With permission from Elsevier.

Source: [84].

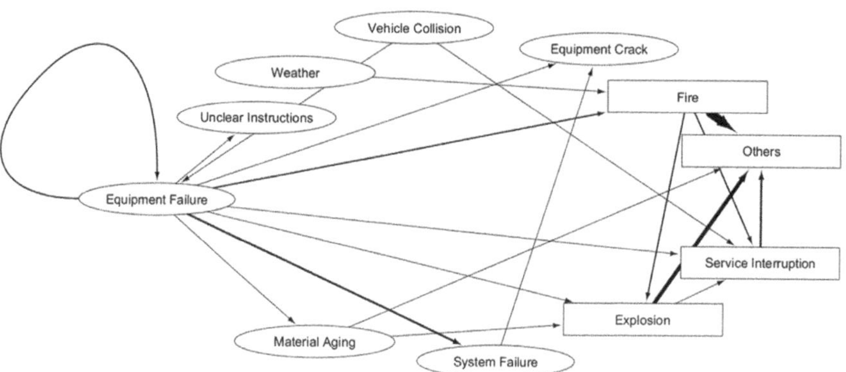

FIGURE 4.16 Network diagram with nodes of inappropriate operation (factors) and leakage (effects) removed. With permission from Elsevier.

Source: [84].)

precautions. In this study, a network diagram with 13 nodes was constructed as shown in Figure 4.15. After removing all factor nodes of inappropriate operation and leakage, the cluster network of effects has been remarkably reduced, as shown in Figure 4.16.

Sathiah et al. [84] propose a CFD methodology that is suitable for the analyses of hydrogen deflagrations in a nuclear reactor containment. They found that when their methodology is applied in an LWR, the turbulent flame acceleration of hydrogen, and consequently the resulting pressure, depend strongly on the initial turbulence in the ignition region.

Furthermore, these authors verified that dynamic pressure loads resulting from hydrogen combustion can be detrimental to the structural integrity of the reactor, safety systems, and reactor containment. Therefore, accurate prediction of these pressure loads is an important safety issue. Experiments performed with a small fan-stirred explosion bomb were used for this purpose. From the presented validation analyses, it was concluded that the maximum pressures were predicted with 13% accuracy, while the rate of pressure rise, *dP/dt*, was predicted to within about 30% [86].

Pasman [87] stresses the need for adequate risk control when dealing with large-scale distribution and the use of hydrogen. This control requires adequate risk assessment, yet despite many years of experience, methods to determine risk lack robustness, since results seem to be too dependent on choices made by the analyst due to uncertainties, lack of data, and different views. This can create frustration amongst people dependent on results. Pasman's study reviews existing methodological weaknesses and current improvements, e.g. the context of developing risk-informed standards, and emphasizes the challenges to further raise quality.

Thus, Pasman [87] raises some suggestions for how HySafe and other groups could work in the right direction. For instance, HySafe pays considerable attention to the fact that because of a much higher diffusivity of hydrogen and the embrittlement effect on some materials, hydrogen leaks cannot be considered similarly to hydrocarbon leaks.

Recently, there has been increasing interest in the application of 3D CFD codes to improve steam and hydrogen distribution and combustion predictions in nuclear power plants. These codes are also used as a supplement to, or in combination with, system codes to identify specific issues that would need further detailed investigation. Lumped-parameters codes are used to analyze the entire accident more globally, while the 3D codes are used to analyze specific phenomena (such as jets, stratification, and flame acceleration simulation) over limited periods. The higher precision of CFD predictions with respect to hydrogen distribution and combustion processes can be explained not only by the nature of the equations used (3D, fully compressible Navier–Stokes equations, including momentum balance and a complete set of fluid dynamic terms) but also by the much finer spatial resolution (typically 1 m^3 versus 1000 m^3 with a lumped-parameter method) and the corresponding mixing effects. Currently, 11 codes are being used by many organizations for hydrogen analysis, namely ASTEC, MAAP/MAAP-CANDU, MELCOR, SPECTRA, COCOSYS, TONUS, GOTHIC, GASFLOW, ANSYS CFX, ANSYS FLUENT, and ANSYS AUTODYN. These codes are continuously under development and review and are continuously validated against new experimental data. The steady increase in computer power over recent years and the advancement of numerical methods has allowed CFD codes to become of practical use and has enabled nuclear engineers to model reacting flows in a realistic geometry with a good spatial resolution [78, 81]. None of the codes is fully validated for the entire list of hydrogen generation, distribution, combustion, and mitigation phenomena, either due to a lack of experimental data or due to a limited application range [81].

A summary of these codes' capabilities and validation statuses for modeling hydrogen generation, distribution, combustion, and mitigation is shown in Table 4.19 [81].

TABLE 4.19

Summary of Codes' Capabilities and Validation Statuses for Modeling Hydrogen Generation, Distribution, Combustion, and Mitigation (Courtesy of the OECD Nuclear Energy Agency (NEA)

Category	ID	Phenomenon Description	Lumped-Parameter Codes					3D Codes					
			ASTEC	MAAP/ MAAP-CANDU	MELCOR	SPECTRA	COCOSYS	TONUS	GOTHIC	GASFLOW	CFX	FLUENT	AUTODYN
Generation	G1	Water radiolysis in sump											
Generation	G2	Metal (Zr, steel) steam reaction (oxidation within the core)	◆	◆	◆	◆							
Generation	G3	Metal corrosion (zinc, steel, aluminium) [1]			◇	◇							
Generation	G4	Molten corium concrete interaction	◆	◇	◆	◇	◆						
Generation	G5	Oxidation in molten pools or debris beds within the core	◇	◇	◆	◆							
Generation	G6	Metal (Zr, steel, B_4C) oxygen reaction within the core	◆	◆	◆	◆							
Distribution	D1	Stratification [2]	◆	◆	◆	◆	◆	◆	◆	◆	◆	◆	
Distribution	D2	Momentum-induced mixing [2]	◆	◆	◆	◆	◆	◆	◆	◆	◆	◆	
Distribution	D3	Buoyancy-induced mixing [2]	◆	◆	◆	◆	◆	◆	◆	◆	◆	◆	
Distribution	D4	Condensation on surfaces [3]	◆	◆	◆	◆	◆	◆	◆	◆	◆	◆	
Distribution	D5	Turbulent flow characterization[4]						◆	◆	◆	◆	◆	
Distribution	D6	Liquid/water in films and pools[5]	◆	◆	◆	◆	◆		◆	◆		◇	
Combustion	C1	Deflagration	◆	◆	◆	◆	◆	◆	◆	◇	◆	◆	
Combustion	C2	Flame acceleration[6]	◇					◆			◆	◆	
Combustion	C3	DDT[6]											
Combustion	C4	Detonation[6]						◆					◆
Combustion	C5	Quenching of detonations											◇
Combustion	C6	Diffusion flame	◆	◇	◆	◇	◆		◇		◆	◆	

Mitigation												
C7	Strong ignition/jet ignition									◇	
C8	Combustion with droplets										
C9	H_2–CO combustion	◇	◇	◇		◇	◇				
M1	H_2 recombination by PAR[7]	◆	◇	◆	◆	◆	◆	◆	◆	◆	◆
M2	H_2 ignition by PAR[8]	◇		◇		◇	◇				
M3	H_2 ignition by igniter	◇	◆	◇	◆	◇		◆		◇	◇
M4	CO recombination by PAR[9]	◇			◇	◇	◇			◆	
M5	Filtered venting system[10]	◇	◇	◇	◇	◇		◇			
M6	Fan/local air cooler	◆	◆	◆	◆	◆		◆		◆	◇
M7	Spray system[11]	◆	◇	◆	◆	◆	◆	◆	◆	◇	◆

◇ - code has this capability (but no validations exist); ◆ - code has this capability and validations have been performed against relevant experiments.

Notes:

(1) Metal corrosion can be modeled in MELCOR and SPECTRA by defining a user input for the reaction coefficients.

(2) Modeling of D1, D2, and D3 by lumped-parameter codes requires specific and detailed nodalization schemes.

(3) For CFX and FLUENT, the condensation models are mostly developed by users (i.e., France, Germany, the Netherlands, etc.). While CFX provides a basic model for condensation, there is a condensation model available in the general FLUENT package provided by the code developer.

(4) The code developers performed fundamental qualification/verification of the turbulence models of their codes against small-scale tests with turbulence characteristics measured; the users performed benchmarks against large-scale integral tests where turbulence took a significant role in mixing, but the turbulence characteristics were not quantified.

(5) Even though CFX and FLUENT can simulate multiphase flows with heat and mass transfer, to limit the computational effort, containment simulations are performed in a single phase. The water in films and pools is either neglected or modeled using a Lagrangian approach. User models for heat and mass transfer in the sump are under development for CFX in Germany. User models for heat and mass transfer with water films on the wall and in the sump are developed and tested for FLUENT in the Netherlands.

(6) GASFLOW and SPECTRA can predict whether a hydrogen–air–steam mixture is flammable or detonable and if it has the potential risk of flame acceleration (FA). However, the code does not simulate the FA, DDT, and detonation.

(7) Detailed PAR models in CFX and FLUENT are developed by code users (i.e., Germany, the Netherlands) and are not available in the general package obtained from the code developer. Simple models to employ manufacture correlations can be easily implemented in both codes with general packages.

(8) ASTEC, COCOSYS, and MELCOR allow the user to trigger the ignition in zones where PARs exist, but no specific model exists.

(9) All codes except AUTODYN can potentially use a similar approach for H_2 recombination to model H_2–CO recombination, but such application is only done for the labeled codes. Validation of PAR models for CO recombination is only partly done by users as not enough data are publicly available.

(10) Similar models for filtered venting systems are built into these codes, which employ user-defined parameters (i.e., removal efficiency, decontamination factor, etc.). For CFX and FLUENT, a similar approach can be used, but no application has been performed.

(11) A general spray model is available in CFX, but it has not been applied for containment spray modeling yet. Some users (i.e., the Netherlands) have developed their own spray model in FLUENT.

In a study [88], a detailed 3D GOTHIC model was used to analyze the hydrogen distribution and determine the preferential hydrogen pathways and areas of hydrogen accumulation along the containment building during a hypothetical severe accident. The PARs' implementation process addressed the preferential zones of hydrogen flow and accumulation, which were identified in the steam generator towers and the dome. It was found that hydrogen accumulates mostly in the dome due to its fast diffusion and rises to the upper part of the containment.

A methodology and accompanying software toolkit (HyRAM) have also been developed to provide a platform for the integration of state-of-the-art, validated science and engineering models and data relevant to hydrogen safety. The HyRAM software toolkit establishes a standard methodology for conducting QRA and consequence analysis relevant to assessing the safety of hydrogen fueling and storage infrastructure. The HyRAM toolkit integrates fast-running deterministic and probabilistic models for quantifying the risks of accident scenarios, predicting physical effects, and characterizing the impact of hydrogen hazards including thermal effects from jet fires, and thermal and pressure effects from deflagrations and detonations in enclosures [89].

The use of cascaded fuzzy neural networks in the prediction of hydrogen concentrations in nuclear power plant containment under severe accidents has been successful with a root-mean-square error level below 5%, thus assisting operators in preventing a hydrogen explosion [90].

In a study by Chang et al., [91] a dynamic risk analysis approach regarding hydrogen leakage in a hydrogen generation unit was applied to address the leakage risk of the hydrogen generation unit using the dynamic Bayesian network methodology. A case study of the hydrogen generation unit was carried out to demonstrate the applicability and advantages of this approach. The output indicated that the leakage probability of the hydrogen generation unit can be significantly decreased through equipment repair. Furthermore, the failure and repair rates of the overflow alarm and pressure sensor were found to be the most significant factors in the hydrogen generation unit leakage. Finally, some active mitigative suggestions are presented in this paper to further reduce the leakage risk of the hydrogen generation unit.

Although explosion risk analysis (ERA) is an effective method to investigate potential accidents in hydrogen production facilities, it suffers from significant hydrogen dispersion explosion scenario-related parametric uncertainty. To better understand the uncertainty in ERA results, numerous CFD scenarios need to be computed. Yet such a large number of CFD simulations are computationally expensive. Thus, Shi et al. [92], presented a stochastic procedure by integrating a Bayesian regularization artificial neural network (BRANN) methodology with ERA to effectively manage the uncertainty as well as reduce the simulation intensity in the hydrogen explosion risk study. This BRANN method randomly generates thousands of non-simulation data presenting the relevant hydrogen dispersion and explosion physics. The generated data are used to develop scenario-based probability models, which are then used to estimate the exceedance frequency of maximum overpressure. The performance of the proposed approach was verified by analyzing the parametric sensitivity on the exceedance frequency curve and comparing the results against the traditional ERA approach.

4.5 ENVIRONMENTAL CONCERNS

4.5.1 METHODS FOR PRODUCING HYDROGEN

Hydrogen is often presented by the media as an environmentally clean fuel because its only combustion product is harmless water. Nevertheless, hydrogen first must be produced, and the effects of hydrogen production on the environment depend on the production process selected, which determines the products [93]. Europe is leading the global revival of this energy carrier, which has its origins back in World War II when it was originally used by the Nazis to produce synthetic fuels from coal due to a lack of liquid fossil fuels [94]. In the future mutually dependent energy system of the EU, hydrogen will play a vital role, as well as sustainable electrification and more efficient and circular use of all resources. Vigorous large-scale development of clean hydrogen is a crucial factor for the EU to control climate change by cost-effectively reducing greenhouse gas (GHG) emissions by a minimum of 50% and toward 55% by 2030 [95].

Hydrogen production is based on a variety of feedstocks such as oil, coal, natural gas, biomass, and organic waste. The main constituents of all these feedstocks are hydrogen and carbon. Furthermore, hydrogen can also be produced through water electrolysis resulting in hydrogen and oxygen. Costs, gas emissions, land use, and environmental issues related to hydrogen production depend on the feedstock and the chosen process [96].

Steam reformation of methane is usually used to produce hydrogen. Yet to produce green hydrogen with minimum global warming emissions, renewable energy should be used, such as wind or solar energy. Other resources such as coal and hydrocarbons may be used to produce blue hydrogen, but the co-produced carbon dioxide should be captured and stored to prevent global warming. Other methods such as thermochemical, biochemical, and thermal splitting of water using heat from nuclear or strong solar energy and hydrogen-producing algae are not sufficiently developed yet to be used on an industrial scale [96].

4.5.1.1 Steam Methane Reforming Using Natural Gas

Despite the possibility of reforming other hydrocarbons to produce hydrogen, methane from natural gas still remains the most suitable. Nevertheless, global resources of natural gas are limited and, for geopolitical reasons, its price has significantly increased recently. In any case, although it co-produces lesser amounts of carbon dioxide compared to other hydrocarbons, this should still be captured and stored. In addition, the relatively recent method of its extraction from shale considerably pollutes the environment, with enormous amounts of water and chemicals used in the process [96].

4.5.1.2 Gasification of Coal and Other Hydrocarbons

In addition to light hydrocarbons, coal and many heavy liquid hydrocarbons, including heavy residual oils and other low-value refinery products, can be used for hydrogen production with the gasification process. However, the production of hydrogen from coal or heavy oils generates large quantities of carbon dioxide which if not captured and stored, will heavily contribute to global warming. Furthermore, coal mining contributes to significant land and water deterioration [96].

4.5.1.3 Electrolysis

The environmental impacts of electrolysis are determined by the types of fuels and technologies utilized to generate electricity. The use of grid power generated by polluting fuels to produce hydrogen with the electrolysis process would result in more global warming pollution than the production of hydrogen with steam reforming of natural gas. Nevertheless, electricity produced by the use of environmentally friendly renewable processes would result in near-zero-emissions cycles. Concerning energy security, electrolysis has the advantage of using surplus electricity from renewable power. At the same time, it may utilize domestically produced electricity, thus offering the advantage of capturing and storing electricity for the production of hydrogen in remote areas [96].

4.5.1.4 Gasification of Biomass

Gasification of biomass deals with heat (thermochemical processes) or anaerobic digestion using bacteria (biochemical processes). Many domestic biomass feedstocks may be used for hydrogen production, including dedicated energy crops and livestock wastes or residues. These resources are usually redundant all over a country. This approach to hydrogen production may result in low to zero greenhouse gas emissions provided the feedstocks are sustainably cultivated. Nevertheless, large-scale production of dedicated feedstocks and collection and transport of crops and residues may create other air, land, and ecosystem issues [96].

4.5.1.5 Nuclear Power

Nuclear power can contribute in many different ways to hydrogen production. It can produce electricity to be used in electrolytically derived hydrogen. It can also supply heat for the energy requirements of steam reformation of natural gas or for a thermochemical process to split water molecules into hydrogen and oxygen. However, considerable impediments hinder the utilization of nuclear power for hydrogen production related to high capital costs, concerns about waste management and disposal, reactor safety, weapons proliferation, and lack of public acceptance. It also raises other severe environmental and health issues concerning the mining and processing of uranium, the risk for accidents, and radioactive waste issues [96].

4.5.2 Environmental Concerns Related to Hydrogen Production Methods

Table 4.20 summarizes some of the likely advantages and disadvantages of the primary hydrogen feedstock and production pathways that have been identified to date.

4.5.3 Hydrogen Emissions in the Atmosphere

Since hydrogen is a small molecule, it can easily leak into the atmosphere, resulting in a lack of knowledge about the total amount of emissions (e.g., leakage, venting,

TABLE 4.20
Pros (+) and Cons (–) of Hydrogen Production Security and Environmental Considerations (Modified from Herzog and Tatsutani, [96]). With permission from NRDC

Production Method	Energy Security Considerations	Environmental Considerations
Steam methane reforming using natural gas as feedstock	(+) A mature technology is available, already used extensively worldwide. (–) Availability of supplies is very sensitive to international relationships.	(–) Low contribution of methane reforming to global warming but high contribution of inevitable methane leakage. (–) Extracting and transporting natural gas could harm sensitive landscapes.
Reforming of coal and other hydrocarbons	(+) Relatively well-established and low-cost process. (+) Process can be used with a variety of heavier hydrocarbons, including low-value refinery products.	(–) Production of hydrogen from coal or heavy oils would generate large amounts of carbon emissions. (–) Coal mining can degrade land and water quality.
Electrolysis using conventional grid or renewable power	(+) Multiple sources available. (+) Use of conventional grid electricity would provide energy security benefits since currently oil is not widely used for power production. (+) Using renewably produced electricity to produce H_2 could offer a means of capturing and storing this energy in remote areas or during off-peak hours. (–) Costs remain relatively high.	(+) Use of renewable power would produce low to zero emissions. (–) Use of conventional grid power would generate more global warming pollution than steam methane reforming with natural gas. (–) Near-term benefits of using renewable power may be greater if used to displace other sources of electricity.
Gasification of biomass	(+) Feedstocks are abundant worldwide.	(+) If feedstocks are sustainably cultivated, process has low to no net global warming emissions. (–) Large-scale production of feedstocks and collection and transport of crops and residues may raise air, land, and ecosystem concerns.
Nuclear power	(+) Uranium to power nuclear plants is relatively abundant worldwide. (–) High capital costs and concerns about reactor safety and weapons proliferation.	(–) Issues of waste management and disposal and extraction and processing of uranium. (+) Low global warming emissions.

and purging) from hydrogen installations. So, the effectiveness of hydrogen as a decarbonization player is not known. In a paper, Ocko and Hamburg [97] evaluated the climate impacts of hydrogen emissions over all timescales by employing already published data to assess the net warming impacts of replacing fossil fuel technologies with their clean hydrogen alternatives. They found that green hydrogen applications in the higher-end emission range of about 10% may only reduce climate impacts from fossil fuel technologies by 50% over the first two decades. This does not justify the notion that green hydrogen applications are climate-neutral. Nevertheless, by employing green hydrogen over a 100 year period, climate impacts could be reduced by about 80%. In contrast, when considering lower-end emissions (1%), green hydrogen could result in limited impacts on the climate over all timescales. For blue hydrogen, related methane emissions mean that this method is more damaging to the climate than fossil fuel technologies over several decades, yet even blue hydrogen offers climate benefits over a 100 year period.

Liquid energy carriers, such as LH_2 carriers, have an unavoidable problem when they are stored at cryogenic temperatures, lower than the boiling temperature of the liquid carrier. This problem is called *boil-off gas (BOG)* and occurs due to temperature differences between the liquid and the environment. BOG generation occurs when energy is stored in tanks or injected into or out of tanks, and it appears in different phases of a liquid energy carrier supply chain, including storage, loading, unloading, and transportation. BOG can be handled by a re-liquefaction facility or flared to the environment [98]. The handling pathway of BOG is decided based on the economics of a project since the BOG re-liquefaction facility is an energy-intensive process. When BOG is flared into the environment, these emissions negatively affect the environment and increase GHG emissions [99].

In any case, hydrogen is on the rise, and the data show an increase over time of about 4% in the past 25 years[100], whereas over the twentieth century, hydrogen levels rose by around 70% to 550 ppb [101].

Investigating GHG emissions from the full life-cycle of energy carriers covering production, transportation, and utilization phases by accounting for leaks and BOG generations can provide a full environmental assessment. The study performed by Breiky and Bicer [102] shows a comparison of GHG emissions from the full life-cycles of energy carriers, namely, LNG, dimethyl ether (DME) [103, 104], methanol, LH_2, and liquefied ammonia (LNH_3) produced either from natural gas or renewables. The main findings of Breiky and Bicer's study are listed as follows:

- *Methanol and DME are less harmful to the environment compared to LNG if considering production and transportation phases only. However, the utilization phase has an important contribution to overall emissions. Considering the full life-cycle, methanol and DME from the reforming of natural gas emits about 95.8 g CO_2 eq/MJ for methanol and 93.7 g of CO_2 eq/MJ for DME. These values are 30% and 27% higher compared to the full life-cycle of LNG. However, the production of methanol and DME from renewables (biomass) results in a reduction of GHG emissions by 75% and 80% respectively in their production phase, which makes DME and methanol alternative fuels alongside LNG.*

- *Regarding generated emissions during the transportation phase, even though DME starts with greater GHG emissions at the production phase compared to LNG, as ocean distance length exceeds 14,000 nmi (nautical miles), emissions from LNG become greater.*
- *The desire to minimize GHG emissions in LH_2 production methods is likely achievable using renewably generated electricity. When electricity is produced from natural gas, the production of LH_2 by steam reforming with carbon capture and storage (CCS) emits about 5.89 kg CO_2 eq/kg LH_2. Using renewables (solar photovoltaics) to produce electricity instead of natural gas results in a reduction of emissions by 40%.*
- *Producing LNH_3 by reforming natural gas without CCS emits around 2.24 kg CO_2 eq/kg LNH_3, however, implementing CCS results in a GHG reduction of about 24%. For the complete life-cycle of LNH_3, total GHG emissions of LNH_3 by reforming of natural gas is 1.87 kg CO_2 eq/kg LNH_3, which is lower compared to LNG, DME, and methanol full life-cycles.*

Thus, the conversion of natural gas into LNH_3, methanol, and DME for long storage could reduce GHG emissions compared to the storage of natural gas as LNG, which emits more GHG emissions during the storage phase. Furthermore, increased utilization of LNH_3 and LH_2 in combustion processes for power generation and transportation could be environmentally beneficial.

Analyzing literature data, Fan et al. [105] found insufficient information on hydrogen leakage from existing hydrogen applications, whereas some information comes from assessment studies and extrapolation but not from on-site measurements. The data available concerning hydrogen leakage in the extension and use of hydrogen in new areas of the hydrogen economy are even more limited. On this issue, applications of hydrogen are expected to be mainly concentrated in certain processes such as green hydrogen production, delivery, road transport, and chemical production, but other future applications are not known without data on actual hydrogen leakage. In a rather uncertain study by the International Energy Agency (IEA) [106] the hydrogen leakage rate for the net-zero scenario (528 million tonnes by 2050) is estimated to be 5.6% compared with an estimated 2.7% in 2020.

Mitigation measures to address the hydrogen leakage issue as proposed by Fan et al. [105] are:

- *Gathering and analysis of data, especially cross-referenced data from actual measurements, to better understand current and future hydrogen leakage rates.*
- *Research and development to improve hydrogen leakage detection, prevention, and mitigation. Hydrogen sensors must be able to detect leakage at much lower detection thresholds than existing sensors and among various types of applications. Such technologies are available but need to be transferred and developed at scale.*
- *Regulations that look beyond safety concerns related to hydrogen's flammability to include the use of hydrogen in different parts of the energy system and tackle leakage of substances more broadly, including hydrogen. It is important that the regulating entities involved co-ordinate at national and international levels to achieve coherent regulations.*

4.5.4 Impacts of Hydrogen Emissions in the Atmosphere

Although hydrogen is considered environmentally friendly compared to hydrocarbons as a transportation fuel, a study by researchers from the California Institute of Technology (Caltech) has shown that if we move to the so-called hydrogen economy, the hydrogen leakage from its extended use will be as much as 10 to 20% [107]. This would result in the rapid escape of huge quantities (estimated as 60,000 to 120,000 tonnes) of this extremely light gas to the ozone layer, resulting in doubling or tripling of hydrogen input into the atmosphere from all current natural or human sources. The output will then be the creation of additional water which would cool and dampen the stratosphere, eventually *thinning the stratospheric ozone layer* by as much as 10%. In addition, the combination of hydrogen with oxygen to form water would create increased noctilucent clouds (high wispy tendrils) appearing at dawn and dusk, which would accelerate global warming.

If the Caltech study is verified, hydrogen impact in the environment would resemble the catastrophic effect chlorofluorocarbons (CFCs) had on the stratospheric ozone layer, as evidenced by the annual holes over the south and north poles. Since no one would like to repeat the errors of the past, there is still time to fully investigate this eventuality and develop cost-effective technologies to minimize leakage before entering a global hydrogen economy.

Nevertheless, there still remain many uncertainties about the hydrogen cycle in the atmosphere. Moreover, the anticipated accumulation of hydrogen in the air is questioned by other scientists and organizations such as the US Department of Energy (Office of Energy Efficiency and Renewable Energy) and the Fuel Cell and Hydrogen Energy Association, claiming that the increase in the total hydrogen concentration would be at least one order of magnitude less than the Caltech researchers estimate.

In a more recent study, leakage of hydrogen into the atmosphere was investigated, and it was found that this would have a broader impact on atmospheric composition, which would partially offset some of the gains obtained in a switch to hydrogen. The net benefit also depends on the extent to which other emissions, such as carbon dioxide, methane, carbon monoxide, volatile organic compounds (VOCs), and nitrogen oxides (NO_x), are reduced [108]. In this work, the authors have focused on the consequences of hydrogen leakage using the state-of-the-art chemistry-climate model UKESM1 to calculate changes in atmospheric composition due to hydrogen leakage into the atmosphere. They have also investigated the impact of these changes on radiative forcing. In addition, they have evaluated the global warming potential (GWP) of hydrogen. Hydrogen by itself is not a GHG, yet changes in its atmospheric concentration can influence the concentrations of important GHGs, namely methane, water vapor, and ozone, in the troposphere and stratosphere as well. Thus, they have used modeled changes to these gases in response to changing atmospheric hydrogen mixing ratios to produce the most comprehensive assessment to date of the hydrogen GWP.

Concerning atmospheric composition, the authors of this study [108] have modeled the impact of increases in hydrogen from its current background concentration of about 0.5 ppm to values up to 2 ppm that span much of the uncertainty range of any future hydrogen economy as well as previous estimates[107, 109].

4.5.5 Reactions of Hydrogen in the Atmosphere and Global Warming

Substituting the current fossil fuel-based energy system with hydrogen in a global hydrogen economy, and considering a leakage rate of 1%, would result in an additive climate impact of 0.6% of the current fossil fuel-based system. Thus, attention should be given to minimizing the leakage of hydrogen from the synthesis, storage, and use of hydrogen in a future global hydrogen economy if we want to achieve the targeted zero net global warming gas emissions [110].

Hydroxyl (OH) radicals are strong atmospheric oxidants and are created in the atmosphere from ozone and water vapor through photocatalytic reactions [100]. Since hydrogen reacts with OH radicals, deactivating them, the increased presence of hydrogen will result in a reduction in tropospheric OH, which in turn results in an increase in the *methane lifetime* in the atmosphere. Thus, for every 1 ppm increase in hydrogen, the methane lifetime increases by about a year, resulting in a methane increase in the atmosphere of around 12%, provided methane emissions remain constant [108].

Regarding *tropospheric ozone* concentration in the atmosphere, in addition to hydrogen concentration, methane, VOCs, and NO_x also play a significant role. Comparing hydrogen and methane, the latter has a much stronger effect on tropospheric ozone. On the other hand, tropospheric ozone decreases with decreases in the concentrations of carbon monoxide, NO_x, and VOCs [108].

Stratospheric ozone is controlled by different chemical processes than those in the troposphere because an increased flux of hydrogen to the stratosphere would lead to increased water vapor. This result combined with methane increases in the stratosphere would lead to increased production of OH radicals, which would cause ozone decreases in the stratosphere. Although these changes are small in the lower to middle stratosphere, larger decreases are expected in the upper stratosphere, driven by the increase in OH radicals, and, to some extent, in polar latitudes, where increased water vapor plays a significant role. Nevertheless, no significant negative impacts on global stratospheric ozone are expected if a global hydrogen economy is adopted [86], despite the pessimistic claims that a hydrogen economy could widen the ozone hole found in previous articles [111].

Regarding *radiative forcing*, in an anticipated hydrogen economy there will be considerable benefits for the climate. Yet increases in atmospheric concentration of hydrogen would also result in increases in tropospheric ozone, water vapor, and methane, which will partially offset the climate benefits of the widespread use of hydrogen. Thus, the hydrogen leakage rate will determine the atmospheric radiative forcing change, as well as the warming effect. Warwick et al. [108] found that an increased leakage of hydrogen into the atmosphere will increase radiative forcing, thus exacerbating global warming. Consequently, climate stability depends not only on hydrogen itself but also on reductions of all other emissions, including carbon monoxide, NO_x, and VOCs.

Arrigoni and Bravo Diaz [112] also agree that the indirect global warming by increased amounts of hydrogen would be attributed to hydrogen leakage. To address this situation, the Clean Hydrogen Joint Undertaking and the US Department of Energy are co-operating with the European Commission, Hydrogen Europe, Hydrogen Europe Research, the Hydrogen Council, and the International Partnership for Hydrogen and Fuel Cells in the Economy to organize a targeted expert workshop. All parties agreed that the hydrogen economy would be directly beneficial to lowering global warming, yet hydrogen would prolong the lifetime of other greenhouse gases, namely methane, ozone, and water vapor, thus indirectly contributing to the increase of the Earth's temperature.

Although hydrogen by itself does not cause climate heating, it aggravates global warming by reacting with other gases in the atmosphere, thus being a *strong indirect greenhouse gas*. This is caused by the stabilization of methane in the atmosphere and the creation of tropospheric ozone and stratospheric water vapor. It is estimated that the climate impact of hydrogen is about 34 times higher than that of CO_2 over a 20 year period. Regarding its impact over 100 years, the global warming potential reduces to between eight and 13 times higher than that of CO_2 [113].

Paulot et al. [114] and Derwent et al. [115] claim that hydrogen is destined to play a crucial role in the decarbonization of the global energy system. As many other authors also stress, they state that hydrogen, since it has the smallest molecule of all elements, can easily permeate through many materials, including metals and alloys, thus posing safety concerns. They also add that up to now limited attention has been paid to the contribution of hydrogen leakage to climate change, especially the participation of hydrogen in global warming effects through mechanisms related to the lifetime of methane and other GHGs in the atmosphere. Derwent et al. [115] estimated that hydrogen has a *Global Warming Potential* over a 100 year time horizon *(GWP100)* equal to 5 ± 1.

However, Warwick et al. [108] found that the hydrogen *GWP100* equals 11 ±5 for a 100 year time horizon, a value which is more than twice the one estimated by Derwent et al. [115] (5±1). Around a third of this figure would originate from increased stratospheric water vapor following an increase in atmospheric hydrogen. Uncertainty in the GWP figures comes from uncertainties in estimations of atmospheric hydrogen, where the extent of the soil sink for hydrogen is the most uncertain factor. A future hydrogen economy would nonetheless have greenhouse consequences and would possibly cause climate disruptions.

Sand et al. [116] used five global atmospheric chemistry models to estimate the *GWP100* of hydrogen as equal to 11.6 ± 2.8. The uncertainty range covers soil uptake, the photochemical production of hydrogen, the lifetimes of hydrogen and methane, and the OH radical feedback on methane and hydrogen. A figure of this magnitude is not negligible, so reducing hydrogen leaks as much as possible is mandatory to benefit from a hydrogen economy. Yet, for the time being, the technology to monitor and detect hydrogen leaks at the scale needed is still under development.

The challenges on the issue of global warming are severe, but we need to be optimistic. As John Quigley (Senior Fellow at the Kleinman Center for Energy Policy) states [117], the hydrogen economy might be our last chance:

2023 was the hottest year in recorded history by a large margin, and the planet is likely to surpass the internationally agreed-to target of 1.5-degree Celsius warming above pre-industrial levels in the next month. But the Biden administration's proposed clean hydrogen rules offer real hope for urgently needed, effective climate policies.

REFERENCES

1. Fact Sheet Series No. 1.008, *Hydrogen Safety*, National Hydrogen Association, Washington, DC, 2006.
2. Sharma, S., Jindal, P., and Khandelwal, K.K., Comparison of hydrogen and gaseous fuel with other conventional fuels on the basis of safety and toxicity, *International Journal of New Innovations in Engineering and Technology*, 5 (2), 52–56, 2016.
3. Crowl, D.A. and Jo, Y.D., The hazards and risks of hydrogen, *Journal of Loss Prevention in the Process Industries*, 20, 158–164, 2007.
4. Karri, V., Ho, T., and Madsen, O., Artificial neural networks and neuro-fuzzy inference systems as virtual sensors for hydrogen safety prediction, *International Journal of Hydrogen Energy*, 33, 2857–2867, 2008. doi:10.1016/j.ijhydene.2008.02.039.
5. Kim, S., Song, Y., Hwang, T.Y., Lim, J.H., and Choa, Y.H., Facial fabrication of an inorganic/organic thermoelectric nanocomposite based gas sensor for hydrogen detection with wide range and reliability, *International Journal of Hydrogen Energy*, 44, 11266–11274, 2019.
6. Liu, Q., Yao, J. Wu, Y., Wang, Y., and Ding, G., Two operating modes of palladium film hydrogen sensor based on suspended micro hotplate, *International Journal of Hydrogen Energy*, 44, 11259–11265, 2019.
7. ANSI, *Guide to Safety of Hydrogen and Hydrogen Systems,* American Institute of Aeronautics and Astronautics, Reston, VA, American National Standard ANSI/AIAA G-095-2004, 2004, Chap. 2. chrome-extension://efaidnbmnnnibpcajpcglclefindmkaj/https://webstore.ansi.org/preview-pages/AIAA/preview_ANSI+AIAA+G-095A-2017.pdf
8. Zuettel, A., Borgschulte, A., and Schlapbach, L. (Eds.), *Hydrogen as a Future Energy Carrier*, Wiley-VCH Verlag, Berlin, Germany, 2008, Chap. 4.
9. MedicineNet.com. www.medterms.com
10. LaChance, J., Tchouvelev, A., and Engebo, A., Development of uniform harm criteria for use in quantitative risk analysis of the hydrogen infrastructure, *International Journal of Hydrogen Energy,* 36, 2381, 2011.
11. CPR, Methods for the determination of possible damage, In *CPR 16E*, The Hague: Directorate-General of Labour of the ministry of Social Affairs and Employment. III. The Netherlands Organization of Applied Scientific Research, 1989. chrome-extension://efaidnbmnnnibpcajpcglclefindmkaj/http://docenti.ing.unipi.it/m.carcassi/materialedidattico/Lez%20Sic%20e%20Analisi%20del%20Rischio/26-VULNERABILITA'.pdf
12. Eisenberg, N.A., et al. *Vulnerability Model: A Simulation System for Assessing Damage Resulting from Marine Spills*, Final Report SA/A-015 245, Document prepared for U.S.

Coast Guard, and available to the public through the National Technical Information Service, Springfield, Virginia 22151, 1975. chrome-extension://efaidnbmnnnibpcajpc glclefindmkaj/https://apps.dtic.mil/sti/tr/pdf/ADA015245.pdf

13. Tsao, C.K. and Perry, W.W., *Modifications to the Vulnerability Model: A Simulation System for Assessing Damage Resulting from Marine Spills*, Report ADA 075 231, Document prepared for U.S. Coast Guard, Office of Research and Development, Washington, D.C. 20590, 1979. extension://efaidnbmnnnibpcajpcglclefindmkaj/https://apps.dtic.mil/sti/pdfs/ADA075231.pdf

14. Opschoor, G., van Loo, R.O.M., and Pasman, H.J., Methods for calculation of damage resulting from physical effects of the accidental release of dangerous materials. in *International Conference on Hazard Identification and Risk Analysis, Human Factors, and Human Reliability in Process Safety*, Orlando, Florida, January 15–17, 1992.

15. Lees, F.P., The assessment of major hazards: a model for fatal injury from burns, *Transactions of the Institution of Chemical Engineers*, 72 (Part B), 127–134, 1994.

16. Jeffries, R.M., Hunt, S.J., and Gould, L., Derivation of probability of fatality function for occupant buildings subject to blast loads, *Health & Safety Executive,* 1997.

17. Simon J.M., Brady, S., Lowell, D., and Michael Quant, M., *Guidelines for Use of Hydrogen Fuel in Commercial Vehicles*, DOT F 1700.7, Report No. FMCSA-RRT-07-020, U.S. Department of Transportation, Federal Motor Carrier Safety Administration, Technology Division (MC-RRT), 1200 New Jersey Ave. SE, Washington, DC, 2007. https://rosap.ntl.bts.gov/view/dot/65

18. ANSI, *Guide to Safety of Hydrogen and Hydrogen Systems*, American Institute of Aeronautics and Astronautics, 12700 Sunrise Vally Drive, Reston, VA 20191, American National Standard ANSI/AIAA G-095-2004, Chap. 3. chrome-extension:// efaidnbmnnnibpcajpcglclefindmkaj/https://webstore.ansi.org/preview-pages/AIAA/ preview_ANSI+AIAA+G-095A-2017.pdf

19. Wang, T., Zhang, H., and Liang, W., Hydrogen embrittlement fracture mechanism of 430 ferritic stainless steel: The significant role of carbides and dislocations, *Materials Science & Engineering A*, 829, 142043, 2022.

20. Chandler, W.T. and Walter, R.J., Testing to determine the effect of high pressure hydrogen environments on the mechanical properties of metals, in ASTM special technical publications (eds.) *Hydrogen Embrittlement Testing,* Edited by ASTM special technical publications, ASTM 543, American Society for Testing and Materials, Philadelphia, 1974, 170. www.osti.gov/biblio/4152947

21. Groenvald, T. D. and Elcea, A. D. Hydrogen stress cracking in natural gas transmission pipelines. in *Hydrogen in Metals: Proceedings of an International Conference on the Effects of Hydrogen on Materials Properties and Selection of Structural Design*, Edited by I. M. Bernstein and A. W. Thompson, ASM International, Champion, Pennsylvania, 23–27, September, 1973.

22. Rowe, M.D., Nelson, T.W., and Lippold, J.C., Hydrogen-induced cracking along the fusion boundary of dissimilar metal welds, *Welding Research*, 31–37, February 1999. chrome-extension://efaidnbmnnnibpcajpcglclefindmkaj/http://files.aws.org/wj/sup plement/AREFAE_1/ARTICLE1.pdf

23. Nguyen, T.T., Bae, K.O., Jaeyeong, P., Nahm, S.H., and Baek, U.B., Damage associated with interactions between microstructural characteristics and hydrogen/methane gas mixtures of pipeline steels, *International Journal of Hydrogen Energy*, 47, 31499–31520, 2022.

24. Lei, Y., Hosseini, E., Liu, L., Scholes, C.A., and Kentish, S.E., Internal polymeric coating materials for preventing pipeline hydrogen embrittlement and a theoretical model of hydrogen diffusion through coated steel, *International Journal of Hydrogen Energy*, 47, 31409–31419, 2022.

25. Ogawa, Y. and Iwata, K., Resistance of pearlite against hydrogen-assisted fatigue crack growth, *International Journal of Hydrogen Energy*, Article in Press, 2022. https://doi.org/10.1016/j.ijhydene.2022.07.074

26. Zhou, C., Yang, Z., Chen, G., Zhang, Q., and Yang, Y., Study on leakage and explosion consequence for hydrogen blended natural gas in urban distribution networks, *International Journal of Hydrogen Energy*, Article in press, 2022.

27. Rigas, F. and Sklavounos, S., Evaluation of hazards associated with hydrogen storage facilities, *International Journal of Hydrogen Energy*, 30, 1501, 2005.

28. Hansen, O.R., Liquid hydrogen releases show dense gas behavior, *International Journal of Hydrogen Energy*, 45, 1343–1358, 2020.

29. Ekoto, I.W., Houf, W.G., Merilo, E.G., and Groethe, M.A., Experimental investigation of hydrogen release and ignition from fuel cell powered forklifts in enclosed spaces, *International Journal of Hydrogen Energy*, 37, 17446–17456, 2012.

30. Mogi, T. and Horiguchi, S., Experimental study on the hazards of high-pressure hydrogen jet diffusion flames, *Journal of Loss Prevention in the Process Industries*, 22, 45–51, 2009.

31. Zalosh, R.G. and Short, T.P., *Compilation and Analysis of Hydrogen Accident Reports*, C00-4442-4, Department of Labor, Occupational Safety and Health Administration, Factory Mutual Research Corp., Norwood, 1978.

32. Williams, G.P., Causes of ammonia plant shutdowns, *Chemical Engineering Progress,* 74, 9, 1978.

33. Ordin, P.M., A review of hydrogen accidents and incidents in NASA operations, in *9th Intersociety Energy Conversion Engineering Conference*, American Society of Mechanical Engineers, New York, 1974.

34. EHSP, European Hydrogen Safety Panel, Fuel Cell and Hydrogen Joint Undertaking (FCH 2 JU), Statistics, lessons learnt and recommendations from the analysis of the Hydrogen Incidents and Accidents Database (HIAD 2.0), HIAD 2.0 Database developed by the Joint Research Centre of the European Commission (JRC) in the frame of HySafe, an EC co-funded Network of Excellence in the 6th Framework Programme, 2021. chrome-extension://efaidnbmnnnibpcajpcglclefindmkaj/https://www.clean-hydrogen.europa.eu/system/files/2021-09/Lessons%2520learnt%2520from%2520HIAD%25202.0-Final.pdf

35. Thompson, J.D. and Enloe, J.D., Flammability limits of hydrogen-oxygen-nitrogen mixtures at low pressures, *Combustion and Flame*, 10 (4), 393–394, 1996.

36. Fisher, M., Safety aspects of hydrogen combustion in hydrogen energy systems, *International Journal of Hydrogen Energy*, 11 (9), 593–601, 1986.

37. BARPI, Ministere de la Transition Ecologique, Flash Aria, June 2020. www.aria.developpement-durable.gouv.fr/

38. Zhang, C., Wen, J., Shen, X., and Xiu, G., Experimental study of hydrogen/air premixed flame propagation in a closed channel with inhibitions for safety consideration, *International Journal of Hydrogen Energy*, 44, 22654–22660, 2019.

39. Li, Y., Yan, C., Liu, Q., Zhou, Y., and Gao, W., Inerting effect of carbon dioxide on confined hydrogen explosion, *International Journal of Hydrogen Energy*, 44, 22620–22631, 2019.

40. McHale, E.T., Geary, G., von Elbe, G., and Huggett, C., Flammability limits of H_2-O_2-fluorocarbon mixtures, *Combustion and Flame*, 16, 167, 1971.

41. Creitz, E.C., Inhibition of diffusion flames by methyl bromide and trifluoromethyl-bromide applied to the fuel and oxygen sides of the reaction zone. *Journal of Research for Applied Physics and Chemistry*, 65, 389, 1961.

42. Duan Q., Xiao, H., Gao, W., Wang, Q., Shen, X., Jiang, L., and Sun, J., An experimental study on shock waves and spontaneous ignition produced by pressurized

hydrogen release through a tube into atmosphere, *International Journal of Hydrogen Energy*, 40, 8281, 2015. http://dx.doi.org/10.1016/j.ijhydene.2015.04.097

43. Wionsky, S.G., Predicting flammable material classifications, *Chemical Engineering*, 79, 81, 1972.

44. Beach, R., Preventing static electricity fires, *Chemical Engineering*, 71, 73, 1964.

45. Beach, R., Preventing static electricity fires, *Chemical Engineering*, 72, 63, 1965a.

46. Beach, R., Preventing static electricity fires, *Chemical Engineering*, 72, 85, 1965b.

47. *FM Approval CN 5700 – Approval Standard for Explosion Suppression Systems*, FM Approvals, Norwood, Massachusetts, January 2022. chrome-extension://efaidnbmn nnibpcajpcglclefindmkaj/https://www.fmapprovals.com/-/media/Feature/Approval-Standards/5700-pdf.pdf

48. Baker, W.E. and Tang, M.J., *Gas, Dust and Hybrid Explosions*, Elsevier, Amsterdam, 1991, 42.

49. Kim, W.K., Mogi, T., and Dobashi, R., Flame acceleration in unconfined hydrogen/air deflagrations using infrared photography, *Journal of Loss Prevention in the Process Industries*, 26, 1501–1505, 2013a.

50. Kim, W.K., Mogi, T., and Dobashi, R., Fundamental study on accidental explosion behavior of hydrogen-air mixtures in an open space, *International Journal of Hydrogen Energy*, 38, 8024–8029, 2013b.

51. Molkov, V., Fundamentals of hydrogen safety engineering, Part II, 2012, www.bookb oon.com

52. Abbasi, T. and Abbasi, S.A., The boiling liquid expanding vapour explosion (BLEVE): Mechanism, consequence assessment, management, *Journal of Hazardous Materials*, 141, 489–519, 2007. dx.doi.org/10.1016/j.jhazmat.2006.09.056

53. Abbasi, T. and Abbasi, S.A., The boiling liquid expanding vapour explosion (BLEVE) is fifty…and lives on! *Journal of Loss Prevention in the Process Industries*, 21, 485–487, 2008. dx.doi.org/10.1016/j.jlp.2008.02.002

54. CCPS, Center for Chemical Process Safety, *Guidelines for Evaluating the Characteristics of Vapor Cloud Explosions, Flash Fires, and BLEVEs*, American Institute of Chemical Engineers, New York, 1994.

55. Lide, D.R. (Ed.), *Handbook of Chemistry and Physics*, 75th ed., CRC Press, Boca Raton, FL, Chap. 6, 1994.

56. Salla, J.M., Demichela, M., and Casal, J., BLEVE: A new approach to the super-heat limit temperature, *Journal of Loss Prevention in the Process Industries*, 19, 690, 2006.

57. Bjerketvedt, D., Bakke, J.R., and Wingerden, K.V., Gas explosion handbook. *Journal of Hazardous Material*, 52, 1, 1997.

58. Ustolin, F., Song, G., and Paltrinieri, N., The influence of H2 safety research on relevant risk assessment, *Chemical Engineering Transactions*, 74, 1393–1398, 2019.

59. Wingerden, K., Kluge, M., Habib, A.K., Ustolin, and F., Paltrinieri, Medium-scale tests to investigate the possibility and effects of BLEVEs of storage vessels containing liquified hydrogen, *Chemical Engineering Transactions*, 90, 547–552, 2022.

60. Philips, H., *Explosions in the Process Industries*, Institute of Chemical Engineers, Warwickshire, UK, 1994, 5.

61. Lee, J. H. et al., Hydrogen-air detonations, in *Proceedings of the 2nd International Workshop on the Impact of Hydrogen on Water Reactor Safety*, edited by M. Berman, SAND82-2456, Sandia National Laboratories, Albuquerque, NM, 1982.

62. Gordon, S., and McBride, B.J., *Computer Program for Calculation of Complex Chemical Equilibrium Compositions and Applications*, developed at the NASA Lewis Research Center, NASA Reference Publication 1311, 1994. https://ntrs.nasa.gov/citati ons/19950013764

63. Bull, D.C., Ellworth, J.E., and Shiff, P.J., Detonation cell structures in fuel/airmixtures, *Combustion and Flame,* 45, 7, 1982.

64. Explosion Protection Specification, Kidde Fire Protection, Oxfordshire, UK. www.astrasecurity.cz/download/SHZ_Explosion_Protection_EN.pdf

65. Bauwens, C.R., Chao, J., and Dorofeev, S.B., Effect of hydrogen concentration on vented explosion overpressures from lean hydrogen-air deflagrations, *International Journal of Hydrogen Energy,* 37,17599–17605, 2012.

66. U.S. DOT, *Clean Air Programme – Use of Hydrogen to Power the Advanced Technology Transit Bus (ATTB): An Assessment.* Report DOT-FTA-MA-26-0001-97-1, U.S. Department of Transportation, Washington DC, 1997.

67. EIGA, European Industrial Gases Association, *Safety in Storage. Handling and Distribution of Liquid Hydrogen,* Report DOC 06/02/E, Brussels, 2002. chrome-extension://efaidnbmnnnibpcajpcglclefindmkaj/https://h2tools.org/sites/default/files/Doc6_02SafetyLiquidHydrogen.pdf

68. Lee, J., Initiation of gaseous detonation, *Annual Review of Physical Chemistry,* 28, 75, 1977.

69. Eichert, H., Hydrogen-air deflagrations and detonations: Numerical calculation of 1-d tube combustion processes, *International Journal of Hydrogen Energy,* 12, 171, 1987.

70. Molkov, V., Fundamentals of hydrogen safety engineering, Part I, 2012, www.bookboon.com

71. Hord, J., Is hydrogen a safe fuel? *International Journal of Hydrogen Energy ,* 3, 157, 1978.

72. Tretsiakova-McNally, S., *Dealing with hydrogen explosions, HyResponse,* Grant agreement No: 325348. https://cordis.europa.eu/project/id/325348

73. Saeid, M.H.S., Khadem, J., Emami, S., Ghodrat, M., Effect of diffusion time on the mechanism of deflagration to detonation transition in an inhomogeneous mixture of hydrogen-air, *International Journal of Hydrogen Energy,* 47, 23411–23426, 2022.

74. Zhang, X., Yu, J., Huang, T., Jiang, G., Zhong, X., and Saeed, M., An improved method for hydrogen deflagration to detonation transition prediction under severe accidents in nuclear power plants, *International Journal of Hydrogen Energy,* 44, 11233, 2019.

75. Gharari, R., Kazeminejad, H., Mataji Kojouri, N., and Hedayat, A., A review on hydrogen generation, explosion, and mitigation during severe accidents in light water nuclear reactors, *International Journal of Hydrogen Energy,* 43, 1939, 2018.

76. Tong, L.L., Hydrogen risk for advanced PWR under typical severe accidents induced by DVI line break, *Annals of Nuclear Energy,* 94, 325–331, 2016.

77. OECD, Organisation for Economic Co-Operation and Development Committee on the Safety of Nuclear Installations, Nuclear Energy Agency, Organization for Economic Co-Operation and Development, Status Report on Hydrogen Management and Related Computer Codes NEA/CSNI/R(2014) 8, 2014. http://refhub.elsevier.com/S0360-3199(17)34637-2/sref14

78. IAEA, International Atomic Energy Agency, *Mitigation of Hydrogen Hazards in Severe Accidents in Nuclear Power Plants,* TECDOC-1661, 2011.

79. Corium (Nuclear Reactor), Wikipedia. https://en.wikipedia.org/wiki/Corium_(nuclear_reactor)

80. Choi, Y.S., Lee, U.J., Lee, J.J., and Park, G.C., Improvement of HYCA3D code and experimental verification in rectangular geometry, *Nuclear Engineering and Design,* 226, 337, 2003.

81. NEA, Nuclear Energy Agency, *Status report on hydrogen management and related computer codes,* Committee on the Safety of Nuclear Installations, NEA/CSNI/R(2014) Nuclear Energy Agency, Organisation for Economic Co-operation and Development,

Paris, France, 8, 2014. chrome-extension://efaidnbmnnnibpcajpcglclefindmkaj/https://www.oecd-nea.org/upload/docs/application/pdf/2021-02/csni-r2014-8.pdf

82. Lee, S.D., Suha, K.Y., and Jae, M., A framework for evaluating hydrogen control and management, *Reliability Engineering and System Safety*, 82, 307, 2003.

83. Tong, L., Zou, J., and Cao, X., Analysis on hydrogen risk mitigation in severe accidents for pressurized heavy water reactor, *Progress in Nuclear Energy*, 80, 128–135, 2015. http://dx.doi.org/10.1016/j.pnucene.2014.12.011

84. Lam, C.Y., Fuse, M, and Shimizu, T., Assessment of risk factors and effects in hydrogen logistics incidents from a network modeling perspective, *International Journal of Hydrogen Energy*, 44, 20572–20586, 2019.

85. Sathiah, P., Komen, E., and Roekaerts, D., The role of CFD combustion modeling in hydrogen safety management – Part I: Validation based on small scale experiments, *Nuclear Engineering and Design*, 248, 93–107, 2012a.

86. Sathiah, P., van Haren, S., Komen, E., and Roekaerts, D, The role of CFD combustion modeling in hydrogen safety management—II: Validation based on homogeneous hydrogen–air experiments, *Nuclear Engineering and Design*, 252, 289–302, 2012b.

87. Pasman, H.J., Challenges to improve confidence level of risk assessment of hydrogen technologies, *International Journal of Hydrogen Energy*, 36, 2407–2413, 2011.

88. Lopez-Alonso, E., Papini, D., and Jimenez, G., Hydrogen distribution and Passive Autocatalytic Recombiner (PAR) mitigation in a PWR-KWU containment type, *Annals of Nuclear Energy*, 109, 600–611, 2017.

89. Groth, K.M. and Hecht, E.S., HyRAM: A methodology and toolkit for quantitative risk assessment of hydrogen systems. *International Journal of Hydrogen Energy*, 42, 7485–7493, 2017.

90. Choi, G.P., Kim, D.Y., Yoo, K.H., and Na, M.G., Prediction of hydrogen concentration in nuclear power plant containment under severe accidents using cascaded fuzzy neural networks, *Nuclear Engineering and Design*, 300, 393–402, 2016.

91. Chang, Y., Zhang, C., Shi, J., Li, J., Zhang, S., and Chen, G., Dynamic Bayesian network based approach for risk analysis of hydrogen generation unit leakage, *International Journal of Hydrogen Energy*, 44, 26665–26678, 2019.

92. Shi, J., Chang, B., Khan, F., Chang, Y., Zhu, Y., Chen, G., and Zhang, C. Stochastic explosion risk analysis of hydrogen production facilities, *International Journal of Hydrogen Energy*, 45, 13535–13550, 2020.

93. Nowotny, J. and Veziroglu, T.N., Impact of hydrogen on the environment, *International Journal of Hydrogen Energy*, 36, 13218–13224, 2011.

94. van Renssen, S., The hydrogen solution? *Nature Climate Change*, 10, 799–801, 2020. https://doi.org/10.1038/s41558-020-0891-0.

95. EC, European Commission, *A Hydrogen Strategy for a Climate-Neutral Europe*, Brussels, 2020. chrome-extension://efaidnbmnnnibpcajpcglclefindmkaj/https://energy.ec.europa.eu/system/files/2020-07/hydrogen_strategy_0.pdf

96. Herzog, A. and Tatsutani, M., *A Hydrogen Future? An Economic and Environmental Assessment of Hydrogen Production Pathways*, NRDC (Natural Resources Defense Council), New York, United States, 2005. chrome-extension://efaidnbmnnnibpcajpcglclefindmkaj/https://www.nrdc.org/sites/default/files/hydrogen.pdf

97. Ocko, I.B. and Hamburg, S.P., Climate consequences of hydrogen emissions, *Atmospheric Chemistry and Physics*, 22, 9349–9368, 2022. https://doi.org/10.5194/acp-22-9349-2022.

98. Kwak, D.H., Heo, J.H., Park, H., Seo, S.J., and Kim, J.K., Energy-efficient design and optimization of boil-off gas (BOG) re-liquefaction process for liquefied natural gas (LNG)-fuelled ship, *Energy*, 148, 915–929, 2018.

99. Liu, C., Zhang, J., Xu, Q., and Gossage, J.L., Thermodynamic-analysis-based design and operation for boil-off gas flare minimization at LNG receiving terminals, *Industrial & Engineering Chemistry Research*, 49, 7412–7420, 2010.

100. Pearman, G. and Prather, M., Don't rush into a hydrogen economy until we know all the risks to our climate, *The Conversation*, 2021. https://theconversation.com/dont-rush-into-a-hydrogen-economy-until-we-know-all-the-risks-to-our-climate-140433

101. Patterson et al., H2 in Antarctic firn air: Atmospheric reconstructions and implications for anthropogenic emissions, *PNAS*, 118(36), 1–8, 2021. https://doi.org/10.1073/pnas.2103335118

102. Breiki, M.A. and Bicer, Y., Comparative life cycle assessment of sustainable energy carriers including production, storage, overseas transport and utilization, *Journal of Cleaner Production*, 279, 123481, 2021.

103. Semelsberger, T.A., Borup, R.L., and Greene, H.L., Dimethyl ether (DME) as an alternative fuel, *Journal of Power Sources*, 156, 497–511, 2006.

104. Park, S.H. and Lee, C.S., Combustion performance and emission reduction characteristics of automotive DME engine system, *Progress in Energy and Combustion Science*, 39, 147–168, 2013.

105. Fan, Z., et al., *Hydrogen Leakage: A Potential Risk for the Hydrogen Economy*, *Columbia/SIPA*, Center on Global Energy Policy, July 2022. www.energypolicy.columbia.edu/publications/hydrogen-leakage-a-potential-risk-for-the-hydrogen-economy/

106. IEA, International Energy Agency, Global Hydrogen Review 2021, October 21, 2021. https:// doi.org/10.1787/39351842-en.

107. Tromp, T.K., Shia, R.L., Allen, M., Eiler, J.M., and Yung, Y.L., Potential environmental impact of a hydrogen economy on the stratosphere, *Science*, 300(5626), 1740–1742, June 13, 2003. doi:10.1126/science.1085169.

108. Warwick, N., Griffiths, P., Keeble, J., Archibald, A., and Pyle, J., *Atmospheric implications of increased Hydrogen use*, GOV.UK, Department for Energy Security and Net Zero and Department for Business, Energy & Industrial Strategy, Open Government Licence v3.0, Crown copyright 2022. chrome-extension://efaidnbmnnn ibpcajpcglclefindmkaj/https://assets.publishing.service.gov.uk/media/624eca7fe90e0 729f4400b99/atmospheric-implications-of-increased-hydrogen-use.pdf

109. Warwick, N.J., Bekki, S., Nisbet, E.G., and Pyle, J.A., Impact of a hydrogen economy on the stratosphere and troposphere studied in a 2-D model, *Geophysical Research Letters*, 31, L05107, 2004.

110. Derwent, R., Simmonds, P., O'Doherty, S., Manning, A., Collins, W., and Stevenson, D., Global environmental impacts of the hydrogen economy, *International Journal of Nuclear Hydrogen Production and Application*, 1 (1), 57–67, 2006.

111. Ball, P., Hydrogen fuel could widen ozone hole, *Nature*, 13 June 2003. https://doi.org/10.1038/news030609-14; www.nature.com/articles/news030609-14

112. Arrigoni, A. and Bravo Diaz, L., *Hydrogen emissions from a hydrogen economy and their potential global warming impact*, EUR 31188 EN, Publications Office of the European Union, Luxembourg, JRC130362, 2022, ISBN 978-92-76-55848-4. https://doi:10.2760/065589

113. Kurmayer, N.J., *Scientists reiterate concerns about climate-warming hydrogen leaks*, Euractiv, November 21, 2023. www.euractiv.com/section/energy-environment/news/scientists-reiterate-concerns-about-climate-warming-hydrogen-leaks/

114. Paulot, F., Paynter, D., Naik, V., Malyshev, S., Menzel, R., and Horowitz, L.W., Global modeling of hydrogen using GFDL-AM4.1: Sensitivity of soil removal and radiative forcing, *International Journal of Hydrogen Energy*, 46, 13446–13460, 2021.

115. Derwent, R.G., et al. Global modelling studies of hydrogen and its isotopomers using STOCHEM-CRI: Likely radiative forcing consequences of a future hydrogen economy, *International Journal of Hydrogen Energy*, 42, 9211–9221, 2020.

116. Sand, M., Skeie, R.B., Sandstad, M., Krishnan, S., Myhre, G., Bryant, H., Derwent, R., Hauglustaine, D., Paulot, F., Prather, M., and Stevenson, D., A multi-model assessment of the global warming potential of hydrogen, *Communications Earth & Environment*, 4, 203 2023. https://doi.org/10.1038/s43247-023-00857-8

117. Quigley, J., 2024 *Starts with a Grim Climate Milestone, but Hydrogen Policy Hope*, Kleinman Center for Energy Policy at the University of Pennsylvania, January 12, 2024. https://kleinmanenergy.upenn.edu/news-insights/2024-starts-with-a-grim-climate-milestone-but-hydrogen-policy-hope/

5 Hazards in Hydrogen Storage Facilities

5.1 STORAGE OPTIONS

The storage options for hydrogen (including those under investigation) are the following [1]:

- Compressed hydrogen gas (GH_2) in cylinders or tanks.
- Tethered balloons, "bags", or water displacement tanks (low-pressure GH_2).
- Hydrogen adsorbed into metal to form a metal hydride (MH).
- Liquid hydrogen (LH_2) in cryogenic tanks.
- Adsorption on high-surface-area carbon powder in tanks.
- Encapsulation in glass microspheres (experimental).
- Adsorption on carbon nanotubes (experimental).
- In water (H_2O; not a "fuel").
- In ammonia (NH_3).
- In liquid hydrocarbons: gasoline, diesel fuel, alcohol, liquid natural gas (LNG), propane or butane (LPG), etc.
- In gaseous hydrocarbons: compressed natural gas (CNG), biogas, etc.

The first three hydrogen storage options mentioned, namely, GH_2, MH, and LH_2, are the "state-of-the-art" methods that are the most frequently applied in vehicular and stationary applications. The last two hydrogen storage options are the fossil fuels (liquid and gaseous hydrocarbons) that currently dominate our global fuel production and consumption.

The principal groups of hydrogen storage technologies are either physical-based or material-based, as shown in Figure 5.1. Physical-based technologies include storing hydrogen as compressed gas, cold/cryo-compressed, and liquid hydrogen storage. Material-based storage is divided into two main sub-groups, chemical sorption/chemisorption, and physical sorption/physisorption [2, 3].

Ren et al. [4] reviewed research efforts for hydrogen storage in both physical and chemical storage materials. In physical storage, they cited current research on new porous materials with increased surface-to-volume ratio and also collected information for works dealing with reaction enthalpies. In chemical storage, they evaluated

DOI: 10.1201/9781003313007-5

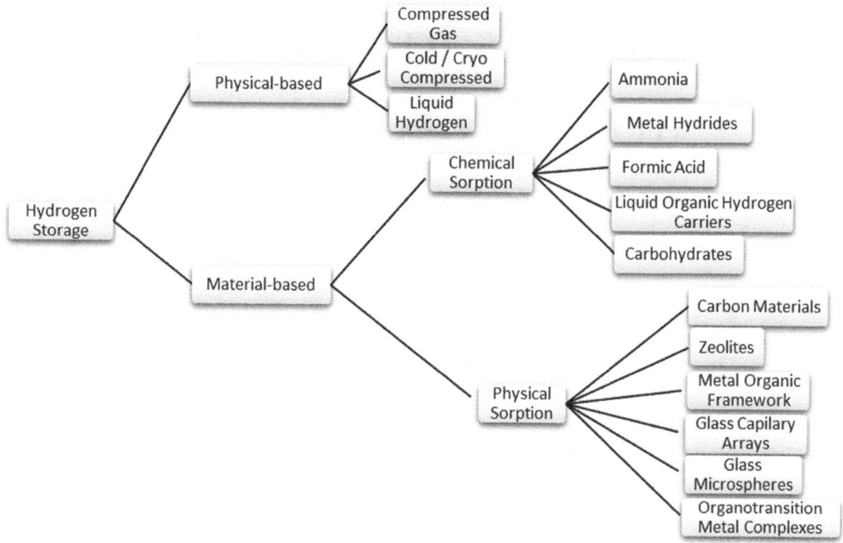

FIGURE 5.1 Hydrogen storage methods. With permission from Elsevier.

Source: Adapted from [2].

research on the kinetics and thermodynamics of relevant reactions including composite hydride systems and nanoconfinement of hydride materials.

Moradi and Groth [2] reviewed the current state-of-the-art on storage, delivery, and the related safety and reliability open questions. They found that the hazards are well known, and there is much progress in material compatibility, storage capacity, and simulation accuracy. Yet issues like the production cost of pressure vessels, liner blistering, fire resistance, contamination and leakage control, and large-scale utilization of hydrogen have not been investigated in depth and need further research. Moradi and Groth [2] finally concluded that:

- *Cryogenic and compressed storage are the most mature technologies. Cryogenic has a poor energy efficiency, and compressed storage requires large volumes due to low density of hydrogen.*
- *Composite storage vessel's reliability needs to be further improved. Damage mechanisms, inspection methods, and maintenance policies have to be determined.*
- *Material-based storage methods are still in their early development stage and need more time to prove themselves as viable long-term solutions.*

Assessing the status of methods for delivering hydrogen, Moradi and Groth [2] found that:

- *Gaseous delivery with tube trailers is not economically viable for long distances and high demands, liquid hydrogen tankers are a better option in that sense but their energy loss is high (up to 40%). Pipeline is the most reasonable solution but for it to be used, huge amount of resources and high market demand is needed.*

- *Selection of the delivery method should be based on regional specifications and potentials, demand, and economics.*
- *Central hydrogen production would require careful selection of delivery method. On the other hand, the idea of distributed hydrogen production is another pathway that is being investigated by many researchers in the field and could potentially lessen difficulties of delivering hydrogen to far destinations.*

5.1.1 Hydrogen Storage Materials

Hydrogen storage materials can store hydrogen in the form of hydrides or as molecular hydrogen. Three kinds of hydrogen atoms, protide (hydride) H^-, protium H^o, and proton H^+, exist in the hydrides. Boron and aluminum form negatively charged molecular hydrides (B–H, Al–H) based on the electronegativity difference. Carbon and nitrogen form positively charged molecular hydrides. Transition metals such as Ti, V, Zr, Pd, La, and Ce form hydrides with protium [5, 6].

Structural models of hydrogen storage materials are shown in Figure 5.2. These materials can safely store higher densities of hydrogen compared with the gaseous and liquid hydrogen storage systems at room temperature [6]. Therefore, the systems using hydrogen storage materials are considered for various kinds of applications including the adoption of renewable energy on a global scale.

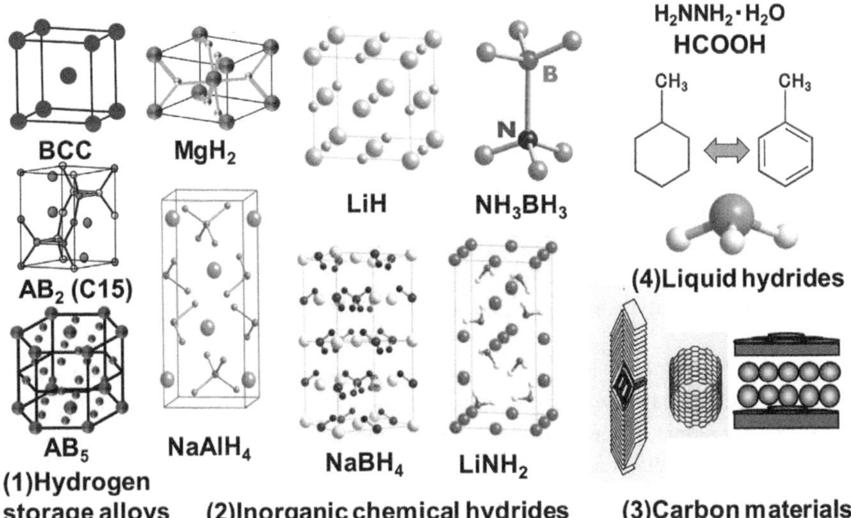

FIGURE 5.2 Structural models of hydrogen storage materials. With permission from Elsevier.

Source: [6].

5.1.2 Hydrogen Storage Alloys

Hydrogen storage alloys composed of the hydride-forming transition metals A and the non-hydride-forming metals B are considered attractive hydrogen storage materials. $LaNi_5$ is a typical AB_5-type hydrogen storage alloy. This alloy can reversibly store 1.4 wt% of hydrogen between 3 and 0.1 MPa at 293 K under a hydrogen dissociation pressure of 0.14 MPa. AB_5-type alloys are used as negative electrodes of nickel–metal hydride batteries [6]. BCC (body-centered cubic structure) alloys have a large amount of hydrogen density [7], and an alloy such as $Ti_{0.22} Cr_{0.39} V_{0.39}$ (Ti-Cr-V) absorbs and desorbs 2.2 wt% of hydrogen at 296 K [6].

Zhang et al. [8] investigated the use of low-temperature hydrides in submarines. These storage materials combined with fuel cells are preferable compared to the noisy liquid fuel combustion engines. This application is considered superior even when compared with the low-noise internal combustion engines of the Dutch Stirling system, the MESMA turbine, and the closed loop diesel systems. Thus, very competent submarines have been designed and commercialized recently, such as the export class submarine U214 in 2004 by Thyssen Krupp Marine Systems, which is based on metal hydride storage materials for hydrogen and fuel cell-based propulsion systems.

5.1.3 Inorganic Chemical Hydrides

Various kinds of chemical hydrides, such as ionic, covalent, and complex hydrides composed of light elements, are expected to realize high gravimetric and volumetric densities of hydrogen from 3 to 20 wt% (volumetric H_2 density: 2–8 kg H_2/100 L) [9–13]. Ionic hydrides and complex hydrides are stable, except $NaAlH_4$. It has been reported that MgH_2 is an intermediate hydride between an ionic hydride and a covalent hydride [6]. Covalent hydrides such as AlH_3 and NH_3BH_3 have been found to be unstable. The gravimetric hydrogen densities of these materials increase with a decrease in the density. There are three methods to desorb hydrogen: hydrolysis, ammonolysis, and thermolysis.

Hydrolysis. This is applied to lightweight hydrides (chemical hydrides: ionic hydrides, complex hydrides, covalent hydrides) consisting of protide H^- or negatively charged molecular hydrides that generate large amounts of hydrogen at room temperature. Ionic hydrides can hydrolyze to yield hydrogen gas and hydroxides.

Ammonolysis. Ammonia is a molecular hydride and electron-rich compound whose hydrogen becomes slightly positively charged [6]. Alkali metal hydrides (LiH, NaH, KH) react with ammonia at room temperature to generate hydrogen and metal amide [14]. Metal amide is recycled back to hydride and NH_3 under hydrogen flow at 373–573 K.

Thermolysis. Thermolysis of chemical hydrides is an attractive method for hydrogen storage and refers to reversible hydrogen desorption reactions, usually of alanates (e.g., $NaAlH_4$) [15–17], borohydrides (e.g., $LiBH_4$) [9, 18], lithium nitride (Li_3N) [6], alane (AlH_3) [19], and ammonia borane (NH_3BH_3) [6].

5.2   STORAGE IN POROUS MEDIA

5.2.1   Porous Materials

Increased safety accompanies the storage of hydrogen in porous media, as well as lower pressure storage and design flexibility, but this technology is not ready to be used yet. Reversible sorption in porous media may be physisorption (van der Waals forces) or chemisorption (as in metal hydrides). The materials that are extensively studied as sorbents are [20]:

- Carbon-based materials, nanotubes, nanofibers, activated carbons, activated fibers, carbons from templates, powders, doped carbons, and cubic boron nitride alloys.
- Organics, polymers, zeolites, silicas (aerogels), and porous silicon.

Comparing various types of *storage in carbon-based materials*, nanotubes appear to have higher storage capacity than activated carbons. Yet research data for nanotubes are sometimes conflicting, probably due to uncertainties in the material used.

Storage in other noncarbon materials includes:

- Self-assembled nano-composites/aerosols. These are nanostructured open-celled foams with very low densities, inexpensive, and safe to be used for hydrogen storage by physisorption.
- Zeolites. These are crystalline nanoporous materials, available at low cost, environmentally friendly, and safe to be used.
- Metal organic materials. These are usually zeolitic-type materials with backbone made of carbon possessing tailored properties and high potential.
- Other materials, such as glass microspheres, hydride slurries, boron nitride nanotubes, bulk amorphous materials (BAMs), hydrogenated amorphous carbon, hybrids, metal–organic frameworks (MOFs), and sodium borohydride.

5.2.2   Nano-composite Materials

The high working temperature of chemical hydrides due to the slow reaction rate (high activation energy) and the difficulty in controlling thermodynamic stability limits the practical application of chemical hydride systems. These properties can be improved by synthesizing nano-composite materials. The nano-composite materials encompass a catalyst and composite chemical hydrides at the nanometer scale. The catalyst increases the reaction rate by allowing a different mechanism to occur, which has a lower activation energy. The thermodynamic stability of the nano-composite materials, standard enthalpy difference, and standard entropy difference can be controlled by the combination of chemical hydrides. In addition, by the nano-size effects, the kinetics of hydrogen absorption and desorption are improved, and the thermodynamic stability of the materials may be modified [6, 8, 21, 22].

Mosquera-Vargas et al. [23] studied the hydrogen storage properties of purified multi-walled carbon nanotubes (MWCNTs) at room temperature. Nanotubes were exposed to hydrogen at pressures in the range of 0.39 to 13.33 kPa. It was found that the hydrogen adsorption capacity depends principally on the morphological and structural characteristics of the carbon nanotubes and their specific surface area. The adsorption kinetics from various types of purified MWCNTs were also investigated as a function of the gas pressure and the performance of these MWCNTs on the reversibility of the hydrogen loading/unloading process of successive cycles of gas sorption/desorption.

5.2.3 CARBON MATERIALS

Materials with very high specific surface area seem promising for hydrogen storage. Super-activated carbon having a specific surface area of about 3000 m²/g can reversibly adsorb and desorb 5 wt% of hydrogen at 77 K and 1.3 wt% of hydrogen at 296 K. Gravimetric and volumetric H_2 densities of super-activated carbon at 20 K are 10 wt% and 7.5 kg H_2/100 L, respectively [24–27]. In comparison to compressed or liquid hydrogen storage techniques in cylinders, physical adsorption is safer, especially for mobile systems, and enables better volumetric capacities [28–29].

Ozturk [28] investigated novel porous graphene framework structures to study various hydrogen adsorption media to meet the storage targets set. The lithium atoms' dispersion was studied, aimed at the enhancement of hydrogen storage. It was found that the structures with small pores partly lost their storing abilities with a dopant addition like lithium, especially for low-temperature sorption. In addition, Ozturk observed that the pillar destruction between graphene layers was prevented by using more suitable linkers. Hence, the pillars used in this study achieved sufficient strength to keep graphene layers distant with bonding monodentate from tree tips in the opposite site architecture. According to the author, the possible utilization of these materials in hydrogen storage containers could help to the prevalence of the hydrogen economy.

In a study, Shiraz and Tavakoli [30] found that heteroatoms, especially nitrogen as a dopant in graphene nano-sheets, enhance the gravimetric hydrogen density. Graphene decorated by both alkali and transition metals improved the results. Nitrogen-doped Pd-decorated graphene proved to be one of the best hybrid systems tested. Moreover, density functional theory has been used, thus confirming the results. The roles of interlayer spacing and functional groups on the hydrogen storage properties of both graphene oxide and reduced graphene oxide were also investigated [31]. It was found that the oxygen functional groups function as spacers between the graphene layers, thus increasing the interlayer space that, in turn, adsorbs many more hydrogen molecules on the surface of carbon nano-sheets.

Jung et al. [32] investigated a Pt-decorated graphene foam using bio-inspired polydopamine that acts as a reducing agent and strong metal binder, which they suggested as a successful hydrogen storage material. It was found that the extent of functionalized polydopamine on graphene could control the specific surface area of the three-dimensionally interconnected graphene foam. The degree of polydopamine functionalization could be used to maximize the hydrogen storage in the Pt-decorated graphene.

5.3   REACTIVITY OF STORAGE MATERIALS

Complex metal hydrides and chemical hydrides are promising materials to store compressed hydrogen gas or liquefied hydrogen for onboard vehicular applications. Nevertheless, the public should first accept these onboard storage materials, since the safety of these systems in vehicles seems to still have some issues of concern compared to the well-known fossil fuels already used for more than a century in the transportation sector. Khalil [33–36] discusses the results of an experimental program dealing with the metal hydrides' reactivity, their mechanical impact sensitivity, and ignition in contact with hot surfaces. In Figure 5.3, the $NaAlH_4$ mechanical impact test results from this program are shown. The wafer ignited upon the first impact. The powder compact of $3Mg(NH_2)_2 \cdot 8LiH$ was also found to be sensitive to mechanical impact where the test sample ignited upon first impact. The NH_3BH_3 powder compact, however, did not ignite during the impact tests.

These tests were performed to identify the safety-critical characteristics of selected solid-state hydrogen storage materials. Based on the results of those tests, risk mitigation methods were proposed by Khalil [33] to eliminate or mitigate the identified risks. The effectiveness of the proposed risk mitigation methods was also examined. The insights gained from the experimental program could be useful to ongoing research efforts aimed at identifying the best hydrogen storage materials from a safety standpoint as well as supporting current and future risk-informed hydrogen safety standards. The main conclusions derived from this experimental program are the following:

- Powder compaction was effective in reducing the risks associated with the reactivity of the examined hydride materials. However, this risk mitigation method failed to suppress the sensitivity of the examined hydrides to mechanical impact and hot surface contact.

FIGURE 5.3   A 4 g $NaAlH_4$ wafer ignited upon first impact (free fall height of a 10 kg weight = 1 m). With permission from Elsevier.

Source: [33].

- Solid ammonia borane (NH_3BH_3) is highly reactive when dispersed in air as a dust cloud.
- Some complex metal hydrides such as $3Mg(NH_2)_2 \cdot 8LiH$ and chemical hydrides such as AlH_3 are more reactive in the discharged phase compared to the charged phase.
- Some of the complex metal hydrides such as charged $NaAlH_4$ and charged $3Mg(NH_2)_2 \cdot 8LiH$ powder compacts are sensitive to mechanical impact. The effectiveness of selected flame-retardant chemical additives was examined as a means of mitigating this risk.

5.4   LIQUID HYDRIDES

Liquid hydrides are promising hydrogen storage materials due to the convenience of integration into existing transportation systems in contrast to compressed or liquid hydrogen. The principal liquid hydrides under investigation are ammonia, formic acid, methyl-cyclohexane, and hydrous hydrazine [37]. The gravimetric density of hydrogen-storing ammonia is 17.8 wt%, which is sufficiently high. Ammonia's density is also beneficial: 0.682 g/cm^3 at 0.1 MPa and 240 K, and 0.603 g/cm^3 at 1 MPa and 298 K. Its explosive limits in the air are limited (15–28 vol%), and the permissible exposure limit (PEL) is 50 ppmv [6].

Furthermore, liquid ammonia has a high volumetric hydrogen density of 0.107–0.121 kg H_2/L. A potential liquid hydride is also methyl-cyclohexane (C_7H_{14}) which has a moderate hydrogen storage density of 6.2 wt%, volumetric hydrogen density of 0.047 kg H_2/L, explosive limits in the air of 1.2–6.7 vol% and PEL equal to 500 ppmv). Its PEL value is ten times higher than ammonia's, so it is ten times less toxic, which is a significant advantage over ammonia and permits its handling in the same way as gasoline under ambient conditions. An alternative is toluene with explosive limits in the air of 1.2–7.1 vol% and PEL of 200 ppmv. This is produced by the dehydrogenation of methyl-cyclohexane but has to be reshipped to the place producing hydrogen [38]. Hydrous hydrazine is a liquid with a hydrogen content of 8.1 wt%, and a volumetric hydrogen density of 0.083 kg H_2/L. That is also an interesting candidate for hydrogen generation at ambient conditions. Finally, formic acid, with explosive limits in the air of 18–54 vol% and PEL of 5 ppmv, is a liquid hydride with hydrogen contents of 4.4 wt% and volumetric hydrogen density equal to 0.053 kg H_2/L [6].

Liquid organic hydrogen carriers (LOHCs), in general, have a hydrogen storage capacity of 6–8 wt% and can store a large amount of hydrogen for a prolonged time at ambient temperature and pressure. Kwak et al. [39] investigated the dehydrogenation of four homocyclic LOHCs, including methyl-cyclohexane, hydrogenated biphenyl-based eutectic mixtures, perhydro-monobenzyl-toluene, and perhydro-dibenzyl-toluene, using a 0.5 wt% Pt/Al_2O_3 heterogeneous catalyst.

In conclusion, the hydrogen densities of liquid hydrides have sufficiently large values in the range of 4.4–17.8 wt% and volumetric hydrogen densities in the range of 0.053–0.121 kg H_2/L. From a safety perspective, the liquid hydrides all have narrower explosive limits in the air than pure hydrogen, although their toxicity ranges from low to high, which poses some issues of concern.

5.5 AMMONIA AS A HYDROGEN CARRIER AND STORAGE MATERIAL

The following properties are required to apply to hydrogen storage materials for hydrogen carriers: high gravimetric and volumetric densities of hydrogen, small heat of formation (high hydrogen conversion efficiency), liquid, durability, safety, easy handling, and abundant resources. Liquid NH_3 satisfies those properties except for the safety.

Ammonia burns without the emission of CO_2 and can produce electricity when burnt in gas turbine generators [40]; [41]. Ammonia is recognized by the International Energy Agency (IEA) as a hydrogen carrier for fuel cells and as a fuel for direct combustion [6]. However, there are several issues in ammonia combustion, such as low flammability, high nitrogen oxides (NO_x) emissions, and low radiation intensity. Thus, further research into ammonia flame dynamics and chemistry is required to overcome these challenges [41].

The low flammability of ammonia in air could be improved by mixing it with other fuels such as methane and hydrogen, preheating gases, and supplying additional oxygen. However, it is difficult and expensive to inject pure oxygen into an ammonia combustion system for power generation. The fuel co-firing with ammonia increases expense because it requires an additional gas supply facility and a new safety system. Thus, an advanced combustion technology needs to be developed that could reduce NO_x emissions without using costly equipment such as selective catalytic reduction (SCR) devices. SCR is an active emission control technology that can considerably lower the emissions of NO_x in modern diesel-powered vehicles and gas turbines. The concentration of NO_x emissions currently formed when burning ammonia in a gas turbine is several hundred ppm. These challenges counterbalance the advantages of using ammonia as a carbon-free fuel [42]. Hence, it is necessary to develop a technology to reduce NO_x emission, for instance by using selective non-catalytic reduction (SNCR) devices during the combustion of ammonia. In a more practical approach, methods such as multi-stage (rich-lean) combustion have been developed to considerably reduce NO_x emissions [43].

Ammonia is a very promising fuel to support the decarbonization efforts of the energy sector, thus demanding intense research and development to overcome the issues of using ammonia as a fuel. Elbaz et al. [44] reviewed recent R&D accomplishments in increasing ammonia reactivity and reducing NO_x emissions. They spotted new chemical kinetics mechanisms of ammonia combustion that could help to control NO_x emissions from turbine operation and other industrial applications and address the low propagation speed, high ignition delay time, narrow flammability limits, and low flame radiation and temperature. In this review paper, dual-fuel or co-firing was found to be the conventional approach to using ammonia as a fuel, as well as various combustion promoters added to ammonia, such as hydrogen, methane, carbon monoxide, syngas, dimethyl ether, diethyl ether, and dimethoxymethane. Hydrogen has been proven the most efficient promoter, with methane being the worst among the tested fuels. Ammonia evaporation before combustion is an issue of concern due to the cost and complexity of the combustors, but it seems to have been overcome by spraying ammonia into the combustion chamber. However, this incurs extra costs of cooling the reactants due to the large heat of vaporization of liquid ammonia.

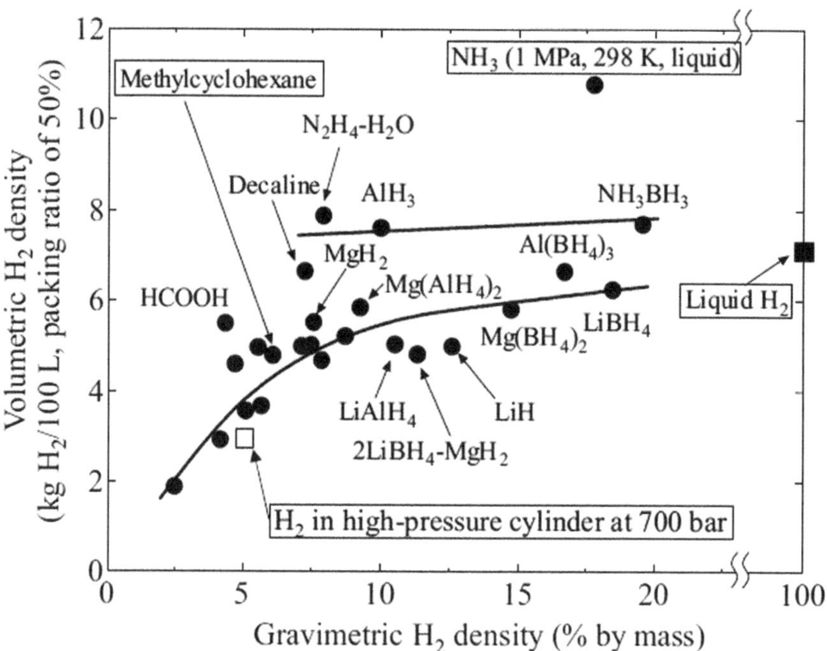

FIGURE 5.4 Gravimetric and volumetric H$_2$ densities of various hydrogen carriers. With permission from the NH3 Fuel Association.

Source: [41].

Figure 5.4 shows the gravimetric and volumetric hydrogen densities for various hydrogen carriers [41]. Besides pure hydrogen, all other compounds require energy to absorb and release hydrogen. Ammonia has a very high hydrogen density and can be used both as a fuel in internal combustion engines and as a fuel in solid oxide fuel cells (SOFCs)[45].

Although the issues raised by the UK's Health and Safety Executive (HSE) concerning the use of ammonia in electricity/power generation are less significant than those raised for transportation (Table 5.1), several concerns need to be considered. Furthermore, the increase in industrial incidents involving ammonia, and human injuries following ammonia release [40], should be taken into account. The National Fire Protection Association (NFPA), USA, has classified ammonia as a toxic substance, thus making it a chemical of high risk for health [46]. However, due to its low reactivity, the hazards from accidental combustion or explosions are much lower compared to other fuel gases and liquids (Figure 5.5). It should be noted that ammonia is rated as a toxic substance but *not* as a poison (NFPA Health: 3). Ammonia is a toxic substance with an exposure limit equal to 25 ppm for 8 h of exposure (time-weighted average), and a lethal concentration level (LC$_{50}$) for 20 min of exposure equal to 29,000 ppm [46].

TABLE 5.1
Comparison Between Ammonia in Transportation and Electricity/Power Generation. Courtesy of the Ammonia Energy Association

	Transportation	Electricity Generation
Safety	Very critical	Not as critical
Cracking	Cracking reactors heavy/ expensive	Easily done
Storage tank weight	Critical	Not an issue
Storage tank robustness	Need to be "indestructible"	Existing storage tanks are suitable
Distribution	Complicated	Relatively simple
Start-up	Problematic	Not many start-ups
Operational	Pumps operated by non-professionals	Delivered/handled by professionals

Source: [46].

Substance	Health	Flammability	Reactivity
Ammonia	3	1	0
Hydrogen	0	4	0
Gasoline	1	3	0
LPG	1	4	0
Natural Gas	1	4	0
Methanol	1	3	0

0=No hazard, 4=Severe hazards

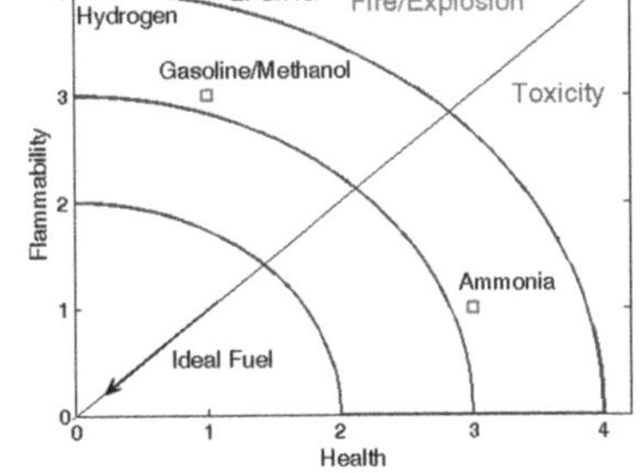

FIGURE 5.5 Toxicity and fire/explosion comparison of different fuels. Courtesy of the Ammonia Energy Association.

Source: [46].

Human exposure limits to ammonia depend on the legislation and exposure time. The limit is set between 25 and 50 ppm, with dangerous consequences for exposure to concentrations above 300 ppm. Table 5.2 provides some exposure guidance according to the NFPA. The variances clearly show that there is a need for additional research establishing more accurate values for industries and the exposed specific potential users.

In conclusion, care should be taken when handling ammonia because of its toxicity and corrosivity, which are critical factors for living organisms and construction materials. The high hydrophilicity of ammonia to water can destroy skin with severe burns to respiratory tracts in humans and animals. This is accompanied by intense corrosion of copper-based alloys, and lower corrosivity to nickel-based alloys. It is recommended that high-strength steel (>450 MPa) not be used because this is

TABLE 5.2
Exposure Guidance for Ammonia. Courtesy of the Ammonia Energy Association

Effect	Ammonia concentration in air (by volume)
Readily detectable odor	20–50 ppm
No impairment of health for prolonged exposure	50–100 ppm
Severe irritation of eyes, ears, nose, and throat. No lasting effect on short exposure	400–700 ppm
Dangerous, less than half an hour of exposure may be fatal	2000–3000 ppm
Serious edema, strangulation, asphyxia, rapidly fatal	5000–10,000 ppm

Source: [46, 47].

susceptible to low-stress corrosion. It is also recommended that spray coatings (i.e., zinc) be used in welding areas. Finally, good maintenance programs are required to ensure the integrity of components [42].

Regarding the shipping industry, the target set is for a 50% reduction in greenhouse gas emissions by 2050, compared to the 2008 levels. To this end, ammonia seems to be a promising candidate to replace fossil fuels for limiting global warming. Thus, ammonia-powered fuel-cell technology is proposed as a reliable and efficient power-generation alternative as well as an environmentally friendly fuel, not forgetting its high toxicity. Trivyza et al. [48] holistically investigated the safety, operability, and reliability of an ammonia-fuel-cell-powered ship, as well as fuel bunkering and other issues such as hazard spotting resulting from the presence and use of ammonia on ships. They concluded that the principal hazard from this innovative onboard use of ammonia would be the probability of exposure of the crew to ammonia, due to the toxicity of the fuel. A proposed potential safeguard is placing the bunkering station at a distant safe location while enabling the crew to control the procedure remotely. Other measures are ample ventilation and sufficient outlets in hazardous areas and proper identification of hazardous zones and exits. Special attire is also necessary for the ship crew to minimize exposure to ammonia risk.

Ammonia is a commodity chemical with the second largest production volume worldwide (behind only sulfuric acid), and with over 100 million tons transported per year it has a well-established distribution system. Its production is estimated to account for approximately 2% of global energy production. This does not happen with hydrogen, which is lacking proper and sufficient infrastructure. Thus, the ammonia economy looks easier to attain than the hydrogen economy, since its infrastructure is already established with similar environmental benefits [49].

The hydrogen contained in ammonia (17.65% by weight) can be recovered through thermal catalytic decomposition or electro-oxidation. On the other hand, ammonia can be used as a fuel in fuel cells without the need for a reactor to separate hydrogen. The life-cycle production cost of energy from ammonia is estimated at USD1.2/

kWh, compared to USD3.8/kWh for methanol and USD25.4/kWh for hydrogen. Consequently, ammonia is beneficial in many aspects and presents a sustainable and cost-effective fuel for fuel cells [50, 51].

Despite its toxicity and corrosivity, ammonia is a well-known commodity chemical in which a broad experience and expertise in synthesis, transportation, and utilization have been invested for over a century. Well-tested and effective health and safety standards already exist for the whole spectrum of its industrial applications. As a result, and due to the strict environmental constraints set for power generation from fossil fuels, the "hydrogen through ammonia" economy looks in many ways more profitable and sustainable, with increasing industrial support compared to other alternatives [40].

Sanchez et al. [52] have evaluated ammonia as a chemical for hydrogen storage and as a green fuel for power generation provided it has been generated from green hydrogen. In this respect, the production of ammonia should start from renewable energy sources. The authors consider a feed based on a mixture of ammonia and hydrogen to be used as fuel, with the hydrogen produced through ammonia decomposition. A simulation and an optimization approach are presented in their study to determine the optimal path and the conditions of ammonia-based power production. Several technical and economic evaluations are analyzed for various alternatives. They estimated energy efficiencies of around 40% for the transformation of ammonia into power, including not only the gas turbine but also the entire process.

By analogy to hydrogen, the ammonia industry has also adopted the color spectrum already applied to hydrogen to describe the carbon burden of the existing methods to produce ammonia. *Grey*, also called brown, ammonia is the conventional ammonia that has been produced in the same way for over 100 years by the Haber–Bosch process of combining hydrogen and atmospheric nitrogen. *Blue* ammonia is the ammonia for which the byproduct CO_2 is captured and stored, thus reducing the climate impact compared with grey ammonia. *Green* ammonia is produced from hydrogen produced in turn from the electrolysis of water but powered by alternative energy. *Turquoise* ammonia is produced by the pyrolytic conversion of methane into solid carbon and hydrogen, which then reacts with nitrogen to make ammonia. Environmentally speaking, turquoise ammonia lies somewhere between green and blue ammonia [53].

Fasihi et al. [54] performed a study on the global potential of green ammonia based on hybrid photovoltaic–wind power plants. In this study, a model is presented based on cost-optimized hourly power supply from hybrid PV–wind power plants with data from on-site and coastal power-to-ammonia plants. The model estimates the least-cost ammonia production based on efficiency and cost projections for years 2020, 2030, 2040, and 2050. They found that in an islanded installation, photovoltaics will be the predominating technology by 2030 in most parts of the world except for far southern and far northern regions. Gas turbines would not have a significant role in the designed system, and batteries would rather balance the electricity requirements of the ammonia plant than increase the performance of electrolyzers.

Service [55] claims that *reverse fuel cells* can use renewable power to make ammonia from air and water. This is a far more environmentally friendly method than the century-old industrial Haber–Bosch process. A reverse fuel cell could use renewable electricity for the chemical reaction that makes ammonia. According to this study, water will react at the anode to make hydrogen ions (H^+), which will migrate to the cathode to react with nitrogen to create ammonia. This reaction is efficient but slow.

In the past two decades, several alternatives to the existing Haber–Bosch process have been proposed, including electrochemical ammonia synthesis. Garagounis I. et al. [56, 57] investigated the electrochemical synthesis of ammonia in solid electrolyte cells and reviewed the literature on this alternative. The research data were classified according to the engaged electrolyte in aqueous, molten salt, or solid electrolyte cells, and the operating temperature, which is determined by the electrolyte used. The performance of these systems was also evaluated.

Lim et al. [58] have produced high-purity hydrogen by the thermal-electrochemical decomposition of ammonia at an intermediate temperature of 250°C. The process is enabled by the use of a solid acid-based electrochemical cell in combination with a bi-layered anode, thus comprising a thermal-cracking catalyst layer and a hydrogen electro-oxidation catalyst layer.

The process of conversion of hydrogen into ammonia, only to convert it back to hydrogen, appears illogical. Nevertheless, liquefaction of hydrogen requires a lot of energy to freeze it to temperatures below –253°C, thus consuming a third of its energy content. In contrast, ammonia liquefies easily at –10°C under low pressure. The energy needed to convert hydrogen to ammonia and recover it afterward is about the same as to liquefy hydrogen. Furthermore, since sufficient infrastructure is already available for handling and transporting ammonia, the conversion of hydrogen to ammonia and back is the process of choice [55].

Ammonia is a feasible hydrogen carrier that allows the storage and transport of hydrogen using well-established infrastructure while maintaining high hydrogen storage density. However, cracking ammonia into hydrogen needs a lot of energy. Thus, direct Joule-heating of the NiCrAl foam catalyst support was suggested and demonstrated by Badakhsh et al. [59], to minimize heat transfer and achieve lower reactor volume and higher efficiency and power density than previously reported reformers. The proposed system shows the highest hydrogen production rate per unit reactor volume ($1.56 \text{ cm}^3_{H2}/\text{s/cm}^3_{Reactor}$) reported for ammonia decomposition reactors to date. Compared with external heating, the specific activity of the foam catalyst was enhanced by up to ten times using the Joule-heating phenomenon. Thus, a reforming efficiency of about 69.2% was obtained for the proposed ammonia decomposition reactor.

In a feasibility study, Jackson et al. [60] proposed a feasible cracking technology for ammonia to be applied in the UK and other countries aiming at aiding the implementation of hydrogen technology. Due to environmental conditions, they conclude that the cost of solar energy in areas such as the Middle East is much lower than in the UK. Thus, it would be cheaper to import hydrogen in the form of green ammonia and crack it in the UK than produce green hydrogen in the UK. Furthermore, the lithium imide catalyst proposed by these authors shows high potential to lower the cost of ammonia decomposition to produce hydrogen with higher performance than the current catalysts used. Thus, the lithium amide–imide system showed significantly higher catalytic activity than sodium amide and ruthenium, reducing the temperature of 90% conversion by around 50°C. In addition, improving the economics of the ammonia decomposition process, the feedstock elements are far more readily available than the current catalytic elements like ruthenium. Altogether, the catalytic system based on lithium imide is likely to result in cheaper catalyst formulations.

Green ammonia, created by nitrogen from the air and hydrogen from water, and supported by renewable energy, is a carbon-free energy vector with many applications, including green electricity. Because of the low cost of storing and transporting ammonia of any kind, green ammonia could be delivered to every country, without the geological storage limitations of carbon capture and storage. In a study, Cesaro et al. [61] performed a techno-economic assessment to estimate the cost of electricity from ammonia based on anticipated technological improvements up to 2040. They found that green ammonia could be available for less than 400 USD/t in 2040, with the possibility of it being as low as 300 USD/t if electrolyzers obtain sufficient cost reductions or if more beneficial renewable resources are available to supply a global green ammonia market.

It is anticipated that ammonia will play a significant role in the future energy mix. Current analyses at all levels and sectors ensure that ammonia has already obtained a deep penetration in the energy market, thus enabling the use of scattered energy sources and decarbonization of the discontinuous, renewable grid. Nevertheless, many issues still exist before ammonia can be commercially developed as a main energy vector. Challenges and studies in the areas of synthesis, use, safety, and economics are all under careful examination at various levels, thus ensuring that our understanding of the use of ammonia and its practical implementation will be accomplished within the following decades [42].

Nevertheless, liquid hydrocarbons are still widely used in transportation because of their extremely high *energy density* (Table 5.3). The values of properties in this table were determined using the WolframAlpha Computational Knowledge Engine [62]. It is interesting to note that although the gravimetric energy density of LH_2 is three times greater than that of gasoline, its volumetric energy density is less than three times lower than gasoline's. Furthermore, liquid hydrogen has much lower H_2 density in kilograms per cubic meter than other chemical hydrogen storage media that are currently used or are intended to be used as energy carriers. Among them, ammonia and hydrazine possess the greatest H_2 densities, but their use in common applications is limited because of their high chemical reactivity and toxicity. Nevertheless, hydrazine has been used effectively since World War II as a rocket fuel or monopropellant in space exploration (e.g., *Viking* and *Phoenix* landers), in military aircraft (e.g., F-16 fighter aircraft), and other industrial uses (e.g., in polymers and pharmaceuticals and as a gas precursor in air bags).

However, water possessing very high H_2 density is the hydrogen storage reservoir most likely to be used in the future after the development of reasonably priced methods for hydrogen separation from water.

TABLE 5.3

Hydrogen Densities in Different Forms of Hydrogen Storage

Fuel	Formula	Density* (kg/m^3)	Gravimetric Energy Density* (MJ/kg)	Volumetric Energy Density (MJ/L)	H_2 Density $(kg\ H_2/m^3)$
GH_2	H_2	0.09	142	0.013	0.09
LH_2	H_2	71	142	10.2	71
LNG (methane)	CH_4	424	55.5	23.5	106
LPG (propane)	C_3H_8	582	50.1	29.2	106
Gasoline	C_8H_{18}	737	47.3	34.9	118
Methanol	CH_3OH	791	22.7	18	99
Ethanol	C_2H_5OH	789	29.7	23.4	103
Cyclohexane	C_6H_{12}	779	46.7	36.4	111
Methyl-cyclohexane	C_7H_{14}	770	46.6	35.9	95
Ammonia (liquid at BP)	NH_3	683	18.6	12.7	121
Hydrazine	N_2H_4	1011	19.2	19.4	126
Water	H_2O	1000	—	—	111

* WolframAlpha Computational Knowledge Engine was used for the determination of property values (www.wolframalpha.com).

5.6 STORAGE AS LIQUID HYDROGEN

There are cases where liquid hydrogen is beneficial, for instance, when high purity is needed. Disadvantages are boil-off losses, the temperature stability necessary to avoid overpressure, and, mainly, the liquefaction energy requirement. Technology for storage of LH_2 is commercially available in vessels with sizes ranging from 0.1 to thousands of cubic meters. The main concern in LH_2 tank design is the construction of an effectively insulated container. This is normally obtained with vacuum-jacketed, double-walled containers. The preferred shape is spherical due to its minimum surface-to-volume ratio and its more uniform distribution of stresses and strains[63].

Special attention is given to the multi-layer construction of the insulation, which may consist of 60 to 100 reflective foils fixed on the outside of the inner vessel, with spacers between each layer acting as thermal barriers and a total thickness of at least 20 mm for tank sizes up to 300 m^3. The void volume between the two vessel walls may be filled with reflective perlite powder or hollow glass microspheres that considerably reduce heat loss and allow for less demanding vacuum than normally needed (1.3 Pa). For safety reasons, care is taken with regard to the potential shifting of insulation particles during contraction of the inner vessel, which leads to compaction upon re-expansion that may rupture the supporting structures. Suitable materials for construction of cryogenic tanks are carbon steel for the outer vessel and stainless steel or aluminum for the inner vessel. Tubing is generally made from stainless steel [63, 20].

In addition to reflective materials, liquid nitrogen (LN_2) is usually used for safety reasons in large LH_2 tanks to fill the space created by an additional outer wall. Up to now, NASA has constructed the largest LH_2 tanks, including two identical storage tanks of 3800 m^3 capacity each, at the Kennedy Space Center in Florida, for its space shuttle program. The construction materials are austenitic stainless steel for the inner wall, with an inside diameter of 18.75 m, and carbon steel for the outer wall, with an inside diameter of 21.34 m. The ullage is about 15%, permitting each tank to be filled up to 3218 m^3. Operation pressure is 620 kPa, with a boil-off rate equal to 0.025% per day, or about 800 L per day.

van Wingerden et al. [64] performed three tests to investigate the eventuality and effects of BLEVEs of storage vessels containing LH_2. The tests were executed to determine the consequences of storage vessels containing LH_2 being engulfed in a fire. The work was implemented at the Test Site for Technical Safety of the Bundesanstalt für Materialforschung und -prüfung (BAM) in Germany, and it was a research co-operation between BAM and Gexcon as part of the SH2IFT program. The vessels tested had a capacity of 1 m^3 and were insulated with double-walled vacuum insulation. The orientation of the vessels varied as well as the insulation material used (perlite or multi-layer insulation). The vessels were filled up to 35% and the fire was provided by a propane burner positioned homogeneously under the storage vessel. In one of the tests, a BLEVE of the storage vessel occurred, resulting in a fireball and fragments. The conditions in the vessel (e.g. temperature and pressure) during the implementation of the tests were monitored in all tests.

5.7 HAZARD SPOTTING

5.7.1 HAZARDOUS ZONES

In atmospheric conditions, hydrogen is a colorless, odorless gas, much lighter than air. The low density in conjunction with the small particle size allows the penetration of hydrogen molecules into some metals and alloys such as cast iron and high-carbon steel [65]. The penetration may end in small hydrogen leaks or, in the presence of cracks within the wall, contribute to crack spread, material strength decrease, and subsequent fracture.

Hydrogen reacts violently with oxidizers such as nitrous oxide, halogens (especially with fluorine and chlorine), and unsaturated hydrocarbons (i.e., acetylene) with intensely exothermic reactions. Hydrogen gas forms combustible or explosive mixtures with atmospheric oxygen over a wide range of concentrations in the air in the range 4.0–75 vol% and 18–59 vol%, respectively. Since hydrogen has much broader flammability and detonability limits than any other fuel, it should never be stored unless it is below the lower flammability limit (LFL). Thus, industry standards for storage safety are set well below 0.25 of LFL, that is, less than 1% oxygen with the hydrogen [1].

Saffers and Molkov [66] constructed a nomogram (Figure 5.6) for the calculation of hydrogen jet flame length and maximum flame width based on the experimental best-fit correlation for flame length and the application of the under-expanded jet theory by Molkov et al. [67]. The ratio of the maximum visible hydrogen flame

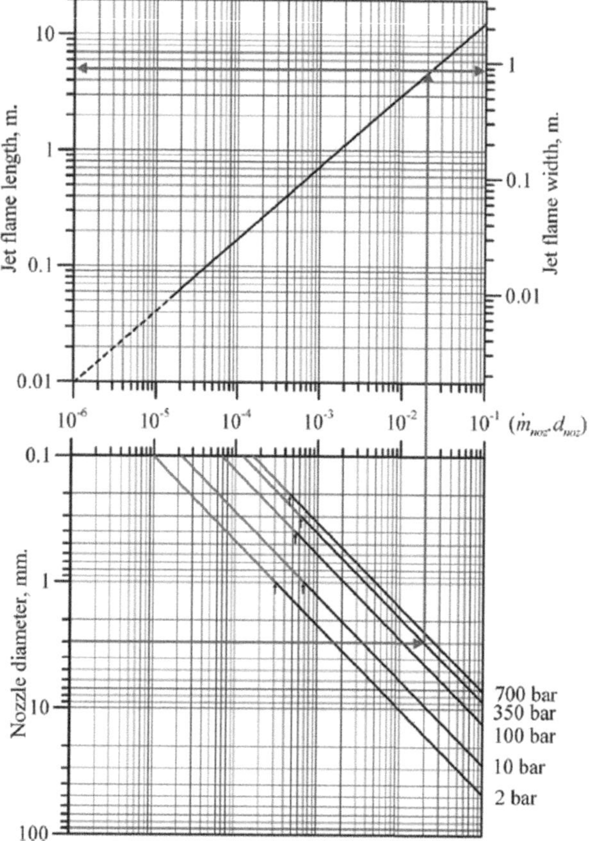

FIGURE 5.6 Nomogram for the determination of hydrogen jet flame length and width. With permission from Elsevier.

Source: [66].

width to hydrogen flame length is 0.17 [68]. Only physical leak diameter and storage pressure are necessary to determine the jet flame length and width (the losses are neglected, thus giving a conservative approach). It is the simplest known tool to assist different stakeholders in determining the jet flame length and width. The nomogram also encompasses the blowoff phenomenon of hydrogen flames at small nozzle diameters. For each storage pressure in the nomogram, the spouting pressure was calculated using the under-expanded jet theory [67], and the corresponding diameter to blowoff was determined from the experimental data given by Mogi and Horiguchi [69]. The solid lines correspond to the stable flames, while the dashed lines represent the blowoff zone, where a hydrogen flame is stabilized only in the presence of a permanent ignition source at the adequate location. The black arrows indicate the transition from the stable flames to the blowoff zone. When considering a release from a 350 bar storage tank through a 3 mm diameter nozzle (see blue arrowed lines on the nomogram) the nomogram gives a jet flame length of about 5 m and a width of about 0.85 m (the best-fit line).

Jet fires are visually not detectable due to the invisibility of hydrogen flames, especially in daylight, so a hydrogen fire is very difficult to spot, and this has resulted in delayed action and serious injuries of carelessly approaching persons.

Hydrogen is nontoxic; however, besides the hazards of flash fire, jet fire, and gas cloud explosion in the event of an accidental release, hydrogen may act as an asphyxiant at sufficiently high concentrations by depleting the oxygen available in atmospheric air.

5.7.2 Refrigerated Storage Conditions

Hydrogen is stored at normal temperature either in a gaseous state under medium (4.1 to 8.6 bar) or high pressure (140 to 400 bar), or in the liquid state under low temperature and moderate pressure. When stored at medium pressure, hydrogen should be kept in low-carbon steel or other materials unaffected by hydrogen embrittlement. High-carbon steel tanks are not suitable for storing hydrogen under pressure. To prevent hydrogen tank embrittlement, cold-rolled or cold-forged steels and those having weld hard spots in excess of about Vickers hardness number 260 should be avoided. Nonmetal tanks such as composite-fiber tanks avoid hydrogen embrittlement concerns and derating.

Medium-pressure GH_2 storage tanks usually have smaller size and greater weight, for a given storage capacity, relative to low-pressure GH_2 tanks. Storage tanks for hydrogen should be hydrostatically tested to at least twice the operating pressure, equipped with a pressure release valve, and always installed outdoors for safety reasons. In addition, inlet and outlet lines should be equipped with flash-back arrestors [1].

In any case, there is a significant hazard potential for mechanical explosion of hydrogen vessels (tanks and cylinders) when exposed to high temperature or thermal radiation. Usually, the cause of overheating is a neighboring fire (primary event) that causes temperature increase in the vessel shell and its content (Figure 5.7). The vessel finally bursts (secondary event) spilling its content, which, if flammable, is usually immediately ignited and burns in the form of a jet fire or a fireball. Typical ignition sources for the spilling fuel may be electrostatic sparks developed during content discharging or the flames of the neighboring fire. This situation is known as a domino effect since an initial accident gives rise to another, thus generating a chain of accidents with escalating impact on the surroundings [70, 71]).

In contrast to storage vessels for high-pressure gaseous hydrogen, liquefied hydrogen vessels operate at moderate-to-low pressures that normally do not exceed 20 bar. As a result, it is reasonable for the walls to be designed with lower pressure resistance than that for storage of hydrogen gas (which may have operation pressures of up to 400 bar, thus being liable to increased risk of failure). In the event that the container is engulfed in a fire, its metal is heated and loses mechanical strength. While the liquid phase absorbs significant amounts of heat, vapors possess far lower specific heat capacity. Therefore, the heat supplied to that part of the container where the

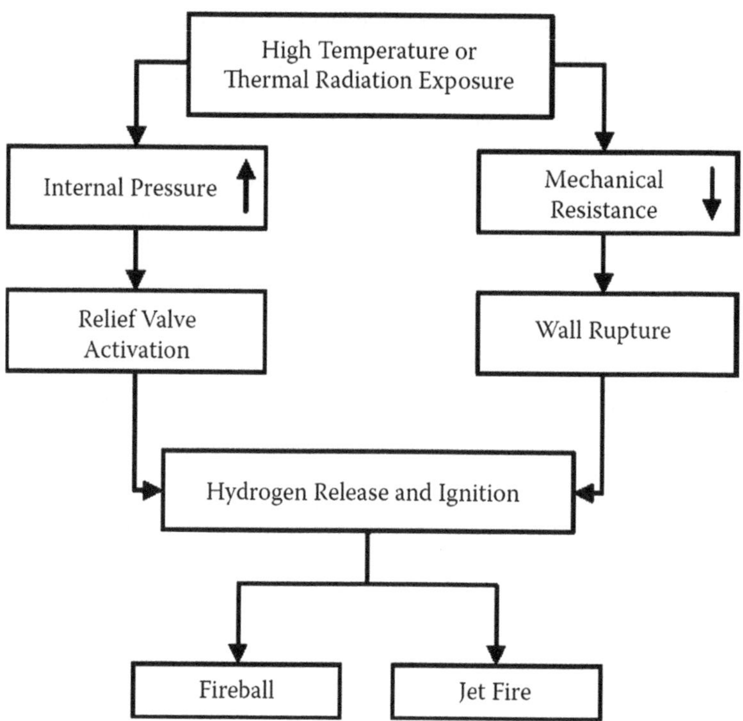

FIGURE 5.7   Thermal radiation effects on pressurized hydrogen vessels. With permission from Elsevier.

Source: [83].

vapor phase exists will raise the local wall temperature much more, thus weakening its strength [70].

Concerning liquefied gas storage, vessel overheating may result in internal temperatures higher than the boiling point of the content, even without vaporization initiation of the liquid phase, and then the liquid is superheated. This phenomenon is observed when there is a shortage of nucleation sites (i.e., impurities, crystals, or ions) in the bulk of the liquid. However, there is a temperature limit above which the fluid cannot remain in the liquid state (the homogeneous nucleation limit or superheat limit temperature). In this limit, random molecular density fluctuations within the bulk of a liquid produce hole-like regions of such molecular dimensions that they may act as bubbles [72]. The end result is an explosive flash of the liquid accompanied by a strong shock wave that propagates through the fluid and ruptures the container, spilling the content into the atmosphere (a boiling liquid expanding vapor explosion, or BLEVE; see Chapter 4) [73]. Missiles of the ruptured walls may travel hundreds of meters, whereas the flammable content ignites, forming a sphere that burns from the outer to the inner layers and is known as a *fireball*. The overall process is shown in Figure 5.8.

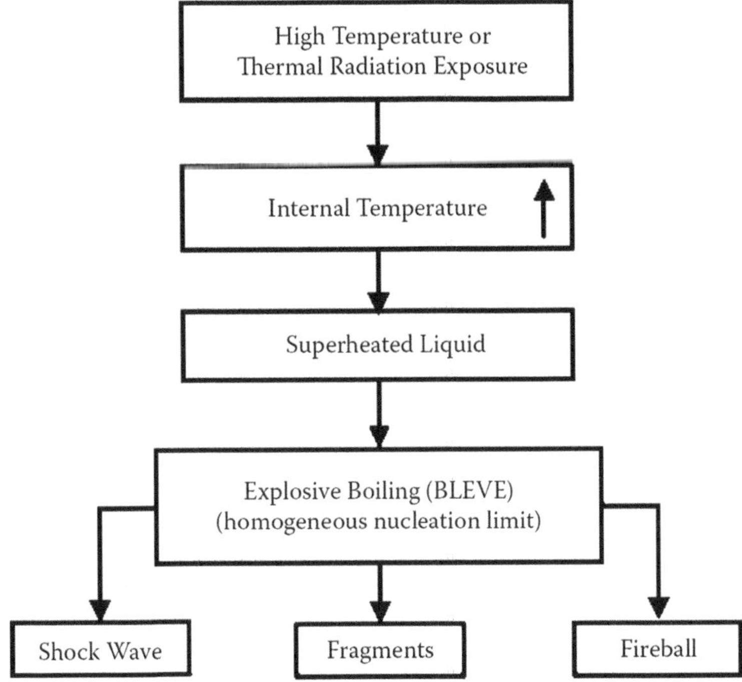

FIGURE 5.8   Thermal radiation effects on liquefied hydrogen vessels. With permission from Elsevier.

Source: [83].

Research on the modeling of fireball phenomena was recently carried out by Sklavounos and Rigas, [74] with reasonable quantitative estimations of the thermal load received in the vicinity. A good qualitative simulation was also accomplished for the formation and evolution of the fireball after initiation of burning (Figure 5.9). Fireball development was based on the release and ignition of 1708 kg of propane. The burning cloud moved upward as expected, obtaining positive buoyancy due to high temperatures, whereas air inflow from the left (inlet boundary condition) caused horizontal fireball shift to the right.

5.7.3 Cryogenic Storage Conditions

At hydrogen refilling stations and in vehicles, hydrogen has to be compressed at high pressures (up to 400 bar), due to its low energy content per unit volume. Nevertheless, in some applications, hydrogen, as well as other gases (e.g., carbon dioxide, nitrogen, helium, and methane), need to be stored in a liquid state at very low temperatures for volume restriction. These temperatures are often lower than −73°C, so that storage conditions are characterized as cryogenic (as distinguished from the refrigerated condition). These conditions are preferred, for example, in uses such as rocket propulsion,

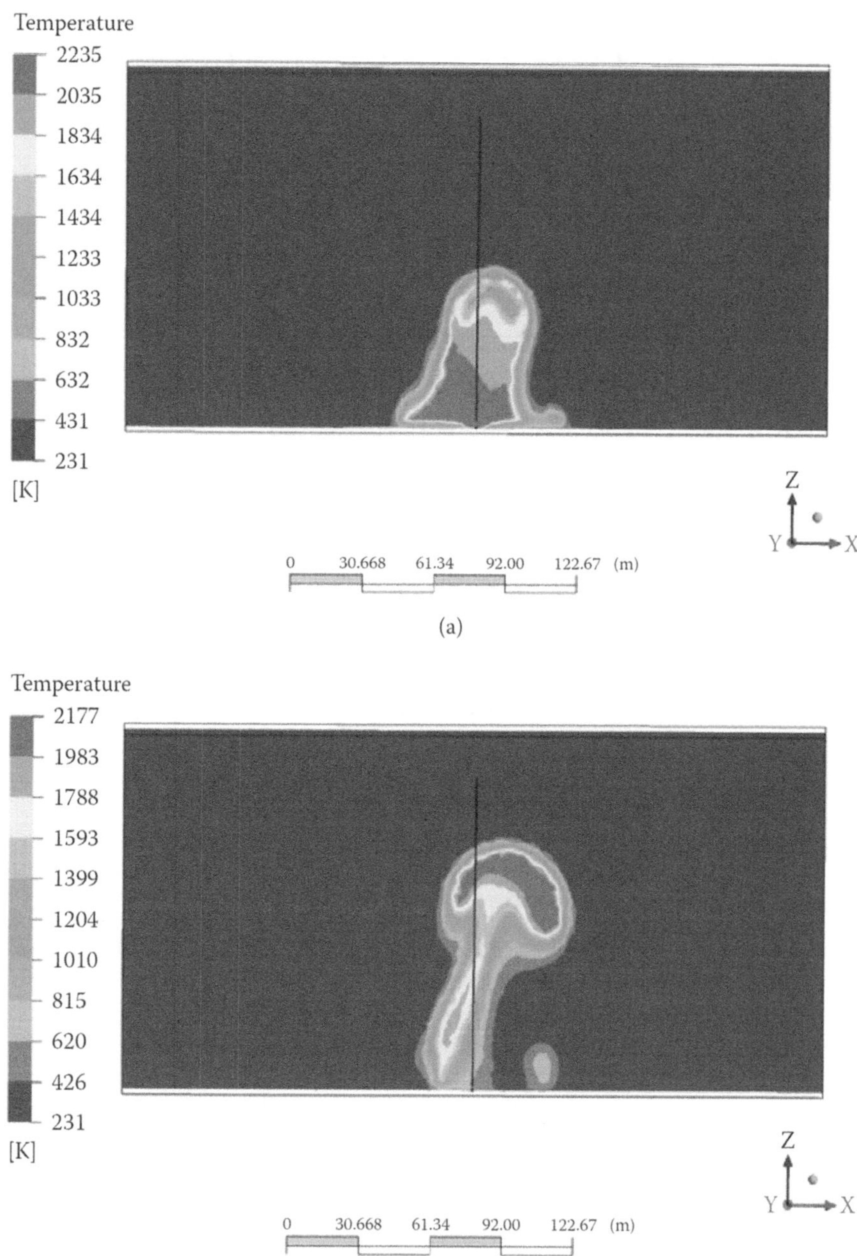

FIGURE 5.9 Fireball development after release and ignition of 1708 kg of propane. The burning cloud moves upward, obtaining positive buoyancy due to high temperatures, whereas air inflow from the left (inlet boundary condition) causes horizontal fireball shift to the right. (a) 1.2 s after ignition, (b) 3.2 s after ignition. With permission from Elsevier.

Source: [74].

as well as in warehousing for convenience and economy. A crucial factor in liquefaction is the critical temperature above which the gas cannot be liquefied by pressure application only.

In practice, hydrogen is kept liquefied at extremely low temperatures, below −240.2°C, and moderate pressures (20–30 bar). Major hazards related to cryogenic storage stem from:

- *Embrittlement of service materials.* The low temperatures inside storage tanks and transmission pipelines may cause significant susceptibility of the structural material to vibrations and shocks. Mild steel and most iron alloys at liquid hydrogen temperatures lose their ductility, being liable to increased risk of mechanical failure [75]. Figure 4.1 in Chapter 4 presents the dramatic decrease of impact strength with a decrease in temperature. The same risk exists with other elements of equipment (e.g., control valves, gauges) incorporated in vessels and pipes.
- *Liquid hydrogen spills.* Liquefied hydrogen releases yield dramatically larger volumes of combustible clouds (1 L of liquid produces on evaporation 851 L of gas). Therefore, the consequences of a fire or explosion will be more extensive than those of a pressurized hydrogen release.
- *Extremely low temperature.* Low temperature can cause severe tissue frostbite if the material comes into contact with the human body. Flesh may adhere quickly to cold, insufficiently insulated pipes or vessels and tear on attempting to withdraw it [76].
- *Hydrogen cloud dispersion.* A hydrogen spill that originates from cryogenic storage results in cloud formation that disperses in a manner similar to a heavier-than-air gas, thus increasing the risk of accidental fires and explosions.

5.8 HAZARD EVALUATION

5.8.1 Methodology

Hazard evaluation in hydrogen storage facilities aims to determine all credible accident scenarios. A variety of methods (event tree, failure modes and effects analysis, what-if, fault tree) can be found in the relevant literature [77–79]. (See also Section 8.2.4.) Of these, event tree analysis (ETA) is a formal technique and one of the standard approaches used for industrial incident investigation. ETA is a logic model that graphically portrays the combination of events and circumstances in an accident sequence. It is an inductive method, which begins with an initiating undesirable event and works toward a final result (outcome). The general procedure for ETA involves the following steps:

1. Determination of the initiating events that can result in certain types of accident.
2. Identification of the critical factors that may affect the evolution of the initiating event.

3. Construction of the event tree taking into account the interaction between critical factors and the initiating event.
4. Designation and evaluation of resulting accidental events.

Applying ETA in fuel gas releases, the critical factors that may substantially affect the final outcome are the time of ignition of the resulting cloud and the confinement provided by the surroundings. The former is related to the mixing of escaping fuel gas with air. When immediate ignition occurs, gas cloud mixing with atmospheric oxygen is still limited; thus, ignition takes place on the outer layer that is between the flammable limits, whereas the inner core of the cloud is too rich in fuel to ignite. As the buoyancy forces of the hot gases begin to dominate, the burning cloud rises and becomes more spherical in shape, forming a ball of flames. This elevation gradually causes further mixing of the gas with oxygen, which brings new volumes of gas into the flammable limits, thus sustaining the fire. In contrast, when delayed ignition occurs, the fuel cloud may have been adequately mixed with air, so that after ignition it flashes back. This differs from a fireball scenario since it proceeds faster and can burn from inner to outer flammable layers, provided a proper ignition source is found there. Thus, a deflagration or detonation may occur, the latter necessitating a more uniform (and in narrower concentration limits) mixing of hydrogen with air, and in addition some increased degree of confinement.

5.8.2 ETA Method Application

Rigas and Sklavounos [71] have applied the ETA method with regard to accidental hydrogen release as shown in Figure 5.10. It is obvious from this figure that, unless an immediate ignition takes place, there is some time of dispersion that intervenes between release and ignition. Generally, if hydrogen flammability zones were known, it would be possible to take preventive measures and to prepare emergency response planning against fires and explosions.

Consequently, a major issue arises regarding the computation of the dispersion succeeding an accidental hydrogen release. Furthermore, even if no ignition takes place, escaping hydrogen may accumulate in closed spaces adjacent to the source, posing an asphyxiation hazard for the people there. Hydrogen dispersion may be considered safe only when no ignition occurs and no confined space exists [63].

In this section, the main hazards associated with hydrogen storage procedures were analyzed. In addition, possible accidental events that hydrogen may yield were indicated by performing an ETA of a hydrogen release. The analysis showed that, unless an immediate ignition occurs, determination of the LFL distance of the fuel–air mixture is of major importance for the purpose of loss prevention, such as ignition source elimination.

5.9 QUALITATIVE PREDICTION OF CLOUD TRAVEL

Generally, the behavior of a gas (or vapor) cloud during dispersion can be either buoyancy or gravity driven. The former is relevant with lighter-than-air gases, and the latter with heavier-than-air gases.

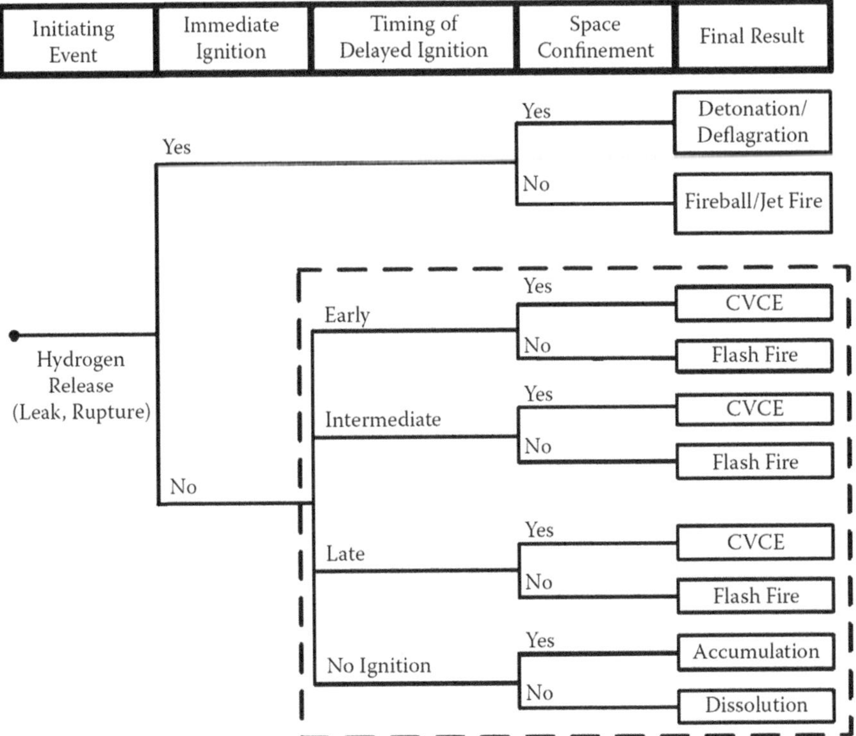

FIGURE 5.10 Event tree analysis adapted to accidental hydrogen releases. With permission from Elsevier.

Source: [101].

Currently, dispersion estimations are performed via the use of semiempirical, one-dimensional models (the so-called box models) and computational fluid dynamics (CFD) codes. The selection of the appropriate dispersion model in an accidental release scenario requires the behavior of the dispersing gas to be known, since each model is specialized for one type of release (buoyancy or gravity driven).

At atmospheric conditions (20°C, 1 atm), hydrogen is a lighter-than-air gas and preserves the properties of light gas behavior even when releases from high-pressure storage systems occur. However, once hydrogen escapes while stored in a liquid state, its dispersion behavior turns to that of a heavy rather than a light gas.

The different way in which a gas disperses according to the storage conditions has been observed in field-scale trials at Lawrence Livermore National Laboratory (LLNL) installations with another light gas (natural gas) that was left to escape and disperse while stored in a liquid state [80]. In this case, instead of rising, the vapor cloud moved along the ground, being under the LFL, for several tens of meters.

5.10 GAS DISPERSION SIMULATION

5.10.1 CFD MODELING

Generally, CFD methods follow a deterministic procedure to approximate a problem. They consider the fundamental Navier–Stokes equations for mass, momentum, and heat transfer processes, in conjunction with other partial differential equations for describing further processes such as turbulent mixing between the dispersing gas and the atmospheric air.

The first step in a CFD simulation attempt includes the definition of the three-dimensional geometry (computational domain), in addition to the subdivision of the entire domain into a number of smaller control volumes (cells) that form a mesh. The geometry comprises a unified set of parametric surfaces built in an appropriate CAD interface.

In CFX (a CFD code), the solution process begins with discretizing the spatial domain into finite control volumes using the mesh constructed in the preprocessor stage. Integration of the subsequent balance equations over each control volume is based on the finite volume method [81]. The integral equations obtained are converted to a system of algebraic equations using a high-resolution numerical scheme and are solved iteratively at nodal points inside each cell, aiming at minimization of the residuals to acceptable convergence levels. As far as their solution is concerned, CFX uses a coupled solver that solves the hydrodynamic equations (for pressure and velocity components) as a single system. This solution approach uses a fully implicit discretization of the equations at any given time step, reducing the number of iterations needed for convergence.

5.10.2 CFD SIMULATIONS OF GASEOUS AND LIQUEFIED HYDROGEN RELEASES

Sklavounos and Rigas [82], working on large-scale gas dispersions, have used the general-purpose CFD code named CFX, first validating it against experimental data from large-scale gas dispersion trials for isothermal heavy gas releases. In addition, these authors have also dealt with non-isothermal dense gas releases [83]. The reliability of CFD simulation of cryogenic gas release was examined by Sklavounos and Rigas against field-scale experimental data including hydrogen and natural gas releases. A characteristic example of natural gas experiments is the Burro series trials of LNG spills. The project was sponsored by the United States Department of Energy (US DOE) and was conducted by LLNL and the Naval Weapons Center (NWC) at the China Lake site in California during the summer of 1980 to determine the transport and dispersion of vapor from spills of LNG on water. A detailed description of the experiments and analysis of the results are given in a report by LLNL [84]. Hydrogen dispersion experiments with data available to the public are the LH_2 spills conducted in 1980 by Witcofski and Chirivella [85] at the NASA Langley Research Center at the White Sands Test Facility in New Mexico. The objective of the program was to study the phenomena associated with the rapid release of large amounts of LH_2, with emphasis on the generation and dispersion of the flammable hydrogen cloud.

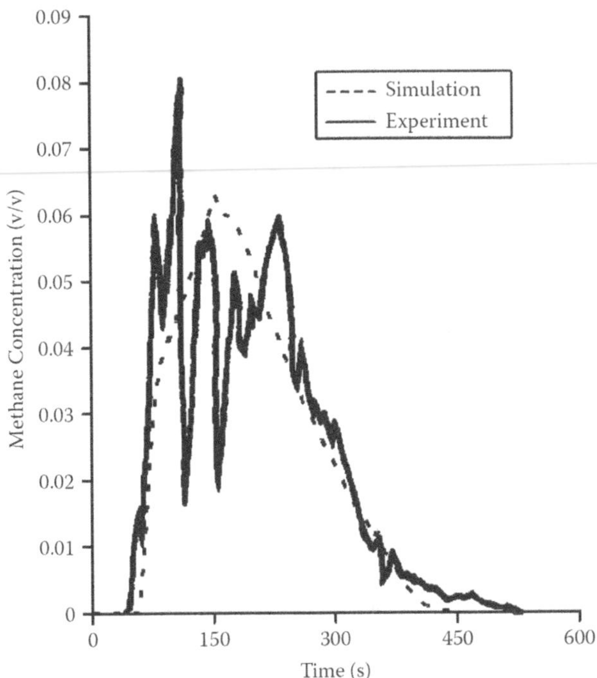

FIGURE 5.11 Computational and experimental methane concentration histories at point (140, 0, 1) for the Burro 8 trial. With permission from Elsevier.

Source: [83].

Figures 5.11 and 5.12 show the experimental gas concentration histories versus the computational ones for the LNG Burro 8 trial and the LH_2 Burro trial 6 obtained by Sklavounos and Rigas [83]. The experimental observations of the concentration change are found statistically to be in good agreement with the computationally predicted histories, showing that the CFD predictions are reasonable. Further elaboration of simulation results and the calculation of specific statistical performance measures led to the conclusion that the predicted values were lying within a factor of two of the observed values of cloud passage duration, maximal concentration, and total dose; therefore, it was inferred that the predictions of the CFX code were consistent with the data from the field tests.

The behavior of hydrogen during its atmospheric dispersion has also been examined by Rigas and Sklavounos [71] considering two different cases: isothermal release under pressurization conditions, and cryogenic (non-isothermal) release under liquefaction conditions. Three-dimensional simulations were run for a typical accidental release scenario, which included hydrogen discharge under pressurization and liquefaction conditions in an imaginary computational domain with dimensions 160 m × 40 m × 30 m. Hydrogen release was modeled as a ground-level area source with inflow rate equal to 2 kg/s for 5 s, with typical atmospheric conditions (1 atm, 20°C), ground temperature equal to 15°C, and wind speed equal to 3 m/s.

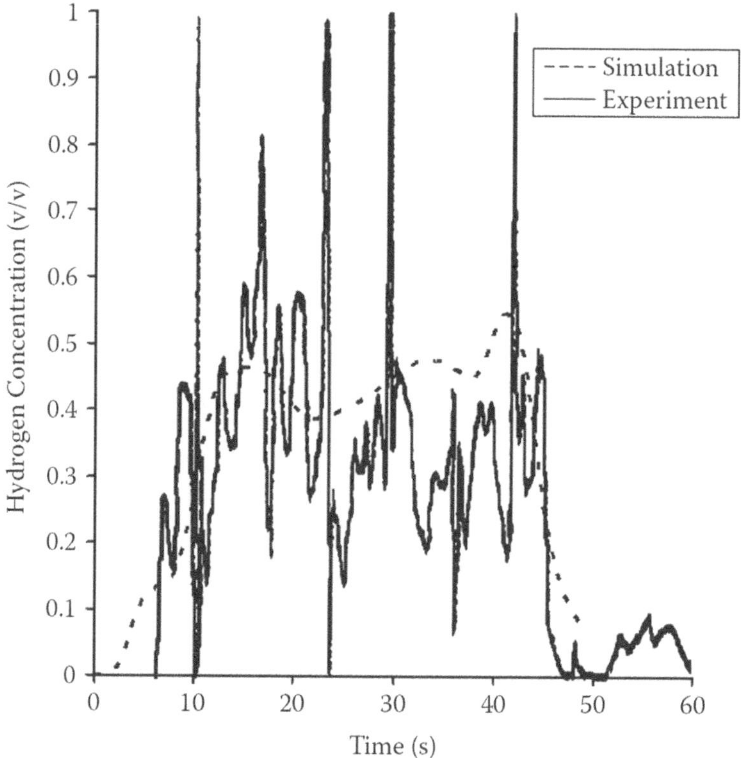

FIGURE 5.12   Computational and experimental hydrogen concentration histories at point (9.1, 0, 1) for the LH_2 trial 6. With permission from Elsevier.

Source: [83].

The results demonstrated substantial differentiation between pressurized and cryogenic hydrogen release as far as cloud dispersion is concerned. In the former case, the cloud lifts immediately, moving almost vertically (Figure 5.13), whereas in the latter, it travels downwind almost horizontally, remaining close to the ground (Figure 5.14).

It can be seen from Figure 5.15 that the spreading of a cryogenic-origin hydrogen cloud simulated by Rigas and Sklavounos [71] is similar to that of LNG that has been experimentally observed in field-scale trials at LLNL installations [80], namely horizontal cloud shift at low height. This is the result of the greater inertia of the heavy cryogenic gas that reduces the rate of turbulent mixing, thus leading to delayed dissolution times. In addition, the greater density of the cloud produces a gravity-driven flow that tends to reduce the cloud height and increase its width. Thus, in contrast to lighter-than-air gas, the persistence of the heavy cloud at low height significantly increases the risk of ignition.

The simulation of atmospheric dispersion of clouds from liquefied *cryogenic hydrogen releases* by Rigas and Sklavounos [71] showed that hydrogen disperses as a heavier-than-air gas when discharged in cryogenic conditions, in contrast with a

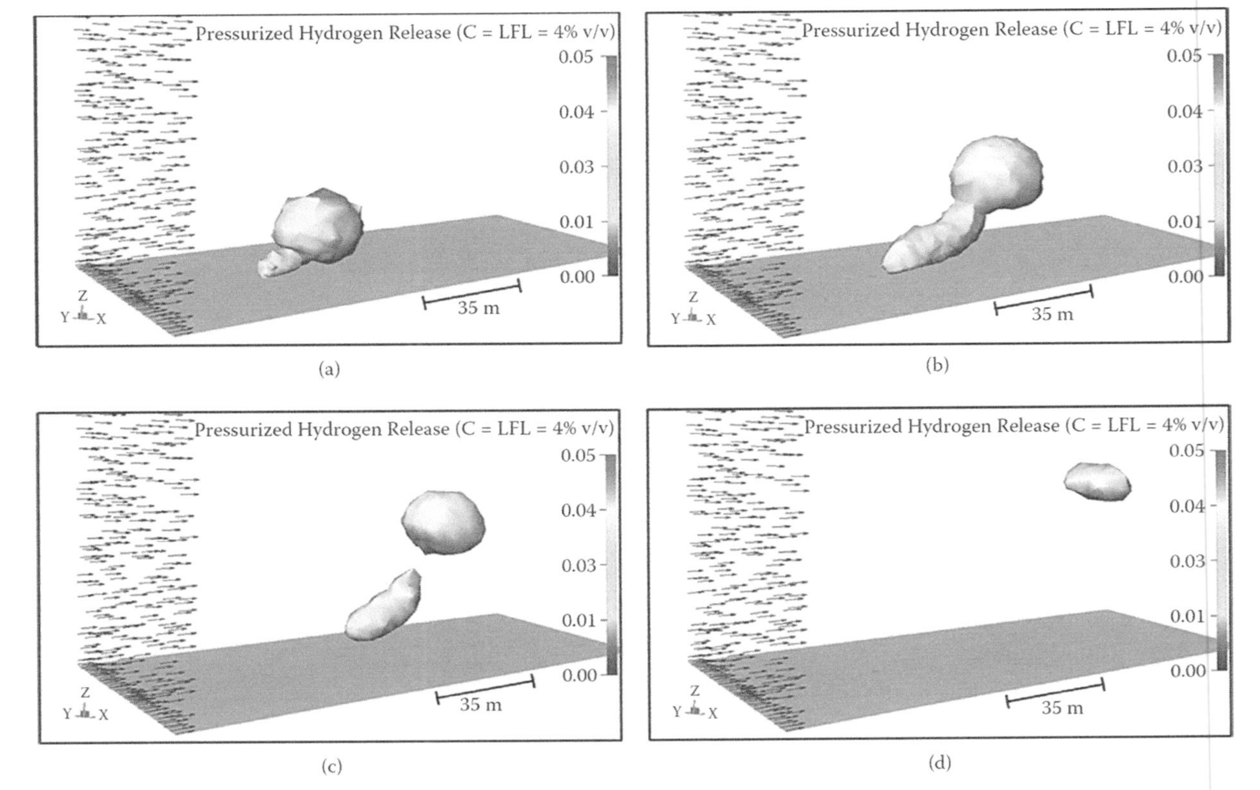

FIGURE 5.13 Simulated dispersion snapshots following pressurized hydrogen release at times (a) 2.4 s, (b) 6.2 s, (c) 9.6 s, and (d) 12.4 s after start of release. With permission from Elsevier.

Source: [101].

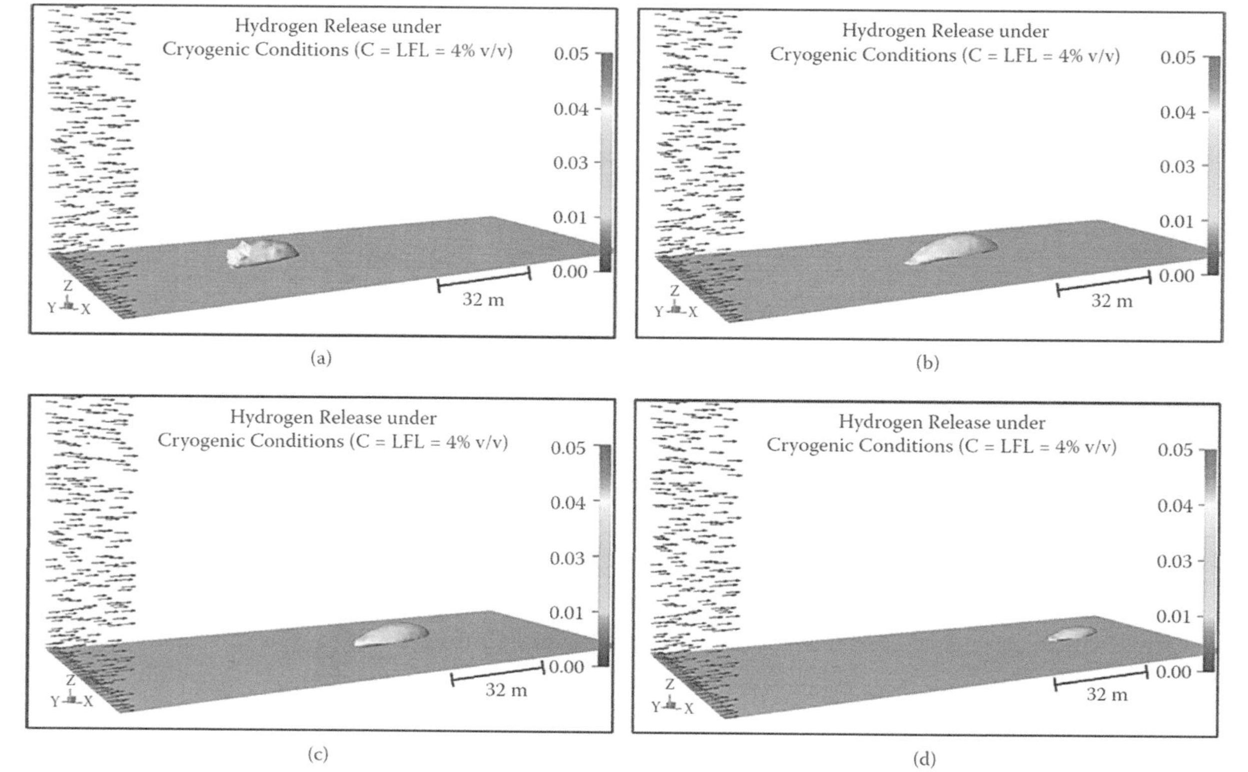

FIGURE 5.14 Simulated dispersion snapshots following cryogenic hydrogen release at times (a) 3.5 s, (b) 10 s, (c) 13 s, and (d) 18.5 s after start of release. With permission from Elsevier.

Source: [71].

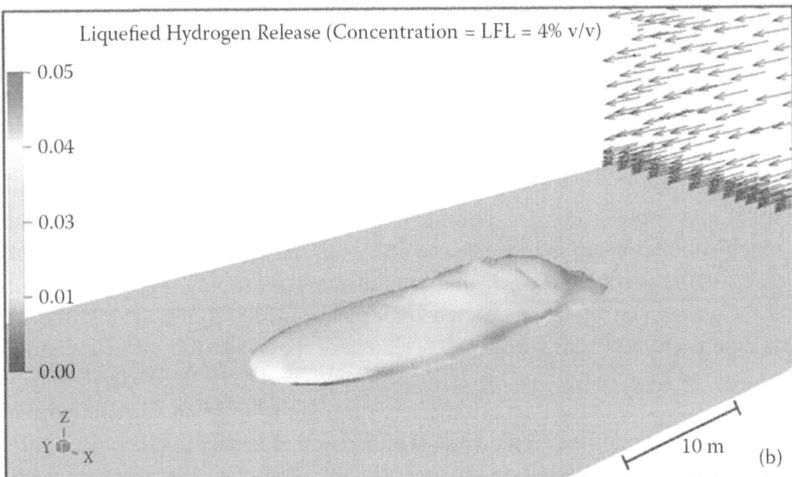

FIGURE 5.15 Qualitative comparison of dispersing cloud following (a) liquefied natural gas release; experimental record from Burro series trials; and (b) liquefied hydrogen release; computational estimation 7 s after release initiation.

Source: [121, 71]. With permission from Elsevier.

high-pressure gas release that produces a lighter-than-air cloud. In the former case, hydrogen remains within the LFL range close to the ground for some time, increasing the risk for various types of fires and explosions. Visualization of computer-simulated cryogenic hydrogen dispersions demonstrated that hydrogen cloud behavior in this case is similar to that observed during LNG experimental releases. Consequently, accidental liquefied hydrogen release scenarios should be addressed as heavier-than-air releases, using the appropriate dispersion models for calculating heavy and not light gas dispersion. Moreover, this change of the behavior of a light gas on dispersion during cryogenic release should be kept in mind in hazard evaluation procedures and when establishing safety measures.

In practical applications, dispersion calculations need to be performed in *topographically complex environments*. Thus, solid obstacles intervening in the dispersion area should be considered in the computations. In previous validation work, it has been proven that CFD codes constitute powerful tools for complex terrain dispersion simulation, providing very accurate results with excellent visualization capabilities, which can be helpful in quantitative risk analysis applications [82]. The dominating mixing mechanism between gas and air, which is the turbulent flow developing in the vicinity of the solid obstacle, was then computationally approached satisfactorily by several turbulence models: the k–ε and shear stress transport (SST) models showed improved robustness during solver processing. The SSG model entailed increased central processing unit (CPU) time without significant enhancement of accuracy of results. The SSG, k–ε, and SST models appeared to overestimate maximum concentrations recorded in the trials, whereas the k–ω model underestimated them.

Generally, the results obtained through the numerical simulations showed good agreement compared with the experimental data, leading to the conclusion that CFD techniques can be effectively used in consequence assessment procedures concerning toxic/flammable dispersion scenarios in real terrains, where box models have limited capabilities.

Venetsanos and Bartzis [86] have also investigated CFD modeling of large-scale LH_2 spills in an unobstructed environment using the ADREA-HF code. The authors performed simulations to investigate the effects of the source model (jet or pool), the modeling of the earthen sides of the pond around the source, and the ground heat transfer. Comparison of the predicted hydrogen concentrations against the experimental results from NASA's trial-6 experiment at the available sensor locations was made. Generally, simulation jet modeling was not found to be in good agreement with experimental data, with the exception of the simulation based on modeling the source as a two-phase jet pointing downward, which resulted in predicted concentrations in much better agreement with the experiments. Modeling the source as a pool resulted in overestimation of concentration levels. An illustration of the predicted contours of hydrogen concentration (by volume) on a symmetry plane at 21 s from a hydrogen pool is shown in Figure 5.16.

The ADREA-HF CFD code has also been investigated as a prediction tool for consequence assessment of hydrogen applications by Venetsanos et al. [87]. To this end, various types of hydrogen release scenarios were considered, including gaseous and liquefied releases, open, semiconfined, and confined environments, and sonic (underexpanded) and low momentum releases.

5.10.3 RELEASES IN CONFINED SPACES

Swain et al. [88] have proposed the hydrogen risk assessment method (HyRAM) to be used for establishing venting in buildings that contain hydrogen-fueled equipment or for determining optimum hydrogen sensor locations. The method can be used to

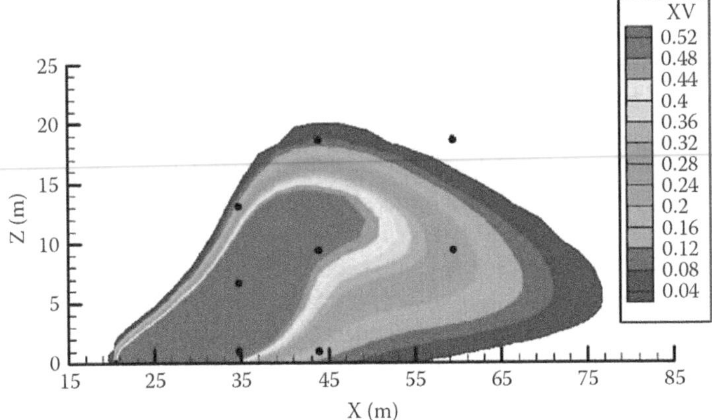

FIGURE 5.16 Predicted contours of hydrogen concentration (by volume) on symmetry plane at $t = 21$ s (Case 5: pool, with fence and contact heat transfer). With permission from Elsevier.

Source: [86].

reduce CFD modeling and utilize helium as a hydrogen surrogate to limit hazardous validation tests for hydrogen leakage into simple geometric enclosures.

An intercomparison exercise on the capabilities of CFD models to predict the short- and long-term distribution and mixing of hydrogen in confined spaces, as in a garage, was performed by Venetsanos et al. [89]. The rectangular volume studied was 78.4 m³. The project engaged 12 different organizations, ten different CFD codes, and eight different turbulence models. The authors found large variations in predicted results in the first phase of the benchmark among the various modeling approaches, and this was attributed to differences in turbulence models and numerical accuracy options (time/space resolution and discretization schemes). Nevertheless, during the second phase of the benchmark, the variation between simulated and predicted results was reduced. (See also Chapter 9.)

Papanikolaou et al. [90] have studied numerically the release of hydrogen inside a naturally ventilated residential garage with an internal volume of 66.3 m³. For validation purposes they used past experiments performed by Swain et al. [24] with helium as a surrogate gas. This was, in fact, an inter-laboratory study checking results from four different CFD packages, namely, ADREA-HF, FLACS, FLUENT, and CFX, with the experimental measurements inside the garage. Agreement between the partners' predictions and the experimental data was generally good. Yet it was determined that there was a tendency to overestimate the results of the upper sensors for the small and medium vent sizes and underestimate for the large vent size. The results of the lower sensors were generally over-predicted for the small and medium vent sizes, whereas for the larger vent size the four codes either over-predicted or under-predicted at the sensor close to the leak and over-predicted at the sensor close to the vent.

In another article on hydrogen release and dispersion in enclosed spaces, Zhang et al. [91] compared experimental data with CFD-computed results and the calculations of an analytical method. They found good agreement of all three approaches, provided a suitable value is used for Smagorinsky's constant to obtain "fine tuning" of the CFD application. The investigated space was a rectangular room with a volume of 78.4 m^3. The Fire Dynamics Simulator (FDS) code (a large-eddy simulation-based code) available from NIST (National Institute of Standards and Technology) was used for CFD simulations [92].

Papanikolaou et al. [93] have obtained CFD simulations on small hydrogen releases inside a ventilated facility providing an internal volume of 25 m^3 and, thus, assessing ventilation efficiency. The study focused on the safety assessment of a facility hypothetically hosting a small H_2 fuel cell system. Dispersion experiments to validate the simulations were also performed. Simulation was accomplished by using the ADREA-HF CFD code. In general, good agreement was found between predicted and experimentally measured concentration/time histories for all simulated cases.

Venetsanos et al. [94] summarized the results from the InsHyde project on the use of hydrogen in confined spaces. (Again, see also Chapter 9.) The project investigated realistic small-to-medium indoor hydrogen leaks and provided recommendations for the safe use and storage of indoor hydrogen systems using experimental and simulation approaches. The authors used hydrogen and nitrogen for their experiments to evaluate short- and long-term dispersion patterns in garage-like settings. In addition, they performed combustion experiments to evaluate the maximum amount of hydrogen that could be safely ignited indoors, and they extended their study to the ignition of larger amounts of hydrogen in obstructed environments outdoors. Their work included pre-test simulations, validation of the available CFD codes against previously performed experiments with significant CFD code intercomparisons, and CFD application to investigate specific realistic scenarios.

Skjold et al. estimated the consequences of vented hydrogen deflagrations for homogeneous [95] and inhomogeneous [96] mixtures in 20 foot ISO containers considering the impacts of vented hydrogen deflagrations. Both papers compare experimental and simulation results and discuss the implications of the findings for safety issues related to various hydrogen applications. Other authors have predicted with sufficient accuracy the stratification of hydrogen inside a container during release. Nevertheless, the results of the model predictions sometimes differ significantly, especially in estimations of the maximum reduced explosion pressure. The results of this study, performed as part of a HySEA project, offer a strong motivation for those who work on simulations of consequence models to improve their models, implement automated procedures, and update training and guidelines for users of the models.

In a continuation of their previous studies, Skjold, [97] investigated the effect of congestion in vented hydrogen deflagrations in containers for homogeneous and inhomogeneous mixtures. Other parameters investigated include hydrogen concentration, vent area, type of venting device, ignition position, and level and type of congestion inside the container. The results confirmed that internal congestion can significantly increase the maximum reduced explosion pressure in vented deflagrations, compared to vented deflagrations in empty enclosures. As such, it is important to incorporate the effect of congestion in the theoretical and/or empirical correlations recommended in standards and guidelines for explosion protection.

Chen et al. [98] used the commercial code FLACS to analyze the mechanism and influence factors of the overpressure development during vented hydrogen deflagrations. The code was validated by previous experimental data found in the literature, and the effects of concentration, ignition location, and vent area on the vented overpressure were discussed in detail. It was found that three overpressure peaks were formed at the moments of the vent failure, the external explosion, and the occurrence of the maximum flame surface area in the vessel. The relationship between overpressure and vent area fits the power law with a negative exponent, while the larger vent area may lead to a stronger effect of the external explosion on the internal overpressure.

5.10.4 THERMAL HAZARDS FROM HYDROGEN CLOUDS

Atmospheric releases of flammable gases such as hydrogen may lead to major fires with extensive effects on the surroundings, mainly due to the intense thermal load emissions. In activities where hazards are associated with cloud fires, there is the need for societal risk assessment that involves the estimation of hazardous zones due to the resulting thermal radiation. However, up to now only limited work has been done on modeling the effects of flash fires; available techniques have been judged insufficient by the Center for Chemical Process Safety (CCPS) [70].

A methodology for modeling *flash fire radiation* was developed by Raj and Emmons and proposed by CCPS [70], in which a flash fire is modeled as a two-dimensional turbulent flame propagating at constant speed. Yet the model requires simplifying assumptions (e.g., stationary location of burning cloud) and presents certain deficiencies (e.g., ignoring of ignition site location). Moreover, a multipoint grid-based approach for radiation impact analysis was adopted by Pula et al. [99], providing estimations of radiation effects at different locations in the area of interest. A three-dimensional computational approach based on fluid dynamics techniques was attempted, aimed at the estimation of resulting thermal radiation emissions and overpressure in large-scale cloud fires [100–101].

Experimentation in the field of gas cloud fires appears to be limited. A unique set of large-scale experiments that involve the release, dispersion, ignition, and combustion of flammable natural gas clouds in the open air does, however, exist, with the code name Coyote. Coyote series trials conducted by LLNL in 1983 at the Nevada Test Site provided an integrated dataset for use in validation studies [102–103]. The objective of the experiments was to determine the transport and dispersion of vapors from LNG spills and in addition to investigate the damage potential of vapor cloud fires. Transient simulations of the flammable cloud combustion accomplished by Sklavounos and Rigas [100–101] and performed using the CFX (CFD code referred to on page XX) provided reasonable estimations of the thermal flux histories (Figure 5.17), while the predictions of the thermal impulses proved to be statistically valid and consistent with the experimental results obtained at LLNL installations [102], as verified by strict statistical tests.

In addition, post-processing of the computational results resulted in visualizations of the combustion progress within the cloud. Thus, comparisons between the computational and the experimentally observed combustion regions were obtained, revealing

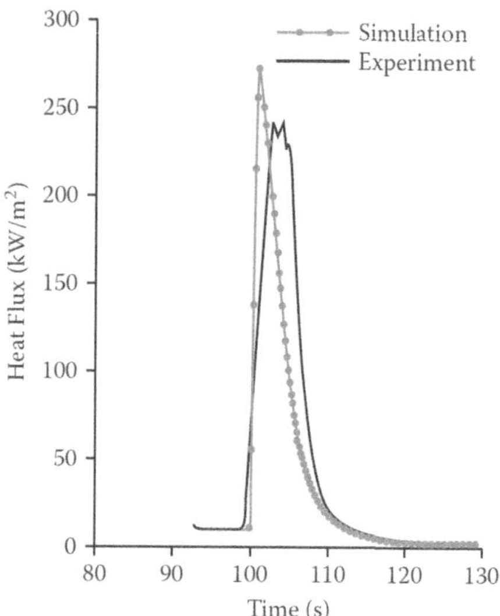

FIGURE 5.17   Experimental versus computational heat flux histories for Coyote trial 3 at position (65, 57, 1). With permission from Elsevier.

Source: [101].

good agreement between the corresponding snapshots (Figure 5.18). Generally, it was concluded that the evolution of a gas cloud fire and the oncoming consequences may be efficiently simulated utilizing CFD techniques.

5.10.5   SIMULATIONS IN THE CHEMICAL INDUSTRIES

Chemical industries (including the petroleum industry) produce and consume huge quantities of hydrogen for their needs, mainly for upgrading fossil fuels and for the production of ammonia. As a result, many accidents have occurred in these industries due to the use of hydrogen. To counteract these occurrences, many publications have recently stressed the use of CFD tools to simulate complicated chemical and physical phenomena with the aim of preventing loss of control of chemical reactors.

Zhai et al. [104] have proposed CFD simulation with the detailed chemistry of steam reforming of methane (SRM) for hydrogen production in an integrated microreactor. The authors claim that SRM in microreactors has great potential to develop a low-cost, compact process for hydrogen production via a shortening of reaction time from seconds to milliseconds. They simulated a microreactor design for SRM reaction with the integration of a microchannel for Rh-catalyzed endo-thermic reactions, a microchannel for Pt-catalyzed exothermic reactions, and a wall in between with a Rh- or Pt-catalyst-coated layer. They accomplished CFD modeling by adopting elementary reaction kinetics for the SRM process in the CFD model and describing combustion in the channel by global reaction kinetics. The model predictions were validated against experimental data from the literature. For the

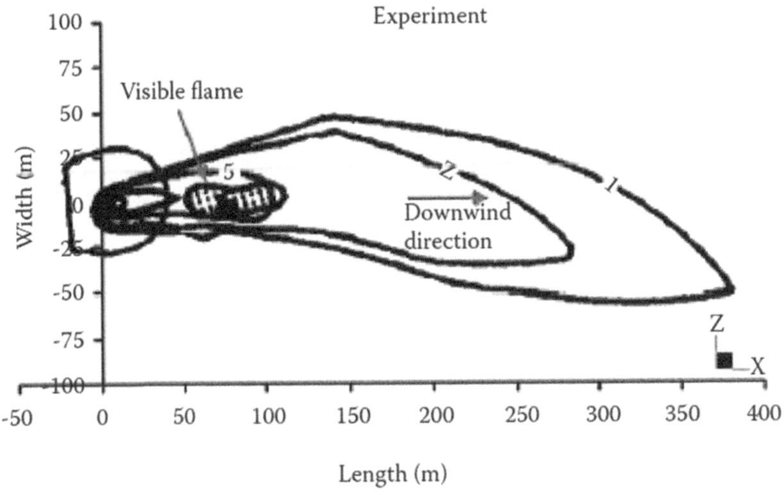

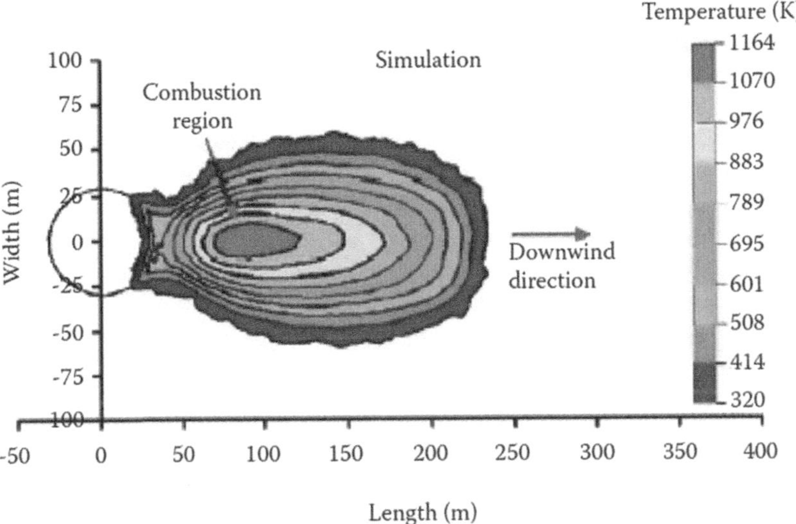

FIGURE 5.18 Experimental and simulated burning cloud snapshot for Coyote trial 3 (1 m height, 103 s). With permission from Elsevier.

Source: [101].

extremely fast reactions in both channels, the simulations indicated the significance of the heat conduction ability of the reactor wall, as well as the interplay between the exothermic and endothermic reactions (e.g., the flow rate ratio of fuel gas to reforming gas). This approach provides a potential way of efficiently investigating and controlling these highly exothermic and hazardous reactions, aiming at developing more economical hydrogen production.

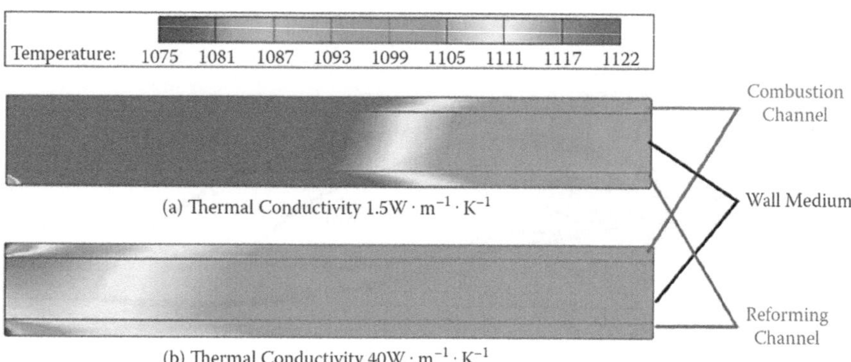

FIGURE 5.19 The temperature profiles inside the reactors with metallic and ceramic walls (residence time 10 ms, fuel gas:reforming gas [2:3, inlet temperature 1073 K, H_2O:CH_4 [3:1, absolute pressure 1 atm]). With permission from Elsevier.

Source: [104].

The reactor performance was investigated using a metallic wall with thermal conductivity of 40 W m⁻¹ K⁻¹ and a ceramic wall with thermal conductivity of 1.5 W m⁻¹ K⁻¹, as shown in Figure 5.19.

Successful CFD simulation has also been accomplished by Dou and Song [105] on hydrogen production from steam reforming of glycerol in a fluidized bed reactor. This conversion is of great environmental value considering the high surplus of glycerol now produced from the conversion of vegetable oils or animal fats to biodiesel fuel. Thus, glycerol is becoming a cheap market product and is expected to become a waste problem. The FLUENT code was used in these simulations with the following three-step reaction scheme:

$$C_3H_8O_3 + H_2O \leftrightarrow CH_4 + 2CO_2 + 3H_2$$

$$CH_4 + H_2O \leftrightarrow CO + 3H_2$$

$$CO + H_2O \leftrightarrow CO_2 + H_2$$

Chemical thermodynamics and experimentation revealed that to avoid coke deposition on the Ni-based catalyst, the reaction temperature should be raised to 600°C. Thus, this CFD simulation is a valuable tool to predict and prevent hazardous conditions when applying this new chemical process industrially.

To counteract hydrogen deflagrations in the process industries, Baraldi et al. [106] utilized the results of the experiments performed by Pasman and Groothuisen [107] on explosions of stoichiometric hydrogen–air mixtures in a cylinder-type combustion vessel of volume 0.95 m³ equipped with pressure relief vents. The authors simulated experimental results with a discrepancy below 20% for the case of a 0.2 m² vent, whereas the discrepancy with a 0.3 m² vent was much higher (up to 45%). The flame propagation for the case with a 0.3 m² vent is illustrated in Figure 5.20.

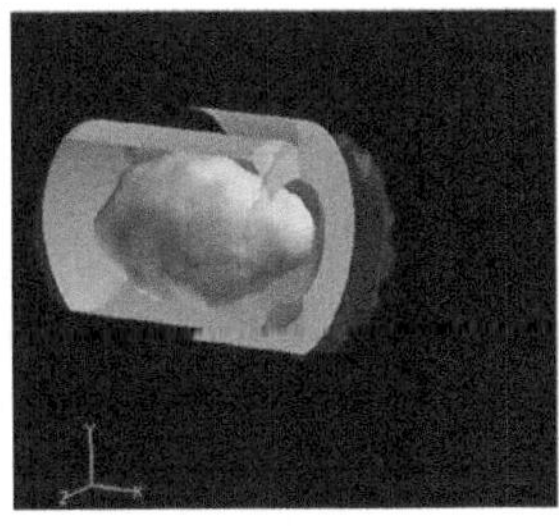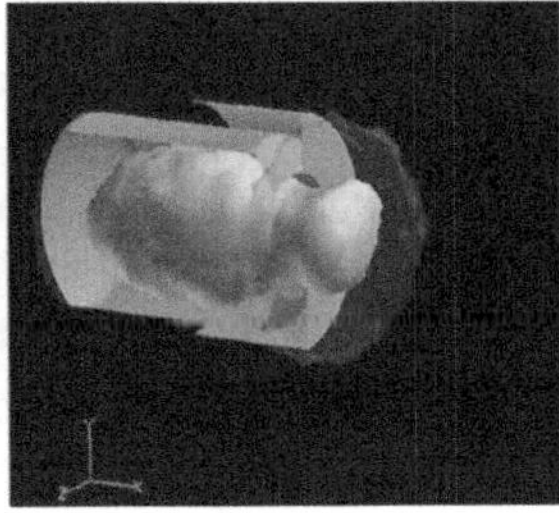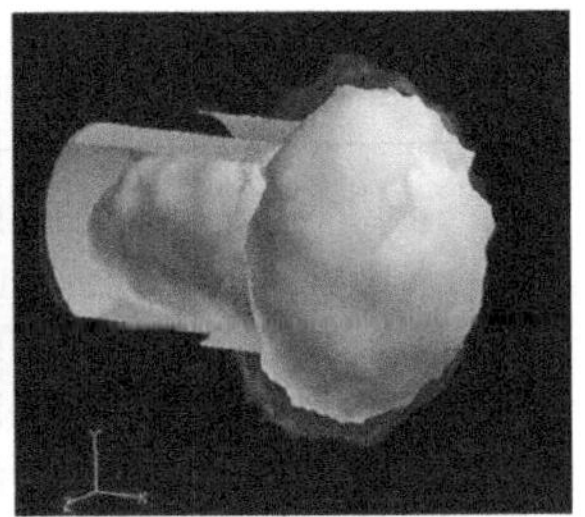

FIGURE 5.20   Flame propagation of a stoichiometric mixture of hydrogen–air through a 0.3 m^2 vent. With permission from Elsevier.

Source: [106].

Xu et al. [108] proposed a heat transfer model based on a multi-region and multi-physics approach for modeling the transient heat transfer behavior of composite cylinders subjected to fire impingement. The fire scenario was modeled using the fireFOAM software. Three-dimensional governing equations, based on the finite volume method, were written to simulate the heat transfer through the regions of composite laminate, liner, and pressurized hydrogen, respectively. The governing equations were solved sequentially with temperature-dependent material properties and coupled interface boundary conditions. The proposed conjugate heat transfer model was validated against a bonfire test of a commercial Type 4 cylinder, and its transient heat transfer behavior was also discussed.

Bauwens and Dorofeev [109] simulated the impacts of an accidental release of hydrogen within installations of a capacity of more than 15,000 m^3. A Large Eddy Simulation (LES) Navier–Stokes Computation Fluid Dynamics (CFD) solver, called fireFOAM, was utilized to model the hydrogen release. The CFD model results were evaluated through a study that used experimental results from a 1/6-scale hydrogen release from the literature and a grid sensitivity study. Using the model, a parametric study was performed by varying release rates and enclosure sizes and examining the concentrations that developed. The hydrogen dispersion results were then used to calculate the corresponding pressure loads from hydrogen–air deflagrations in the facility.

5.10.6 Simulations in the Nuclear Industry

A severe hazard in nuclear facilities, in the event of loss of control, is the accidental production of large quantities of hydrogen via various mechanisms. Fast reactions, for example, in a few minutes, that may contribute to this phenomenon include zirconium–steam reactions, steel–steam reactions, molten core–water reactions, and molten core–concrete reactions. On the other hand, slow reactions may produce considerable quantities of hydrogen. Such reactions include water radiolysis and corrosion of aluminum, zinc-based paint, and galvanized steel. Mixing of hydrogen with air in a confined space will result in flammable or even detonable mixtures taking into account the wide flammability and detonability limits of hydrogen–air mixtures.

This eventuality has occurred in many cases of nuclear accidents, including the catastrophic hydrogen explosion at the Fukushima Daiichi nuclear power complex in Japan on March 14, 2011.

In an effort to prevent such accidents, Baraldi, Heitsch, and Wilkening [110] performed CFD simulations of hydrogen deflagrations in a simplified European pressure reactor (EPR) containment, aiming to construct prediction models for the overpressures and high temperatures generated from these events. In this multiparametric analysis, the authors investigated pressure differences between the two side walls, as well as various geometrical configurations considering the number of vents, size, and position. Single and multiple ignition points were also investigated. CFX and REACFLOW codes were used in this work, and useful results were obtained.

Redlinger [111] validated DET3D (a CFD tool) against experimental and theoretically calculated results, aiming to investigate and improve nuclear reactor safety. The author simulated hydrogen detonations in complex three-dimensional geometries occurring in assumed accident scenarios in nuclear reactors. The potential of hydrogen–air mixtures for flammability, flame acceleration, deflagration-to-detonation transition, and detonability was checked using the σ (expansion ratio) and λ (detonation cell size) criteria of mixtures, as described by Dorofeev et al. [112]. Then the detonability conditions were used as initial data to run the DET3D code. Good agreement was found between computations and experiments in all studied cases.

The GASFLOW CFD code was applied by Xiong, Yang, and Cheng [113] in an effort to investigate hydrogen risk at the Qinshan-II nuclear power plant in China. The authors analyzed the effect of water spray on hydrogen risk in the containment considering a large break loss of coolant accident (LBLOCA). They selected three different spray strategies, namely, without spray, with direct spray, and with both direct and recirculation spray. They determined a significant influence of spray modes on hydrogen distribution inside the containment. The efficiency of the passive autocatalytic recombiners (PARs) was also investigated. No substantial influence by spray modes was observed. The results obtained by a CFD analysis of a new PAR model were in good agreement with the experimental results.

Heitsch et al. [114] performed simulations of the Paks nuclear power plant in Hungary using GASFLOW, FLUENT, and CFX codes by modeling a defined severe hydrogen release accident scenario. The authors investigated the risk of losing integrity of containment after hydrogen deflagrations. The mitigation measure proposed and assessed was the implementation of catalytic recombiners in the plant. Both unmitigated and recombiner-mitigated simulations were performed, and the results were compared. Simulations of a full containment subject to a severe accident were not possible until recently due to the prohibitive computing times and the limited capabilities of the codes; yet this work proved that this is now possible with improved codes and more powerful computers. All the same, the wall clock time needed for each of the simulations in this work was around 40 days, even when six to eight processors were used in parallel. The analysis showed that at least 30 recombiners are needed if it is desirable to avoid significant flammable clouds with more than 8% hydrogen being formed. The main compartments of the containment modeled in the CFD simulations are shown in Figure 5.21. Figure 5.22 illustrates the reduction of hydrogen in the containment for both unmitigated and recombiner-mitigated cases.

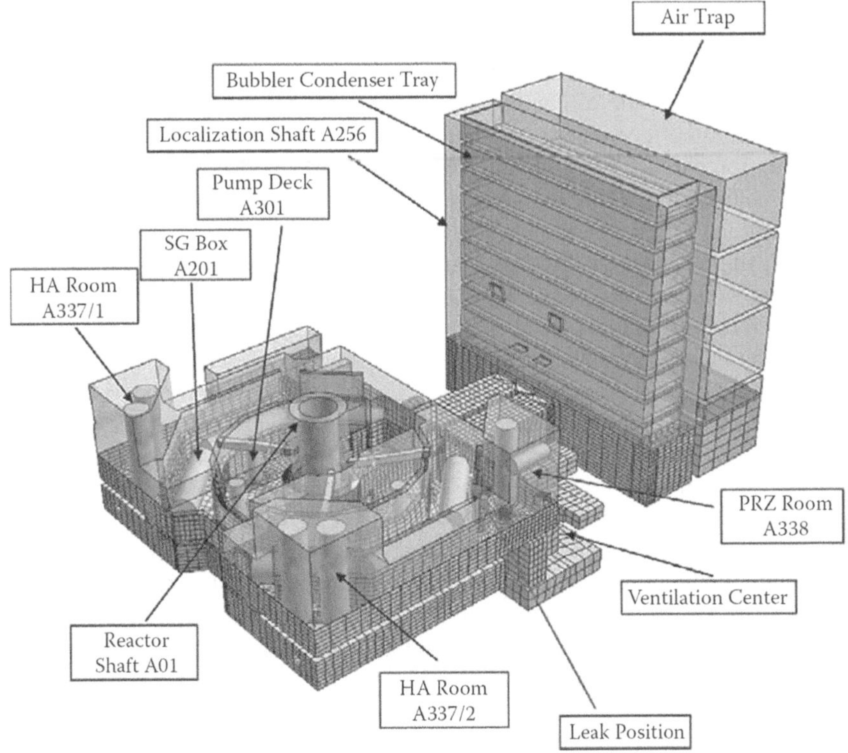

FIGURE 5.21 Regions of the Paks nuclear power plant modeled in the CFD simulations. With permission from Elsevier.

Source: [114].

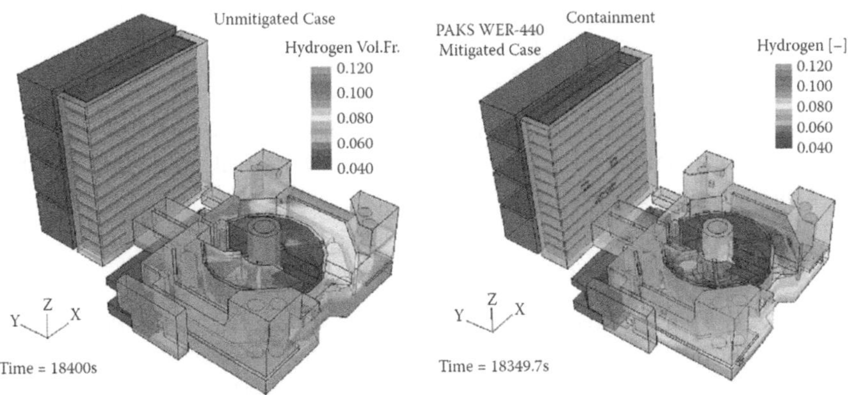

FIGURE 5.22 Hydrogen concentrations in the regions modeled, at around 18,400 s. With permission from Elsevier.

Source: [114].

Prabhudharwadkar and Iyer [115] formulated a model compatible with a CFD code to simulate hydrogen distribution using a passive catalytic recombiner in a nuclear power plant containment aiming to prevent hydrogen–air explosions. This was an effort to counteract the difficulties in modeling resulting from the much smaller size of a catalytic recombiner (needing very fine mesh size inside the recombiner channels) compared to the containment. After conducting a parametric study, the authors modeled the recombiner channels as single computational cells. This avoids full resolution of these channels, allowing the use of a one-node recombiner channel modeling approach, thus offering a significant reduction in computational time. Predictions with the single-node approach were within 5% accuracy for steady-state cases when compared to the detailed simulations.

Ferng and Chen [5] analyzed the thermal-hydraulic characteristics and hydrogen generation from graphite–water reactions after a steam generator tube rupture in the HTR-10 reactor (a graphite-moderated and helium-cooled reactor with a pebble-bed core). They developed a transient three-dimensional compressible CFD model to investigate four postulated accidents in HTR-10, considering ruptures in one or two steam generator tubes which cause hydrogen and carbon monoxide generation. The computations showed that the quantity of hydrogen produced from the reaction of water with graphite is not sufficient to reach the flammability limits of hydrogen–air mixtures. In fact, the maximum quantity of hydrogen that may be produced from two tube ruptures is a volume fraction of about 0.5%, which is much less than the lower flammability limit of 4%. In addition, the maximum temperature of the nuclear fuel in the reactor core was computed to be 1218 K, much less than the limiting temperature (1873 K) for graphite. The authors therefore concluded that the nuclear fuel would not be damaged under the worst-case conditions.

Wang and Cao [116] proposed a three-dimensional CFD analysis model to model hydrogen behavior during an assumed severe accident induced by a loss-of-coolant accident (LOCA) in an advanced passive pressurized water nuclear reactor (PWR) containment. The hydrogen distribution in the containment was studied by simulating with algebraic and k–ε turbulence models. In hydrogen distribution results simulated with the algebraic model, all hydrogen is gathered in the dome, and the peak concentration reaches 17%. When the k–ε model is adopted, the peak concentration in the dome is 8%, and hydrogen stratification is established in the entire upper space. In addition, the algebraic model cannot simulate turbulence diffusion near the source. Hence, it is reasonable to select the k–ε model instead of the algebraic model to study the hydrogen behavior in the containment.

5.10.7 Recent Advances in Hydrogen Storage Safety

Li et al. [117] have reviewed the safety of hydrogen storage and transportation. In their paper, they present the recent progress on hydrogen leak diffusion characteristics, leak spontaneous combustion mechanisms, and material hydrogen damage mechanisms from the viewpoint of a theoretical experimental approach, as well as numerical simulations. They observe that regardless of the progress achieved on the safety features of hydrogen, there are still some challenges and limitations. Thus, they claim that additional research is needed, such as optimization of the kinetic mechanism of the high-pressure hydrogen leakage reaction and turbulence model, investigation of the expansion and dilution of hydrogen clouds after liquid hydrogen flooding, and further research on the spontaneous ignition of leaked hydrogen and the damaging effects of hydrogen on materials.

Zhou et al. [118] have also reviewed the research progress on the self-ignition of high-pressure hydrogen discharge. Significant research achievements have been obtained in the past 20 years, indicating that the minimum self-ignition discharge pressure of hydrogen is approximately 2 MPa (20 bar). The ruptured discharge tube shape and the type of bursting disc play a significant role in the characteristics of hydrogen self-ignition. In addition, the shock-induced self-ignition mechanism of hydrogen has been extensively studied recently. On this topic, the initial conditions greatly affect the critical self-ignition pressure of hydrogen using formation change, development, and propagation of shock waves.

Le et al. [119] used a quantitative risk assessment methodology to investigate the effects of design parameters such as storage size, mass flow rate, storage pressure, and storage temperature, on the safety of hydrogen storage systems. Their quantitative risk assessment procedure includes data collection and hazard identification, frequency analysis, consequence analysis, and risk analysis. The Millers model and TNO multi-energy were used in the consequence analysis to simulate the jet fire and explosion hazards, respectively. It was found that the storage capacity and pressure are the most critical factors in the risk effect analysis of a hydrogen storage system.

The future of hydrogen energy has been studied by Hassan et al. [120] concerning advancements in storage technologies and implications for sustainability. The findings indicated that with the advancement in hydrogen storage technologies and their sustainability consequences, policymakers, researchers, and industry stakeholders can make the right decisions to promote the transition toward a hydrogen-based energy future that would be clean, sustainable, and resilient. The authors also stress the critical role that hydrogen energy storage technologies could play in steering the transition towards a more sustainable, low-carbon future. The principal safety concern associated with hydrogen storage is the risk of leaks or ruptures in storage tanks or pipelines. Consequently, current research focuses on developing inherently safer storage and transportation technologies that can minimize these risks and prevent fires and explosions. To this end, efforts should be directed at the development of advanced materials and designs for storage tanks and pipelines, and the development of safety protocols, guidelines, and emergency response procedures for hydrogen storage and transportation.

REFERENCES

1. Pyle, W., Hydrogen storage, *Home Power*, 59, 14–15, June/July, 1997.
2. Moradi, R. and Groth, K.M., Hydrogen storage and delivery: Review of the state of the art technologies and risk and reliability analysis, *International Journal of Hydrogen Energy*, 44, 12254–12269, 2019.
3. Niaz, S., Manzoor, T., and Pandith, A.H., Hydrogen storage: Materials, methods and perspectives, *Renewable and Sustainable Energy Reviews*, 50, 457–469, 2015.
4. Ren, J., et al., Current research trends and perspectives on materials-based hydrogen storage solutions: A critical review, *International Journal of Hydrogen Energy* , 42, 289–311, 2017.

5. Ferng, Y.M. and Chen, C.T., CFD investigating thermal-hydraulic characteristics and hydrogen generation from graphite-water reaction after SG tube rupture in HTR-10 reactor, *Applied Thermal Engineering*, 31, 2430, 2011. http://refhub.elsevier.com/S0360-3199(19)31975-5/sref4

6. Kojima, Y., Hydrogen storage materials for hydrogen and energy carriers, *International Journal of Hydrogen Energy*, 44, 18179–18192, 2019.

7. Akiba, E., Hydrogen-absorbing alloys. *Current Opinion in Solid State & Materials Science*, 4, 267–272, 1999.

8. Zhang, Q., et al., Facile synthesis of small MgH2 nanoparticles confined in different carbon materials for hydrogen storage, *Journal of Alloys and Compounds*, 825, 153953, 2020.

9. Rumble, J.R., Editor-in-chief, *CRC Handbook of Chemistry and Physics*, 99th ed., CRC Press, Taylor & Francis Group, Boca Raton, FL; 2018–2019.

10. Wolf G, Baumann J, Baitalow F, and Hoffmann FP. Calorimetric process monitoring of thermal decomposition of B-N-H compounds. *Thermochimica Acta*, 343, 19–25, 2000.

11. Voss, J., Hummelshøj, JS., Łodziana, Z., and Vegge, T. Structural stability and decomposition of $Mg(BH_4)_2$ isomorphs – An ab initio free energy study. *Journal of Physics: Condensed Matter*, 21, 012203, 2009.

12. Milanese C., et al., Complex hydrides for energy storage, *International Journal of Hydrogen Energy*, 44, 7860–7874, 2019.

13. Manickam, K., et al., Future perspectives of thermal energy storage with metal hydrides. *International Journal of Hydrogen Energy*, 44, 7738–7745, 2019.

14. Yamamoto, H., Miyaoka, H., Hino, S., Nakanishi, H., Ichikawa, T., and Kojima, Y., Recyclable hydrogen storage system composed of ammonia and alkali metal hydride. *International Journal of Hydrogen Energy*, 34, 9760–9764, 2009.

15. Bogdanovic, B. and Schwickardi, M., Ti-doped alkali metal aluminium hydrides as potential novel reversible hydrogen storage materials. *Journal of Alloys and Compounds*, 253–254, 1–9, 1997.

16. Bogdanovic, B., Brand, R.A., Marjanovic A, Schwickardi M, and Tolle J., Metal-doped sodium aluminium hydrides as potential new hydrogen storage materials. *Journal of Alloys and Compounds*, 302, 36–58, 2000.

17. Morioka, H., Kakizaki, K., Chung, S.-C., and Yamada, A., Reversible hydrogen decomposition of KAlH4. *Journal of Alloys and Compounds*, 353, 310–314, 2003.

18. Li, H.W., Kikuchi, K., Nakamori, Y., Ohba, N., Miwa, K., Towata, S., et al., Dehydriding and rehydriding processes of well crystallized Mg(BH4)2 accompanying with formation of intermediate compounds. *Acta Materialia*, 56, 1342–1347, 2008.

19. Graetz, J., Reilly, J.J., Kulleck, J.G., and Bowman, R.C., Kinetics and thermodynamics of the aluminum hydride polymorphs. *Journal of Alloys and Compounds*, 446–447, 271–275, 2007.

20. Tzimas, E., Filiou, C., Peteves, S.D., and Veyret, J.B., *Hydrogen Storage: State-of-the-Art and Future Perspective*, European Commission, Directorate General Joint Research Centre, Petten, The Netherlands, 2003.

21. Broom, D.P., et al., Concepts for improving hydrogen storage in nanoporous materials, *International Journal of Hydrogen Energy*, 44, 7768–7779, 2019.

22. Liu, S.S., et al., Improved reversible hydrogen storage of $LiAlH_4$ by nano-sized TiH_2, *International Journal of Hydrogen Energy*, 38, 2770–2777, 2013.

23. Mosquera-Vargas, E., Tamayo, R., Morel, M., Roble, M., and Díaz-Droguett, D.E., Hydrogen storage in purified multi-walled carbon nanotubes: Gas hydrogenation cycles effect on the adsorption kinetics and their performance, *Heliyon*, 7, e08494, 1–13, 2021.

24. Kojima, Y., Kawai, Y., Koiwai, A., Suzuki, N., Haga, T., Hioki, T., et al. Hydrogen adsorption and desorption by carbon materials, *Journal of Alloys and Compounds*, 421, 204–208, 2006a.

25. Noh, J.S., Agarwal, R.K., and Schwarz, J.A., Hydrogen storage systems using activated carbon, *International Journal of Hydrogen Energy*, 12, 693–700, 1987.

26. Chahine, R. and Bose, T.K. Low-pressure adsorption storage of hydrogen, *International Journal of Hydrogen Energy*, 19, 161–164, 1994.

27. Kiyobayashi, T., Takeshita, H.T., Tanaka, H., Takeichi, N., Zuttel, A., Schlapbach, L., et al., Hydrogen adsorption in carbonaceous materials – How to determine the storage capacity accurately, *Journal of Alloys and Compounds*, 330, 666–669, 2002.

28. Ozturk, Z., Lithium decoration characteristics for hydrogen storage enhancement in novel periodic porous graphene frameworks, *International Journal of Hydrogen Energy*, 46, 11804–11814, 2021.

29. Benard, P. and Chahine, R., Storage of hydrogen by physisorption on carbon and nanostructured materials, *Scripta Materialia*, 56, 803–808, 2007.

30. Shiraz, H.G. and Tavakoli, O., Investigation of graphene-based systems for hydrogen storage, *Renewable and Sustainable Energy Reviews*, 74, 104–109, 2017.

31. Rajaura, R.S., et al., Role of interlayer spacing and functional group on the hydrogen storage properties of graphene oxide and reduced graphene oxide, *International Journal of Hydrogen Energy*, 41, 9454–9461, 2016.

32. Jung, H., et al., Bio-inspired graphene foam decorated with Pt nanoparticles for hydrogen storage at room temperature, *International Journal of Hydrogen Energy*, 41, 5019–5027, 2016.

33. Khalil, Y.F., Science-based framework for ensuring safe use of hydrogen as an energy carrier and an emission-free transportation fuel, *Process Safety and Environmental Protection*, 117, 326–340, 2018.

34. Khalil, Y.F., Dust cloud combustion characterization of a mixture of LiBH4 destabilized with MgH2 for reversible H2 storage in mobile applications, *International Journal of Hydrogen Energy*, 39, 16347–16361, 2014.

35. Khalil, Y.F., Risk quantification framework of hydride-based hydrogen storage systems for light-duty vehicles, *Journal of Loss Prevention in the Process Industries*, 38, 187–198, 2015.

36. Khalil, Y.F., Experimental determination of dust cloud combustion parameters of a-AlH3 powder in its charged and fully discharged states for H2 storage applications, *Journal of Loss Prevention in the Process Industries*, 44, 334–346, 2016.

37. Makepeace, J.W., He, T., Weidenthaler, C., Jensen, T.R., Chang, F., Vegge, T., Ngene, P., Kojimam Y., de Jongh, P.E., Chen, P., and David, W.I.F., Reversible ammonia-based and liquid organic hydrogen carriers for high-density hydrogen storage: recent progress, *International Journal of Hydrogen Energy* , 44, 7746–7767, 2019.

38. Alhumaidan, F., Cresswell, D., and Garforth, A. Hydrogen storage in liquid organic hydride: Producing hydrogen catalytically from methylcyclohexane, *Energy Fuel*, 25, 4217–4234, 2011.

39. Kwak, Y. et al., Hydrogen production from homocyclic liquid organic hydrogen carriers (LOHCs): Benchmarking studies and energy-economic analyses, *Energy Conversion and Management*, 239, 114124, 2021.

40. Valera-Medina, A., Xiao, H., Owen-Jones, M., David, W.I.F., Bowen, P.J., Ammonia for power, *Progress in Energy and Combustion Science*, 69, 63–102, 2018.

41. Kobayashi H., Hayakawa, A., Somarathne, K.D., Kunkuma, A., and Okafor, E.C., *Science and technology of ammonia combustion, Proceedings of the Combustion Institute*, 37, 109–133, 2019.

42. Valera-Medina, A., et al., Review on ammonia as a potential fuel: From synthesis to economics, *Energy & Fuels*, 35, 6964–7029, 2021.

43. Lee, H. and Lee, M., Recent advances in ammonia combustion technology in thermal power generation system for carbon emission reduction, *Energies*, 14, 5604, 1–29, 2021. https://doi.org/10.3390/en14185604

44. Elbaz, A.M., Wang, S., Guiberti, T.F., and Roberts, W.L., Review on the recent advances on ammonia combustion from the fundamentals to the applications, *Fuel Communications*, 10, 100053, 2022.

45. Okanishi, T., et al., Comparative study of ammonia-fueled solid oxide fuel cell systems, *Fuel Cells,* 17, 383–390, 2017.

46. Karabeyoglu, A. and Brian, E., Fuel conditioning system for ammonia fired power plants, in *9th Annual NH3 Fuel Association Conference*, NH3 Fuel Association, 2012. https://nh3fuel.files. wordpress.com/2012/10/evans-brian.pdf

47. Thomas, G. and Parks, G., *Potential Roles of Ammonia in a Hydrogen Economy*, U.S. Department of Energy, Washington, DC, 2006.

48. Trivyza, N.I., Cheliotis, M., Boulougouris, E., and Theotokatos, G., Safety and reliability analysis of an ammonia-powered fuel-cell system, *Safety*, 7, 80, 1–21, 2021.

49. Jensen, J.O., Vestbo, A.P., Li, Q., and Bjerrum, N.J., The energy efficiency of onboard hydrogen storage, *Journal of Alloys and Compounds*, 446–447, 723–728, 2007.

50. Cheddie, D., Ammonia as a hydrogen source for fuel cells: A review, Chapter 13 in *Book: Hydrogen Energy – Challenges and Perspectives*, edited by Minic, D., IntechOpen Publ., 2012. http://dx.doi.org/10.5772/47759

51. Aziz, M., Wijayanta, A.T., and Nandiyanto, A.B.D., Ammonia as effective hydrogen storage: A review on production, *Storage and Utilization, Energies*, 13, 3062, 2020.

52. Sanchez, A., Castellano, E., Martin, M., and Vega, P., Evaluating ammonia as green fuel for power generation: A thermo-chemical perspective, *Applied Energy*, 293, 116956, 2021.

53. Tullo, A.H., Is ammonia the fuel of the future? *Chemical & Engineering News, Petrochemicals*, 99(8), 2021. https://cen.acs.org/business/petrochemicals/ammonia-fuel-future/99/i8

54. Fasihi, M., Weiss, R., Savolainen, J., and Breyer, C., Global potential of green ammonia based on hybrid PV-wind power plants, *Applied Energy*, 294, 116170, 2021.

55. Service, R.F., Ammonia—A renewable fuel made from sun, air, and water —could power the globe without carbon, Science, *AAAS*, 12 July 2018. doi: 10.1126/science.aau7489; www.science.org/content/article/ammonia-renewable-fuel-made-sun-air-and-water-could-power-globe-without-carbon.

56. Garagounis, I., Kyriakou, V., Skodra, A., Vasileiou, E., and Stoukides, M., Electrochemical synthesis of ammonia in solid electrolyte cells, *Frontiers in Energy Research*, 2, 1–10, 2014.

57. Garagounis, I., Vourros, A., Stoukides, D., Dasopoulos, D., Stoukides, M., Electrochemical synthesis of ammonia: Recent efforts and future outlook, *Membranes*, 9, 112, 2019.

58. Lim, D.K., et al., Solid acid electrochemical cell for the production of hydrogen from ammonia, *Joule*, 4, 2338–2347, 2020.

59. Badakhsh, A., et al., A compact catalytic foam reactor for decomposition of ammonia by the Joule-heating mechanism, *Chemical Engineering Journal*, 426, 130802, 2021.

60. Jackson, C., et al., *Ammonia to Green Hydrogen Project, Feasibility Study*, Engie, Siemens, Ecuity, and STFC Publishers, Birmingham, B3 1BG, 2020.

61. Cesaro, Z., Ives, M., Nayak-Luke, R., Mason, M., and Banares-Alcantara, R., Ammonia to power: Forecasting the levelized cost of electricity from green ammonia in large-scale power plants, *Applied Energy*, 282, 116009, 2021.
62. WolframAlpha, Computational Knowledge Engine. http://www.wolframal-pha.com
63. Verfondern, K., Safety considerations on liquid hydrogen, Forschungszentrum Juelich GmbH, *Energy and Environment*, 10, 25–27, 2008.
64. Wingerden, van K., Kluge, M., Habib, A.K., Ustolin, F., and Paltrinieri, N., Medium-scale tests to investigate the possibility and effects of BLEVEs of storage vessels containing liquified hydrogen, *Chemical Engineering Transactions*, 90, 547–552, 2022.
65. Warren, P., *Hazardous Gases and Fumes*, Butterworth-Heinemann, Oxford, 1997, 96.
66. Saffers, J.B. and Molklov, V.V., Towards hydrogen safety engineering for reacting and non-reacting hydrogen releases, *Journal of Loss Prevention in the Process Industries*, 26, 344–350, 2013.
67. Molkov, V.V., Makarov, D., and Bragin, M.V. Physics and modelling of under-expanded jets and hydrogen dispersion in atmosphere, in *24th International Conference on Interaction of Intense Energy Fluxes with Matter*, Elbrus, Russia, 1-6 March 2009.
68. Schefer, R. W., Houf, W. G., Bourne, B., and Colton, J., Spatial and radiative properties of an open-flame hydrogen plume. *International Journal of Hydrogen Energy*, 31, 1332–1340, 2006.
69. Mogi, T. and Horiguchi, S., Experimental study on the hazards of high-pressure hydrogen jet diffusion flames, *Journal of Loss Prevention in the Process Industries*, 22, 45–51, 2009.
70. CCPS (Center for Chemical Process Safety), *Guidelines for Evaluating the Characteristics of Vapor Cloud Explosions, Flash Fires and BLEVEs*, American Institute of Chemical Engineers, New York, 1994, 158.
71. Rigas, F. and Sklavounos, S., Evaluation of hazards associated with hydrogen storage facilities, *International Journal of Hydrogen Energy*, 30, 1501, 2005.
72. Lide, R. (Ed.), *Handbook of Chemistry and Physics*, 75th ed., CRC Press, Boca Raton, FL, 1994, Chap. 6.
73. Shebeko, Yu. N., Shevchuck, A.P., and Smolin, I.M., BLEVE prevention using vent devices, *Journal of Hazardous Materials*, 50, 227, 1996.
74. Sklavounos, S. and Rigas, F., Advanced multi-perspective computer simulation as a tool for reliable consequence analysis, *Process Safety and Environmental Protection*, 90 (2), 129, 2012.
75. Pohanish, R.P. and Green, S.A., *Rapid Guide to Chemical Incompatibilities*, Van Nostrand Reinhold, New York, 1997, 434.
76. Carson, P. and Mumford, C., *Hazardous Chemicals Handbook*, Butterworth-Heinemann, Oxford, 1994, 181, 202.
77. Taylor, J.R., *Risk Analysis for Process Plant, Pipelines and Transport*, Chapman & Hall, London, 1994, 102.
78. CCPS, *Guidelines for Hazard Evaluation Procedures*, American Institute of Chemical Engineers, New York, 1992, 69.
79. HSE (Health and Safety Executive), *Application of QRA in Operational Safety Studies*, Health and Safety Executive report 025, Bootle, Merseyside, 2002, 19.
80. LLNL (Lawrence Livermore National Laboratory), *Burro Series Data Report*, LLNL/NWC report UCID-19075 Vol. 1, LLN, Berkeley, CA, 1982.
81. Versteeg, H.K. and Malalasekera, W., *An Introduction to Computational Fluid Dynamics: The Finite Volume Method*, Longman, New York, 1995, 85.
82. Sklavounos, S. and Rigas, F., Validation of turbulence models in heavy gas dispersion over obstacles, *Journal of Hazardous Materials*, 108, 9, 2004.

83. Sklavounos, S. and Rigas, F., Fuel gas dispersion under cryogenic release conditions, *Energy and Fuels*, 19, 2535, 2005.

84. LNL, *Description and Analysis of Burro Series 40-m3 LNG Spill Experiments*, LLNL/NWC report No. UCRL-53186, Lawrence Livermore National Laboratory, Berkeley, CA, 1981.

85. Witcofski, R.D. and Chirivella, J.E., Experimental and analytical analyses of the mechanisms governing the dispersion of flammable clouds formed by liquid hydrogen spills, *International Journal of Hydrogen Energy*, 9, 425, 1984.

86. Venetsanos, A.G. and Bartzis, J.G., CFD modeling of large-scale LH2 spills in open environment, *International Journal of Hydrogen Energy*, 32, 2171, 2007.

87. Venetsanos, A.G., Papanikolaou, E., and Bartzis, J.G., The ADREA-HF CFD code for consequence assessment of hydrogen applications, *International Journal of Hydrogen Energy*, 35, 3908, 2010.

88. Swain, M.R., et al., Hydrogen leakage into simple geometric enclosures, *International Journal of Hydrogen Energy*, 28, 229, 2003.

89. Venetsanos, A.G., Papanikolaou E., Delichatsios, M., Garcia, J., Hansen, O.R., Heitsch, M., Huser, A., Jahn, W., Jordan, T., Lacome, J.M., Ledin, H.S., Makarov, D., Middha P., Studer, E., Tchouvelev, A.V., Teodorczyk, A., Verbecke, F., and Van der Voort, M.M., An inter-comparison exercise on the capabilities of CFD models to predict the short and long term distribution and mixing of hydrogen in a garage, *International Journal of Hydrogen Energy*, 34, 5912, 2009.

90. Papanikolaou, E., Venetsanos, A.G., Heitsch, M., Baraldi, D., Huser, A., Pujol, J., Garcia, J., and Markatos, N., HySafe SBEP-V20: Numerical studies of release experiments inside a naturally ventilated residential garage, *International Journal of Hydrogen Energy*, 35, 4747, 2010.

91. Zhang, J., Delichatsios, M.A., and Venetsanos, A.G., Numerical studies of dispersion and flammable volume of hydrogen enclosures, *International Journal of Hydrogen Energy*, 35, 6431, 2010.

92. FDS—Fire Dynamics Simulator, NIST (National Institute of Standards and Technology). http://fire.nist.gov/fds

93. Papanikolaou, E., Venetsanos, A.G., Cerchiara, G.M., Carcassi, M., and Markatos, N., CFD simulations on small hydrogen releases inside a ventilated facility and assessment of ventilation efficiency, *International Journal of Hydrogen Energy*, 36, 2597, 2011.

94. Venetsanos, A.G., Adams, P., Azkarate, I., Bengaouer, A., Brett, L., Carcassi, M.N., Engebø, A., Gallego, E., Gavrikov, A.I., Hansen, O.R., Hawksworth, S., Jordan, T., Kesslerm, A., Kumar, S., Molkov, V., Nilsen, S., Reinecke, E., Stoecklin, M., Schmidtchen, U., Teodorczyk, A., Tigreat, D., and Versloot, N.H.A., On the use of hydrogen in confined spaces: Results from the internal project InsHyde, *International Journal of Hydrogen Energy*, 36, 2693, 2011.

95. Skjold, T., et al., Blind-prediction: Estimating the consequences of vented hydrogen deflagrations for homogeneous mixtures in 20-foot ISO containers, *International Journal of Hydrogen Energy*, 44, 8997–9008, 2019a.

96. Skjold, T., et al., Blind-prediction: Estimating the consequences of vented hydrogen deflagrations for inhomogeneous mixtures in 20-foot ISO containers, *Journal of Loss Prevention in the Process Industries*, 61, 220–236, 2019b.

97. Skjold, T., Vented hydrogen deflagrations in containers: Effect of congestion for homogeneous and inhomogeneous mixtures, *International Journal of Hydrogen Energy*, 44, 8819–8832, 2019c.

98. Chen, Y., Yuan, D., Guo, J., and Liu, X., Research on mechanism and influence factors of the overpressure development in vented hydrogen deflagrations based on numerical simulation, *International Journal of Hydrogen Energy*, 44, 22681–22690, 2019.

99. Pula, R., Khan, F., Veitsch, B., and Amyotte, P., Revised fire consequence models for off-shore quantitative risk assessment, *Journal of Loss Prevention in the Process Industries*, 18, 443, 2005.

100. Sklavounos, S. and Rigas, F., Simulation of Coyote series trials, Part I: CFD estimation of non-isothermal LNG releases and comparison with box-model predictions, *Chemical Engineering Science*, 61, 1434, 2006a.

101. Rigas, F. and Sklavounos, S., Simulation of Coyote series trials – Part II: A computational approach to ignition and combustion of flammable vapor clouds, *Chemical Engineering Science*, 61, 1434, 2006b.

102. LLNL, *Coyote Series Data Report*, UCID: 19953 Vol. 1, 2. LLNL/NWC, Lawrence Livermore National Laboratory, Berkeley, CA, 1983.

103. LLNL, *Vapor Burn Analysis for the COYOTE series LNG spill Experiments*, UCID: 53530, LLNL/NWC, Lawrence Livermore National Laboratory, Berkeley, CA, 1984.

104. Zhai, X., Ding, S., Cheng, Y., Jin, Y., and Cheng, Y., CFD simulation with detailed chemistry of steam reforming of methane for hydrogen production in an integrated micro-reactor, *International Journal of Hydrogen Energy*, 35, 5383, 2010.

105. Dou, B. and Song, Y., A CFD approach on simulation of hydrogen production from steam reforming of glycerol in a fluidized bed reactor, *International Journal of Hydrogen Energy*, 35, 10271, 2010.

106. Baraldi, D., Kotchourko, A., Lelyakin, A., Yanez, J., Gavrikov, A., Efimenko, A., Verbecke, F., Makarov, D., Molkov., V., and Teodorczyk, A., An inter-comparison exercise on CFD model capabilities to simulate hydrogen deflagrations with pressure relief vents, *International Journal of Hydrogen Energy*, 35, 12381, 2010.

107. Pasman, H.J. and Groothuisen, Th.M., Design of pressure relief vents. In *Loss Prevention in the Process Industries*, edited by C.H. Bushman, Elsevier, New York, 1974, pp. 185–189.

108. Xu, B.P., Cheng, C.L., and Wen, J.X., Numerical modelling of transient heat transfer of hydrogen composite cylinders subjected to fire impingement, *International Journal of Hydrogen Energy*, 44, 11247–11258, 2019.

109. Bauwens, C.R. and Dorofeev, S.B., CFD modeling and consequence analysis of an accidental hydrogen release in a large scale facility, *International Journal of Hydrogen Energy* , 39, 20447–20454, 2014.

110. Baraldi, D., Heitsch, M., and Wilkening, H., CFD simulations of hydrogen combustion in a simplified EPR containment with CFX and REACFLOW, *Nuclear Engineering and Design*, 237, 1668, 2007.

111. Redlinger, R., DET3D-ACFD tool for simulating hydrogen combustion in nuclear reactor safety, *Nuclear Engineering and Design*, 238, 610, 2008.

112. Dorofeev., S.B., Kuznetsov, M.S., Alekseev, V.I., Efimenko, A.A., and Breitung, W., Evaluation of limits for effective flame acceleration in hydrogen mixtures, *Journal of Loss Prevention in the Process Industries*, 14, 583, 2001.

113. Xiong, J., Yang, Y., and Cheng, X., CFD application to hydrogen risk analysis, *Science and Technology of Nuclear Installations*, 2009. doi: 1155/2009/213981.

114. Heitsch, M., Huhtanen, R., Techy, Z., Fry, C., and Kostka, P., CFD evaluation of hydrogen risk mitigation measures in a VVER-440/213 containment, *Nuclear Engineering and Design*, 240, 385, 2010.

115. Prabhudharwadkar, D.M. and Iyer, K.N., Simulation of hydrogen mitigation in catalytic recombiner, Part II: Formulation of a CFD model, *Nuclear Engineering and Design*, 241, 1758, 2011.

116. Wang, D. and Cao, X., Numerical analysis of different break direction effect on hydrogen behavior in containment during a hypothetical LOCA, *Annals of Nuclear Energy*, 110, 856–864, 2017.

117. Li, H., Cao, X., Liu, Y., Shao, Y., Nan, Z., Teng, L., Peng, W., and Bian, J., Safety of hydrogen storage and transportation: An overview on mechanisms, techniques, and challenges, *Energy Reports*, 8, 6258–6269, 2022.

118. Zhou, S., Luo, Z., Wang, T., He, M., Li, R., and Su, B., Research progress on the self-ignition of high pressure hydrogen discharge: A review, *International Journal of Hydrogen Energy* , 47, 9460–9476, 2022.

119. Le, S.T., Nguyen, T.N., Linforth, S., and Ngo, T.D., Safety investigation of hydrogen energy storage systems using quantitative risk assessment, *International Journal of Hydrogen Energy* , 48, 2861–2875, 2023.

120. Hassan, Q., Sameen, A.Z., Salman, H.M., Jaszczur, M., and Al-Jiboory, A.K., Hydrogen energy future: Advancements in storage technologies and implications for sustainability, *Journal of Energy Storage*, 72, 108404, 2023.

121. Lees, F.P., *Loss Prevention in the Process Industries*, Butterworth Heinemann, Oxford, 1996.

6 Hazards of Hydrogen Use in Transportation

6.1 GENERAL CONSIDERATIONS

As early as 1978, Professor Trevor Kletz introduced the principles of inherent safety (see Chapter 7 for a detailed presentation). One of his ideas was that "What you do not have can't leak" [1]. Nevertheless, as long as we need a source of energy to move our cars, we have to choose among gasoline, diesel fuel, liquid propane gas, natural gas, hydrogen, or some other source. Once we choose hydrogen as a motor fuel, we need to identify principles of inherent safety to cope with the hazards stemming from this new fuel; for instance, minimization of stored quantities and the length of stay in the hazardous zone, especially in refueling stations.

It should be noted also that, although hydrogen has been proven relatively safe in the chemical and aerospace industries, its anticipated widespread application in the automotive industry and, consequently, in our everyday lives will necessitate the handling of large quantities of hydrogen in gaseous or liquid states by unskilled personnel. This will considerably increase the frequency of small to medium-sized accidents. Obviously, significant attention should be paid to the design of inherently safer hydrogen-fueled cars, safer storage systems onboard, and refueling stations that take into account human error.

Another safety issue is the transport of huge quantities of hydrogen by tankers that have proven quite unsafe on roads, as expressed by their high accident rates. Alternatively, pipeline systems look much safer, and these are to be used in the near future for the mass transport of hydrogen, either in gaseous or liquid state. Hydrogen pipelines with operating pressures up to 100 MPa have been in use in Europe for many decades with no accident reports. Of these pipelines, the longest ones have been operating in Germany (215 km) since 1938 and in France (290 km) since 1966 [2].

Bethoux [3] states that a pipeline is the least expensive means of transmission in terms of excessive flow rates but is best for modest distances because the auxiliaries consume 1.4% of the lower calorific value of hydrogen every 150 km. This should be compared with 2.3% of losses inside the entire French transmission and distribution energy network according to RTE (the power transmission system operator of today's France).

Lins and Almeida [4] propose a multicriteria decision tool for assessing the risk in hydrogen pipelines. By way of incorporating the decision maker's options into

DOI: 10.1201/9781003313007-6

the decision process, the model allows sections of the pipeline to be ranked in terms of likely risk. The proposed model is an evolution of the preceding research in the field.

Regarding overland transport of hydrogen, this can be done both by way of the street (trucks) or rail (trains) via racks of metal cylinders of compressed gaseous hydrogen (200 bar) and tanks of liquid hydrogen, with an alternative pressure of as much as 700–900 bar. The cylinders are more costly but their use is greener. This has an enormous effect on the price of the hydrogen distributed and also a prime environmental effect because hydrogen truck transport emits about 1 kg CO_2 eq/kg H_2 transported 100 km, in line with Bethoux [3]. As a consequence, it is recommended to stay well within a threshold distance of 100 km. Therefore, road transport must be reserved for the final kilometers of delivery to the provider stations.

Hydrogen long-distance shipping is still in the undertaking and demonstration stage. This third option of maritime shipping targets transporting liquid hydrogen over long distances, intending to link low-density areas suitable for low-carbon power manufacturing (photovoltaics and wind electricity) with regions with growing energy demand. A cost-lowering option would be to use the hydrogen losses from cryogenic tanks as fuel to propel the ship [3].

6.2　GROUND TRANSPORTATION

6.2.1　Hydrogen Systems in Vehicles

Consistent with Verhelst [5], there are three potential options to permit carbon-free ground transportation: battery electric vehicles (BEVs), fuel cell electric vehicles (FCEVs or FCVs), and internal combustion engines (ICEs) through the usage of a combination of renewable fuels and/or carbon capture and storage. Hydrogen-fueled internal combustion engines (H_2-ICEs) have been thoroughly studied for many decades. Nevertheless, Verhelst [5] reviews the merits of H_2-ICEs but also what makes them doubtlessly unsuitable as a practical option. Also, the weight and recharging time of BEVs make them unfit for long-haul programs, leaving FCEV and ICE automobiles as the most effective answers with long-term business capability [6].

According to Kalghatgi [7], the BEV and the FCEV could replace ICEs, and gaseous fuels such as natural gas and biofuels are viable alternatives to conventional liquid fuels crafted from crude oil. However, these options for the present device all start from a very small base, face essential barriers to unrestrained and quick growth [8], and even by 2040 are not projected to account for more than around 10% of global transport. It is crucial that these alternatives undergo a life-cycle assessment to evaluate their environmental and other advantages, and that the impacts are not displaced from the engine tailpipe to someplace else. A possible viable alternative to the ICE is the very efficient fuel cell powered with the aid of hydrogen. However, this requires an extensive international hydrogen infrastructure to be built first. The process of refueling hydrogen is quicker and the range is longer for an FCEV in comparison to a BEV [8].

Kalghatgi [9] also reports that the UK authorities are thinking about banning the sale of any new car with an internal combustion engine beginning in 2035 and maybe even by 2030. Therefore, starting from the date to be set, only full electric vehicles, namely BEVs and FCEVs running on hydrogen, will be allowed to be produced and sold.

Weger et al. [10] investigated the anticipated effects on greenhouse gas (GHG) and air pollutant emissions due to a probable transition toward a hydrogen economy in German road transport. They studied the impacts of an anticipated whole transition from fossil-fueled to hydrogen-powered German delivery, investigating various scenarios. They concluded that German emissions range between −179 and +95 Mt CO_2 eq yearly, depending on the scenario adopted, with renewably powered electrolysis leading to the greatest emissions reduction, while electrolysis based on the fossil-fuel scenario to produce electricity leads to the greatest increase.

Jayakumar et al. [11] studied the probable techno-economic advantages of utilizing hydrogen for cars in preference to battery electric-powered vehicles or conventional internal combustion engine cars. Their assessment demonstrates that hydrogen production, storage, and distribution costs are the predominant challenges, and extensive development is still necessary for hydrogen to compete with either the internal combustion engine automobile or the battery electric-powered automobile. The secondary demanding situations have additionally been presented in this study, including the cost of the fuel cell stack (the assembly of fuel cells that use hydrogen and oxygen to produce electricity) and the changes associated with internal combustion engine automobiles, in addition to regulatory and safety issues, which obstruct the contemporary sizable utilization of hydrogen.

The objective of the study performed by Sinigaglia et al. [12] is to review the feasibility and impacts of all stages from production to the very last use of hydrogen as a substitute for fossil fuels for transport purposes. This article presents a general discussion of the route hydrogen goes through from its production until its utilization, examining the related technological, economic, and environmental issues for the sustainability of the hydrogen economy. Furthermore, it provides the primary demanding situations and research fields that need greater engagement by researchers and political decision makers. The results imply that hydrogen production strategies are not sufficiently advanced to be competitive. The production methods displaying the best results are hybrid production methods, accompanied by thermal and photonic. The main difficulty in terms of storage is obtaining a sufficient volumetric density. The volumetric density of the current compression storage is about 5.7 wt% capacity. According to Sinigaglia et al. [12], fuel cells need to attain extended durability (current durability is approximately 120,000 km) and lower production costs to be competitive with fossil-fueled mobility.

Milojevic [13] states that the usage of compressed gaseous hydrogen as an alternative fuel is an effective, currently available way to assist in solving environmental issues. As regards safety, hydrogen has advantages compared to gasoline and diesel because it is not toxic, neither is it carcinogenic nor a corrosive gas, and it does not contaminate ground or water in the event of a release. The anticipated use of hydrogen in cars should be based on suitable hydrogen engine technology. The lean-burn heavy-duty hydrogen engines examined are promising because of their lower NO_x emissions and higher fuel efficiency compared to stoichiometric engines. Nevertheless, to meet the most stringent Euro VI emission standards for NO_x, it is vital to convert to stoichiometric combustion combined with exhaust gas recirculation and a three-way catalyst after treatment.

Jeerh et al. [14] have reviewed recent progress in ammonia fuel cells and their probable applications. They deal with different types of ammonia fuel cells, namely oxygen anion-conducting electrolyte-based solid oxide fuel cells (SOFC-Os), proton-conducting electrolyte-based solid oxide fuel cells (SOFC-Hs), alkaline fuel cells (AFCs), and alkaline membrane-based fuel cells (AMFCs), all of which have been assessed against the effects of electrolytes, electrocatalysts, and operating temperature. The proficiency of each technology and its eventual integration into real applications have been also briefly reviewed. These authors found that much research has been carried out for SOFCs, specifically SOFC-Hs, which are the most promising alternative to ammonia fuel cell technology up to this point. Lately, low-temperature ammonia fuel cells have also been studied which may be compatible with a variety of small–medium power range devices. However, progress within the commercialization stage remains lagging and there are several challenges that need to first be addressed before this technology is fully appreciated. The power density and stability of low-temperature ammonia fuel cells mainly need to be improved before broad commercial application. Nevertheless, ammonia fuel cells provide a safer and dependable energy supply, which can outweigh many of the limitations associated with direct hydrogen use.

Of the numerous choices for alternative clean fuels that are both energy efficient and environmentally friendly, hydrogen has continually been considered the best clean alternative. Ravi and Aziz [15] reviewed many available options for using hydrogen for mobility applications and discussed their relative advantages and weaknesses. In addition to properly studied methods like FCEVs, hydrogen-based ICEs, and dual-fuel operation with hydrogen, this work also assesses the technical and economic feasibilities of using hydrogen in the generation of electrofuels and their implications for current infrastructure and future energy demands.

Electrofuels, also known as e-fuels or synthetic fuels, are considered replacement fuels. Captured carbon dioxide or carbon monoxide and green hydrogen are used to produce electrofuels, thus resulting in a low carbon footprint. The most widely known electrofuels are butanol, other alcohols, and biodiesel. The European Federation for Transport and Environment has suggested e-kerosene to be used in the aviation sector aiming at considerably limiting the climate impact of aviation [16].

6.2.1.1 Hydrogen in Internal Combustion Engines

According to the US Department of Transportation [17], hydrogen may be used in one of three ways to power vehicles:

- As a replacement for gasoline or diesel fuel in an internal combustion engine,
- As a supplement to gasoline or diesel fuel used in an internal combustion engine, or
- To produce electricity in a fuel cell.

Internal combustion engines based on the Otto and Diesel cycles can be operated with hydrogen or hydrogen mixtures with other liquid fuels. Ricardo (1924) and Burstall (1927) were the first to carry out investigations on using hydrogen as a fuel in vehicles, but it was Erren (1930) who performed intensive studies on hydrogen–air and hydrogen–oxygen motors. The first internal combustion engine car in the United States fueled by hydrogen was presented by Billings in Utah in 1966. Today, significant developments are underway in Germany and Japan [17].

Hydrogen cannot be used directly as a fuel for replacement of gasoline; therefore, engine modifications are necessary. The problematic hydrogen properties to be taken into account are its low ignition energy and high flame propagation speeds that can cause self-ignition during the mixture preparation or flame flashback. The octane number of hydrogen is much lower than that of gasoline and this can cause low performance and fast wearing out of the engine. Nevertheless, the wide ignition range of hydrogen allows burning of lean fuel–air mixtures, and this gives a large control range. Uncontrolled premature ignition or even flashback into the intake manifold is usually prevented by adding water as a ballast and also by timed individual port injection of hydrogen close to the cylinder intake. The only noxious gases found in hydrogen-fueled engine emissions are nitric oxides, and many papers have been published on this issue.

The schedules developed for hydrogen admixture in fossil fuels of ICEs include port fuel injection (PFI) and direct injection (DI) [18, 19]. The PFI option offers excellent load efficiency with extremely low emissions, however, transforming a fuel spark ignition engine into an H_2-ICE lowers its power output. Furthermore, the DI concept leads to excellent efficiency with slight emissions [20].

Xu et al. [21] obtained low emissions on a hydrogen-fueled spark ignition engine at the cold initial period under rich combustion via ignition timing management. The average NO_x emissions were successfully reduced by 90.2%, and unburnt hydrocarbons (HCs) and carbon monoxide (CO) emissions created by the evaporated lubricant oil reduced respectively by 33.8% and 19.7% in the first 6 s throughout the cold start procedure with the delay of the ignition timing control.

According to Fromm et al. [22], another option for using hydrogen in an internal combustion engine is the *opposed-piston (OP) engine*. These engines were first manufactured in the 1890s and have been used since then in many ground, marine, and aviation programs. The most noteworthy OP engines have been the Junkers Jumo 205 (Wilkinson 1940) and Jumo 207. These are German engines that were used before and all through World War II on planes known for their exceptional power and performance.

In OP engines there are two pistons in every cylinder, with the piston crowns opposite to each other. The engines have two crankshafts connected by a gear train to combine torque and relate piston motion. The piston crowns form the combustion chamber when the pistons approach the middle of the cylinder. OP engines function usually on the two-stroke concept. The intake ports are placed at one end of the cylinder and the exhaust ports are at the other end of the cylinder. The motion of the pistons opens and closes the ports, thus enabling the gas exchange with efficient, uniflow scavenging. This configuration results in the avoidance of cylinder

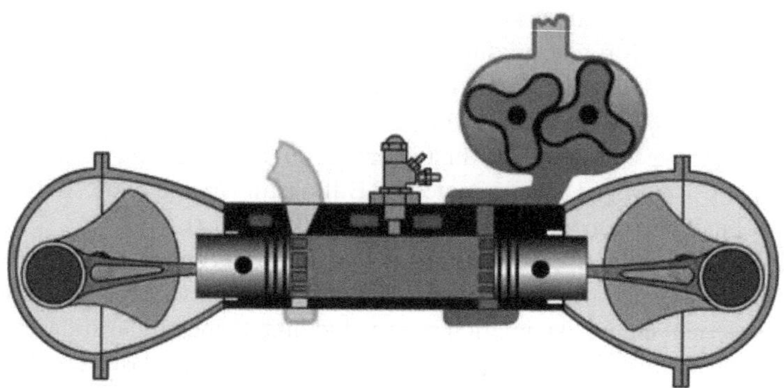

FIGURE 6.1   Opposed-piston (OP) engine schematic. Open access article distributed under the terms of the Creative Commons Attribution License 4.0.

Source: [23].

valves. Consequently, valves, valve springs, valve keepers, camshafts, pushrods, rocker arms, and other parts necessary to control gas exchange in traditional engines are all removed. Most OP engines function with diesel fuel and compression ignition. The OP engine has lower heat loss (approximately 30% less) than similar four-stroke engines due to a reduced surface area/volume ratio of the combustion chamber. Other benefits are the reduced power loss to cooling fans and the improved car aerodynamics as a result of smaller radiators. An example of opposed-piston architecture is shown in Figure 6.1 [22].

Achates Power and other ICE manufacturers long ago started state-of-the-art research on hydrogen combustion in OP engines. In these engines, an air/fuel ratio (λ, lambda number) of 2.2 or greater creates limited quantities of nitrogen oxides, and λ is related to the stoichiometric air/fuel ratio (see also Chapter 4). When $\lambda=1$, ideally balanced combustion conditions result in neither oxygen starvation nor excess (zero oxygen balance). The ability of an OP engine to function with lean fuel mixtures, producing negligible NO_x, particulate matter (PM), HC, and CO_2 makes it a clean engine with near-zero emissions of hydrogen combustion. Hydrogen opposed-piston engines have an inherent advantage over conventional internal combustion engines, with an efficiency approaching or exceeding that of fuel cells [6, 23]. Another advantage of the OP engines is their ability to operate with 98% to 99% hydrogen purity level, whereas 100% purity is needed in fuel cells. Consequently, the hydrogen used in fuel cells is much more expensive than that accepted by OP engines.

6.2.1.2   Electric and Fuel Cell Vehicles

The conversion of the chemical energy of hydrogen to mechanical energy is obtained through fuel cells to generate heat, water, and electricity with the lowest environmental impact [24]. Comparing FCEV technology with internal combustion engines, fuel cells are more efficient and less polluting, with no vibration and noise [25]. The major fuel cell types are, according to Kirubakaran et al. [26]:

- Proton exchange membrane fuel cell (PEMFC).
- Alkaline fuel cell (AFC).
- Solid oxide fuel cells (SOFC).
- Molten carbonate fuel cell (MCFC).
- Direct methanol fuel cell (DMFC).
- Phosphoric acid fuel cell (PAFC).

The pressures usually applied to deliver hydrogen are 350 and 700 bar, in vessels wrapped with carbon fibers. Small cars store hydrogen gas onboard in cylinders with a capacity of 4–6 kg and at 700 bar pressure. Concerning bigger vehicles using FCEV technology, such as trucks and forklifts, the pressure used is more often 350 bar. Vehicles using FCEV technology are more expensive compared to those with ICEs and also BEVs [24].

As Field [27] states, electric vehicles equipped with batteries, and FCEVs, are quite different, having different characteristics, advantages, and disadvantages. The batteries of electric cars are too heavy to be carried far while the as-yet insufficiently developed hydrogen cars are currently too expensive.

As stated by Thompson et al. [28] and Bethoux [3], proton exchange membrane fuel cell (PEMFC) technology has developed to be used in fuel cells (FCs) and fuel cell systems. Among the known applications of PEMFCs are space flight auxiliary power units, silent electric generators, hydrogen-powered domestic boilers and electric micro-cogeneration, and forklifts. This new technology is considered sufficiently mature now to be applied to road cars, and many prototype FCEVs have been designed and tested successfully. Based on 500,000 FC systems/yr, their cost was USD124/kW$_{elec}$ in 2006, dropping to USD45/kW$_{elec}$ in 2017. This latter value is quite competitive with the ICE vehicle cost (USD30/kW$_{elec}$), as shown in Figure 6.2.

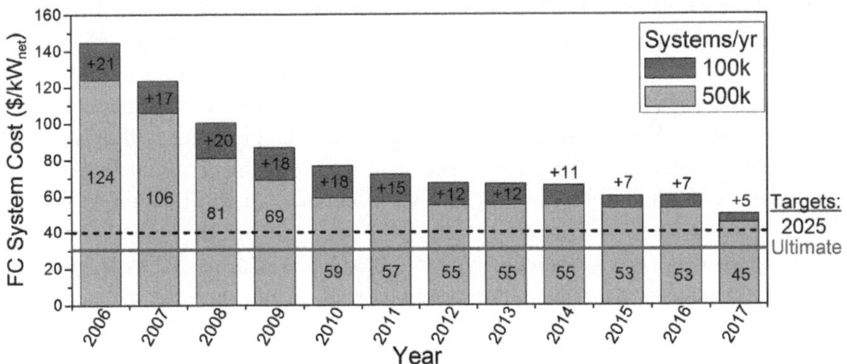

FIGURE 6.2 Total 80 kW$_{net}$ automotive FC system cost estimates at high annual production volumes (100,000 and 500,000 systems/yr). The 2025 and ultimate system cost targets are USD40 and USD30/kW$_{net}$. Previous and current year values in nominal USD; target values in out-year USD. With permission from Elsevier.

Source: [28].

TABLE 6.1
Pros and Cons of Hydrogen Fuel Cell Cars. With permission

Hydrogen Car Pros	Hydrogen Car Cons
Greener compared to conventional cars	High purchase price
Use of renewable resources	High depreciation
Efficient power	Charging issues
No noise pollution	Lack of infrastructure
Short fueling time	Limited selection of cars
Long range	Technology not mature yet
Absence of fumes	Large investments for R&D necessary
No air pollution	Hydrogen fuel is quite expensive
Less greenhouse gas emissions	Safety issues
Suitable alternative for commuting	Not entirely green yet

Source: [30].

Considering the drawback of the high purchase cost of a hydrogen car currently, we should recall that the first microwave oven on the market cost around USD12,000 in 1955, while today you can buy one for USD50, as *Bill Gates* stresses [29]. So, he suggests that a research and development acceleration of advanced clean-energy technologies would result in cost-competitive clean products, for instance, less expensive FC systems. To this end, he believes that a new pattern has to be created, which he calls *Breakthrough Energy Catalyst*, consisting of philanthropists, companies, and governments aiming at lowering production costs of the most important innovative clean technologies.

Andreas [30] has summarized the main pros and cons of hydrogen fuel cars, as shown in Table 6.1, and concluded that hydrogen FC cars could replace ICE cars at some point in the future. Yet, since this technology is still lagging, greater efforts are needed to accomplish the target set. This is why electric cars will dominate the market in the near future. For the time being, battery and FC technologies will exist side-by-side taking into account their similarities, with BEVs being more cost-effective for short-range and small vehicles, and FCEVs more suitable for medium-to-large and long-range vehicles [31–33].

As regards *hydrogen's efficiency problem*, Baxter [34] claims that hydrogen cannot be efficient because, in this case, the energy must go from electricity to hydrogen gas and back to electricity if we want to move a car. This is known as *energy vector transition*, and to support this opinion he uses the following example:

Let's take 100 watts of electricity produced by a renewable source such as a wind turbine. To power an FCEV, that energy has to be converted into hydrogen, possibly by passing it through water (the electrolysis process). This is around 75% energy-efficient, so around one-quarter of the electricity is automatically lost.

The hydrogen produced has to be compressed, chilled and transported to the hydrogen station, a process that is around 90% efficient. Once inside the vehicle, the hydrogen needs converted into electricity, which is 60% efficient. Finally, the electricity used in the motor to move the vehicle is around 95% efficient. Put together, only 38% of the original electricity – 38 watts out of 100 – are used.

And Baxter continues as follows:

With electric vehicles, the energy runs on wires all the way from the source to the car. The same 100 watts of power from the same turbine loses about 5% of efficiency in this journey through the grid (in the case of hydrogen, I'm assuming the conversion takes place onsite at the wind farm).

You lose a further 10% of energy from charging and discharging the lithium-ion battery, plus another 5% from using the electricity to make the vehicle move. So you are down to 80 watts... In other words, the hydrogen fuel cell requires double the amount of energy.

On the other hand, as Arnold [35] supports, hydrogen has the enormous advantage of production during an oversupply of electricity grids when the energy is not utilized by other consumers. In addition, the surplus hydrogen that is produced as a byproduct by several industries, and is a burden to them, can be used by hydrogen fuel cell vehicles after proper cleaning. Considering also the risks of hydrogen fuel cell cars, the hydrogen stored in liquid form in thick-walled tanks is particularly safe. Numerous crash tests have confirmed the safety of hydrogen storage vessels onboard and the related safety systems.

As stated conclusively by Goyal [36], hydrogen fuel cells are among the best decarbonization means to address global warming provided the hydrogen is produced from renewable energy. Nevertheless, there are still some significant issues, including the cost to be addressed before the market accepts it. When these problems are solved, hydrogen fuel cells will offer a truly renewable and clean power source both for stationary and mobile applications in the coming future.

6.2.1.3 Hydrogen Storage in Vehicles

The current possibilities for storage of hydrogen in vehicles are in the gaseous state inside pressure vessels, in the liquid state in vacuum-insulated tanks, or absorbed in metal hydride storage tanks. Unfortunately, none of these hydrogen storage systems offers the irrefutable advantages of gasoline as a fossil energy carrier. Pressurized gas storage vessels are heavy and occupy a large volume on a vehicle. In addition, the high storage pressure of >20.0 MPa raises serious safety concerns.

Liquid hydrogen tanks are larger than gasoline tanks and require complicated construction in order to maintain very low storage temperatures. Safety problems arise in this case too, stemming from the cryogenic conditions and evaporation loss.

Hydride-forming metals may be either of the low-temperature type, such as FeTi, with a reversible hydrogen absorption temperature range of 25°C to 100°C, or of the high-temperature hydride material type, such as Mg_2Ni, desorbing hydrogen at much higher temperatures (about 350°C). The latter materials have a higher hydrogen capacity, but this comes at the expense of more energy to desorb the stored hydrogen and with increased safety considerations due to a much higher operating temperature.

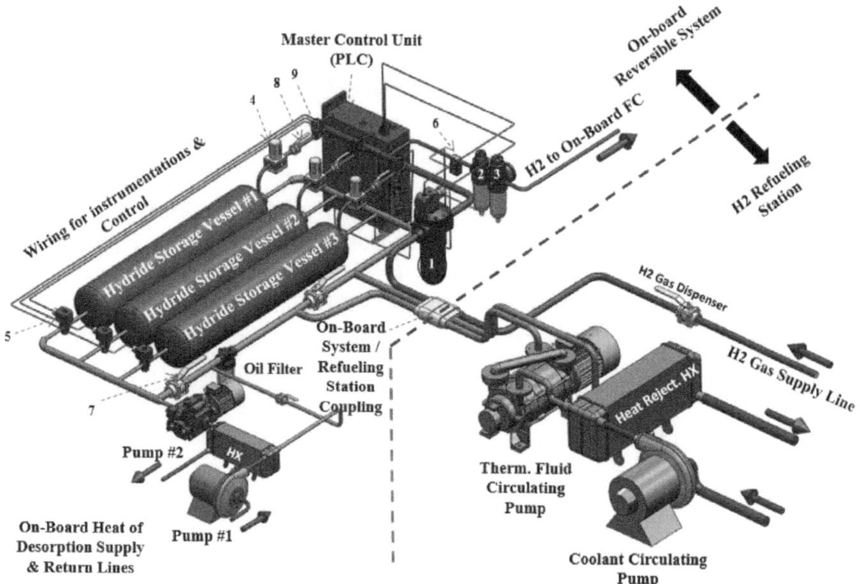

FIGURE 6.3 Baseline design of an onboard reversible hydrogen storage system. With permission from Elsevier.

Source: [37].

Khalil [37] proposed a set of materials tests (based on ASTM and United Nations testing standards) aimed at examining hydrogen storage materials such as $NaAlH_4$, AlH_3, $2LiBH_4+MgH_2$, $3Mg(NH_2)_2\cdot8LiH$, NH_3BH_3, and activated carbon (Maxsorb AX-21). These tests were scheduled to imitate conditions that the hydrogen storage materials may come up against during hypothesized accident scenarios, such as hot-surface touch, mechanical impact, and rapid depressurization. The test results revealed possible fire and explosion hazards under these conditions, which could be useful for risk scenario quantification, the characterization of a material's hazard class, and preparation of new hydrogen safety codes and standards. Concerning hazard mitigation, it was found that powder compaction could suppress the pyrophoricity of hazardous hydride powders used to store hydrogen. It was also found that (NH_4) H_2PO_4 is an effective flame retardant if added to the hydrides such as $NaAlH_4$ powder before compaction. The activated carbon Maxsorb AX-21 was proven a safer hydrogen storage medium in comparison with hydrides. Some hydrogen storage media like sodium alanate (sodium aluminum hydride, $NaAlH_4$) can be recharged safely onboard, whereas others such as ammonia borane (NH_3BH_3) and alane (AlH_3) should be recharged offboard the vehicle. Finally, Khalil [37] deployed a meaningful design for an onboard reversible rechargeable hydrogen storage system (Figure 6.3).

6.2.1.4   Hydrogen Tanks Onboard

In the United States, the Motor Vehicle Fire Research Institute (MVFRI) has contracted with Southwest Research Institute (SwRI) to perform testing of fuel systems. Zalosh,

TABLE 6.2
Different Types of Pressure Vessels for Hydrogen Storage. With permission from Elsevier

Types of Pressure Vessels	Weight	Cost	Pressure
Type I: Full metal pressure vessels	Heaviest	The lowest cost	up to 200 bar
Type II: Steel vessel with a glass-fiber composite layer added around the steel	30–40% lighter than Type I	50% more than Type I	300 bar
Type III: Fully wrapped vessels with composite and metal liner	70% lighter than Type I	About twice the cost of Type II	350–700 bar
Type IV: Full composite	80% lighter than Type I	Higher cost than Types I–III	up to 1000 bar
Type V: Linerless fully composite pressure vessel	85% lighter than Type I	--	--

Source: [24].

of the MVFRI, conducted a study on compressed natural gas (CNG) and hydrogen vehicle fuel tank failure incidents, in which he gives useful recommendations for both fuels when stored onboard in cylindrical pressure vessels [38].

According to Abohamzeh et al. [24], delivery cost in ground transportation of hydrogen storage vessels is lower due to their low weight and great capacity. Hydrogen is stored as a gas in five types of pressure vessels, as presented in Table 6.2. Type I vessels, manufactured from steel or aluminum, have the lowest cost but are the heaviest. Type II vessels, manufactured by a combination of a composite and a metal, are 50% more expensive than Type I vessels, but their weight is 30–40% lower. Type III vessels are made with a metal liner and their cost is about twice the cost of type II, but their weight is approximately 50% lower. Type IV vessels are even more improved and cost much more. They are manufactured from carbon-glass composites and a high-strength polymeric liner. Type V vessels are made without a liner and are fully composite pressure vessels. They are as yet still experimental [39].

6.2.1.5 Hydrogen Permeation

Considering safety, permeation of hydrogen through materials, which is caused by the small size of its molecules that can migrate through various storage materials or deteriorate their properties by embrittlement, may become a critical hydrogen property. As regards compressed gaseous hydrogen storage systems, permeation is determined as the flow rate of hydrogen passing through the walls of vessels, interface materials, or piping. Permeation increases with increasing material aging, temperature, and storage pressure. The permeation rate in pressure vessels of Types I, II, and III is insignificant. Losses of hydrogen might happen within several years, and this does not usually pose any safety hazards for common applications. However, it is a safety issue for pressure vessels with polymeric (nonmetallic) liners (pressure vessels of Type IV), which present higher rates of hydrogen permeation [24]. Efforts are made to determine an upper limit for hydrogen permeation in vehicles. To estimate the allowable permeation rate, several factors should be taken into account, such as safety, environmental issues, vehicle scenarios, and hydrogen dispersion patterns. In general, the permeation rate is significantly lower in new containers, whereas in older vessels, possible micro-cracks due to material aging affect the resin/carbon fiber matrix, thus increasing the permeation rate of hydrogen.

6.2.1.6 Liner Blistering in Pressure Vessels

The Type IV pressure vessels are manufactured by an overwrap of carbon fiber composites, a polymeric liner, and at least one boss. Under high pressures, hydrogen gas may be absorbed by the polymeric liner, and if the depressurization rate becomes higher than the rate of desorption of the absorbed gas, liner blistering may occur [40]. A prediction model for blistering in a Type IV hydrogen storage tank has been developed by Yersak et al. [41]. Their model has shown good agreement with experimental results.

A setup has been built by Pepin et al. [42] to reproduce liner blistering that can result in liner failure. Liner deformation as well as composite cracks were observed with the tomographic examination of collapsed samples. The authors found that the collapses occur at the liner/composite interface and in regions that are not sufficiently bonded. It was also found that blistering in the Type IV hydrogen tank depends on working conditions such as including dwell time at residual pressure level before emptying initial filling pressure, the emptying rate, and the initial filling pressure. The main collapse after the decompression test resulted in a void 1.54 mm high, which after seven days had decreased to 0.51 mm [42].

6.2.1.7 Embedded Emissions

Generating a car calls for numerous strategies. Let us observe the cylinder block of an ICE as an example. If virgin aluminum is needed, bauxite ore needs to be mined first. Next, the ore should be transported to a refining facility. It would be treated by the Bayer and Hall-Héroult process to transform the bauxite ore into aluminum oxide and in the end into aluminum via the electrolytic reduction technique. Then, the aluminum will be transported to the casting installation to produce cylinder blocks before being machined, assembled into the ICE, and then fixed into position in the car. Each step of this cycle calls for energy, and normally there will be a carbon footprint produced during each step. This carbon footprint is otherwise referred to as *embedded emissions*. These are emissions produced to bring the car into the showroom in advance of driving a single passenger mile, as stressed by Conway [43].

The majority of life-cycle analysis (LCA) studies show that hybrid vehicles (in particular hybrid electric vehicles, which amalgamate an ICE system with an electric-powered system), FCEVs, and BEVs all have higher embedded emissions than an equivalent ICE, mainly because of the lithium-ion battery pack. This is attributed to the high energy requirements to produce a lithium-ion battery pack, specifically the cathode production. The cathode chemistry usually utilized in battery packs of early BEVs is lithium–nickel–manganese–cobalt–oxide chemistry with a high cobalt cost and low metal availability worldwide. Although BEVs and plug-in hybrid electric vehicles move toward different cathode chemistries, graphite anodes remain the dominant choice [43].

6.2.1.8 Safety Comparisons of Hydrogen, Methane, and Gasoline

Substitution of conventional fuels by alternative energy carriers has been implemented to some extent by the introduction of natural gas as a generalized fuel in the world market. Its use is not limited to industry and the home, but extends to public means of transportation, especially in Europe. The prospects for hydrogen use are similar to those of natural gas, and a proposal for their combined use has been made. Besides the techno-economic and environmental advantages discussed for hydrogen application, another significant issue is that of the comparative safety between natural gas and hydrogen concerning the application, transport, and storage procedures.

Thermophysical, chemical, and combustion properties of hydrogen, methane, and gasoline are given for comparison in Table 3.1. Of these fuels, gasoline is certainly the easiest and perhaps the safest fuel to store because of its higher boiling point, lower volatility, and narrower flammability and detonability limits. This generalized consideration is based on our previous discussion concerning fire and explosion hazards. Nevertheless, hydrogen and methane (the principal ingredient of natural gas) can also be safely stored using current technology.

Despite its volumetric energy density, hydrogen has the highest energy-to-weight ratio of any fuel. Unfortunately, this weight advantage is usually overshadowed by the high weight of the hydrogen storage tanks and associated equipment. Thus, most hydrogen storage systems designed for transport applications are considerably bulkier and heavier than those used for liquid fuels such as gasoline or diesel. A comparison of the hazards posed by hydrogen, methane, and gasoline follows [44, 45]:

- *Size of molecules.* Since the hydrogen molecule is the smallest of all, it will leak through permeable materials, whereas methane and gasoline will not, but the difference in leakage rates is very low. Hydrogen has approximately three times the energy of methane by mass, but one-third the energy of methane by volume. Therefore, at the same pressure, three times the volume of hydrogen will have the same total energy content as methane. For pinhole-size leaks from high-pressure systems, about three times the volume of hydrogen will leak over methane.
- *Fuel spills.* In the event of a fuel spill, it is expected that a fire hazard will develop most rapidly, in descending order, for hydrogen, methane, and gasoline. With regard to fire duration, gasoline fires last the longest and hydrogen fires are the shortest, while all three fuels burn at nearly the same flame temperature. In fact, for spillage of identical liquid fuel volumes, hydrocarbon fires will last five to ten times longer than hydrogen fires.
- *Odorization.* Natural gas is odorized so that leaks can be detected, whereas gasoline is normally odorous. Since natural gas distribution piping exists in so many places and natural gas is piped into buildings and homes, odorization is a prudent although not entirely effective safety measure. Leaks will only be detected if someone is present to smell them and respond. Hydrogen as an

industrial gas or fuel cell vehicle fuel is not odorized because sulfur-containing substances (mercaptans) contaminate the catalysts of a fuel cell.

- *Buoyancy.* Hydrogen is 14.5 times lighter than air at normal temperature and pressure, whereas methane is 1.8 times lighter and gasoline vapor is heavier than air. Thus, hydrogen will rise much more quickly causing greater turbulent diffusion, which reduces its concentration to below the lower flammability limit (LFL) more rapidly. Moreover, hydrogen diffuses into air approximately four times faster than methane and 12 times faster than gasoline, thus causing rapid concentration decrease to safe levels.

- *Energy of explosion.* The energy of explosion values given in Table 3.1 should be considered as the theoretical maxima, and yield factors of 10% are considered reasonable for fuel–air explosions. For equivalent volume storage, hydrogen has the least theoretical explosive potential of the three fuels considered, although it has the highest heat of combustion and explosive potential on a mass basis.

- *Flammability and detonability limits.* The broader flammability and detonability limits of hydrogen coupled with its rapid burning velocity render hydrogen a greater explosive threat than methane or gasoline. The LFLs for hydrogen and methane are similar (4% for hydrogen and 5% for methane). However, hydrogen has a much wider range between the LFL and the lower detonability limit (LDL) than methane (4 to 18% for hydrogen versus only 5 to 5.7% for methane). This means that a hydrogen gas concentration of more than three times that of methane is required to produce a detonable mixture. For safety reasons, LFL is used normally in place of LDL for hydrogen, which provides an additional safety factor. A gas concentration in air equal to 25% of the LFL for hydrogen corresponds to 1% hydrogen in air, but 25% of the LDL is 4.5% hydrogen in air. Therefore, gas detection using LFL gives an earlier warning for a hydrogen detonable mixture than a methane detonable mixture.

- *Ignition energy.* At concentrations up to approximately 10% hydrogen and methane in air, hydrogen has the same ignition energy as methane. As the hydrogen concentration increases toward a stoichiometric mixture of 29% hydrogen in air, the ignition energy drops to about one-fourteenth of that for methane and one-twelfth of gasoline. Since we are generally concerned with the prevention of ignitable mixtures, the LFL is the important property. However, the energy levels required for the ignition of these fuels are so low that common ignition sources, such as static electricity discharges from a human body, will ignite any of these fuels in air.

- *Autoignition temperature.* Hydrogen and methane possess unusually high autoignition temperatures (585°C and 540°C, respectively), whereas gasoline, with autoignition temperature ranging from 227°C to 477°C, appears more hazardous.

- *Deflagrations.* A confined deflagration of hydrogen–air or methane–air will produce a static pressure rise ratio of less than 8:1. Explosion pressures for confined deflagrations of gasoline–air are about 70 to 80% of those for hydrogen–air. Unconfined deflagration overpressures are usually <7 kPa. However, a pressure of 3 to 4 kPa is sufficient to cause structural damage to buildings; therefore, unconfined large volume gas-phase explosions can be destructive. Thus, it is apparent that confined deflagrations with up to 8 atm (811 kPa) of explosion pressure can be devastating, and even unconfined deflagrations can cause slight-to-moderate structural damage, and injure people via fire and window-glass shrapnel.

- *Detonations.* Pressure rise ratios of ~15:1 for hydrogen–air or methane–air detonations and a ratio of ~12:1 for a gasoline–air detonation are normally expected. The impulse created by the explosion pressure profile should be taken into account in evaluating explosion damage and in the design of barricades or other structures constructed to mitigate explosion consequences. Hydrogen burning speed (time to peak pressure) is ten times greater than that of methane. This indicates that a hydrogen detonation will be of much greater severity, yet with shorter positive phase duration and peak explosion overpressure close to that of methane. As a result, material constructions should respond to the same overpressure in a shorter time period.
- *Shrapnel hazard.* This depends on explosion overpressure and with ordinary enclosures *(L/D* <30) is about the same for hydrogen–air and methane–air and somewhat less severe for gasoline–air mixtures. Nevertheless, in long structures such as tunnels or pipes, hydrogen poses a greater explosion risk than the other two fuels owing to its greater tendency for a deflagration-to-detonation transition (DDT). Thus, hydrogen presents the greatest hazard of shrapnel damage.
- *Radiant heat.* Owing to the heat-absorbing water vapor created during hydrogen combustion and the absence of a carbon combustion reaction, the radiant heat from a hydrogen fire is significantly less than that of a hydrocarbon fire, and this reduces the risk of secondary fires. Combustible materials may actually be placed closer to a hydrogen flame than a methane flame. The reduced radiant heat means less heating of adjacent equipment in case of a major fire and hence a lower probability of a domino effect that leads to escalating damage and losses.
- *Hazardous smoke.* The potential for smoke inhalation damage is judged to be most severe, in descending order, for gasoline, methane, and hydrogen fires.
- *Flame visibility.* Unlike visible methane and gasoline flames, hydrogen burns with a near-invisible flame in daylight, although contaminants in air generally add some visibility. Nevertheless, hydrogen flames are visible at night, and modern detection equipment can detect them even in daylight.
- *Firefighting.* Normally, hydrogen and methane fires should be allowed to burn until gas flow is stopped or until liquid spills are consumed because of the potential explosive hazard resulting from extinguishing these fires. However, the fire should be controlled by cooling the storage tanks with water in all situations. Dry chemicals and high-expansion foams can be used to extinguish methane and gasoline fires.

In conclusion, hydrogen has been used and stored safely in industry for quite a long time as compressed gas or liquefied hydrogen, and it seems that metal hydride storage will be equally safe or even safer. Consideration of future hydrogen applications reveals apparently manageable safety problems in the industrial and commercial markets. Although hydrogen safety problems have been efficiently controlled in industry, additional safety analyses are needed in the transportation and residential fuel markets.

6.2.1.9 Resistance of Hydrogen Storage Vessels to High Temperatures and Fire

Polymeric materials are drastically more vulnerable to high temperatures in comparison to metallic materials, with the maximum operating temperature being the crucial factor for their use in pressure vessels. Therefore, their resistance to high temperatures and fire is extremely important, above all for onboard utilization. Ruban et al. [46] performed a bonfire test on fully composite vessels and found that the pressure increase in the vessels prior to leakage or bursting was relatively small (up to 12.7%), but the bursting delay was between 6 and 12 min, which was not acceptable.

Hupp et al. [47] investigated the effect of fire intensity, fire impingement area, and internal pressure on the fire resistance of composite pressure vessels constructed to be used onboard cars. These authors also performed bonfire tests on 7.5 L pressure vessels. The evaluation of the results showed that the time the vessels resisted fire was shorter with higher vessel surface temperature, larger fire impingement area, and higher initial hydrogen pressure.

Saldi and Wen [48] used both a finite volume CFD and a finite element model they developed for the simulation and evaluation of a Type IV pressure vessel response to an external propane fire and succeeded in simulating a published relevant experiment.

Magneville et al. [49] derived a thermo-mechanical law describing the behavior of a composite used to simulate the failure of a pressurized vessel. This behavior law took into account several physical phenomena such as fiber and matrix failure and plastic deformation. Validation of the law on wound carbon-epoxy structures was obtained via experimental tests. Moreover, this material law was also analyzed by a finite element model of the tank at three different temperatures, to investigate the thermal effect on the burst phenomenon.

A further study conducted by Zheng et al. [50] applied a three-dimensional CFD model to evaluate the characteristics of heat transfer in 700 bar hydrogen composite vessels exposed to localized fire. They observed that during 600 seconds of exposure to the localized fire, the rise in pressure and temperature was slow. The intensity of convective heat transfer in the internal hydrogen was higher than the conductive heat transfer in the walls of vessels when the area of flame impingement was far from the pressure-relief device. Furthermore, the heat transfer in Type III vessels was faster compared to Type IV vessels.

Zheng et al. [51] studied a Type III vessel experimentally and numerically to evaluate the air and hydrogen effect in the filling media. They developed a three-dimensional CFD model of the whole process of localized fire testing on the activation of a thermally activated pressure-relief device (TPRD) and observed that by extending the time of exposure to the localized fire, the activation time of the TPRD increases.

Hong et al. [52] evaluated the mechanical response to elevated temperatures of 70 MPa Type III hydrogen storage vessels manufactured with especially thick (around 40 mm) carbon/epoxy composite layers and an aluminum liner, as well as the material thermal degradation of composites. They used the ply-based modeling technique taking into consideration the variations in the winding angle and the composite thickness of the dome sections. Heat transfer analysis was applied to calculate the temperature distribution of composite layers during the pressurization process. The temperature-affected system properties were computed for each finite element to determine the stress distribution of the pressure vessel. The results showed that under the elevated temperatures, the winding pattern proved capable of resisting the 70 MPa pressure applied to the vessel, thus satisfying all design regulations for Type III pressure vessels.

Leh et al. [53] performed a failure analysis of a Type IV hydrogen composite vessel pressurized at 700 bar. The authors investigated burst simulations of hydrogen pressure vessels made using the filament winding technique. Their study aims to improve the manufacture of safer and cost-effective hydrogen storage vessels and consequently ease the inception and acceptance of hydrogen technology for transport, satisfying environmental requirements as well.

Type IV composite pressure vessels may burst if they are exposed to fire due to the destruction of the outer layers. Yet, if a pressure critical value is not exceeded, a leak may happen rather than a blowout. This occurs when the liner melts due to heat transfer through the composite shell. To investigate this failure mode, Halm et al. [54] performed fire impingement tests aiming to interpret and simulate the fire proficiency of Type IV hydrogen storage vessels. The phenomenon was approached both experimentally (exposing the vessels to contact with fire) and numerically via a finite element model especially deployed to simulate the combined effects of mechanical failure and temperature. It was found that both approaches may predict precisely the time either to burst of the composite vessel or its leak.

6.2.1.10   Resistance of Storage Vessels to Impact

The structures of composite pressure vessels are complicated, and their characteristics and integrity depend on several factors. Thus, the design of a dependable pressure vessel as regards safety requirements, and especially the resistance against impact, needs an in-depth knowledge of all influencing factors such as matrix strength, delamination behavior, fiber strength, and dome geometry. Wu et al. [55] analyzed numerically and experimentally the damage mechanisms of carbon fibers. A finite element model was developed and applied, and the computed results were compared against experimental data from tests conducted by these authors on the composite cylinders under impact from a variety of impactors. The simulation proved successful since the computed and experimental results were found to be in good agreement with each other.

Further studies have focused on the influence of repetitive and single impacts on burst pressure of glass fiber-reinforced vessels by Demir et al. [56]. These authors found a considerable decrease in the burst pressure of the vessels under repeated loading compared to single impacts. They also found that the burst pressure of empty pressure vessels under impact was higher than the burst pressure of the vessels under impact filled with water.

However, the properties of composite materials change remarkably with fiber density and stacking procedure, among other factors. Thus, probabilistic approaches for the prediction of fiber failures can be an acceptable path forward according to Ramirez et al. [57]. The finite element simulations applied by these authors were found to be able to predict the total behavior of a 700 bar Type IV hydrogen storage vessel as regards radial and axial displacements due to increasing inner pressure, as well as the failure modes. Thus, they were capable of explaining phenomena such as the non-linearity due to the presence of a gap between the liner and the composite shell or the possible leaks due to delamination in the dome.

Han and Chang [58] performed a finite element (FE) analysis to simulate the burst of a Type III hydrogen pressure vessel subjected to an impact load induced by free fall. The FE code was checked through a comparison of the computed results with those from experimental results. The Type III hydrogen pressure vessel tested had an aluminum liner and thick composite layers. The composite laminates were simulated by use of the ply-based modeling technique. It was found that even when some layers had failed by delamination or after a matrix failure in the transverse direction, the overall structure was still safe after the impact under the service conditions.

Alcantar et al. [59] presented two methods for the weight optimization of 70 MPa composite Type IV hydrogen storage vessels with a polymeric liner. Assuming a safety factor of 2.25, they first applied in their study the classical laminate theory and subsequently constructed an objective function based on the Tsai–Wu criterion, composite thickness, a safety factor, and a penalization factor. The optimized results showed that their methodology could decrease the Type IV hydrogen storage tank weight by up to 11.2% compared to a traditional design.

6.2.2 Accidents Caused by Hydrogen Use in Vehicles

6.2.2.1 Hydrogen Vehicle Hazards

Hydrogen is a promising fuel gas for transport; therefore, hazards associated with this application should be thoroughly investigated. These hazards should be considered in situations when a vehicle is inoperable, in normal operation, and in collisions.

Potential hazards are commonly related to fire, explosion, or toxicity as in any other case. *Toxicity* hazards can be ignored because neither hydrogen nor its combustion products are toxic, as mentioned earlier. *Fire and explosion* of hydrogen used for transport may come from the fuel storage in the vehicle, the fuel supply lines, or from the fuel cell, if such a system is used. Among them, the fuel cell is the least hazardous, although hydrogen and oxygen in the existing technology are separated by a thin polymer membrane (20 to 30 μm). The initiating event in this case would be the membrane rupture resulting in the combination of hydrogen and oxygen, yet in such an event the fuel cell would lose its potential, which can be easily detected by a control system. The supply lines can then be immediately disconnected. The operating temperature of a fuel cell (60°C to 90°C) is too low to act as a thermal ignition source, but the two reactants may combine on the catalyst surface to create ignition conditions. Nevertheless, the damage potential is limited since the amount of hydrogen in the fuel cell and the fuel supply lines is small.

The greatest damage potential is located in the fuel gas tank because the largest amount of hydrogen is found there. The *failure modes* that should be considered in normal operation, and in collision as well, are the following:

- *Catastrophic rupture of a tank.* This may be the result of a manufacturing defect, abusive handling or stress fracture, puncture by a sharp object, or external fire combined with failure of a pressure-relief device to open.
- *Massive leak.* This may be due to faulty operation of a pressure-relief valve, chemically induced failure of the tank wall, puncture by a sharp object, or normal operation of a pressure-relief valve in case of fire.
- *Slow leak.* Probable reasons may be stress cracking of the tank liner, faulty operation of a pressure-relief valve, faulty coupling from the tank to the feed line, or failure of the fuel line connection.

Because a catastrophic rupture is rather unlikely, failure modes resulting in large or slow hydrogen release have been identified experimentally. Countermeasures to avoid the failure modes mentioned could be, for instance:

- *Leak prevention* by a suitable safety design, allowing for shock and vibration tolerance of high-pressure lines.
- *Leak detection* by a proper detector or by adding an odorant in the fuel, which with the existing technology is a problem for fuel cells.
- *Ignition prevention* by automatic disconnection of the battery in an accident, separation of fuel supply lines and electrical systems, and design of proper systems for both active and passive ventilation, such as openings to allow hydrogen to escape upwards.

Based on these failure modes, a detailed risk assessment of the most probable or most severe *hydrogen accident scenarios* was conducted by Swain, Shriber, and Swain [60], including fuel tank fire or explosion in unconfined spaces and in tunnels, fuel line leaks in unconfined spaces, fuel leak in a garage, and refueling station accidents. These studies have determined that:

- A well-designed hydrogen fuel cell car should be safer than a natural gas or a gasoline car in collisions in open spaces.
- A hydrogen fuel cell car should be nearly as safe as a natural gas car, and both should be safer than a gasoline or propane car in a tunnel collision.
- The greatest risk appears in the scenario of a hydrogen leak in a garage, which in the absence of passive or active ventilation may result in fire or explosion.

In conclusion, hydrogen fuel presents similarities and differences from other transportation fuels, with some parameters tending to make accidents more severe and others making accidents less severe. Thus, it is not clear whether hydrogen would entail more or less risk in transportation. The good safety records of trucks carrying compressed and liquefied hydrogen by road, for instance in the United States and

FIGURE 6.4 Type IV hydrogen fuel tank test setup on propane burner. With permission from the School of Engineering, University of Edinburgh.

Source: [64].

Germany, add confidence that there are not large, unknown risks associated with hydrogen fuel use. Despite public perception, the use of hydrogen as a fuel for transport may be safer than gasoline or natural gas in some respects [61].

Because no relevant hydrogen incidents had been reported at the time of his study, Zalosh suggested that hydrogen tanks behave with the same failure modes as CNG tanks. The MVFRI funded two tests to determine the consequences of fire on these tanks. Descriptions of those tests have been reported by Zalosh and Weyandt [62] and Weyandt [63].

In these experiments, propane burner fires were set under cylinders that were filled with hydrogen at a pressure of 32 to 34 MPa without pressure-relief devices. The first test was done with a 72 L Type IV cylinder (Figure 6.4), while the second one used an 88 L Type III cylinder installed under a sports utility vehicle (SUV) (Figure 6.5). The Type IV cylinder ruptured after six minutes and 27 seconds of fire exposure. The burner flame ignited the SUV before the Type III cylinder ruptured after 12 minutes and 18 seconds of fire exposure. In both cases, the cylinder wraps prevented the hydrogen temperature and pressure from increasing significantly above their pretest values.

The type IV hydrogen cylinder primary remains were found about 82 m away from the burner (Figure 6.6), whereas the largest fragment of the Type III hydrogen cylinder was found about 41 m away from the burner (Figure 6.7).

In these tests, some blast overpressure measurements were performed as shown in Figures 6.8 and 6.9. The ideal blast wave pressures calculated for blast energies of 13.4 MJ and 15.2 MJ are also shown for comparison. The blast energies were calculated by Zalosh [64] from the hydrogen isothermal expansion of the gas inside the pressurized tank during rupture, based on the methodology of Baker et al. [65,

FIGURE 6.5 Burner below type-IV cylinder under an SUV. With permission from the School of Engineering, University of Edinburgh.

Source: [64].

FIGURE 6.6 Type IV hydrogen cylinder fragment. With permission from the School of Engineering, University of Edinburgh.

Source: [64].

FIGURE 6.7 Type III hydrogen cylinder fragment. With permission from the School of Engineering, University of Edinburgh.

Source: [64].

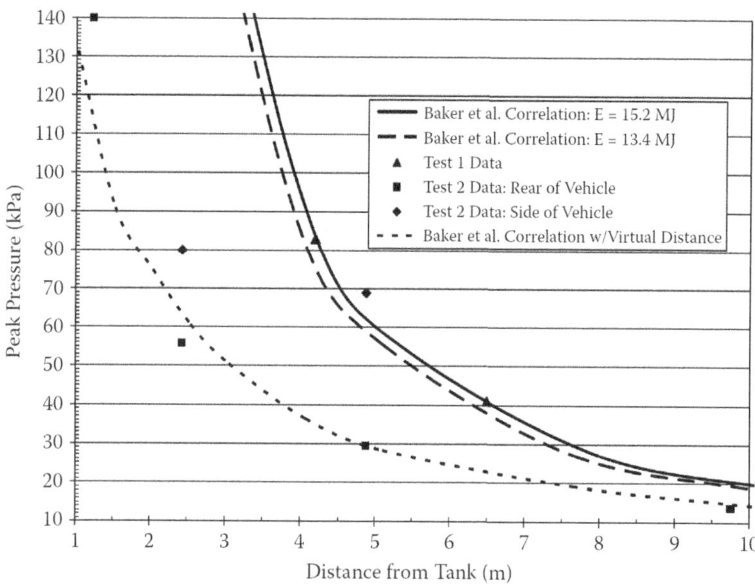

FIGURE 6.8 Pressures of fuel tank blast waves versus distance from tank. With permission from the School of Engineering, University of Edinburgh.

Source: [64].

FIGURE 6.9 Hydrogen fireball about 170 m/s after tank rupture in Test 2 (cylinder installed under an SUV). With permission from the School of Engineering, University of Edinburgh.

Source: [64].

66]. The agreement with the data demonstrates the validity of blast wave calculations based on the cylinder pressure and volume.

In addition to blast waves, fireballs were formed in both tests after the cylinders ruptured (Figure 6.9). In the cylinder-only test, the maximum diameter of the fireball was 7.7 m, whereas, in the cylinder-under-SUV test, the maximum diameter of the fireball was about 24 m. The larger fireball diameter in the SUV test indicates that with respect to formation of the fireball, not only the cylinder rupture, but also the SUV fire contributed to the fireball size.

In conclusion, the results of these tests suggest that the hazard zone in an anticipated hydrogen tank explosion extends to a radius of more than 100 m from the burning vehicle.

Stephenson [67] investigated the following principal crash-induced fire safety issues related to hydrogen-fueled vehicles:

- Crash forces and exposure to fire,
- Fuel tanks,
- Fuel lines,
- Hydride devices,
- Reformers,
- Fuel cells,
- Regulator failures,
- Venting from various sources,
- Mechanical energy from tank rupture.

Aiming at a reasonable mass fraction of a vehicle, the compressed hydrogen tank should preferably be a Type III (aluminum liner) or Type IV (plastic liner) carbon fiber-wrapped tank. Typical pressures in these tanks are 345 to 690 bar (34.5 to 69 MPa). Although these pressures seem rather hazardous, the tanks are so well constructed as to withstand that much pressure in a way that they are unlikely to burst in a crash. In contrast, the tanks may be more vulnerable at low pressures when the tank wall is less supported from the inside.

With regard to *regulators*, many suppliers construct regulators that are placed at least partially inside the tank. In this way, the regulators are well protected, and it is quite unlikely that they will be torn off in a crash. This event would cause a more severe accident owing to the rapid evacuation of the entire contents of the tank and its ignition either by existing ignition sources or by autoignition due to static electricity discharges. A countermeasure to avoid such high-pressure hydrogen releases is to have an in-tank solenoid shut-off valve that will stop the high-pressure hydrogen release as soon as possible.

A worst-case scenario is the exposure of the hydrogen tank to a liquid hydrocarbon *pool fire* from another vehicle following a crash. In this case a fiber composite tank is a good insulator that may prevent the temperature and pressure from rising considerably. Although a partially filled tank may not exceed its normal operating pressure, the fire will gradually weaken the carbon fiber or fiberglass overwrap and eventually the tank will burst if the hydrogen is not vented. This is usually accomplished by a thermally activated pressure-relief device, provided that the relief valve receives the same heat flux from the fire as the tank. Consequently, the use of both thermally activated and pressure-activated pressure-relief devices would be preferable.

Another issue is the *mechanical energy* stored in the compressed gas tank which could see an impulse of about 6000 Newton-seconds if a tank ruptures and leaks and produces a horizontal jet in a preferred direction. This impulse would accelerate a 1360 kg car on a friction-free surface to a velocity of about 16 km/h [68].

Aging of tanks is also a severe issue. Thus, all compressed gas tanks should have an expiration date to protect against fatigue and corrosion failures (for instance, at the most 15 years of tank life). A high-pressure regulator and several meters of tubing are needed to lower the supply pressure to about 10 bar (1 MPa). Since all of these components may be severely damaged in a crash and with ignition sources nearby, it is reasonable to assume that leaking hydrogen will ignite. Consequently, keeping the trapped volumes low so that the amount of energy release from such small fires is less than that required to ignite nearby materials is indispensable. Furthermore, materials in the vicinity of hydrogen tubes should be carefully selected, accompanied by wise spacing as well.

Sometimes, a *hydride storage device* of some kind may be part of the system. If this device is housed in a pressure vessel, this should hold the full pressure in the event the whole quantity of hydrogen is released from the hydride; thus, a pressure-relief device is necessary in this case. Furthermore, all of the hydrogen should be driven out if the hydride system were to be exposed to an electrical fire or liquid hydrocarbon pool fire.

Fuel cells also contain small quantities of hydrogen that could easily cross the thin membrane and escape in the event of a crash; hence, it will be necessary to safely vent the releasing hydrogen. *Reformer systems* also have some inventory of flammable and hot gases, careful management of which is needed.

In summary, the most important countermeasures according to Stephenson [67] include:

1. The location and protection of major components and routing of electrical wires and fuel lines.
2. The selection of low flammability materials for parts that might be exposed to minor electrical or hydrogen fires.
3. The rapid disconnection of electrical and hydrogen sources after detection by the vehicle crash sensors of crashes over a specified severity. The hydrogen can also be shut off via a variety of system sensors, such as low or high pressures or temperatures, in various lines or components.

According to Muradov [69], a serious danger lies in the frequent use of high pressures to ensure a high gravimetric and volumetric density. The possible self-ignition of compressed hydrogen following its sudden release into the air has been demonstrated by Bethoux [3]. This spontaneous ignition of hydrogen is caused by the shock wave at the boundary between the low ambient pressure and the high hydrogen pressure while escaping from the vessel. Therefore, spontaneous ignition is a phenomenon experienced in hydrogen vessels. The general idea is to limit the leakage occurring when the strength of the hydrogen vessel wall weakens due to elevated temperature. To address this issue, international safety regulations demand any onboard hydrogen tank be equipped with a thermal pressure-relief device (TPRD) that releases the compressed hydrogen as soon as the 110°C temperature set point is reached [70]. The purpose of this device is to set free the pressurized hydrogen at a controlled flow rate in order to exploit both the extreme diffusivity and the high buoyancy of hydrogen. Therefore, if the TPRD functions suitably, it will be able to both keep away from fire and explosion eventualities. The efficiency of this system depends on the environmental and operating conditions and reduces the consequences of a catastrophic tank rupture at a cost of not more than a few percent of the tank's content [70, 71].

The pressure effects of an ignited release from a hydrogen-powered car in a garage with a single outlet were investigated by Brennan et al. [72]. The authors conducted a safety study considering the hazards arising from both the remaining non-ignited gas and also the ignited release through a 3.34 mm diameter orifice from an onboard hydrogen storage tank at 700 bar in a small garage with only a small escape window. This is equivalent to the release rate of hydrogen from a TPRD. It has been estimated using a CFD approach and an analytical model that the resulting overpressure due to this non-ignited release is about 0.55 kPa, which is not likely to cause any damage in the garage. In contrast, the overpressure calculated for an eventual ignition of the release is two orders of magnitude greater, exceeding 55 kPa in less than 1 s, a situation that can cause destruction of the structure. Furthermore, the ignition of this explosive atmosphere would limit the oxygen to unbreathable air for humans in the garage.

Tamura et al. [70] experimentally investigated the fire spread behavior between gasoline and hydrogen fuel cell vehicles (HFCVs) and also between many HFCVs. They found that the principal cause of fire spread from the fire source vehicle to the adjacent vehicle was the flames from the burning vehicle and not the hydrogen flame from the functioning TPRD. Nevertheless, things are not the same in ferries carrying parked HFCVs with little or no space in between, when an activated TPRD of an HFCV car could generate hydrogen flames which in turn may trigger the ignition of the underfloor TPRD of adjacent HFCVs. Thus, damage minimization from an HFCV fire can be obtained by early detection and extinguishing a fire before the TPRD activates.

Figure 6.10 shows how the fire is transmitted from an HFCV to an adjacent gasoline vehicle in time intervals in the experiments conducted by Tamura et al. [70]. The HFCV was ignited first (time 0 min), and when the temperature reached 110°C at time 30 min at the TPRD, the relief valve opened to release hydrogen from the storage vessel through the vent pipe directing downwards, as indicated by the flames observed underneath the HFCV. Yet, despite this hazardous situation, the gasoline vehicle did not catch fire at this time point, but finally caught fire at time 57 min 26 s, or about 28 min after the HFCV's TPRD activation.

These tests and some incidents involving hydrogen reported by Adomaitis [73] and Jin and Chung [74] have shown that safety remains an open topic despite the precautions taken by car manufacturers and hydrogen producers and distributors to obtain demonstrators that are attractive to the general public.

Dadashzadeh et al. [75] report that, considering improvements to HFCV safety, two significant projects are currently being carried out. The first project deals with the design of thermic barriers to limit the temperature rise of hydrogen vessels and TPRDs onboard. The start time of the hydrogen pressure vessel burst (normally between 6–12 min) is thus increased, giving time to onboard persons to move away. Other projects work intensively to optimize TPRD performance without increasing their cost too much. The concept, in this case, is to invent proper means to mitigate the effect of the local fire hazard resulting from TPRD ignition due to hydrogen release through it that could endanger passengers or nearby persons close to the HFCV [76]. Adaptation of the TPRD via a rotary release device would direct the releasing hydrogen into an empty space and would allow the passengers to get out quickly from the damaged car. To accomplish the set goal, the emergency system requires many temperature sensors to monitor the hydrogen storage vessel, the passenger compartment, and the surroundings of the HFCV. Then, collecting all this information, the TPRD processor would be able to decide in real time the best hydrogen release direction.

Thus, according to Li and Sun [76], a well-designed monitoring and control system should be installed in a hydrogen car. The installed rotatable TPRD device should be regulated by a controller to decide when and in which direction the hydrogen should be released. The control system should be able to gather information from both the inside and the outside of the vehicle. The system should take information from pyroelectric infrared sensors inside the car to detect whether there is any person onboard. Ultrasonic radars should also be installed to detect the outdoor environment and inform the system which direction is void to canalize the released hydrogen. Furthermore, temperature sensors are required to initiate through the TPRD the activation process of hydrogen release. The technical parameters of the sensors applied in the control system are listed in Table 6.3.

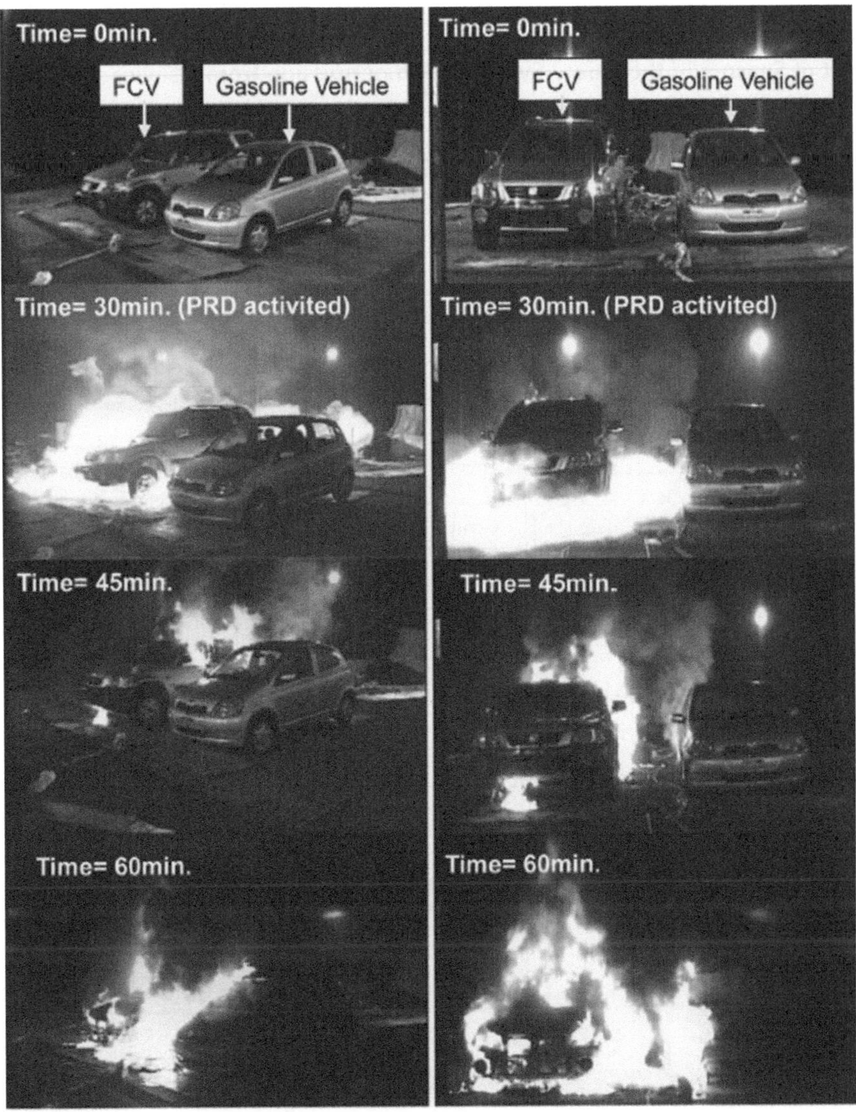

FIGURE 6.10 Photographic view of Test 1. With permission from Elsevier.

Source: [70].

Li and Sun [76] also suggest a central control unit to process the information and send instructions to the motor and the alarm device according to the flowchart shown in Figure 6.11.

Li and Sun [76] depicted the proposed system layout as shown in Figure 6.12. To detect the temperature in the vicinity of the TPRD, two temperature sensors are

TABLE 6.3
List of Sensors of the Safety System. With permission from Elsevier

	Pyroelectric Infrared Sensors	Ultrasonic Radars	Temperature Sensors
Detection parameters	Distance <7 m Detection angle 100°	Distance 0.05 m ~12 m Detection angle 85°	Distance 0.5~1 m Temperature range −40~115°C
Function	Detect people inside the vehicle	Detect void open space outside the vehicle	Detect ambient temperature around TPRD

Source: [76].

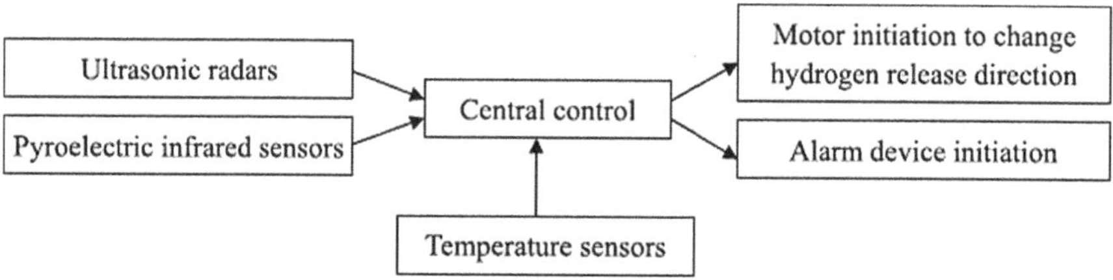

FIGURE 6.11   Flowchart of the system. With permission from Elsevier.

Source: [76].

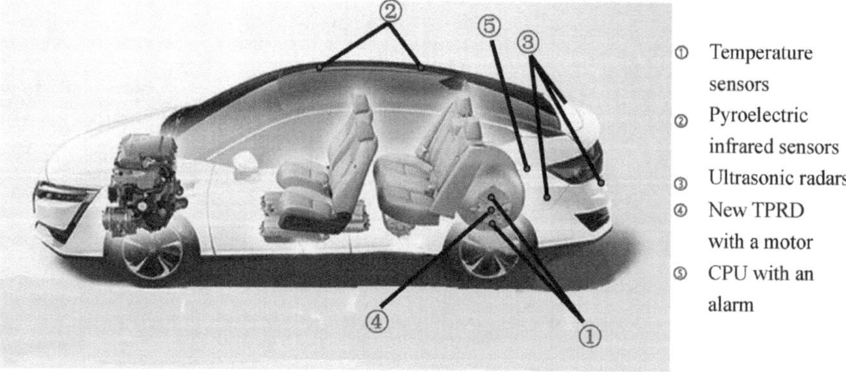

FIGURE 6.12   System layout of the mitigation measures. With permission from Elsevier.

Source: [76].

located near the TPRD. Two pyroelectric infrared sensors are placed inside the car, one above the front seats and the other above the rear seats to cover all of the space in the vehicle. There are also three ultrasonic radars, one at the rear of the vehicle to detect the backward space outside the vehicle, and the other two are placed on the left rear side and the right rear side of the vehicle. The central control unit and the alarm are installed in the rear region close to the temperature sensors.

Fuel cells use protection containers (enclosures) that should be able to ventilate in order to set free hydrogen emitted due to normal function or accidental leaks. In general, passive ventilation via buoyancy-driven release is preferable to forced mechanical ventilation. Proper function depends on factors such as vent number, size, shape, and position. Vents are usually rectangular openings with louvers for external protection. The influence of louvers on ventilation has been investigated by Ghatauray et al. [77]. These authors compared louver and plain vents with the same opening area using helium as a surrogate gas to simulate hydrogen leaks and found that louvers increased stratified level helium concentrations by typically in excess of 15%. The CFD Flow Simulator (SolidWorks) was also used, which showed a good qualitative agreement with experimental results.

Fuel cell electric vehicles (FCEVs) have given good grounds for expecting them to be a sustainable alternative for exploiting surplus energy produced by renewables, which would render the transition to zero-carbon emissions feasible. Nevertheless, this technology has not yet fully developed, especially in issues related to hydrogen storage vessels. Manufacturers of hydrogen vessels normally use carbon fibers to reinforce the vessels due to their superb mechanical strength, yet the production of carbon fibers requires highly energy-consuming processes that increase the carbon footprint. For that reason, Benitez et al. [78] performed a life-cycle assessment study on FCEVs, targeting the production processes of both the hydrogen storage vessel and carbon fibers. The outcome of this study is that the tank has a negative impact on climate change but is not responsible for other toxic-related environmental issues.

For the Hydrogen and Fuel Cell (H2FC) systems to be acceptable to the public there needs to be a persuasive safety engineering design and a sufficiently educated and trained workforce developing this new technology. According to Saffers and Molkov [79], this can be obtained through the creation and development of a relatively new profession, that is Hydrogen Safety Engineering (H2SE). H2SE is defined by these authors as:

An application of scientific and engineering principles to the protection of life, property and environment from adverse effects of incidents/accidents involving hydrogen.

Saffers and Molkov [79] claim that inherent safety principles should be incorporated in the design of hydrogen systems, as well as resourceful safety approaches in dealing with upcoming safety and feasibility challenges. To this end, new disciplines should be introduced into the curricula of universities, in addition to the recognition of H2SE as a new occupation that will promote global hydrogen fuel cell technology.

6.2.2.2 Accidental Release of Hydrogen from Vehicles

With regard to the safety concerns raised on the use of hydrogen as a motor fuel, Swain [68] produced a video demonstrating the characteristic features of a fuel leak and ignition for a hydrogen-fueled car and a gasoline-fueled car in the open air. Figure 6.13 shows some of the well-known footage of a hydrogen car fire and a gasoline car fire in the open air one minute after ignition. It is evident from this

FIGURE 6.13   Photo from a video that compares fires from an intentionally ignited hydrogen tank release to a small gasoline fuel line leak. Hydrogen-fueled car (left), gasoline-fueled car (right). Time: One minute after ignition when hydrogen flow starts to subside while gasoline vehicle fire begins to enlarge. After 100 seconds, all of the hydrogen was gone and the car's interior was undamaged. With permission.

Source: [68].

photo that the hydrogen can generate very long vertical jet flames and high flame temperatures. Nevertheless, the body of the hydrogen-fueled car was not ignited, and the flame lasted only 100 seconds. It is noteworthy that the maximum temperature inside the back window was only 19.5°C. In contrast, the leak from the gasoline-fueled car formed a pool fire, which engulfed the whole car. The car continued to burn for several minutes and was completely destroyed. This demonstration showed that a hydrogen-fueled car is less hazardous than a gasoline-fueled car in the open air, at least under the specific test conditions. Yet the consequences of a hydrogen-fueled vehicle fire would be much more severe inside a semiconfined space such as a road tunnel, or even worse in an enclosed space such as a garage. This is why a great many studies have been conducted recently dealing with this issue, as subsequently discussed here and also in Chapter 9.

A comparison safety study on gas release from an activated TPRD between hydrogen vehicles and compressed natural gas (CNG) vehicles was conducted by Li and Luo [80]. This study aimed to estimate the differences in hazard distances and release durations of TPRD activation between these types of vehicles during an accident in an open environment. The outline of the assumed accident was that of impinging fire jets set free from a TPRD with 4.2 mm diameter of cars with comparable driving range (4 kg hydrogen storage at 35 MPa and 20 kg methane storage at 25 MPa). It was estimated that the release duration for the CNG vehicle is over two times longer than that for the hydrogen vehicle, which results in a longer time for the firefighters to safely proceed toward the damaged vehicle. For both hydrogen vehicles and CNG vehicles, the maximum hazard distance occurs a few seconds after the opening of the TPRD. Soon after that, the flames start shrinking and the hazard distances fall off. The minimum safety distance that firefighters equipped with bunker gear may approach is 6 m and 14 m from the hydrogen vehicle and CNG vehicle, respectively. For approaching people, a safe perimeter of 12 m and 29 m should also be arranged around the accident scene for hydrogen vehicles and CNG vehicles, respectively.

As with all safety systems, the failure probability of a high-pressure storage tank rupture in a fire cannot be adjusted to zero, and this is the case also with TPRD activation failure in a vehicle incident. A smoldering fire is also a case in which TPRDs may fail to operate due to exposure to insufficient heat. A relevant accident occurred when a CNG-powered garbage truck exploded in Hamilton, NJ, USA on January 26, 2016, and blasted a hole in the front of a nearby house. A total of four houses were damaged in the explosion, though no injuries were immediately reported [81, 82].

Molkov et al. [83] investigated the dynamics of shock waves and fireballs after a hydrogen vessel ruptured when it was set on top of a fire. Experiments on the burst of vessels put in a fire with pressures of 35 MPa and 70 MPa were first performed. The results were then utilized to prove the validity of a CFD model applied to analyze the related physical phenomena. The applied CFD model successfully imitated the experiments. The simulated fireball is similar to the experiment and displays typical growth before buoyancy forces it to rise.

Underground tunnels play an important role in modern road transportation systems. Thus, in a hydrogen economy, hydrogen cars will regularly circulate in ordinary underground tunnels. Wu [84] investigated potential fire hazards and fire scenarios of hydrogen cars in road tunnels, as well as the implications for fire safety measures and ventilation systems in existing tunnels. The CFD simulations carried out by this author concerned two fire scenarios showing that the jet flame hazard can be unique for hydrogen cars. For high release rate, the flame might cause an oxygen deficit inside the tunnel. Impingement of the hydrogen jet flame on the tunnel ceiling would produce high ceiling temperatures, causing damage to tunnel infrastructure. In addition, the oxygen deficit caused by the hydrogen fire may also pose a flash fire hazard inside the tunnel and ventilation ducts. In Figure 6.14, the smoke flow in a

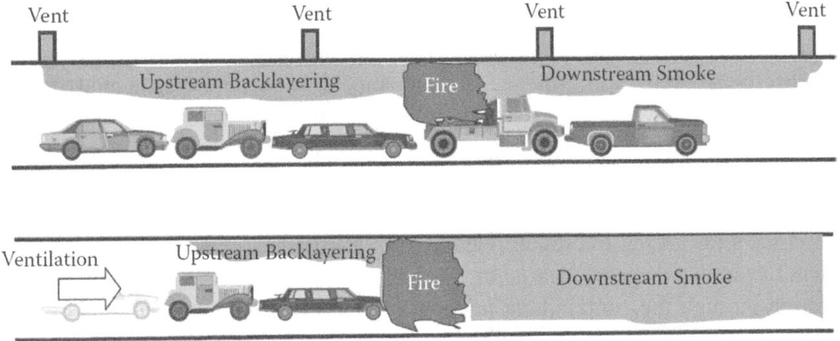

FIGURE 6.14 A tunnel fire and smoke flow under the influence of longitudinal ventilation. With permission from Elsevier.

Source: [84].

tunnel controlled by longitudinal ventilation is shown. The upstream smoke layer is called "backlayering" and is sensitive to ventilation velocity. The "critical velocity" is the velocity that just about completely eliminates the upstream backlayering and prevents smoke flow from traveling against the ventilation, thus forcing the smoke to move in only one direction.

CFD modeling of hydrogen release, dispersion, and combustion for various automotive scenarios was performed by Venetsanos et al. [85], aimed at investigating the potential effects of releases from compressed gaseous hydrogen systems on commercial vehicles in urban and tunnel environments. The authors also performed comparative releases from compressed natural gas systems. Typical nonarticulated single-deck city buses were analyzed in this study. The analysis showed that in worst-case scenarios in tunnels a deflagration may occur, which, in very unfavorable conditions causing intense turbulence (e.g., multiple obstacles), may evolve into a deflagration-to-detonation event.

A CFD simulation study to investigate the risk from hydrogen-fueled vehicles inside tunnels was performed by Middha and Hansen [86]. The study concerned cars containing 700 bar gas tanks releasing either upward or downward, or liquid hydrogen tanks releasing only upward. Buses were also studied containing 350 bar gas tanks and releasing upward in two different tunnel layouts and a range of longitudinal ventilation conditions. A probabilistic risk assessment study was also performed resulting in a maximum pressure level of 0.1 to 0.3 bar(g), which represents a limited human fatality risk, yet with somewhat higher pressures in specific places (e.g., under the vehicles) due to wave enhancement from reflections.

Baraldi et al. [87] investigated the scenario of a large-scale accidental release and subsequent ignition and deflagration of hydrogen in a 78.5 m long tunnel using numerical analysis in their simulation. Experiments have shown that explosions in such semiconfined spaces can result in severe consequences due to the entrapment of hydrogen for some time inside the structure; this increases the probability of ignition and also leads to the enhancement of pressure waves caused by the tunneling effect. Thus, 1 kg of hydrogen at stoichiometric ratio generated an overpressure peak of about 150 kPa in a tunnel environment, whereas the same quantity and composition resulted in only 10 kPa when ignited in the open. It is noteworthy that the presence of vehicles inside the tunnel did not increase considerably the peak pressure. Numerical simulations using five different CFD codes were partly successful with regard to pressure peak prediction, whereas the agreement between experimental data and simulation results was not satisfactory for the pressure rise rate. This discrepancy was attributed to the mesh resolution, the numerical scheme, and some inaccuracy in the physical description of the flame acceleration.

A potential risk associated with the use of hydrogen in automotive applications is a slow, long-lasting hydrogen release from a vehicle's components into an inadequately ventilated enclosed structure such as a garage. This leak may stem from permeation through the containment materials found in compressed gaseous hydrogen storage systems and facilitated by the extremely small size of hydrogen molecules. Consequently, this is an issue that requires special consideration for containers, especially when constructed of nonmetallic liners (mainly polymers). The molecular diffusion through the walls or interstices of a container vessel, piping, or interface

TABLE 6.4
Comparative Permeability of Materials to Hydrogen [(mol H_2) $\cdot$ s^{-1} $\cdot$ m^{-1} $\cdot$ MPa$^{-\frac{1}{2}}$]

Material	$T = 20°C$	$T = 55°C$
Iron	5.47×10^{-11}	2.38×10^{-10}
High density polyethylene (HDPE)	9.30×10^{-13}	3.17×10^{-12}
Carbon fiber/epoxy composite	1.85×10^{-13}	5.79×10^{-13}
Stainless steel 303	4.09×10^{-16}	7.69×10^{-15}
Aluminum	2.47×10^{-24}	7.03×10^{-22}

Source: Data from References [88] and [89].

material has been established by SAE International [88] and cited by Makarov and Molkov [89] as shown in Table 6.4.

Adams et al. [90] investigated the same problem and developed a methodology to estimate an allowable upper limit for the hydrogen permeation rate from road vehicles into enclosed spaces. After this study, the worst garage ventilation rate was determined by SAE to be 0.03 ac/h (air changes per hour) and the maximum prolonged material temperature was 55°C. In addition, the authors concluded that stratification inside these spaces is insignificant and, therefore, it would be valid to assume homogeneous distribution of hydrogen at the flow and ventilation rates considered. They also concluded that since the effects of aging on the permeation behavior of complete containers is uncertain, a factor of two should be used for calculating lifetime and replacing of containers.

6.2.2.3 Consequences of Storage Vessel Failure

Kashkarov et al. [91] constructed nomograms for fast estimation of safety distances from a blast wave generated either by the rupture of stationary (e.g., at refueling stations) hydrogen storage tanks or by onboard vessels engaged in a fire. The nomograms are to be used by first responders to estimate safety distances, and they are based on both overpressure and impulse of the created destructive wave from a hydrogen leak. The validated model of a burst wave decline published by Molkov and Kashkarov [92] was used to construct the nomograms which are destined for both the first responders to use on-site and safety engineers to use as tools to design onboard hydrogen safety systems and safe infrastructure systems.

These authors use three damage threshold criteria and four hazard zones around the hydrogen vessel endangered by a diminished high-pressure wave resulting from an eventually exploding hydrogen vessel in a fire. The four hazard zones are "Fatality", "Serious injury", "Slight injury", and "No harm" (see Figure 6.15). The "No harm" distance or the evacuation perimeter is defined as the distance with a "temporary threshold shift" (TTS), for instance, temporary loss of hearing. This corresponds to an overpressure threshold equal to 1.35 kPa and an impulse above 1 Pa s. "Slight injury" is the zone where burst overpressure is in the range 1.35–16.5 kPa, "Serious injury"

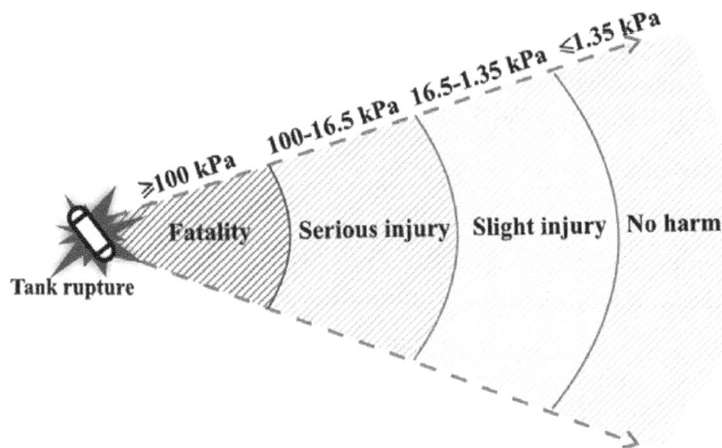

FIGURE 6.15   Hazard zones for people based on selected thresholds. With permission from Elsevier.

Source: [91].

is the zone where the overpressure is in the range 16.5–100 kPa, and "Fatality" is the zone extending from the vessel to a distance where overpressure equals 100 kPa [91].

Safety distances for humans are designed for hydrogen storage tank volumes of a capacity from 10 L (e.g., bike storage tanks) to 10,000 L (e.g., hydrogen refueling station tanks) pressurized to 200 bar, 350 bar, 700 bar, or 1000 bar. A hydrogen refueling station may host several hydrogen storage vessels, the volume of which may be, for example, 300 L each, and the storage pressure may be as high as 1000 bar. This example is graphically presented in the nomogram shown in Figure 6.16 for a stand-alone tank rupture to graphically estimate the hazard distances for the selected harm criteria to people.

The nomogram in Figure 6.17 allows calculation of safety distances from a blast wave generated by an under-vehicle onboard tank rupture in a fire [91].

The first persons dealing with an accident of a hydrogen fuel cell vehicle (HFCV) should be properly informed on how to provide emergency action. In a safety study performed by Tamura et al. [70], a hydrogen leak at a rate of 2000 NL/min from an HFCV was dealt with an equal or faster enforced wind of 10 m/s leeward from the blower area. Soon after, the hydrogen concentrations around the vehicle were measured and ignition tests were performed to check the effectiveness of this artificial wind blow and the effectiveness of this emergency action. They found that this emergency safety action for a hydrogen-leaking vehicle can be successfully applied to be able to approach the vehicle after a while.

6.2.3   Fueling Stations

Although the first commercial hydrogen-fueled cars are now available in the marketplace (although at very high prices), the principal problem for using such a car today

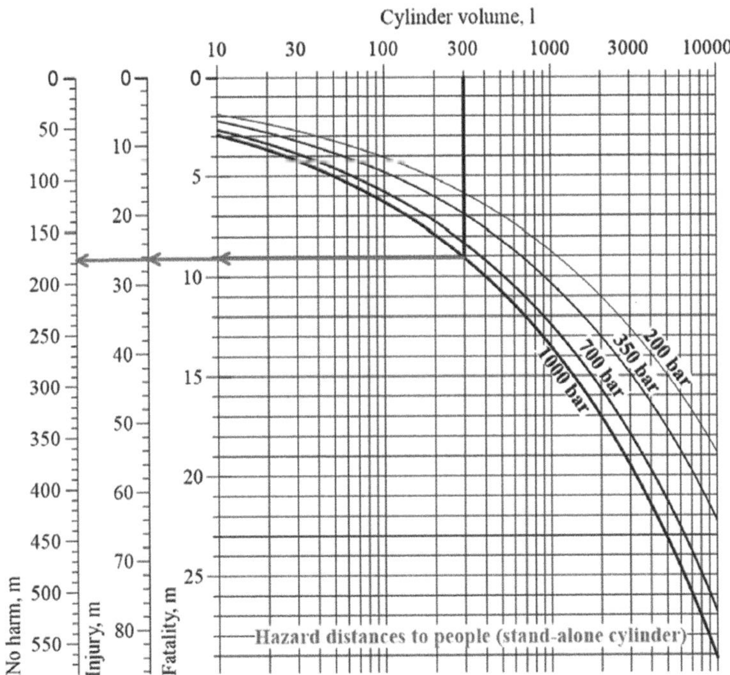

FIGURE 6.16 Nomogram for first responders showing hazard distances for people (stand-alone tank rupture). With permission from Elsevier.

Source: [91].

is the absence of a network of refueling stations. Thus, the owner of a hydrogen-fueled vehicle must limit travel to distances where the vehicle can be refueled securely; clearly, the lack of hydrogen infrastructure significantly hinders the emergence of the so-called hydrogen economy. The declarations for a national policy to be prepared for the hydrogen economy, first by the prime minister of Iceland in 2001 and shortly thereafter by the European Commission President Romano Prodi in 2002, were very promising, and the efforts toward this aim were substantial. Then, in January 2003, President George W. Bush shared the same sentiment in his State of the Union Address. In January 2004, Governor Arnold Schwarzenegger declared in his State of the State Address that California would have a hydrogen highway [93]. Consequently, the safety of hydrogen refueling stations should be investigated in parallel with other hydrogen safety issues.

Markert et al. [2] analyzed a number of safety concerns related to various potential future hydrogen infrastructure projects to be built for a sufficient supply of refueling stations within an urban region. The authors stress the need for a rapid change of building codes that concern specifically the use of fuels giving off vapors heavier than air, such as gasoline vapor and LPG, but not for hydrogen, which is much lighter than air. For instance, it is required that ventilation openings in garages should be near the

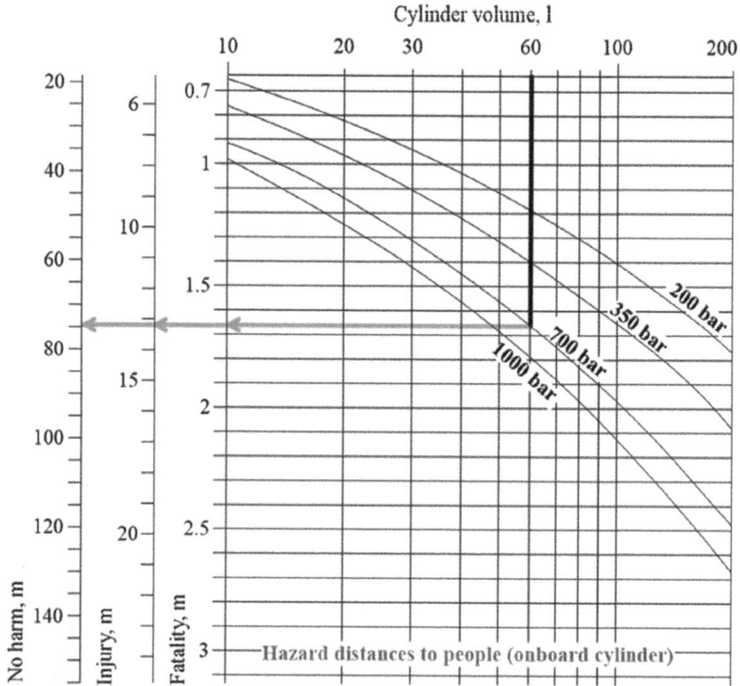

FIGURE 6.17   Nomograms for hazard distances to people from a blast wave after onboard (under-vehicle) tank rupture in a fire (example of tank 60 L, 700 bar) for first responders. With permission from Elsevier.

Source: [91].

ground with often no openings at higher levels. Evidently, such enclosed spaces are prone to destructive hydrogen explosions.

The safety of hydrogen refueling stations could be increased by reducing the amount stored at the station according to inherently safer design principles (Chapter 7). This could be accomplished by pipeline transmission or even by small on-site hydrogen production facilities. In the case of truck delivery, a measure to increase safety could be just leaving the trailer at the refueling station instead of unloading the tankers into buffer storage. With regard to cars, refueling safety could be greatly improved by exchanging whole car tanks at refueling stations, which could be easily and safely recharged as cartridges at remote sites.

Kikukawa [94] conducted a consequence analysis and safety verification of hydrogen fueling stations using CFD simulation. The study was validated against the results of horizontal squirt tests of hydrogen leaking from a pinhole (0.2 mm) in high-pressure gas facilities, such as the piping in a hydrogen refueling station. Ignition by both static electricity (autoignition) and a flame outside the station were analyzed.

Baraldi et al. [95] performed CFD simulations of hydrogen dispersion and combustion in accident scenarios in a mock-up liquid hydrogen refueling station,

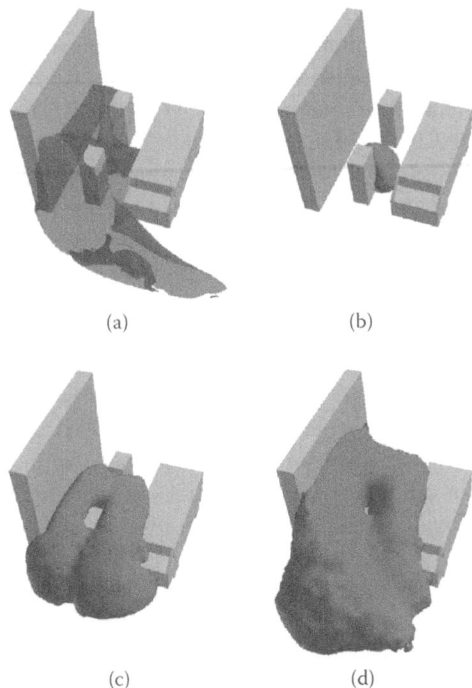

FIGURE 6.18   Non-mitigated west wind scenario. Ignition is located between the dispensers. The molar fraction (0.04) in blue color (a) shows the initial hydrogen flammable cloud. The isosurface of temperature (1000 K) in red color (b, c, d) shows the propagation of the flame. With permission from Elsevier.

Source: [95].

investigating accident scenarios caused by a hose break during LH_2 vehicle refueling. Both a non-mitigated and a mitigated accident scenario were analyzed assuming five different ambient conditions. It was found that the expectation that the wind plays a positive role in increasing the dispersion of the flammable cloud, thus reducing the overpressures in eventual explosions, is not justified for all wind directions. The propagation of the flame for the non-mitigated scenario with the wind coming from the west is shown in Figure 6.18.

Toward the same goal, and utilizing the same mock-up hydrogen refueling station, Makarov et al. [96] simulated a hydrogen explosion involving hydrogen stored in a gaseous state. A total of seven partners obtained nine simulation results using various CFD tools. The paper details the models and numerical codes used and presents the simulated pressure transients obtained in comparison with the experimental pressure records. The comparative model analysis was based on simulation results achieved. The simulated maximum overpressure and the characteristic rate of pressure rise were treated as major output parameters. The simulations had, in general, a reasonable agreement with experimental data, although some codes predicted relatively large overpressures in the recirculation area behind the car.

Regarding human error analyses, Castiglia and Giardina [97] have dealt with errors of operators occurring in hydrogen storage systems of fueling stations during the maintenance of safety vents. In their case study they used first- and second-generation human reliability assessment (HRA) methods. The likelihood of wrong actions was estimated by the HEART (Human Error Assessment and Reduction Technique) methodology. The results were compared with other results obtained from the cognitive reliability and error analysis method (CREAM, second-generation) methodology. The critical analyses of the totality of results ended up with suggestions on the right maintenance process and the most suitable safety equipment to reduce the accident risk.

The safety of a hydrogen fueling station was assessed by Al-shanini et al. [98] via an accident simulation scenario and design. The impending measures proposed and taken into account were related to human factors. The failure likelihoods of these measures were estimated using fault tree analysis (FTA) and event tree analysis (ETA), respectively. The results from this study showed that the proposed estimation method is suitable for estimating the probabilities of specific outcomes. In addition, the method could provide dynamic fine-tuning by way of updating the inputs with new data.

Qian et al. [99] performed a numerical study of accidental hydrogen release from a pressure vessel in a fueling station. They investigated hydrogen release in a momentum-buoyancy-controlled arrangement and also in a momentum-controlled arrangement. The profile of the hydrogen gas cloud in release direction, buoyancy/ gravity direction, and natural convection direction, that is the y direction, z direction, and x direction, is plotted in this study quantitatively, and the two arrangements are compared. The results indicate that during the continuous diffusion process in the momentum-buoyancy-dominated arrangement the profile of the flammable gas cloud moves upward, whereas in the momentum-dominated arrangement, the flammable gas cloud rather accumulates near the ground.

Quantitative risk assessment (QRA) of a predefined hypothetical gaseous hydrogen refueling station was investigated by Papanikolaou et al. [100]. In total, 15 scenarios were simulated. Five of them concerned hydrogen releases in confined ventilated spaces (inside the compression and the purification/drying buildings) and ten scenarios concerned releases in open and semiconfined spaces (in the storage cabinet, storage bank, and refueling hose of one dispenser). The simulations of the open and semiconfined scenarios showed that the wind velocity has an insignificant effect on the size of the flammable clouds. The limited effect of the wind velocity is attributed to the presence of the surrounding buildings that cause a wind velocity field within the refueling station. The risk assessment parameters (such as maximum flammable hydrogen mass and mixture volume and maximum horizontal and vertical distance of LFL cloud) are mainly influenced as expected by the release rate (i.e., the leak diameter size).

A benchmarking exercise on QRA methodologies for hydrogen safety in a virtual refueling station was conducted by Ham et al. [101] (see also Chapter 9). The benchmark exercise showed that big differences exist in the approach of QRA and in the nature of the results obtained, probably owing to limitations in the codes used. The scatter is wider in the computations of dispersion parameters (e.g., dimension of flammable cloud and explosive mass). Nevertheless, this scatter does not appear in a clustering of numerical results versus analytical results.

The safety of hydrogen refueling stations can be greatly improved by the installation of monitoring systems. By these means, not only the safety of their operations but also the fueling performance may be optimized. Indeed, monitoring systems in hydrogen refueling stations are indispensable if the temperature and pressure in vehicle tanks are required to be controlled and set under certain limits in a way that the safety standards of operation are met. In a study, Wang et al. [102] proposed a black-box machine learning model to be used in order to learn the relationship between the initial operating conditions and the final parameters of fueling processes. Using this model, the final temperature of a fueling process can be kept below 85°C and the pressure under 70MPa, with the final state-of-charge (SOC) set between 95% and 100%.

There are cases where hydrogen fueling stations are not compliant with acceptable norms concerning safe design. For those cases, LaFleur et al. [103] performed a study based on a performance design that could be an alternative for them to comply with established safety codes. To this end, they approached the issue by applying QRA techniques combined with other risk estimation techniques to ensure compliance. The methodology makes use of HyRAM (Hydrogen Risk Analysis Model), a QRA tool created by Sandia National Laboratories, US, which is suitable for estimating the risk values needed for risk-equivalent designs. HyRAM combines specific deterministic models that describe hydrogen releases and resulting flames with probabilistic risk models to finally quantify failure probabilities [104].

Liang et al. [105] simulated and analyzed an eventual leakage and explosion at the Dalian (China) hydrogen refueling station. The authors applied the CFD-based FLACS software to simulate and analyze a leakage and blast from this hydrogen storage installation. They analyzed the influence of wind speed and direction, as well as the leakage direction on the hazardous effects of such a hypothetical accident. The hazardous zones (lethal, farthest harmful distance, and longest lethal zones) resulting from an explosion scenario were estimated. The hydrogen leakage and explosion scenario of the 90 MPa hydrogen storage tank is quite likely to cause the most serious damage. In the event of the same direction of wind and leakage, the farthest harmful distance resulting from an explosion of the leaking gas would be 35.7 m and the farthest lethal distance 18.8 m. Furthermore, it is advised that the hydrogen tube trailer not be parked in the hydrogen refueling station when the amount of stored hydrogen is significant.

Kim et al. [106] performed safety assessments for the on-site and off-site hydrogen fueling stations in Korea using Hazard and Operability (HAZOP) and FTA techniques to analyze the risks in these installations. The HAZOP assessment was applied to four Korean hydrogen fueling stations to spot the inherent hazardous sections, investigate the issues reducing the process efficiency and check the proper installation of the safety devices. All four on-site hydrogen fueling stations in Korea are composed of facilities for storing and filling the hydrogen gas produced through the steam reforming method. The safety assessment was accomplished through the division of the piping and instrumentation diagram (P&ID) of the hydrogen fueling station into 27 suitable study segments which included all processes, namely reforming, shift reaction, adsorption, compression, and dispersion. The critical factors such as flow, pressure, temperature, level, step, time, and adsorption were analyzed at each study segment. The guide words (e.g., more, less, none, other, as well as, reverse) were combined to generate 222 irregular states in total. The risks thus determined were investigated to find out whether the safety systems installed following the acceptable standards were adequate for accident prevention. Finally, according to the HAZOP assessment, no severe problems were spotted in these installations.

An FTA of a hydrogen fueling station that produces hydrogen on-site by natural gas reforming was also performed by Kim et al. [106], as shown in Figure 6.19. The FTA approach was applied to estimate the frequency of important destructions obtained from the previous qualitative HAZOP assessment. The Technique for Human Error Rate Prediction (THERP) was also applied for the quantification of human errors. As the top event of the FTA, a hazardous hydrogen leakage at the installation was chosen. The inputs for this assessment were the reliability data of the basic events (Figure 6.19).

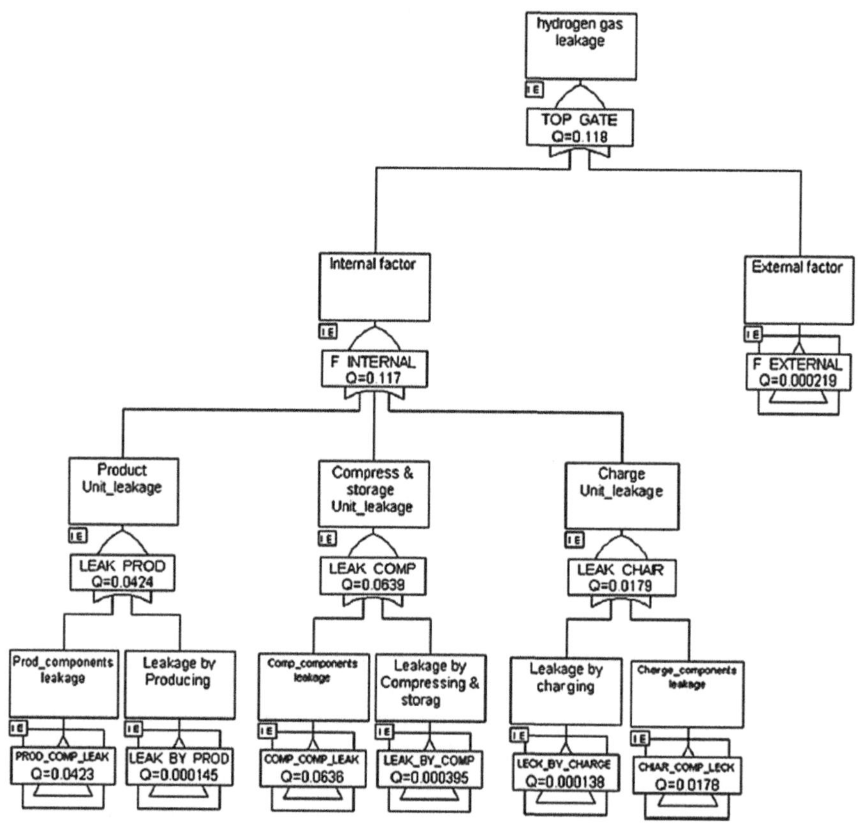

FIGURE 6.19 Fault tree analysis of a hydrogen leak in a hydrogen fueling station. With permission from Elsevier.

Source: [106].

A study by Kim et al. [107] used FLACS (a computational fluid dynamics software) to simulate a hydrogen leak and blast at a hydrogen refueling station, storing hydrogen at 100, 200, 300, or 400 bar through a variety of hydrogen leaking holes equal to 0.5, 0.7, or 1.0 mm in diameter. The 3D study refers to existing stations in Korea and was based on real 3D geometrical data from these installations. Validation of the simulation results was obtained by comparison with experimental results of generated real jets of leaking hydrogen.

The fireFOAM software was used by Yu et al. [108] to investigate the flame mitigation effect of vertical barriers in hydrogen refueling stations. These authors focused on the mitigation effect of obstruction walls against jet flames, a decrease in flame size, and a change in flame direction. A real-gas equation of state was applied to simulate the high-pressure hydrogen expansion through holes. They found that the flame length reduces with higher barrier walls, and the flame deflection angle increases also with the height of the walls. These findings mean that a barrier wall can be beneficial toward the set goal of limiting the hazardous area created by jet flames from hydrogen leaks. Nevertheless, there are drawbacks to too-high barrier walls because a high wall increases the surface area of a flame impinging on the wall, thus increasing flame radiation intensity. They finally concluded that the best results are obtained when the height difference between the barrier wall and the hydrogen jet escape hole is no more than 1 m.

6.2.4 LIFE-CYCLE ANALYSES FOR THE AUTOMOTIVE SECTOR

Life-cycle analysis (LCA) is a useful methodology to quantify all the emissions associated with a sector or process. In the transportation sector, all carbon dioxide emissions are taken into account in specific research. Conway [43] proposes in a study six principal phases:

1. Mining and extraction of all raw materials needed to construct a vehicle, including fuels.
2. Making of all parts.
3. Treatment and transportation of fuels from refineries to refueling stations.
4. Emissions from combustion processes.
5. Component and fluid services.
6. End-of-life disposal and recycling.

Various approaches are taken into consideration in an LCA. For instance, a "tank-to-wheel" analysis considers only phase 4, a "well-to-wheel" analysis considers phases 3 and 4, and a "cradle-to-grave" analysis considers phases 1–6.

The following excerpt from Conway's study [43] provides incident background and clarifies the differences between various approaches:

Life-Cycle Analysis (LCA) is becoming an increasingly important topic in the automotive industry due to the rise of electric and electrified vehicles and alternative fuels. These new powertrains reduce emissions generated in phase 4 and move them into other phases, thus, reducing the relevance of the current *"tank-to-wheel" approach taken by regulators. A key example of why broader LCA is necessary can be seen with a simple "tank-to-wheel" approach case. A horse breathes in air and turns part of the oxygen into CO_2, which it then exhales. If the CO_2 were measured from the horse, like we currently measure emissions from a vehicle's tailpipe, then it would be seen that at a 4-mph canter, a horse emits 530 g/mile CO_2, which is equivalent to a 2019 Corvette. Thus, under the current "tank-to-wheel" regulation, the horse is not compliant with fuel economy standards, and we would conclude that their use should be limited. However, if a "well-to-wheel" approach is considered, it is possible to see that most of the fuel used in the vehicle comes from fossil sources while the energy used to power the horse comes from low carbon, biogenic sources. Biogenic CO_2 essentially means that at some point in the cycle, bio-matter has absorbed atmospheric CO_2. In this example, the grass/hay used by the horse as its energy source absorbed CO_2 during its growth. This means that although the horse breathes out the same amount of CO_2, it has a much lower overall carbon footprint. Furthermore, several arguments exist for a complete "cradle-to-grave" analysis due to CO_2 emissions associated with battery pack production, leading to a significant carbon footprint before the vehicle has driven a single mile. However, until at least a "well-to-wheel" approach is adopted in regulation, an inaccurate analysis will be performed, leading to incorrect conclusions and possible actions.*

As regards environmental impact, Bicer and Dincer [109] conducted an LCA of methanol, hydrogen, and electric vehicles to compare the impacts of vehicles powered by these means on the environment and humans. For every one of them, all the stages were investigated from generation of raw materials to scrapping of the vehicles. The environmental impact groups chosen to perform this LCA were global warming, human toxicity, and ozone layer depletion. As fuels, hydrogen and methanol were directly utilized in internal combustion engine vehicles. The results of this study were astonishing since they indicate that electric vehicles generate higher human toxicity costs stemming from the previous making and maintenance stages. Furthermore, the fact that hydrogen's energy density is much higher than that of methanol means a hydrogen-powered vehicle is an environmentally more benign option considering global warming and ozone layer depletion issues.

6.2.5 COMPARISON OF HYDROGEN AND OTHER ALTERNATIVE FUELS WITH AMMONIA

6.2.5.1 General Considerations

It seems that some countries have already decided to select the energy transition to the hydrogen economy using ammonia as a hydrogen carrier. In particular, in 2021, Japan ordered and received three cargoes of blue liquefied ammonia from the United Arab Emirates since it is a strong proponent of decarbonization. Moreover, the Japanese shipping company NYK Line has started research and development of international collaborations aiming to establish ammonia-powered vessels for liquid ammonia transport. As an indication of Japan's preference for hydrogen, the 2021 Tokyo

Olympic flame burned hydrogen [110–112]. Yet it should be taken into account that the conversion of hydrogen to ammonia results in up to 25% energy loss, as Sadik-Zada says [113].

Ammonia has the capacity to become a zero-emissions fuel, thus possessing the possibility of being a crucial factor in the future global energy economy. Mostly in shipping, its easy handling and storage conditions offer a great advantage against the alternatives, compressed and liquid hydrogen. Moreover, according to Gerlitz et al. [114], ammonia has the benefit of existing infrastructure to some extent onshore as a chemical with enormous production for other uses, thus needing much fewer investments for storage installations than hydrogen.

It seems that ammonia represents a promising alternative fuel for shipping. In any case, the final acceptance of ammonia as a hydrogen carrier will be decided by its final price in the market. In addition to ammonia's handling and storage benefits, ammonia provides long-lasting sea transport without considerable loss of ship space for a fair price. On the other hand, ammonia-propelled ships are still under development, and the ammonia global infrastructure for production and distribution is not ready yet [114].

For ammonia to be used as a fuel in the maritime industry, specific hazards have to be faced, namely toxicity, corrosiveness, and flammability, which pose severe challenges during ship-to-ship bunkering. The corrosion can be prevented by selecting suitable construction materials whereas, for the impacts of toxic gas dispersion and fire hazards, risk assessment guidelines for ammonia bunkering should be prepared. Quantitative risk assessments for the toxicity and flammability of ammonia in bunkering based on Bayesian networks (BN) have been performed to investigate these potential risks. The efficient quantification of risks for ammonia ship-to-ship bunkering using BN shows that toxicity has the greatest impact on the risks of ammonia bunkering compared with flammability. Thus, risk mitigation measures should focus on actions that minimize the risk associated with toxicity [115].

Concerning safety, Valera-Medina et al. [116] claim that ammonia has been considered a relatively safe chemical that is now positioned as the second most distributed and commercialized worldwide. Nevertheless, it is clear that ammonia needs to be handled with care because its toxicity and corrosivity are critical aspects for living species (including humans) and materials. The affinity of ammonia to water can lead to skin death, while burning respiratory tracts in humans and animals. This is accompanied by severe corrosion toward copper-based materials, and some corrosivity to nickel-based alloys. It is recommended that high-resistance steel (>450 MPa) is not used, because this can be susceptible to corrosion under low stress. To keep away from these hazards it is recommended that spray coatings (e.g., zinc) are used in welding areas. Also, good maintenance programs are required to ensure the integrity of components.

Regarding air pollution, although ammonia will mitigate carbon-containing emissions, several factors need to be considered before employing this chemical as a fueling source, for example, the production of other unwanted emissions (like NO_x) and the overall impact of slip ammonia toward the atmosphere if systems are not clean enough. Particulate matter increases in the air due to the use of ammonia are another detriment. Thus, there are still many challenges before ammonia is commercially deployed as a main energy vector. Challenges and works in the areas of synthesis, use, safety, and economics are all under scrutiny at various levels, thus ensuring that our understanding of the use of ammonia and its practical implementation is attained within the coming decades [116].

6.2.5.2 Comparison of Ammonia with Other Alternative Fuels

From the point of view of safety, comparing ammonia with other fuels we may conclude that the use of ammonia as a fuel in transportation is not more dangerous than the currently used fuels provided that the hazards are controlled by both technical and regulatory standards. According to Duijm et al. [117], the most significant demands are:

- Adequate safety systems.
- Sufficient safety measures and standards to avoid accidental releases.
- Delivery of ammonia to fueling stations in refrigerated form.
- Adequate safety zones between refueling stations and public areas.

A comparison of the risks of various fuels, including the risk of flash fire, explosions, and intoxication by substances, is depicted in Figure 6.20 [117]. On the assumption that the systems are made as reliable as needed, it is concluded from this picture that ammonia poses fewer hazards than LPG (liquefied petroleum gas) and hydrogen-powered vehicles. The sizable effect distances for hydrogen are due to the great energy density of the released hydrogen, even when released from tiny holes. The difference between the risk levels for LPG and ammonia stems from the more severe reliability requirements for the ammonia system than for the LPG systems, in addition to the assumption that carbon fiber ammonia vessels would not disastrously fail in a fire. Considering the fire and explosion risks for gasoline and methanol, these

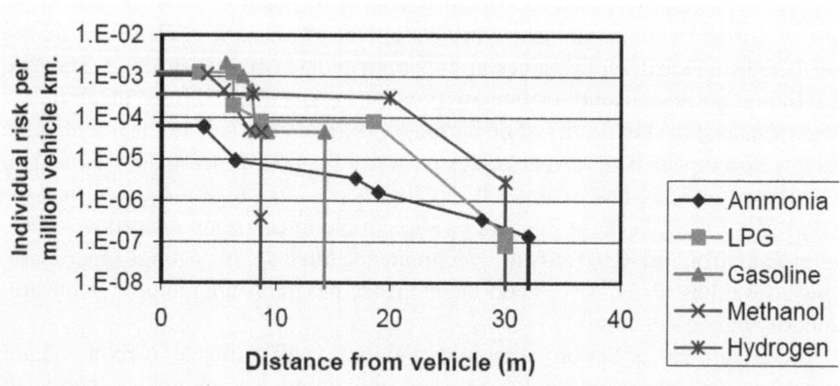

FIGURE 6.20 Vehicle risk related to its fuel. Comparison of individual risk as a function of distance from the vehicle. With Permission from Duijm, N.J.

Source: [117].

are quite close to those for LPG at short distances. However, the effect distances for the worst events are smaller for methanol. It should be reckoned that comparing methanol and gasoline, the latter has a much higher energy content. The energy content of the ammonia, methanol, and hydrogen tanks is similar.

The comparison of individual risk, including the conditional probabilities of flash fire, explosions, and injury by toxic substances, versus the distance from a refueling station for fueling equal energy amounts of ammonia, LPG, and gasoline is depicted in Figure 6.21. Due to its toxicity, even small ammonia releases have relatively sizable fatality distances. Using standard handling of ammonia, a safety distance to the individual risk level of 10^{-6} per year would be more than 150 m. Nevertheless, by applying extra safeguards, this distance can be reduced to around 70 to 120 m. However, this is still considerably larger than that for gasoline and LPG (around 30 and 40 m, respectively). As shown in Figure 6.21, there is not much difference between LPG and ammonia in the largest effect distances, mainly the effects of catastrophic releases due to the catastrophic failure of a storage vessel. By using refrigerated ammonia delivered by road tankers, the consequences of the worst events are reduced to distances less than for LPG, but this is at insignificant risk levels [117].

Concerning environmental impact, the carbon-free fuels ammonia and hydrogen were proposed by Bicer and Dincer [118] to replace heavy fuel oils in the engines of maritime transportation. These authors also proposed hydrogen and ammonia as dual fuels to reduce greenhouse gas emissions. To this end, they conducted a study to assess the environmental impacts of this fuel substitution on transoceanic tankers

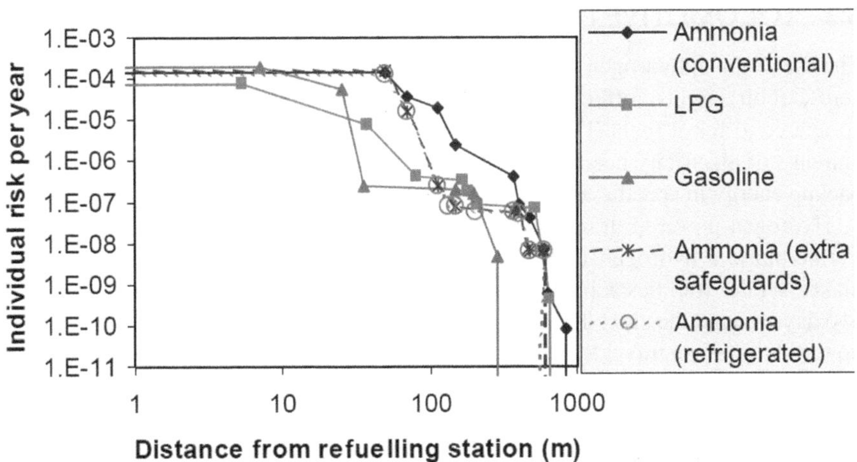

FIGURE 6.21 Individual risk with distance from a refueling station for fueling of equal energy amounts of ammonia, LPG, and gasoline. The options "ammonia (extra safeguards)" and "ammonia (refrigerated)" include both extra safeguards to mitigate hose ruptures. With Permission from Duijm, N.J.

Source: [117].

and freight ships. The life-cycle analysis considered the whole transport life-cycle of ships from their manufacture to fuel production, transportation, and use of hydrogen and ammonia in maritime vessels. Several hydrogen and ammonia production routes, from municipal waste to geothermal options, were analyzed to investigate environmentally benign routes. The examination of various effects on the marine environment included marine sediment and aquatic ecotoxicity. It was found that the ecotoxicity impacts on the marine environment highly depended on the production route of hydrogen and ammonia. The results indicated that by utilizing hydrogen and ammonia as dual fuels, the global warming potential would be considerably lower compared with heavy fuel oil. Thus, using geothermally produced hydrogen and ammonia in tankers would release around 0.98 g and 1.65 g CO_2 eq/tonne-km, respectively, while a heavy fuel oil tanker releases about 5.33 g CO_2 eq/tonne-km in greenhouse gas emissions.

Deniz and Zincir [19] conducted a comparative assessment of alternative fuels that could be used on ships. Thus, methanol, ethanol, liquefied natural gas, and hydrogen were compared environmentally and economically. Eleven criteria were used in the study for comparison from various perspectives, and it was found that methanol and ethanol are not favorable to be used onboard for several reasons. The most highly evaluated and suitable alternative fuel was found to be liquefied natural gas. However, hydrogen proved to be better as regards safety, bunker capability, durability, adaptability to existing ships, and commercial effects criteria. Finally, this study shows that hydrogen could be an alternative to liquefied natural gas as a fuel for ships. Nevertheless, it still requires further studies on compliance with emissions regulations and the effect on engine components issues.

6.3 LOCOMOTIVE TRANSPORTATION

The use of green hydrogen as a locomotive fuel would be very beneficial for meeting zero-carbon targets, offering in addition to zero emissions a higher efficiency compared to fossil fuels. The other alternative, battery power, cannot store the great amounts of electricity needed in trains and consequently can only be used as hybrid backup energy in specific applications [119].

Hydrogen-powered trains can attain significant speeds of up to 140 km/h. Furthermore, a hydrogen fuel-cell passenger train developed by Swiss rail vehicle maker Stadler Rail has achieved a new Guinness World Record, traveling for almost two days around the clock for a distance of 2803 km (1741.7 miles), whereas battery-powered trains can travel less than one-tenth of that distance before recharging again. The hydrogen trains' superiority extends also to the refueling time, which is less than 20 minutes compared to many hours of recharging for the battery trains.

Trains can utilize hydrogen as a fuel in modified internal combustion engines, but the state of the art is its use as a fuel in fuel cells due to their higher efficiency but also their quiet operation. The latter also occurs in battery-powered trains.

Concerning safety, studies have shown that hydrogen-powered vehicles using fuel cells are safer than those using internal combustion engines. Concerns about hydrogen safety stem from its high flammability and explosibility when mixed with air. However, the principal advantage of hydrogen is its supreme buoyancy which takes out hydrogen quickly upwards in case of release. Hydrogen is also a safer alternative to diesel trains. Nevertheless, the energy cost to convert electricity to green hydrogen and then back to electricity is very high unless the initial electricity comes from the surplus electricity of the power grid. The utilization of existing electrified tracks is more efficient than hydrogen-powered and battery-powered trains on routes with more than four trains per hour. On the other hand, the installation of new electrified tracks is quite expensive, and this should also be taken into account. In conclusion, hydrogen trains may not currently be the perfect solution, but they should surely be part of the future locomotive transportation system [119].

Hydrogen fuel cell-powered locomotives have been compared with battery locomotives and it has been found that the former is twice as powerful as the latter [120]. East Japan Railway Company has tested a hybrid diesel generator and passenger rail powered with a hydrogen fuel cell with a power equal to 65 kW. However, fuel cell-powered locomotives still have to cope with some serious issues like the large volumes needed for storage of hydrogen due to its very low density and its flammability hazards. The very high storage pressure of hydrogen is also a severe hazard onboard [24].

Alstom Coradia iLint also set in motion a project on a hydrogen fuel cell-powered locomotive at InnoTrans 2016, and after only two years, in 2018, the iLint™ started commercial service in Germany. Thus, Coradia iLint™ is the first passenger hydrogen fuel cell train worldwide with zero emissions and low noise levels. This train is precisely designed to run on either non-electrified or partially electrified lines with high levels of performance. The iLint™ was designed by the Alstom group in Salzgitter (Germany), a center of excellence for trains, and in Tarbes (France), a center of excellence for traction systems [121].

In the USA, several big companies, namely Progress Rail Inc. (a Caterpillar Inc. Company), BNSF Railway Company, and Chevron USA Inc. (a subsidiary of Chevron Corporation), made known on December 14, 2021 a memorandum of understanding to design and set in action locomotives powered by hydrogen fuel cells. Their task was to investigate the attainability of hydrogen fuel cells compared with conventional fuels for line-haul (transport of goods) rail. There are three principal goals set by the companies in the memorandum of understanding aimed at reaching agreed decisions. First, Progress Rail schedules to design and construct a prototype hydrogen fuel cell locomotive for various types of rail service. Second, Chevron plans to develop the infrastructure to support this project. Third, the prototype hydrogen fuel cell locomotive is scheduled to be given a practical exhibition on BNSF's rail lines at a jointly agreed appropriate time [122].

The first hydrogen-powered locomotive in North America was the prototype hydrogen train which was designed, constructed, and tested on November 30, 2022 by Canadian Pacific Railway utilizing a combination of fuel cells and batteries to supply electrical energy to the traction motors. The project started in late 2020 with the replacement of the diesel tanks by a battery system and of the cooling system and radiator fans by hydrogen storage, while fuel cells replaced the diesel engine and alternator. Meanwhile, a hydrogen storage station has already been constructed on the route. These trains are able to run at a maximum speed of 140 km/h, or 87 mph, although regular speeds on the line are between 80–120 km/h [123–125].

6.4 MARITIME TRANSPORTATION

Various studies have been conducted on the use of different fuel cell technologies for marine applications, for instance:

- Generation of electricity or emergency electrical systems by fuel cells in ships [126].
- Development of a 20 kW hybrid fuel cell power system for small ships [127].
- Energy production on ships by proton exchange membrane fuel cells [128, 129].
- Development of a power system based on polymer electrolyte fuel cells for long-lasting function of sensors [130].
- Proficiency of gas turbine hybrid systems based on solid oxide fuel cells [131].
- Use of fuel cells in ships of the European Maritime Safety Agency [132].
- Development of versatile energy management systems for hybrid fuel cell-powered ships [133].
- Comparison of various plant designs and fuel storage solutions for fuel cell utilization on small ships [134].

These studies confirm that the most promising types of fuel cells for marine applications are low- and high-temperature PEMFCs and SOFCs. Yet, despite significant development in the field, the technology of fuel cells for ships is still at the assessment stage. In addition, the space needed for onboard storage of hydrogen to be used by fuel cells is usually more than that needed for diesel generators. It is expected that fuel cells that utilize liquefied hydrogen will address this issue. Another handicap of fuel cells is the lack of hydrogen storage and delivery system infrastructure in harbors, which prohibits their immediate use in maritime transport.

Kim et al. [135] performed a study on an ammonia-powered ship considering environmental and economic factors. The systems evaluated for the use of ammonia are the polymer electrolyte membrane fuel cell (PEMFC) and the solid oxide fuel cell (SOFC). The SOFC has the tremendous advantage of the direct use of ammonia as a fuel. Yet its main disadvantage is that it cannot quickly increase the fuel delivery rate in case of sudden power demand. To address this problem, a complementary power system is needed to compensate for the delayed response of the SOFC during transient operations. This energy storage system (ESS) would be used to complement (backup) power, and also as a cold-start energy supply. The concept is for the ESS to be charged through the otherwise useless power at a temporary load decrease and to discharge at a subsequent load rise. The ESS can encounter extra load changes, notably in unexpectedly heavy weather. These ammonia-using systems were compared to conventional engines that are fueled by heavy fuel oil, focusing on the economic and environmental viewpoints. The comparison showed that ammonia can be a competitive carbon-free fuel for ships. In addition, the SOFC power system is the most eco-friendly alternative (up to 92.1%) among the proposed systems. Even though this study has some limitations and assumptions, the results indicate that ammonia-fueled ships could offer a meaningful perspective toward solving greenhouse gas issues in the maritime industry.

Concerning the safety of hydrogen versus batteries in ships, Mylonopoulos et al. [136] performed a safety study for a high-speed passenger ferry. The authors compared a battery and a proton exchange membrane hydrogen fuel cell propulsion system. Both versions were properly sized and fitted according to the applicable standards, and their impact on the overall design was evaluated. Hazards for both designs were identified, and their frequency and consequence indexes were identified qualitatively according to the ABS New Technology Qualification and SOLAS Alternative Design and Arrangements. Following this qualitative approach, specific hazard elimination proposals were recommended to reduce the risks of the most common accidental events. The highest-ranked risks were evaluated by quantitative risk assessments using the PyroSim software. The gas dispersion analysis performed for the hydrogen-powered ship indicated that it is crucial for the leakage in the fuel cell room to be stopped within 1 s after detection. This immediate response is expected to prevent the formation of an explosive atmosphere. For the battery propulsion system, the smoke and fire simulation found that the firefighting system could achieve a 30% reduction in fire duration when the fire doors are closed and the ventilation is shut, compared to the scenario without a firefighting system.

Overall, it seems that both hydrogen and battery designs are acceptable concerning safety as long as all the proper risk control measures are taken to mitigate potential accident consequences. Gastight bulkheads and fire-resistant doors are compulsory in hydrogen and battery versions, respectively, and ignition sources should be eliminated in both cases. Hydrogen design can be considered of higher risk due to the wide explosive range of the fuel with air. Appropriate measures, in terms of gas detection and ventilation, should be taken to keep away from the flammability concentration range of hydrogen in the air (4–75%), but especially far away from the detonability range (18–59%). The deflagration-to-detonation transition (DDT phenomenon was assumed to be improbable in a room of volume 19.19 m^3, which means that flame propagation speed would not be significantly increased to reach full detonation effects. In the case of a fire in the fuel cell or battery rooms, the timely use of a firefighting system can significantly reduce the fire duration and minimize the damaging impact [136].

Wang et al. [137] in a study compared three green-powered technology applications on a marine ferry: hydrogen, ammonia, and batteries. The study aimed to compare the fuel consumption, equipment weight, and cost of these systems. The results showed that the difference in the weight of the systems between the battery and the hydrogen version is minor, with the battery system weight being 13,146 kg and the hydrogen system weight 13,646 kg. The system weight of the ammonia-powered version (21,320 kg) is about 1.6 times the battery version system weight. The increased ammonia system weight is due to the weight of ammonia fuel cell stacks in the system, which is 12,100 kg (about 57% of the whole system weight). The total costs for the three designs were estimated in this study to be USD1.24, 3.47, and 7.66 million for the battery, hydrogen, and ammonia versions, respectively. The safety issues are also investigated in the study, including the existing safety design and equipment on the battery-powered version of the ferry, and focus on the importance of ventilation and firefighting systems. For the hydrogen system, special attention was also paid to the detection mechanisms and the safety valves to avoid the flammability range concentrations in case of leakage, due to hydrogen's high flammability. For the ammonia system, safety was mainly focused on its high toxicity which necessitates further developments before the application of ammonia onboard passenger ships.

6.5 HYDROGEN IN SPACE EXPLORATION

6.5.1 LIQUID HYDROGEN: THE FUEL OF CHOICE FOR NASA's SPACE EXPLORATION PROGRAM

The following excerpt is from the Introduction of *Taming Liquid Hydrogen: The Centaur Upper Stage Rocket, 1958–2002* [138]. This report details why the Centaur was so important in NASA history as an upper-stage rocket, being the critical link between its booster stage (Atlas or Titan) and the mission's payload (satellite or spacecraft).

Despite criticism and early technical failures, the taming of liquid hydrogen proved to be one of NASA's most significant technical accomplishments. Hydrogen, being a light and extremely powerful rocket propellant, has the lowest molecular weight of any known substance and burns with extreme intensity (3,038°C or 5,500°F). In combination with an oxidizer such as liquid oxygen, liquid hydrogen yields the highest specific impulse, or efficiency in relation to the amount of propellant consumed, of any known rocket propellant. Because liquid oxygen and liquid hydrogen are both cryogenic gases (liquefied only at extremely low temperatures) they pose enormous technical challenges. Liquid hydrogen must be stored at minus 253°C (minus 423°F) and handled with extreme care. To keep it from evaporating or boiling off, rockets fueled with liquid hydrogen must be carefully insulated from all sources of heat, such as rocket engine exhaust and air friction during flight through the atmosphere. Once the vehicle reaches space, it must be protected from the radiant heat of the Sun. When liquid hydrogen absorbs heat, it expands rapidly. Thus, venting is necessary to prevent the tank from exploding. Metals exposed to the extreme cold of liquid hydrogen become brittle. Moreover, liquid hydrogen can leak through minute pores in welded seams. Solving all these problems required an enormous amount of technical expertise in rocket and aircraft fuels cultivated over a decade by researchers at the National Advisory Committee for Aeronautics (NACA) Lewis Flight Propulsion Laboratory in Cleveland. Today, liquid hydrogen is the signature fuel of the American space program and is used by other countries in the business of launching satellites. In addition to the Atlas, Boeing's Delta III and Delta IV now have liquid-oxygen/ liquid-hydrogen upper stages. This propellant combination was also burned in the main engine of the Space Shuttle. One of the significant challenges for the European Space Agency was to develop a liquid-hydrogen stage for the Ariane rocket in the 1970s. The Soviet Union did not even test a liquid-hydrogen upper stage until the mid-1980s. The Russians are now designing their Angara launch vehicle family with liquid-hydrogen upper stages. Lack of Soviet liquid-hydrogen technology proved a serious handicap in the race of the two superpowers to the Moon. Taming liquid hydrogen is one of the significant technical achievements of twentieth century American rocketry.

At present, NASA's Space Launch System (SLS) is a carrier rocket that can carry the Orion spacecraft, astronauts, and heavy cargo straight to the Moon. Artemis 1 (an uncrewed spacecraft) was sent to the Moon from Kennedy Space Center on November

16, 2022, successfully implemented its mission, and returned to Earth on December 11, 2022. This is the first step following many years of NASA's prioritization of other missions since the Apollo program decades ago. Its main objective was to test the heat shield of the Orion spacecraft. These missions aim to improve current technologies required for future scientific missions, including the exploration of Mars.

The first stage of the carrier rocket is powered by one central engine and two external solid boosters. The Boeing Company, in Huntsville, Alabama, manufactures the SLS central engine system, including the avionics that regulate the vehicle's route during flight. With a height of around 65 m (212 feet) and a diameter of 8.4 m (27.6 feet), the core system storage capacity is 2,763,350 L (730,000 gallons) of liquid hydrogen and liquid oxygen that supplies fuel and oxidizer for the RS-25 engines. This liquid hydrogen and liquid oxygen propulsion system for the core engine is also scheduled for Artemis 2 and 3 as the upper stage of SLS Block 1 [139].

6.5.2 FRANCE'S AND THE EUROPEAN SPACE AGENCY'S SPACE EXPLORATION PROGRAMS

The *French space program* started in 1946 when the Laboratoire de recherches balistiques et aérodynamiques (LRBA, Ballistic and Aerodynamic Research Laboratory) was created in Vernon to develop new rockets based on the German V2 rocket. Astérix was the first French satellite that was successfully launched into space by the Diamant rocket from the Algerian desert on November 26, 1965. France's space launch pads and the French government space agency (CNES) were established in Kourou, French Guiana, in 1965. In 1973, France led the generation of the European Space Agency (ESA) and became its first contributor. With Kourou as the location of the space launch pads of CNES, the French space program receives the advantage of the best ground position for launch sites on Earth. This is because its position (5.3° north of the equator) offers rockets the extra propulsion from the Earth's spinning when the rockets are launched eastward (+460 m/s), thus saving on propellant. No other launch site of other countries benefits from this physical advantage [140].

European Space Agency's space program has developed cryogenic liquid propulsion engines for more than 30 years, with the most recent Vulcain engine being a successful hydrogen/oxygen propulsion system. It is a major technological project by itself and absorbs a quarter of the total investment cost dedicated to the Ariane-5 development program.

In 1957, the French Ministry of Defense decided to make a start on research in cold-temperature propulsion using liquid hydrogen and liquid oxygen, which provide a superior performance to all other fuels in a rocket engine. The major impediment stems from the very difficult control of these two propellants in their extremely cold liquid state (−251°C for hydrogen and −184°C for oxygen). Handling and operating these two liquefied substances in a rocket engine needs sturdy design and construction and special materials as well.

The first cryogenic propulsion engine was made in the United States in 1962, and two years later the French space agency CNES tested its version. The development of these cryogenic engines was expeditious, and by 2005 their performance had improved tremendously. The 1964 prototype engine possessed a one-tonne thrust,

FIGURE 6.22   A Vulcain engine using cryogenic liquid propulsion. Courtesy of the European Space Agency.

Source: [140].

Ariane-4's third stage could release 10 tonnes and the Ariane-5 Vulcain engine could offer 115 tonnes of thrust in its initial version. The latest Vulcain 2 engine provides a thrust of around 130 tonnes (Figure 6.22). This advancement has been accomplished by installing an extremely powerful turbopump into a relatively small space and by successfully controlling the extreme temperatures developed. Parts of the liquid hydrogen turbopump operate at $-251\,^{\circ}$C, being only 50 cm away from the combustion chamber which is at a temperature of $1500\,^{\circ}$C.

The Vulcain engine is a top technological achievement of the Ariane-5 carrier rocket. In total, this engine supplies 8% of the whole thrust required at takeoff and the whole thrust of the propulsion stage after the rejection of boosters. The dimensions of the Vulcain engine are 3 m in height and 1.76 m in diameter, and it weighs 1686 kg. Gimbals are used to fix it on a thrust casing located at the base of the main fuel and oxidizer tanks. Two high-speed turbopumps are used to supply fuel and oxidizer. These pumps inject the cryogenic propellants at high pressure into the combustion chamber at a rate of 235 kg/s. The liquid oxygen pump's power is 3 MW and rotates at a rate of 13,600 rpm. The liquid hydrogen pump revolves at 34,000 rpm and its power is 12 MW, which is the equivalent of the engines of two of France's high-speed TGV trains traveling at maximum speed. The flow rate of liquid hydrogen supplied into the combustion chamber is 41.2 kg/s [141].

6.5.3 Use of Hydrogen in Russia's Space Exploration Program

Russia's space exploration program is based either on kerosene (RP-1) as a fuel and liquid oxygen (LOX) as an oxidizer or the propellant system N_2O_4/UDMH (nitrous oxide/unsymmetrical dimethylhydrazine). The use of the LH_2/LOX system is scheduled for the Angara rocket family which is under development and only for the third stage of the rocket engines [142].

6.5.4 China's Space Exploration Program

China has successfully organized and carried out a final test for its 25 tons of liquid hydrogen–oxygen rocket engine recently. According to its designers, this is the world's greatest testing of its kind and the key technology of China's national program for the development of a super-heavy carrier rocket.

The China Aviation Supplies Holding Company (CAS) announced in December 2021 that this new rocket engine is highly efficient and reliable, performing multiple re-starts and variable thrust adjustments. CAS also claims that this system is capable of highly demanding space tasks such as manned Moon and Mars missions. This new engine is considered the most powerful worldwide, three times stronger than the ones currently used for the upper stage of the Chinese rocket in active service. It is a milestone of China's advancement into the world space powers.

The 25 ton liquid hydrogen–oxygen rocket engine designers of CAS do not provide detailed information about the new engine, and no video testing footage has been shown yet. Nevertheless, space specialists assume that the new engine is probably going to be employed in the third and upper stages of China's Long March-9 super-heavy carrier rocket. The first flight of this super-heavy launch carrier is not anticipated before China's next five-year plan of 2026 to 2030 [143].

China has recently conducted space fuel cell tests on a payload put in orbit by the Tianzhou-5 cargo carrier rocket. This was the first time China organized and carried out such testing according to the China Academy of Space Technology (CAST). The tests aimed to validate the system's performance in extreme conditions such as vacuum, low temperature, and microgravity. The results will yield significant data for developing fuel cell systems to be used in more demanding programs like lunar missions. The carrier rocket Tianzhou-5 launched from the Wenchang Spacecraft Launch Site in the southern island province of Hainan on November 12, 2022. The space center is located in the south (19°12'N north of the equator), yet not as far south as France's space launch pads at Kourou, French Guiana (5.3° north of the equator), which benefits significantly from the Earth's spinning when the rockets are launched eastward (+460 m/s), thus saving propellant [144].

6.6 HYDROGEN-POWERED AVIATION

6.6.1 Decarbonization Through Hydrogen Use in Aviation

Hydrogen as an energy source will play a crucial role in transforming aviation into a zero-carbon system over the next few decades. Novel and groundbreaking aircraft and systems innovations combined with hydrogen technologies can reduce the global warming effect of aircraft by 50 to 90%. Moreover, these innovations could meet the severe reduction goals for aviation emissions set out in the EU Green Deal.

A new independent European study [145] commissioned by Clean Sky 2 and Fuel Cells & Hydrogen 2 Joint Undertakings [146] on hydrogen's potential for use in aviation, was presented at an event on 22 June 2020. It is stressed in the study that hydrogen can be a primary energy source for propulsion, either for fuel cells, direct burn in gas turbine engines, or as a chemical building block for synthetic liquid fuels. Thus, it is concluded in this report that hydrogen could practically power aircraft by 2035 for short-range aviation. This will increase the flight cost by less than €18 extra per person on a short-range flight but will reduce the impact on climate by 50 to 90%.

Such revolutionary innovation will require significant additional aircraft research and development, improvements in fuel cells, and better liquid hydrogen vessels, as well as further investments in hydrogen infrastructure and new regulations and standards to obtain safe, reliable, and economical hydrogen-powered aviation. It is anticipated by experts that these achievements will take ten to 15 years, and therefore intensive research should begin as soon as possible. The study found that the first hydrogen-powered demonstrators may be ready by 2028 provided that sufficient progress on research and innovation could be accomplished by then.

The findings and factors supporting the conclusion of McKinsey & Company's report [145] that hydrogen has the capacity to play a significant role in future propulsion technology are the following:

- *Hydrogen propulsion could have a beneficial effect on climate change.* Hydrogen's extensive use in flight will put an end to CO_2 emissions. With regard also to non-CO_2 emissions, and taking into account the unpredictabilities of these effects, recent studies indicate that hydrogen combustion could reduce climate impact from flights by 50 to 75%.
- *Considering imminent technical developments, hydrogen propulsion is most pertinent to commuter, regional, short-range, and medium-range aviation.* Fuel cell-powered propulsion is the most suitable for commuter and regional aircraft because it is the most energy-efficient, climate-friendly, and economical alternative. Selecting this option in place of conventional aircrafts, the operational costs will increase by only USD5–10 per passenger (PAX) (about 10% per PAX).
- *Long-range aviation needs new aircraft designs for hydrogen.* The use of hydrogen in aviation is practically achievable but less suitable for long-range travel from an economic viewpoint. The bulkier and heavier hydrogen tanks would increase the aircraft's length and energy demand, resulting in 40 to 50% increased costs per PAX.
- *Feasibility and economic studies consider hydrogen as a major factor in future aviation.* If hydrogen-powered aircraft are utilized in the most cost-efficient sections of decarbonization, they could contribute to 40% of all aircraft by 2050, with an increased share after 2050.
- *Refueling infrastructure is an achievable challenge in the near future but will require international co-ordination.* By 2040 aviation's global demand for liquid hydrogen would require 10 million tons per annum, which is 5% of the

projected total global hydrogen demand. This aviation demand is expected to be served by existing supply chains now directed to other industries.

- *A more challenging, yet feasible scale-up after 2040 is required.* By 2050, aviation's demand for liquid hydrogen would grow to 40 million tons per year, thus requiring a substantial scale-up in the hydrogen supply chain and airport refueling infrastructure.
- *Courageous steps are urgently needed to take action on decarbonization through hydrogen.* The aviation industry should move faster, as commercialization and certification of aircraft take more than ten years, and considerable fleet replacement another ten years.

6.6.2 Safety Issues Raised Due to the Use of Hydrogen in Aviation

According to McKinsey & Company [145], safety regulations would need to be re-evaluated for LH_2 use in aviation, considering the radically different properties of hydrogen versus fossil jet fuels. Concerning hydrogen storage, hydrogen can be stored as pressurized gas or in liquid form. Since compressed gaseous hydrogen is commercially available and seems to be suitable for shorter flights, McKinsey & Company concentrates on liquid hydrogen (LH_2) storage tanks as they require roughly half as much volume, and they are significantly lighter than tanks for gaseous hydrogen. This is especially important for short- to long-range segments, where aircraft will carry several tons of hydrogen per flight. Compared to kerosene, LH_2 tanks are still about four times as big. Since LH_2 needs to remain cold and heat transfer must be minimized to avoid the vaporization of hydrogen, spherical or cylindrical tanks are required to keep losses low. To efficiently integrate the tanks into the aircraft's fuselage, the airframe will need to be extended, which increases the aircraft's operating empty weight [147–149].

For small aircraft, the problem of hydrogen storage could be solved by storing it in pods under the wings. This would not influence flight performance and maintenance, and safety could be dealt with more easily. Nevertheless, as regards hydrogen storage under wings, safety and reliability systems have not been developed enough yet, so they should be optimized along with hydrogen propulsion systems following extensive testing and certification for commercial aviation. Hydrogen turbines should also be checked against permissible NO_x emissions considering environmental impact. As regards larger airplanes, the fuselage needs to be elongated by five meters to accommodate the two bulky LH_2 vessels to be put behind the passenger cabin. Yet this design raises safety issues since the LH_2 has to pass by the side of the passenger cabin to fuel the two wing-mounted engines.

Refueling infrastructure is another issue of concern as regards the necessary safety perimeter around the hydrogen storage area as determined by the *Seveso Directive* in the EU. However, regional airports normally have plenty of space around them so they could easily address this issue.

On the other hand, considering large airports, hundreds of thousands of tons of LH_2 per year will be required to supply fuel to aircraft, even for only a 25% switch to hydrogen. Thus, the hundreds of delivery trucks traveling every day on the already congested roads to the airport would pose a severe threat to the area. Another more suitable means of transport would be a pressurized hydrogen pipeline. This may be an existing natural gas pipeline that should be rehabilitated before its use as a hydrogen transmitting pipeline. This option is attractive as regards cost-effectiveness since it uses existing infrastructure, but in case of technical difficulties, a new hydrogen-dedicated pipeline would need to be constructed. There is also the possibility of serving an airport with hydrogen produced directly from adjacent water electrolysis plants when renewable energy is available in the area, which is a much more competitive option both economically and environmentally.

According to McKinsey & Company's report [145], new regulations should be developed to ensure the safe handling of LH_2 during its transport and delivery of fuel with trucks, especially as regards free zones around trucks, which have not yet been determined. Also, the required periphery around refueling trucks might be much smaller than suggested as hydrogen does not create pools on the ground due to its high volatility and it evaporates rapidly moving upwards, as some experts claim. Thus, further research and testing on safety issues are needed before the widespread use of hydrogen as a fuel.

Furthermore, according to McKinsey & Company's report [145], additional research on safety measures during refueling should include:

- Hazard spotting and consequence assessment, including leakages during refueling.
- Leakage management and countermeasures.
- Safety standards and regulations to guarantee safe handling and refueling with LH_2.
- Required ignition-free zone around LH_2 refueling equipment and safety buffer zone of parallel operations.

Sustainable aviation fuels (SAFs) may also be an alternative to help achieve net-zero CO_2 emissions by the middle of this century. Nevertheless, experts do not believe that SAFs can fully support aviation to decarbonization by mid-century due to a lack of economic and environmental sustainability. Thus, we are in search of more radical options.

The findings of the Insight Report for the World Economic Forum, July 2022 [150] show that SAFs are not the only option for decarbonizing the aviation sector. Alternative propulsion options exist and could begin to replace jet-fuel aircraft by 2035. These options are related to the development of battery electric, hydrogen fuel cell electric, and hydrogen combustion aircraft. In addition, more work is needed to determine the infrastructure changes, policy and regulatory frameworks, and industry adaptation that will be required.

As quoted in the World Economic Forum's report [150] in July 2022:

Battery-powered aircraft *would eliminate CO_2 as well as all other in-flight emissions. When powered by fully renewable or low-carbon energy, the electricity used to power these aircraft would have minimal climate impact. There are, however, challenges that will need to be overcome to avoid charging batteries directly from electrical grids, which in many parts of the world will remain* *reliant on fossil fuels for decades to come, as well as managing climate impacts currently associated with the production of batteries.*

Based on forecast battery gravimetric energy densities, the maximum operating range of lithium-ion battery-electric aircraft by 2035 is expected to be around 400 km, rising to 600 km in 2050. Extending the range of these aircraft beyond this limit would require the use of breakthrough battery technologies, the development and commercialization of which are extremely difficult to forecast.

Hydrogen fuel cell electric *would eliminate the majority of emissions but could still release water vapor into the atmosphere, which might lead to significant contrail formation – though there is considerable uncertainty here. Certain design decisions and the low temperatures at which fuel cells operate could create the potential to condense any water vapor into a liquid in the exhaust that would likely eliminate the chance of contrail formation. Still, the overall uncertainty about the impact of contrails from these aircraft makes it especially difficult to assess their full climate impact. Fuel cell technology would allow aircraft to be designed for a much longer range than battery electric aircraft – around 2,000 km by 2030, and a possibility of reaching 4,000 km by 2035.*

Hydrogen combustion aircraft *would eliminate CO_2 and soot emissions in-flight but still produce nitrogen oxides (NO_x) emissions, resulting in increased water vapor compared to jet fuel. As with fuel cell aircraft, assessing the total climate impact of hydrogen combustion aircraft is especially difficult due to the high uncertainty about the impact of hydrogen contrails. Even if the impact of these contrails were greater than those produced from flying on jet fuel, there remains the possibility that changes to aircraft operations could be used to reduce or eliminate contrail formation from both hydrogen and traditional jet fuel aircraft. By 2035, hydrogen aircraft may be able to operate over the same distances as jet engines, meaning they offer the potential to replace current jet fuel-powered aircraft at any range and are a viable option for decarbonizing the longest-range flights.*

This report identifies eight key factors that are necessary for realizing the potential of alternative propulsion for moving us toward an aviation system with a true zero climate impact:

1. Ensuring aviation batteries are charged with renewable energy.
2. Accelerating the introduction of green hydrogen.
3. Improving battery life-cycles and management for aviation.
4. Improving battery-electric aircraft energy density.
5. Developing lighter fuel cell systems.
6. Developing lighter storage tanks for liquid hydrogen.
7. Redesigning aircraft for optimized hydrogen performance.
8. Contrail research and mitigation.

Moreover, Charpentreau claimed in AeroTime Hub-Aviation News, on June 29, 2022 [151]:

*The aviation industry is being pressured to cut aircraft emissions and create a more sustainable future for air transport, with both governments and the general public demanding action. While **a fully electric aircraft seems technologically out of reach** in the foreseeable future, several voices see in **hydrogen a mid-term solution toward zero-emission propulsion**.*

6.6.3 COMMERCIALIZATION OF HYDROGEN-POWERED AVIATION

At Airbus, they schedule the development of the world's first zero-emission commercial aircraft by 2035 using hydrogen as a fuel. The three zero-emission aircraft under design are called ZEROe concept aircraft and are all *hybrid-hydrogen aircrafts*. They use liquid hydrogen as a fuel and oxygen on modified gas turbine engines. Furthermore, these aircraft utilize *hydrogen fuel cells* to produce electrical power that supplements the gas turbine, to obtain an efficient hybrid-electric propulsion system [152].

The first ZEROe demonstrator aircraft was an A380 multimodal platform launched by Airbus in 2022. A demonstrator aircraft is an essential first step in developing new aircraft since it offers valuable information and testing concerning the design, processes, materials, and equipment. This information is then used to perform the required fine-tuning of the systems. After the successful completion of all tests, they hope to have prepared their hybrid-hydrogen aircraft by 2025 [153].

Fuel cells for the aviation industry are also scheduled currently by Airbus. To this end, Airbus has searched for a competitive partner and provider from the automotive industry and finally found ElringKlinger, a top constructor of fuel cells in the market. The two companies came into agreement and created the joint venture Aerostack in 2020. In addition, Aerostack co-operates now with some other specialized companies under the German government's project H2Sky [154].

A new propulsion system has been designed by the Stuttgart-based company H2FLY, which is based on direct fuel cell propulsion. The company's prototype HY4 is a double-fuselage aircraft designed to transport four people. The low-temperature hydrogen fuel cell system was developed by the German Space Agency (DLR) and they are now at the stage of integration of liquid hydrogen tanks into the HY4 test aircraft as reported by Hampel [155].

There is also another novel US–British joint venture concerning hydrogen fuel cells called ZeroAvia which aims to operate the first hydrogen-powered commercial flights by 2024. The company has already succeeded in using the electric-hydrogen propulsion system on a Piper M-series aircraft and hopes to test its ZA-600 hydrogen aircraft engine on a Dornier 228 aircraft as part of the HyFlyer II program [151].

Universal Hydrogen is a company based in California which is scheduled to enter regional aviation as soon as 2025, ten years before Airbus's ZEROe turboprop. To this end, the company designs a "plug-and-play" system utilizing existing aircraft, such as the Dash 8 and the ATR 72. Its conversion kit is composed of an electric powertrain and a fuel cell that takes the place of the existing turboprops. The hydrogen is stored in compact tanks developed especially for the system at the rear of the fuselage.

Concerning Boeing, the company started experimenting with hydrogen-powered aircraft as early as 2008 with the Fuel Cell Demonstrator. This was a two-seat Diamond DA20 light aircraft modified to run on a fuel cell by Boeing Research & Technology Europe which took off for a successful first test flight. Following this successful flight, the company has offered no further information on the program, and neither has its European competitor Airbus on its own hydrogen-powered technology. Moreover, in January 2021, Boeing chief executive David Calhoun announced that the company had dismissed hydrogen as a short-term solution. Instead, it seems that the company focuses on sustainable jet fuels although these are less effective in terms of reducing emissions. *"This is the only answer between now and 2050,"* Calhoun said[151].

As reported by Veenstra [156], the world's first-ever larger passenger aircraft (40–80 seats) powered by liquid green hydrogen is preparing to fly between the Netherlands and London in 2028. This major joint venture is constructing a hydrogen-powered aircraft that is entirely green, from fuel to propulsion. While fuel is normally stored in the wings currently, in this new aircraft the hydrogen tanks will be attached to the tail of the aircraft. Then the electricity produced by the engines will be used to rotate the propellers. A scale-up to larger aircraft is possible.

Finally, Trancossi et al. [157] have also designed a new *Hydrogen Fire-safe Airship* that avoids the safety issues observed in older airships. The main advantage of hydrogen is that it is more effective than helium because of its higher buoyancy and its comparatively low production cost by hydrolysis, especially when abundant renewable energy is available at the production site. The authors propose a new construction concept of the buoyant balloon designed in a way that increases fire safety when utilizing hydrogen as the buoyant gas. The proposed buoyant volume is departmentalized to ensure that hydrogen concentration does not reach the hazardous limits in the central airship balloon. This design will bring in a new era of hydrogen airships, which would be much safer than the previous ones. It would also reduce the transport economic costs, when compared with helium-inflated airships, due to the considerably lower price of hydrogen compared to helium.

Yusaf et al. [158] state that today the efficiency of under-development hydrogen-fueled aircraft has been improved but the main hurdles are the lack of refueling stations in airports, sizable production costs, and the consolidated carbon market that impedes their full commercialization. On the other hand, unmanned aerial vehicles (UAV) form another important sector of the aviation industry in which hydrogen has started to be commonly used as an alternative fuel in heavy-duty drones through fuel cell technology. They emphasize that the fundamental argument against hydrogen has been its exorbitant cost rather than its safety.

REFERENCES

1. Kletz, T.A., What you don't have, can't leak, *Chemistry and Industry*, 287–292, May 6, 1978.
2. Markert, F., Nielsen, S.K., Paulsen, J.L., and Andersen, V., Safety aspects of future infrastructured scenarios with hydrogen refuelling stations, *International Journal of Hydrogen Energy*, 32, 2227, 2007.

3. Bethoux, O., Hydrogen fuel cell road vehicles and their infrastructure: An option towards an environmentally friendly energy transition, *Energies*, 13, 6132, 2020.

4. Lins, P.H.C. and Almeida, A.T., Multidimensional risk analysis of hydrogen pipelines, *International Journal of Hydrogen Energy*, 37, 13545–13554, 2012.

5. Verhelst, S., Recent progress in the use of hydrogen as a fuel for internal combustion engines, *International Journal of Hydrogen Energy*, 39, 1071–1085, 2014.

6. Achates Power. https://achatespower.com/passenger-vehicle/

7. Kalghatgi, G., Development of fuel/engine systems—The way forward to sustainable transport, *Engineering*, 5, 510–518, 2019.

8. Kalghatgi, G., Is it really the end of internal combustion engines and petroleum in transport?, *Applied Energy*, 225, 965–974, 2018.

9. Kalghatgi, G., The challenges of energy transition needed to meet decarbonisation targets set up to address Climate change, *Journal of Automotive Safety and Energy*, 11 (3), 276–286, 2020.

10. Weger, L.B., Leitao, J., and Lawrence, M.G., Expected impacts on greenhouse gas and air pollutant emissions due to a possible transition towards a hydrogen economy in German road transport, *International Journal of Hydrogen Energy*, 46, 5875–5890, 2021.

11. Jayakumar, A., Madheswaran, D.K., Kannan, A.M., Sureshvaran, U., and Sathish, J., Can hydrogen be the sustainable fuel for mobility in India in the global context?, *International Journal of Hydrogen Energy*, 47, 33571–33596, 2022.

12. Sinigaglia, T., Lewiski, F., Martins, M.E.S., and Siluk, J.C.M., Production, storage, fuel stations of hydrogen and its utilization in automotive applications-A review, *International Journal of Hydrogen Energy*, 42, 24597–24611, 2017.

13. Milojevic, S., Optimization of the hydrogen system for city busses with respect to the traffic safety, in *20th World Hydrogen Energy Conference (WHEC 2014)*, vol. 2, Gwangju, South Korea, 15-20 June 2014. ISBN: 978-1-63439-655-4

14. Jeerh, G., Zhang, M., and Tao, S., Recent progress in ammonia fuel cells and their potential applications, *Journal of Materials Chemistry A*, 9, 727, 2021.

15. Ravi, S.S. and Aziz, M., Clean hydrogen for mobility e Quo vadis?, *International Journal of Hydrogen Energy*, 47, 20632–20661, 2022.

16. Electrofuel. https://en.wikipedia.org/wiki/Electrofuel#CITEREFRoyal_Society2019.

17. U.S. Department of Transportation, *Guidelines for Use of Hydrogen Fuel in Commercial Vehicles*, Federal Motor Carrier Safety Administration, Washington, DC, November 2007.

18. Gurz, M., Baltacioglu, E., Hames, Y., and Kaya, K., The meeting of hydrogen and automotive: A review, *International Journal of Hydrogen Energy*, 42, 23334–23346, 2017.

19. Deniz, C. and Zincir, B., Environmental and economical assessment of alternative marine fuels, *Journal of Cleaner Production*, 113, 438–449, 2016.

20. Verhelst, S., Maesschalck, P., Rombaut, N., and Sierens, R., Increasing the power output of hydrogen internal combustion engines by means of supercharging and exhaust gas recirculation, *International Journal of Hydrogen Energy*, 34, 4406–4412, 2009.

21. Xu, P., Ji, C., Wang, S., Bai, X., Cong, X., Su, T., and Shi, L., Realizing low emissions on a hydrogen-fueled spark ignition engine at the cold start period under rich combustion through ignition timing control, *International Journal of Hydrogen Energy*, 44, 8650–8658, 2019.

22. Fromm, L.J., Redon, F., and Salvi, A., *Opposed-Piston Engine Potential: Low CO2 and Criteria Emissions*, Chapter 4 in: *Engines and Fuels for Future Transport*, Book series: Energy, Environment, and Sustainability, edited by Kalghatgi, G., Agarwal, A.K., Leach, F., Senecal, Springer, Singapore, 2022.

23. Czyz, Z., Siadkowska, K., and Sochaczewski, R., CFD analysis of charge exchange in an Aircraft opposed-Piston diesel engine, *MATEC Web of Conferences,* 252, 04002, 2019. https://doi.org/10.1051/matecconf/201925204002

24. Abohamzeh, E., Salehi, F., Sheikholeslami,. M., and Abbasi, R., Review of hydrogen safety during storage, transmission, and applications processes, *Journal of Loss Prevention in the Process Industries,* 72, 104569, 2021.

25. Sharaf, O.Z. and Orhan, M.F., An overview of fuel cell technology: Fundamentals and applications, *Renewable and Sustainable Energy Reviews,* 32, 810–853, 2014.

26. Kirubakaran, A., Jain, S., and Nema, R.K., A review on fuel cell technologies and power electronic interface, *Renewable and Sustainable Energy Reviews,* 13, 2430–2440, 2009.

27. Field, K., Hydrogen fuel cell & battery electric vehicles — Technology rundown, *Clean Technica,* August 11, 2018. https://cleantechnica.com/2018/08/11/hydrogen-fuel-cell-battery-electric-vehicles-technology-rundown/

28. Thompson, ST., et al., Direct hydrogen fuel cell electric vehicle cost analysis: System and high volume manufacturing description, validation, and outlook, *Journal of Power Sources,* 399, 304–313, 2018.

29. Gates, B., Clean jet fuel needs to be more like a microwave oven, *GatesNotes,* July 13, 2021. www.gatesnotes.com/Energy/Introducing-a-new-way-to-invest-in-clean-energy-innovation

30. Andreas, Pros and Cons of Hydrogen Fuel Cell Cars. https://environmental-conscience.com/hydrogen-fuel-cell-cars-pros-cons/

31. Wikipedia, Hydrogen vehicle, https://en.wikipedia.org/wiki/Hydrogen_vehicle

32. Caldwell, A., *Why hydrogen cars will be Tesla's biggest threat,* Insider, July 14, 2020. www.businessinsider.com/hydrogen-fuel-cell-cars-teslas-biggest-threat-2019-12?r=DE&IR=T

33. Hydrogen Fuel Cell Partnership, Cost to Refill. https://cafcp.org/content/cost-refill, (accessed January 2024).

34. Baxter, T., Hydrogen cars won't overtake electric vehicles because they're hampered by the laws of science, *The Conversation,* 3 June, 2020. https://uk.news.yahoo.com/hydrogen-cars-wont-overtake-electric-111749065.html?guccounter=1&guce_referrer=aHR0cHM6Ly9lbi53aWtpcGVkaWEub3JnLw&guce_referrer_sig=AQAAADyLnjrtQ5iAcAY0cNNq9HssOfhmnWOi319sosXdvAcoCEafyV_leb79IATyQV4yQvf0ZW0uxszeHFJlXsCEXosCZxQkqO3F-MWZdXgVDSQEmpmC0TicIvTVSkoNqR2pykT-bLqQ8-7EM91rCjnhuU8MZFkhxsANgZVl7wiNXlDO

35. Arnold, N., Hydrogen Fuel Cell Cars: Everything You Need to Know, August 12 2022. www.bmw.com/en/innovation/how-hydrogen-fuel-cell-cars-work.html

36. Goyal, S., *Hydrogen car: What are the advantages and disadvantages of hydrogen fuel cells?,* Josh, March 31, 2022. www.jagranjosh.com/general-knowledge/hydrogen-car-advantages-and-disadvantages-of-hydrogen-fuel-cell-1648707956-1

37. Khalil, Y.F., Science-based framework for ensuring safe use of hydrogen as an energy carrier and an emission-free transportation fuel, *Process Safety and Environmental Protection,* 117, 326–340, 2018.

38. Zalosh, R., *CNG and hydrogen vehicle fuel tank failure incidents. Testing, and Preventive Measures, Technical Support and Evaluation of the Fuel Tank Tests; Recommendations for Research Priorities of Hydrogen Fueled Vehicles,* Motor Vehicle Fire Research Institute, Wellesley, MA, 2008.

39. Barthelemy, H., Weber, M., and Barbier, F., Hydrogen storage: Recent improvements and industrial perspectives, *International Journal of Hydrogen Energy,* 42, 7254–7262, 2017.

40. Moradi, R. and Groth, K.M., Hydrogen storage and delivery: Review of the state of the art technologies and risk and reliability analysis, *International Journal of Hydrogen Energy,* 44, 12254–12269, 2019.

41. Yersak, T.A., et al., Predictive model for depressurization-induced blistering of type IV tank liners for hydrogen storage, *International Journal of Hydrogen Energy,* 42, 28910–28917, 2017.

42. Pepin, J. et al., Replication of liner collapse phenomenon observed in hyperbaric type IV hydrogen storage vessel by explosive decompression experiments, *International Journal of Hydrogen Energy,* 43, 4671–4680, 2018.

43. Conway, G., Life-cycle analysis for the automotive sector, Chapter 6, in *Engines and Fuels for Future Transport*, Edited by Kalghatgi, G., Agarwal, A.K., Leach, F., Senecal, K., Springer, Oxford, UK, 2022.

44. Rigas, F. and Sklavounos, S., Evaluation of hazards associated with hydrogen storage facilities, *International Journal of Hydrogen Energy*, 30, 1501, 2005.

45. Hord, J., Is hydrogen a safe fuel? *International Journal of Hydrogen Energy*, 3, 157, 1978.

46. Ruban, et al., Fire risk on high-pressure full composite cylinders for automotive applications, *International Journal of Hydrogen Energy,* 37, 17630–17638, 2012.

47. Hupp, N., et al., Influence of fire intensity, fire impingement area and internal pressure on the fire resistance of composite pressure vessels for the storage of hydrogen in automobile applications, *Fire Safety Journal*, 104, 1–7, 2019.

48. Saldi, Z.S. and Wen, J.X., Modeling thermal response of polymer composite hydrogen cylinders subjected to external fires, *International Journal of Hydrogen Energy,* 42, 7513–7520, 2017.

49. Magneville, B., Gentilleau, B., Villalonga, S., Nony, F., and Galiano, H., Modeling, parameters identification and experimental validation of composite materials behavior law used in 700 bar type IV hydrogen high pressure storage vessel, *International Journal of Hydrogen Energy,* 40, 13193–13205, 2015.

50. Zheng, J., et al., Heat transfer analysis of high-pressure hydrogen storage tanks subjected to localized fire, *International Journal of Hydrogen Energy,* 37, 13125–13131, 2012.

51. Zheng, J., et al., Experimental and numerical investigation of localized fire test for high-pressure hydrogen storage tanks, *International Journal of Hydrogen Energy,* 38, 10963–10970, 2013.

52. Hong, J.H., Han, M.G., and Chang, S.H., Safety evaluation of 70 MPa-capacity type III hydrogen pressure vessel considering material degradation of composites due to temperature rise, *Composite Structures*, 113, 127–133, 2014.

53. Leh, D., et al., A progressive failure analysis of a 700-bar type IV hydrogen composite pressure vessel, *International Journal of Hydrogen Energy*, 40, 13206–13214, 2015.

54. Halm, D., et al., Composite pressure vessels for hydrogen storage in fire conditions: Fire tests and burst simulation, *International Journal of Hydrogen Energy,* 42, 20056–20070, 2017.

55. Wu, Q., Chen, X., Fan, Z., Jiang, Y., and Nie, D., Experimental and numerical studies of impact on filament-wound composite cylinder, *Acta Mechanica Solida Sinica*, 30, 540–549, 2017.

56. Demir, I., Sayman, O., Dogan, A., Arikan, V., and Arman, Y., The effects of repeated transverse impact load on the burst pressure of composite pressure vessel, *Composites: Part B*, 68, 121–125, 2015.

57. Ramirez, J.P.B., Halm, D., Grandidier, J.C., Villalonga, S., and Novy, F., 700 bar type IV high pressure hydrogen storage vessel burst e Simulation and experimental validation, *International Journal of Hydrogen Energy,* 40, 13183–13192, 2015.

58. Han, M.G. and Chang, S.H., Failure analysis of a Type III hydrogen pressure vessel under impact loading induced by free fall, *Composite Structures*, 127, 288–297, 2015.

59. Alcantar, Aceves, S.M., Ledesma, E., Ledesma, S., and Aguilera, E., Optimization of Type 4 composite pressure vessels using genetic algorithms and simulated annealing, *International Journal of Hydrogen Energy*, 42, 15770–15781, 2017.

60. Swain, M.R., Shriber, J., and Swain, M.N., Comparison of hydrogen, natural gas, liquefied petroleum gas, and gasoline leakage in a residential garage, *Energy and Fuels*, 12, 83, 1998.

61. Farrell, A.E., Keith, D.W., and Corbett, J.J., A strategy for introducing hydrogen into transportation, *Energy Policy*, 31, 1357, 2003.

62. Zalosh, R. and Weyandt, N., *Hydrogen fuel tank fire exposure burst test*, SAE 2005 World Congress & Exhibition, SAE Paper No. 2005-01-1886, 2005. ISSN: 0148-7191, e-ISSN: 2688-3627, https://doi.org/10.4271/2005-01-1886

63. Weyandt, N., *Intentional failure of a 5000 psig hydrogen cylinder installed in an SUV without standard required safety devices*, SAE World Congress & Exhibition, SAE Paper No. 2007-01-0431, 2007. ISSN: 0148-7191, e-ISSN: 2688-3627, https://doi.org/10.4271/2007-01-0431

64. Zalosh, R., Blast waves and fireballs generated by hydrogen fuel tank rupture during fire exposure, in *Proceedings of the 5th International Seminar on Fire and Explosion Hazards*, Edinburgh, UK, April 23–27, 2007.

65. Baker, W., Kulesz, J., Ricker, R., Westine, P., Parr, V., Vargas, L., and Mosely, P., Workbook for estimating effects of accidental explosions in propellant ground handling and transport systems, *NASA CR 3023*, August 1978.

66. Center for Chemical Process Safety, *Guidelines for Chemical Process Quantitative Risk Analysis*, 2nd ed., American Institute of Chemical Engineers, New York, 2000.

67. Stephenson, R., Crash-induced fire safety issues with hydrogen-fueled vehicles, in *Presented at National Hydrogen Association's 18th Annual U.S. Hydrogen Conference*, Washington, DC, March 2003.

68. Swain, M.R., Fuel leak simulation, in *Proceedings of the 2001 DOE Hydrogen Program Review*, NREL/CP-570-30535, U.S. Department of Energy, Washington, DC, 2001.

69. Muradov, N., Low to near-zero CO_2 production of hydrogen from fossil fuels: Status and perspectives, *International Journal of Hydrogen Energy*, 42, 14058–14088, 2017.

70. Tamura, Y., Takabayashi, M., and Takeuchi, M., The spread of fire from adjoining vehicles to a hydrogen fuel cell vehicle, *International Journal of Hydrogen Energy*, 39, 6169–6175, 2014.

71. Hussein, H., Brennan, S., and Molkov, V., Dispersion of hydrogen release in a naturally ventilated covered car park, *International Journal of Hydrogen Energy*, 45, 23882–23897, 2020.

72. Brennan, S., Hussein, H.G., Makarov, D., Shentsov, V., and Molkov, V., Pressure effects of an ignited release from onboard storage in a garage with a single vent, *International Journal of Hydrogen Energy*, 44, 8927–8934, 2019.

73. Adomaitis, N. and Reuters, Norway Fines Nel Units USD3 mln Over 2019 Blast at Hydrogen Fuel Station, February 17, 2021. www.reuters.com/business/energy/norway-fines-nel-units-3-mln-over-2019-blast-hydrogen-fuel-station-2021-02-16/

74. Jin, H., and Chung, J., and Reuters, Hydrogen Hurdles: A Deadly Blast Hampers South Korea's Big Fuel Cell Car Bet, 2019. www.reuters.com/article/us-autos-hydrogen-southkorea-insight-idUKKBN1W936A

75. Dadashzadeh, M., Kashkarov, S., and Makarov, D., Risk assessment methodology for onboard hydrogen storage, *International Journal of Hydrogen Energy,* 43, 6462–6475, 2018.

76. Li, Z. and Sun, K., Mitigation measures for intended hydrogen release from thermally activated pressure relief device of onboard storage, *International Journal of Hydrogen Energy,* 45, 9260–9267, 2020.

77. Ghatauray, T.S., Ingram, J.M., and Holborn, P.G., A comparison study into low leak rate buoyant gas dispersion in a small fuel cell enclosure using plain and Louvre vent passive ventilation schemes, *International Journal of Hydrogen Energy,* 44, 8904–8913, 2019.

78. Benitez, A., et al., Ecological assessment of fuel cell electric vehicles with special focus on type IV carbon fiber hydrogen tank, *Journal of Cleaner Production,* 278, 123277, 2021.

79. Saffers, J.B. and Molkov, V.V., Hydrogen safety engineering framework and elementary design safety tools, *International Journal of Hydrogen Energy,* 39, 6268–6285, 2014.

80. Li, Z. and Luo, Y., Comparisons of hazard distances and accident durations between hydrogen vehicles and CNG vehicles, *International Journal of Hydrogen Energy,* 44, 8954–8959, 2019.

81. Shea, K. and NJ.com, *WATCH: Garbage truck explodes in fireball, rips hole in nearby house,* NJ.com. https://www.nj.com/mercer/2016/01/garbage_truck_explosion_dam ages_hamilton_house.html, (accessed January 26, 2016).

82. Today, *Caught on camera: Natural-gas powered garbage truck explodes,* today, 2016. www.today.com/video/caught-on-camera-natural-gas-powered-garbage-truck-explo des-609780803613

83. Molkov, V.V., Cirrone, D.M.C., Shentsov, V.V., Dery, W., Kim, W., and Makarov, D.V., Dynamics of blast wave and fireball after hydrogen tank rupture in a fire in the open atmosphere, *International Journal of Hydrogen Energy,* 46, 4644–4665, 2021.

84. Wu, Y., Assessment of the impact of jet flame hazard from hydrogen cars in road tunnels, *Transportation Research Part C,* 16, 246, 2008.

85. Venetsanos, A.G., Baraldi, D., Adams, P., Heggem, P.S., and Wilkening, H., CFD modeling of hydrogen release, dispersion and combustion for automotive scenarios, *Journal of Loss Prevention in the Process Industries,* 21, 162, 2008.

86. Middha, P. and Hansen, O.R., CFD simulation study to investigate the risk from hydrogen vehicles in tunnels, *International Journal of Hydrogen Energy,* 34, 5875, 2009.

87. Baraldi, D., Kotchourko, A., Lelyakin, A., Yanez, J., Middha, P., Hansen, O.R., Gavrikov, A., Efimenko, A., Verbecke, F., Makarov, D., and Molkov, V., An intercomparison exercise on CFD model capabilities to simulate hydrogen deflagrations in a tunnel, *International Journal of Hydrogen Energy,* 34, 7862, 2009.

88. SAE International, Technical Information Report for Fuel Systems in Fuel Cell and Other Hydrogen Vehicles J2579_200901, 6 January 2009.

89. Makarov, D. and Molkov, V., Modelling of dispersion following hydrogen permeation for safety engineering and risk assessment, in *Presented at II International Conference: "Hydrogen Storage Technologies",* Moscow, Russia, October 28–29, 2009.

90. Adams, P., Bengaouer, A., Cariteau, B., Molkov, V., and Venetsanos, A.G., Allowable hydrogen permeation rate from road vehicles, *International Journal of Hydrogen Energy,* 36, 2742, 2011.

91. Kashkarov, S., Li, Z., and Molkov, V., Blast wave from a hydrogen tank rupture in a fire in the open: Hazard distance nomograms, *International Journal of Hydrogen Energy*, 45, 2429–2446, 2020.

92. Molkov, V. and Kashkarov, S., Blast wave from a high-pressure gas tank rupture in a fire: Stand-alone and under-vehicle hydrogen tanks, *International Journal of Hydrogen Energy*, 40, 12581–12603, 2015.

93. Clark, W.W., Rifkin, J., O'Connor, T., Swisher, J., Lipman, T., and Rambach, G., Hydrogen energy stations: Along the roadside to the hydrogen economy, *Utilities Policy*, 13, 41, 2005.

94. Kikukawa, S., Consequence analysis and safety verification of hydrogen fueling stations using CFD simulation, *International Journal of Hydrogen Energy*, 33, 1425, 2008.

95. Baraldi, D., Venetsanos, A.G., Papanikolaou, E., Heitsch, M., and Dallas, V., Numerical analysis of release, dispersion and combustion of liquid hydrogen in a mock-up hydrogen refuelling station, *Journal of Loss Prevention in the Process Industries*, 22, 303, 2009.

96. Makarov, D., Verbecke, F., Molkov, V., Roe, O., Skotenne, M., Kotchourko, A., Lelyakin, A., Yanez, J., Hansen, O., Middha, P., Ledin, S., Baraldi, D., Heitsch, M., Efimenko, A., and Gavrikov, A., An inter-comparison exercise on CFD model capabilities to predict a hydrogen explosion in a simulated vehicle refuelling environment, *International Journal of Hydrogen Energy*, 34, 2800, 2009.

97. Castiglia, F. and Giardina, M., Analysis of operator human errors in hydrogen refuelling stations: Comparison between human rate assessment techniques, *International Journal of Hydrogen Energy*, 38, 1166–1176, 2013.

98. Al-shanini, A., Ahmad, A., and Khan, F., Accident modelling and safety measure design of a hydrogen station, *International Journal of Hydrogen Energy*, 39, 20362–20370, 2014.

99. Qian, J.Y., Li, X.J., Gao, Z.X., and Jin, Z.J., A numerical study of unintended hydrogen release in a hydrogen refueling station, *International Journal of Hydrogen Energy*, 45, 20142–20152, 2020.

100. Papanikolaou, E., Venetsanos, A.G., Schiavetti, M., Marangon, A., Carcassi, M., and Markatos, N., Consequence assessment of the BBC H_2 refuelling station using the ADREA-HF code, *International Journal of Hydrogen Energy*, 36, 2573, 2011.

101. Ham, K., Marangon, A., Middha, P., Versloot, N., Rosmuller, N., Carcassi., M., Hansen, O.R., Schiavetti., M., Papanikolaou, E., Venetsanos., A., Engebo, A., Saw., J.L., Saffers., J.B., Flores, A., and Serbanescu, D., Benchmark exercise on risk assessment methods applied to a virtual hydrogen refueling station, *International Journal of Hydrogen Energy*, 36, 2666, 2011.

102. Wang, Y., Wang, S., and Deces-Petit, C., Less is more: Robust prediction for fueling processes on hydrogen refueling stations, *International Journal of Hydrogen Energy*, 47, 28993–29005, 2022.

103. LaFleur, A.C., Muna, A.B., and Groth, K.M., Application of quantitative risk assessment for performance-based permitting of hydrogen fueling stations, *International Journal of Hydrogen Energy*, 42, 7529–7535, 2017.

104. Groth, K.M. and Hecht, E.S., HyRAM: A methodology and toolkit for quantitative risk assessment of hydrogen systems, *International Journal of Hydrogen Energy*, 42, 7485–7493, 2017.

105. Liang, Y., Pan., Zhang, C., and Xie, B., The simulation and analysis of leakage and explosion at a renewable hydrogen refuelling station, *International Journal of Hydrogen Energy,* 44, 22608–22619, 2019.

106. Kim, E., Lee, K., Kim, J., Lee, Y., Park, J., and Moon, I., Development of Korean hydrogen fueling station codes through risk analysis, *International Journal of Hydrogen Energy,* 36, 13122–13131, 2011.

107. Kim, E., Park, J., Cho, J.H., and Moon, I., Simulation of hydrogen leak and explosion for the safety design of hydrogen fueling station in Korea, *International Journal of Hydrogen Energy,* 38, 1737–1743, 2013.

108. Yu, X., Yan, W., Zhou, P., Li, B., and Wang, C., The flame mitigation effect of vertical barrier wall in hydrogen refueling stations, *Fuel,* 315, 123265, 2022.

109. Bicer, Y. and Dincer, I., Comparative life cycle assessment of hydrogen, methanol and electric vehicles from well to wheel, *International Journal of Hydrogen Energy,* 42, 3767–3777, 2017.

110. Di Paola, A., UAE Sells Another Blue Ammonia Cargo to Japan in Hydrogen Push, August 18, 2021. www.bloomberg.com/news/articles/2021-08-18/uae-sells-another-blue-ammonia-cargo-to-japan-in-hydrogen-push

111. Kelly-Detwiler, P., A Key to the 'Hydrogen Economy' Is Carbon-Free Ammonia, December 16, 2020. www.forbes.com/sites/peterdetwiler/2020/12/16/maybe-the-hydrogen-economy-will-become-the-ammonia-economy/?sh=4de08fe34936

112. Mizra, A., Adnoc makes new blue ammonia shipment to Japan, September 09, 2021. www.argusmedia.com/en/news/2252286-adnoc-makes-new-blue-ammonia-shipment-to-japan

113. Sadik-Zada, E.R., Political economy of green hydrogen rollout: A global perspective, *Sustainability,* 13, 13464, 2021.

114. Gerlitz, L., Mildenstrey, E., and G., Prause, Ammonia as clean shipping fuel for the Baltic Sea region, *Transport and Telecommunication,* 23 (1), 102–112, 2022..

115. Fan, H. Enshaei, Jayasinghe, S.G., Tan, S.H., and Zhang, Ch., Quantitative risk assessment for ammonia ship-to-ship bunkering based on Bayesian network, *Process Safety Progress,* 16, 1, 2021.

116. Valera-Medina, A., et al., Review on ammonia as a potential fuel: From synthesis to economics, *Energy Fuels,* 35, 6964–7029, 2021.

117. Duijm, N.J., Markert, F., and Paulsen. J.L., *Safety assessment of ammonia as a transport fuel,* Risø-R-1504(EN), Risø National Laboratory, Roskilde, Denmark, February 2005.

118. Bicer, Y. and Dincer, I., Environmental impact categories of hydrogen and ammonia driven transoceanic maritime vehicles: A comparative evaluation, *International Journal of Hydrogen Energy,* 43, 4583–4596, 2018.

119. Ridden, P., Fuel-cell train travels more than 1,700 miles on one tank of hydrogen, New Atlas, March 27, 2024. https://newatlas.com/transport/flirt-h2-fuel-cell-train-guinness/

120. Dincer, I., Technical, environmental and exergetic aspects of hydrogen energy systems, *International Journal of Hydrogen Energy,* 27, 265–285, 2002.

121. Alstom, Alstom Coradia iLint – the World's 1st Hydrogen Powered Train. www.alstom.com/solutions/rolling-stock/alstom-coradia-ilint-worlds-1st-hydrogen-powered-train

122. Kenny, K., Caterpillar, *BNSF and Chevron agree to pursue hydrogen locomotive demonstration, alternative fuels,* Chevron, December 14, 2021. www.chevron.com/newsroom/2021/q4/caterpillar-bnsf-and-chevron-agree-to-pursue-hydrogen-locomotive-demonstration

123. Luczak, M., CP's hydrogen locomotive powers up, *Canadian Pacific*, January 25, 2022. www.railwayage.com/mechanical/locomotives/cps-hydrogen-locomotive-powers-up/

124. Stephens, B., *Canadian Pacific's hydrogen-powered locomotive makes first revenue run, Trains*, November 15, 2022. www.trains.com/trn/news-reviews/news-wire/canadian-pacifics-hydrogen-powered-locomotive-makes-first-revenue-run/

125. Clinnick, R., First North American hydrogen locomotive begins testing, *International Railway Journal*, November 30, 2022. www.railjournal.com/fleet/first-north-american-hydrogen-locomotive-begins-testing/

126. de-Troya, J.J., Alvarez, C., Fernadez-Garrido, C., and Carral, L., Analysing the possibilities of using fuel cells in ships, *International Journal of Hydrogen Energy,* 41, 2853–2866, 2016.

127. Shih, N.C., Weng, B.J., Lee, J.Y., and Hsiao, Y.C., Development of a 20 kW generic hybrid fuel cell power system for small ships and underwater vehicles, *International Journal of Hydrogen Energy,* 39, 13894–13901, 2014.

128. Rivarolo, M., Rattazzi, D., Lamberti, T., and Magistri, L., Clean energy production by PEM fuel cells on tourist ships: A time-dependent analysis, *International Journal of Hydrogen Energy,* 45, 25747–25757, 2020.

129. Cavo, M., Gadducci, E., Rattazzi, D., Rivarolo, M., and Magistri, L., Dynamic analysis of PEM fuel cells and metal hydrides on a zero-emission ship: A model-based approach, *International Journal of Hydrogen Energy,* 46, 32630–32644, 2021.

130. Higier, A., Hsu, L., Oiler, J., Phipps, A., Hooper, D., and Kerber, M., Polymer electrolyte fuel cell (PEMFC) based power system for long-term operation of leave-in-place sensors in Navy and Marine Corps applications, *International Journal of Hydrogen Energy,* 42, 4706–4709, 2017.

131. Ahn, J., et al., Performance and availability of a marine generator-solid oxide fuel cell-gas turbine hybrid system in a very large ethane carrier, *Journal of Power Sources*, 399, 199–206, 2018.

132. Tronstad, T. et al., *Study on the Use of Fuel Cells in Shipping*, EMSA European Maritime Safety Agency, Portugal, 2017.

133. Bassam, A.M., Phillips, A.B., Turnock, S.R., and Wilson, P.A., Development of a multi-scheme energy management strategy for a hybrid fuel cell driven passenger ship, *International Journal of Hydrogen Energy,* 42, 623–635, 2017.

134. Dall'Armi, C., Micheli, D., and Taccani, R., Comparison of different plant layouts and fuel storage solutions for fuel cells utilization on a small ferry, *International Journal of Hydrogen Energy,* 46, 13878–13897, 2021.

135. Kim, K., Roh, G., Kim, W., and Chun, K., A preliminary study on an alternative ship propulsion system fueled by ammonia: Environmental and economic assessments, *Journal of Marine Science and Engineering*, 8, 183, 2020.

136. Mylonopoulos, F., Boulougouris, E., Trivyza, N.L., Priftis, A., Wang, H., and Shi, G., Hydrogen vs. batteries: Comparative safety assessments for a high-speed passenger Ferry, *Applied Sciences*, 12, 2919, 2022.

137. Wang, H., Trivyza, N.L., Mylonopoulos, F., and Boulougouris, E., Comparison of decarbonisation solutions for shipping: Hydrogen, ammonia and batteries, in *SNAME 14th International Marine Design Conference*, Vancouver, Canada, June 2022. doi: https://doi.org/10.5957/IMDC-2022-230

138. Dawson, V.P. and Bowles, M.D., *Taming liquid hydrogen: The Centaur Upper Stage Rocket 1958-2002*, The NASA History Series, National Aeronautics and Space Administration Office of External Relations Washington, DC, NASA SP-2004-4230, 2004.

139. NASA, Space Launch System. https://en.wikipedia.org/wiki/Space_Launch_System

140. Wikipedia, French space program. https://en.wikipedia.org/wiki/French_space_program

141. ESA, European Space Agency, Vulcain Engine. www.esa.int/Enabling_Support/Space_Transportation/Launch_vehicles/Vulcain_engine

142. Wikipedia, Angara (Rocket Family). https://en.wikipedia.org/wiki/Angara_(rocket_family)

143. Global Times, China Marks a Milestone in New-Gen Heavy-Duty Rocket Devt with Successful Engine Test, September 06, 2022. www.globaltimes.cn/page/202209/1274765.shtml

144. China.org.cn, China Completes First In-Orbit Space Fuel Cell Tests, November 27, 2022. www.china.org.cn/china/2022-11/27/content_78539535.htm

145. McKinsey & Company, *Hydrogen-Powered Aviation: A Fact-Based Study of Hydrogen Technology, Economics, and Climate Impact by 2050*, Publications Office of the European Union, Luxembourg, May 2020.

146. Clean Sky 2 and Fuel Cells & Hydrogen 2 Joint Undertakings, Hydrogen-Powered Aviation, May 2020. www.clean-aviation.eu/media/publications/hydrogen-powered-aviation

147. Rondinelli, S., Sabatini, R., and Gardi, A., Challenges and benefits offered by liquid hydrogen fuels in commercial aviation, in *Practical Responses to Climate Change (PRCC) 2014. Engineers Australia Convention* 2014 Melbourne, Australia, Vol. Transport (Aviation) *Conference paper*, November 2014. https://doi.org/10.13140/2.1.2658.9764.

148. Tapeinos, I.G., Koussios, S., and Groves, R.M., Design and analysis of a multi-cell subscale tank for liquid hydrogen storage, *International Journal of Hydrogen Energy*, 41, 3676–3688, 2016.

149. Gomez, A. and Smith, H., Liquid hydrogen fuel tanks for commercial aviation: Structural sizing and stress analysis, *Aerospace Science and Technology*, 95, 105438, 2019.

150. Hyde, D., Barker, B., Hodgson, P., Miller, R., and Rapeaunu, M., *Target true zero unlocking sustainable battery and hydrogen-powered flight*, Insight Report for the World Economic Forum, Geneva Switzerland, July 2022.

151. Charpentreau, C., *What are the most ambitious hydrogen aviation projects?* Aerotime Hub- Aviation News, June 29, 2022. www.aerotime.aero/articles/31398-hydrogen-aviation-projects

152. Andriamisaina, M., *The ZEROe demonstrator has arrived*, AIRBUS, February 22, 2022. www.airbus.com/en/newsroom/stories/2022-02-the-zeroe-demonstrator-has-arrived

153. *ZEROe hydrogen combustion demonstrator*, AIRBUS, February 22, 2022. www.airbus.com/en/newsroom/stories/2022-02-the-zeroe-demonstrator-has-arrived

154. AIRBUS, Could hydrogen fuel-cell systems be the solution for emission-free aviation?, November 30, 2022. www.airbus.com/en/newsroom/stories/2022-11-could-hydrogen-fuel-cell-systems-be-the-solution-for-emission-free

155. Hampel, C., *Hydrogen fuel cell aircraft – What for and when?* Electrive.com, Aug 26, 2022. www.electrive.com/2022/08/26/hydrogen-fuel-cell-aircraft-what-for-and-when/

156. Veenstra, A., World's first commercial hydrogen-powered aircraft is 'Made-in-Holland' and zero emission, *Innovation Quarter*, June 13, 2022. www.innovationquarter.nl/worlds-first-commercial-hydrogen-powered-aircraft-is-made-in-holland-and-zero-emission/

157. Trancossi, M., Dumas, A., and Madonia, M., Fire-safe airship system design, *SAE International*, 2012. https://doi:10.4271/2012-01-1512
158. Yusaf, T., Mahamude, A.S.F., Kadirgama, K., Ramasamy, D., Farhana, K., Dhahad, H.A., and Talib, R.A., Sustainable hydrogen energy in aviation – A narrative review, *International Journal of Hydrogen Energy*, 52, 1026–1045, 2024.

7 Inherently Safer Design

As discussed by Amyotte, MacDonald, and Khan [1], inherent safety is a proactive approach in which hazards are eliminated or lessened so as to reduce risk without over-reliance on engineered (add-on) devices and procedural measures. The concepts of inherent safety (or *inherently safer design*, ISD) have been formalized in the process industries over the past 35 years, beginning with the pioneering work of Professor Trevor Kletz (largely in response to the cyclohexane explosion at Flixborough, in the United Kingdom, in 1974). Many publications on ISD are now available: recent texts [2, 3], review articles including Khan and Amyotte [4], and resource articles such as Hendershot [5].

Professor Kletz and others worldwide have formulated a number of principles or guidelines to facilitate inherent safety implementation in industry. Four basic principles have gained widespread acceptance: (1) minimization (or intensification), (2) substitution, (3) moderation (or attenuation), and (4) simplification.

Minimization calls for the use of smaller quantities of hazardous materials when the use of such materials cannot be avoided. It may also involve performing a hazardous procedure as few times as possible when the procedure is unavoidable. *Substitution* calls for the replacement of a substance with a less hazardous material, or a processing route with one that does not involve hazardous material. It could also involve replacing a hazardous procedure with one that is less hazardous. *Moderation* implies the use of hazardous materials in their least hazardous forms, or the identification of processing options that involve less severe conditions, for example, a lower temperature, pressure, or speed of rotation. *Simplification* requires the design of processes, processing equipment, and procedures in a manner so as to eliminate opportunities for errors by eliminating excessive use of add-on safety features and protective devices.

These descriptions indicate that the principles of ISD focus on the underlying chemistry and physics of materials and processing methodologies. In doing so, they directly address the issue of material hazards and the ensuing risk. This approach is preferable to one in which an attempt is made to assess whether a given energy alternative such as hydrogen is *safe* [6]; hence the use of the term inherently *safer* design. ISD concepts, in conjunction with appropriate safety systems including detection, have been recommended as a good starting point for dealing with aspects of hydrogen safety [7].

DOI: 10.1201/9781003313007-7

TABLE 7.1
Safety Advantages and Disadvantages of Hydrogen

Advantages	Disadvantages
Leaks typically disperse at a high rate because of buoyancy effects.	Leaks cannot be identified by senses of sight or smell.
Health effects due to toxicity are not evident (although asphyxiation concern remains).	Ignition energy is low and readily attainable in industry.
Pooling does not occur.	Range between lower and upper flammability limits is wide (and lower flammability limit is relatively low).
	Flames are not visible.
	Significant overpressures can result from explosion.

Source: Adapted from [11].

The chemical and physical properties of hydrogen present both advantages and disadvantages with respect to general safety considerations and those related specifically to inherent safety. Safety bulletins on hydrogen (e.g., [8, 9]) typically comment on the fact that it is colorless, odorless, and tasteless, meaning that most human senses will not help to detect a leak [9]. On the other hand, and with possible safety benefits, hydrogen is lighter than air, diffuses rapidly, and is nontoxic and nonpoisonous [9].

Molkov [10] has demonstrated that the ubiquitous safety concept of *trade-offs* applies to hydrogen as well as hydrocarbons. On the safety asset side, he states that buoyancy – such that the formation of large combustible clouds may be less likely with hydrogen than with heavier hydrocarbons – is the main advantage. Negative aspects include the ability of a hydrogen–air mixture to detonate (which is greater than that of hydrocarbons) and the higher molecular diffusion of hydrogen compared to other fuels [10]. Miller [11] has summarized the safety advantages and disadvantages of hydrogen based on its inherent properties, as shown in Table 7.1.

Examples of application of the various principles of inherent safety to the hydrogen industry are given in this chapter; general material on ISD is drawn from Kletz and Amyotte [3], with relevant excerpts. Attempts to measure the degree or level of inherent safety in a given process are also described, both generally and with specific emphasis on those measurement techniques designed for use with hydrogen. First, the concept of an ordered approach to risk reduction is presented in the following section.

7.1 HIERARCHY OF RISK CONTROLS

The principles of inherent safety work in conjunction with other means of reducing risk, namely, *passive* and *active engineered safety*, and *procedural safety*, within a framework commonly known as the *hierarchy of controls* (or as some call it, the *priority of controls* or the *safety decision hierarchy*) [1, 3]. Inherent safety, being the most effective and robust approach to risk reduction, sits at the top of the hierarchy; it is followed in order of decreasing effectiveness by passive engineered safety devices (such as explosion relief vents), then active engineered safety devices (such as automatic fire suppression systems), and finally procedural safety measures (such as ignition source control by hot-work permits). Figure 7.1 gives a schematic representation of this way of thinking about loss prevention.

Inherent safety is not a stand-alone concept [1]; as just described and as shown in Figure 7.1, ISD works through a hierarchical arrangement in concert with engineered and procedural safety to reduce risk. However, inherent safety is not necessarily a solution for all hazards and all risks. Nor does the hierarchy of controls invalidate the usefulness of engineered and procedural safety measures. Quite the opposite; the hierarchy of controls recognizes the importance of engineered and procedural safety by highlighting the need for careful examination of the reliability of both mechanical devices and human actions. For example, Molkov [10] comments that while venting is the most widespread mitigation technology, its application to confined hydrogen–air deflagrations can possibly enable a transition to detonation to occur, resulting in greatly increased overpressures. This, of course, is the opposite of the intended effect of lessening the severity of such an incident [10].

As illustrated in Figure 7.2, Hendershot [5] described the hierarchy of chemical process safety strategies as a *spectrum of options*. These considerations are readily apparent in the article by Pasman and Rogers [12], which gives several examples of inherent safety features (as discussed in subsequent sections of this chapter), along with passive devices (e.g., fire walls and explosion barriers), active devices (e.g., sensors, blocking valves, ventilation, and recombiners that catalytically oxidize hydrogen), and procedural measures (e.g., emergency response).

Not all hydrogen safety publications recognize the importance of ISD and some seemingly attempt to invert the risk control hierarchy by promoting safety procedures and training as "probably the most important preventive measure" [13]. Others clearly recognize the importance of inherent safety by emphasizing that "the main focus here should be to avoid significant flammable gas clouds" [14]; this publication also implicitly adopts the hierarchy of controls by distinguishing between passive and active measures and commenting that with hydrogen, the use of active measures can be a challenge because of the issues of wide flammability and high reactivity. Closest to ISD in terms of effectiveness, passive measures for facilitating hydrogen safety, including blast walls [15] and vent relief systems [16–19], have been discussed in both the archival literature and the popular media from the perspectives of the success and failure of such devices.

7.2 MINIMIZATION (INTENSIFICATION)

The need to avoid accumulation of flammable clouds of hydrogen is a key motivating factor for adoption of the minimization principle whenever possible. Reference [14] recommends minimizing confinement to achieve this goal and comments that "the optimal wall is no wall". This is not a universal recommendation, however, as large-momentum horizontal leaks might be better directed upwards (thus putting to good

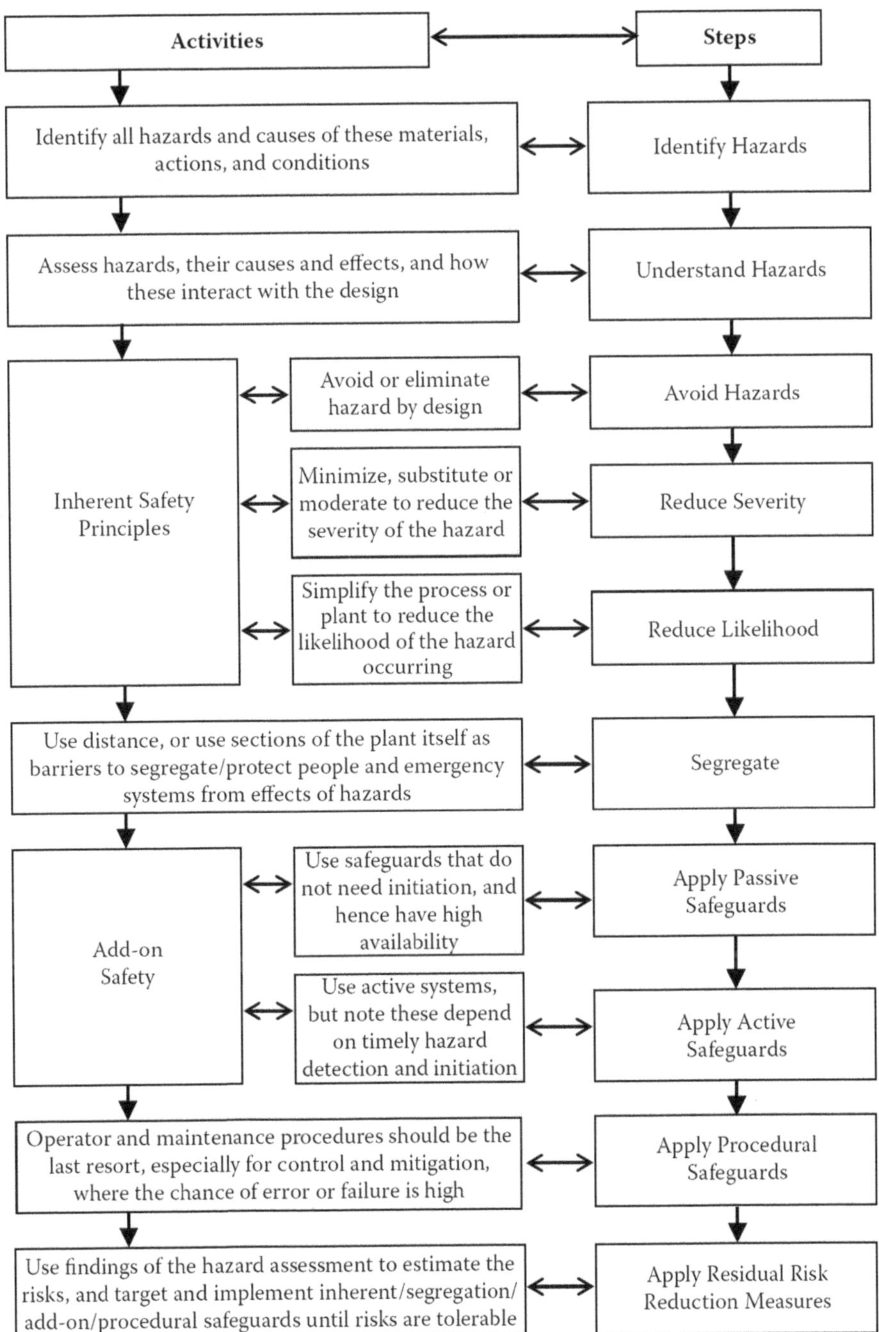

FIGURE 7.1 A systematic approach to loss prevention: the hierarchy of controls. With permission.

Source: [3].

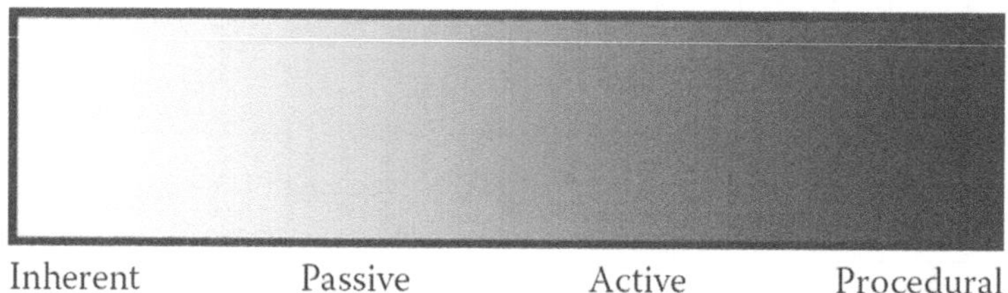

FIGURE 7.2 Representation of safety strategies as a spectrum of options from inherent through to procedural. With permission.

Source: [5].

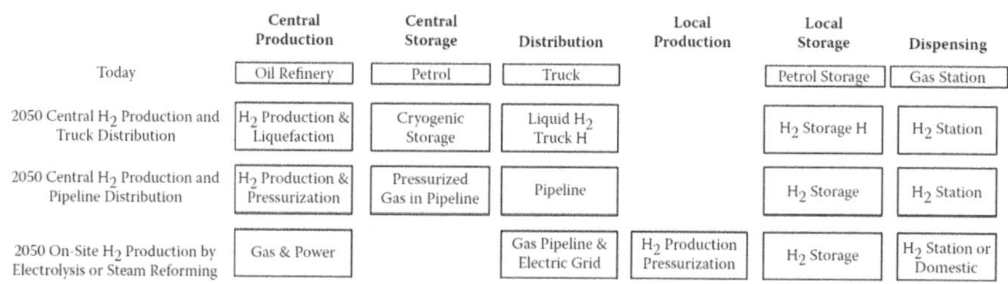

FIGURE 7.3 Comparison of gasoline and hydrogen production, storage, distribution, and dispensing schemes. With permission.

Source: [7].

use the strong positive buoyancy of hydrogen) by the placement of vertical walls around possible leak locations [14]. Minimizing congestion in a process area is also highly beneficial with respect to limiting turbulent flame acceleration and destructive overpressure generation should ignition of a hydrogen cloud occur [14]. Pasman and Rogers [12], on the issue of minimizing release quantities, note the stringent requirement for leak-free connection technologies when handling hydrogen.

In their review of infrastructure options for hydrogen refueling stations, Markert et al. [7] comment that current scenarios are still in the very early stages of development and therefore offer cost-saving opportunities as would be brought about by the ISD approach. This advice acknowledges the fact that inherent safety is typically most effective when considered early in the design sequence. While it is possible to retrofit ISD principles to an existing plant to some extent, incorporation of inherent safety thinking in preliminary hazard and risk assessments can be highly beneficial. This is particularly important with respect to minimization, that is, storing as little hydrogen as possible [7].

Markert et al. [7] present a comparison of three hydrogen refueling alternatives with the current system for gasoline (petrol) production, storage, distribution, and dispensing, as shown in Figure 7.3. The process chain for centralized production and either truck or pipeline distribution (i.e., the first two nonpetroleum options in Figure 7.3) is shown schematically in Figure 7.4. Here one sees the need for careful

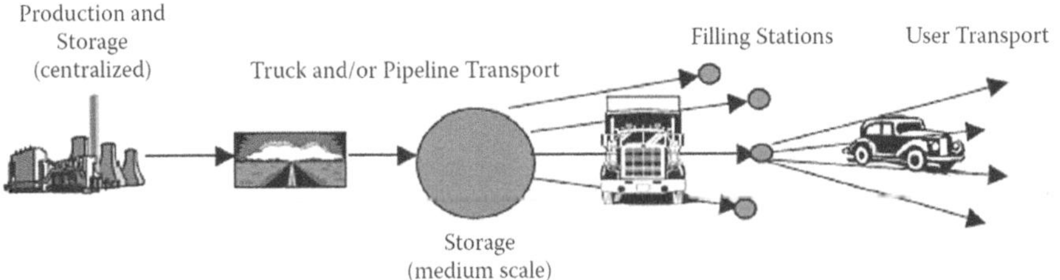

FIGURE 7.4 Process chain for centralized production of hydrogen and distribution by truck or pipeline. With permission.

Source: [7].

consideration of the inventories to be stored both centrally and at medium scale. The on-site option, on the other hand, eliminates the need for medium-scale storage – a case of 100% minimization. There are, of course, a myriad of safety, environmental, and cost factors that must be considered in selecting from the full range of such alternatives. The point being made here is that ISD should be one of these factors.

Another interesting example of minimization is given by the experimental work of Alsheyab, Jiang, and Stanford [20], this time from the perspective of minimizing hydrogen formation during the electrochemical production of ferrate. (The term *ferrate* is often used to mean ferrate(VI), although according to IUPAC naming conventions, it may also refer to other iron-containing oxyanions, such as ferrate(V) and ferrate(IV). Ferrate(VI) refers either to the anion $[FeO_4]^{2-}$, in which iron is in the +6 oxidation state, or to a salt containing this anion.) Ferrate is described by the authors as a strong oxidant used for organic synthesis and water/wastewater treatment and as having several advantages over other commonly used oxidants such as chlorine, hydrogen peroxide, and ozone [20]. They further describe the electrochemical production of ferrate as being preferable to chemical production because of the simplicity of operation, improved safety features, and avoidance of the toxic compound hypochlorite [20]. Using the experimental setup for lab-scale ferrate production shown schematically in Figure 7.5, Alsheyab, Jiang, and Stanford [20] determined that the hydrogen concentrations formed at the cathode were significantly lower than the lower flammability limit for hydrogen in air. Whether this ISD feature brought about by process intensification would hold at the pilot and production plant scales was recognized as a key risk factor requiring further investigation [20].

7.3 SUBSTITUTION

As previously described, one interpretation of the substitution principle is the replacement of a substance with a less hazardous material. This clearly demonstrates the fact that inherent safety must be viewed as hazard specific. For example, while the substitution of nitrogen for natural gas as a purge gas for cleaning pipelines does eliminate the flammability hazard associated with natural gas, it also introduces the asphyxiation hazard that accompanies the use of nitrogen [21]. Thus, if one were to consider

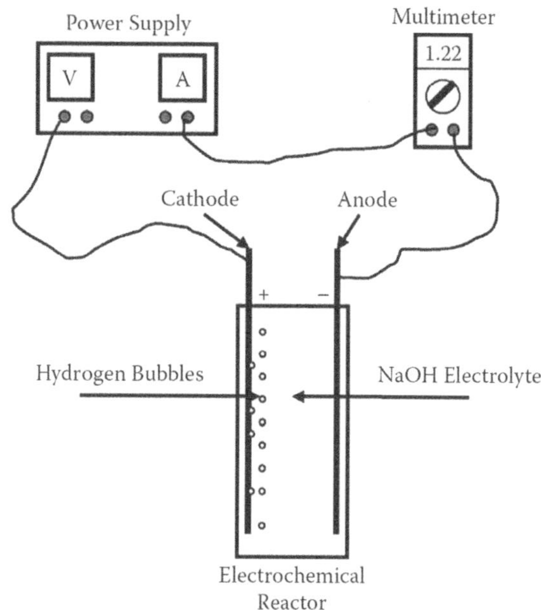

FIGURE 7.5   Ferrate production at the laboratory scale. With permission.

Source: [20].

substituting some other material for hydrogen in a given application, the hazard being avoided must be specified (as in the familiar substitution of nonflammable helium for flammable hydrogen when a lighter-than-air gas is desired).

An extension of the point made in the previous sentence is the use of other substances, not as a substitute for hydrogen but as a means of avoiding hydrogen generation. Pebble-bed nuclear reactors utilizing helium rather than water as a coolant offer the inherently safer advantage of eliminating water from the reactor core. The release of hydrogen gas in the event of an upset is therefore also eliminated [22].

It is, however, the features of hydrogen associated with it being an abundant, cleaner-burning fuel that make it attractive as a substitute for other fuels such as hydrocarbons. It is necessary, therefore, to also invoke the second interpretation of substitution – that of replacing a processing route with one that does not involve hazardous material. A relevant example here is the production of the insecticide carbaryl by two different synthesis routes utilizing the same raw materials but reacting them in a different order [3]. One of the processes produces methyl iso-cyanate as a hazardous intermediate, while the other avoids the production of this chemical that is forever linked with the 1984 tragedy in Bhopal, India (see Section 10.2).

The example in the previous section of minimization of hydrogen inventories in storage also fits with the current discussion on substituting one process route for another. Although it has been shown that hydrogen can be transported by pipeline or truck from centralized plants [7], a more practical, cost-effective production method is on-site catalytic reforming from natural gas [23, 24]. While this latter approach does address the issue of large inventories of hydrogen, it also has a significant environmental impact because of the carbon dioxide generated during the process [24]. To ease this and other problems, Guy [24] comments on the importance of indirect and direct production of hydrogen by renewable energy sources and by sunlight, respectively. These are also examples of synthesis route substitution.

7.4 MODERATION (ATTENUATION)

Pasman and Rogers [12] note that hydrogen can be shipped and stored in various forms: (1) as a compressed gas, (2) absorbed on a substrate material (e.g., as a metal hydride), and (3) as a cryogenic liquid. Each of these options is briefly considered in the following discussion from an inherent safety (moderation) perspective. While not always able to eliminate a given hazard, moderating or attenuating the form of a material or the conditions under which it is processed can be beneficial in terms of risk reduction.

Storing hydrogen as a liquid is attractive because of the high energy density per unit volume [25] (or storage volume efficiency [12]). Potential downsides include the need for heavy, bulky cryogenic tanks [26] (particularly in the case of hydrogen as a transportation fuel), as well as the intrinsic hazards of very low temperatures given that the normal boiling point of hydrogen is −252.9°C (see Chapter 3). Cryogenic temperatures also render traditional detection methods such as marking dyes, radioactive tracing gases, and odorants ineffective [27].

Hydrogen as a gas, of course, presents a significant flammability hazard. Applying the ISD principle of moderation by lowering the gas temperature, moderate reductions in the lower flammability limit of hydrogen can be achieved; the lower limit at hydrogen's normal boiling point (−252.7°C) is 7.8 vol% in air compared with 4 vol% in air at 25°C and atmospheric pressure [28]. Lowering the temperature also brings about a decrease in the upper flammability limit and therefore a narrowing of the flammability range. Varying the gas concentration away from stoichiometric conditions can help to moderate the flammability hazard. The burning velocity of hydrogen is 3.25 m/s at just above its stoichiometric concentration (29.5 vol% in air) at 25°C and atmospheric pressure; the value at the lower flammability limit in these conditions is only 0.04 m/s [28].

Rainer [28] notes that as with other combustible gases, the flammability range of hydrogen can be reduced by the addition of inert gases such as carbon dioxide or nitrogen. Hendershot's point that the hierarchy of controls is in fact a spectrum of options often having indistinct or blurred boundaries [5] is apparent here, especially if the nature of the hazardous material (i.e., hydrogen) is moderated by the addition of nonhazardous materials (i.e., CO_2 or N_2, at least from the perspective of flammability) with the aid of mechanical devices such as sensors and alarms. The question also remains as to the feasibility of using a diluted hydrogen stream when it is the flammability properties of hydrogen that have made it the fuel of choice in a given application. These considerations indicate again the need for trade-offs when attempting to reduce risk via a prioritized ranking of safety measures and also achieve desired operational goals.

A recent article by Middha, Engel, and Hansen [29] addresses the issue of maintaining fuel viability while minimizing hazards and the ensuing risk. The objective of this work was to investigate the addition of hydrogen to natural gas (methane) to create a hybrid fuel, hythane, having a reduced explosion risk compared with the pure component fuels. Because the proposed percentages of hydrogen were less than 50%, this could, strictly speaking, be viewed as moderating the explosion risk of methane as opposed to moderating the similar risk for hydrogen. Additionally, because hythane is a different fuel, this could also be viewed as an example of substitution – thus demonstrating the complementary nature of the various principles of ISD.

The motivation for the work of Middha, Engel, and Hansen [29] is reported to be the need for a stopgap fuel between fossil fuels and hydrogen. Fuel blends of 8 to 30% hydrogen in methane offer the potential for reduced emissions with no significant modifications to existing infrastructure [29]. According to the computational fluid dynamics results obtained in this work, the combined risk of dispersion and explosion for hythane was determined to be comparable to that of methane (or lower in some instances) and lower than that of hydrogen in all cases studied [29].

Metal hydrides provide a means to store hydrogen in solid form [25], and offer an inherently safer approach in some regards to liquid and gaseous storage [12]. Pasman and Rogers [12] further comment, however, that while the hydride itself can be viewed as inherently safer, production of hydrides with acceptable risk and retrieval of the hydrogen at moderate temperatures remain challenging and hence the subject of intensive research.

Yang [30] states that metal hydrides (as well as chemical, carbon-based, and advanced-material hydrides) can potentially store large quantities of hydrogen at low pressures. Low-pressure storage is a classic application of the moderation principle; large inventories are essentially a negation of the minimization principle. The issue of trade-offs appears once more; risk assessments must therefore consider the exposure of hydride systems and their potentially large inventories of hydrogen to high temperatures and the resulting pressure increase brought about by flame impingement from an accidental fire [30].

7.5 SIMPLIFICATION

The ISD principle of simplification is widely applicable in all industrial endeavors. Simplification can be thought of not only in relation to inherent safety but also with respect to passive and active engineered safety devices as well as procedural measures. Barriers and sensors that are simple yet effective provide fewer opportunities for a malfunction to occur. Similarly, an easy-to-understand operating procedure is more likely to be followed than one that is overly complex and seemingly written by people who have never had to perform the given procedure.

One approach to simplification as an ISD measure is to make equipment robust enough to withstand any foreseeable undesired event such as a pressure excursion. This philosophy seems well suited to the hydrogen industry as a possible means of reducing the reliance on explosion relief vents. Cost is an obvious consideration in using thick-walled vessels, which some practitioners would view as a passive rather than an inherent measure.

These points on simplification were considered by Xu et al. [31] in their study of multi-layered stationary high-pressure hydrogen storage vessels (SHHSVs). They give the following list of SHHSV characteristics aimed at enhancing inherent safety [31]:

- As uniform a stress distribution as possible.
- As few welds as possible.
- High fatigue resistance to avoid failure from pressure swings caused by repeated vessel filling and discharging.
- A convenient arrangement for online leak monitoring.
- Hydrogen-compatible materials of construction.

The stipulation of as few welds as possible is also related to the concept of minimizing potential leak and failure locations (welds, flanges, etc.).

Simplification of an overall process can also help to avoid incidents. Guy [24] describes the Chicago Transit Authority Zero Emission Fuel Cell Bus system in which buses store compressed hydrogen in gas cylinders for use in onboard fuel cells. Each bus has a range of 250 miles; what could potentially be a complex, distributed hydrogen refueling system has been simplified by having the buses refuel at a central hub in a time of 15 minutes [24].

Janssen et al. [32] conducted research on the development of a high-pressure electrolyzer with the aim of achieving higher system efficiency for hydrogen production. While conventional electrolytic hydrogen production uses a low-pressure process (an example of ISD through moderation), the process proposed by Janssen et al. [32] employed the ISD principle of simplification to eliminate the need for compressors and buffer tanks from the overall design.

7.6 OTHER EXAMPLES

Two ISD subprinciples provide other examples of inherent safety application to the hydrogen industry: (1) *making incorrect assembly impossible* (simplification) and (2) *limitation of effects* (moderation). As described by Kletz and Amyotte [3], it is not uncommon for the second of these subprinciples to be equated with *avoiding domino or knock-on effects*. This most often happens when limitation of the effects of a hazard is undertaken not only at the hazard source but also with respect to far-field consequences. In this manner and as shown in Figure 7.1, unit segregation provides a bridge in the continuum between ISD and passive barriers.

The US Chemical Safety Board (CSB) has issued a safety bulletin [33] on positive material verification that is pertinent to the matter of taking steps to avoid incorrect assembly [1]. During shutdown of the plant in question, three elbows were removed from a high-pressure, high-temperature hydrogen line for cleaning. Because of different process conditions in the line, two of the elbows were composed of a stronger metal alloy, while the third was constructed of less expensive carbon steel.

During reinstallation, the carbon steel elbow was incorrectly positioned and subsequently corroded, leading to a hydrogen release and fire that caused extensive property damage and minor injury to an employee. A key finding identified by the CSB in relation to this process incident is the following [33]:

> Piping systems can be designed such that incompatible components cannot be interchanged. All three elbows could have been made from the same low alloy steel material, even though this would have meant additional expense. Alternatively, elbow 1 could have been dimensionally different from elbow 2 and 3, although this would have meant additional construction costs.

One approach to limiting the near-field effects of an incident is given by Segal, Wallace, and Keffer [34], who described their fuel supply system to provide gaseous hydrogen to an engine test cell. A key feature of their design was the outdoor storage of 70.5 m^3 of hydrogen in cylinders. While engineered safety devices and presumably procedural measures were also incorporated, separation of the fuel source from operational personnel is an example of simplification through limitation of effects. It should be noted, however, that outside storage of compressed gaseous hydrogen requires consideration of phenomena such as turbulent jet flames arising from a fuel leak and subsequent ignition [12].

Avoidance of domino effects (i.e., damage to adjacent units and harm to people occupying them) is a critical aspect of hydrogen safety [13, 14]. The topic of safety distances (or harm thresholds) with respect to hydrogen dispersion, deflagration, and detonation is therefore an area of intensive research (e.g., [35–37]). A representative example from this body of work is given in Figure 7.6, which shows a nomograph of blast radii corresponding to different damage levels for various masses of hydrogen released and ignited. Figure 7.6 comes from the modeling work of Dorofeev [36] and represents the case of low congestion such as might be experienced within a hydrogen refueling station.

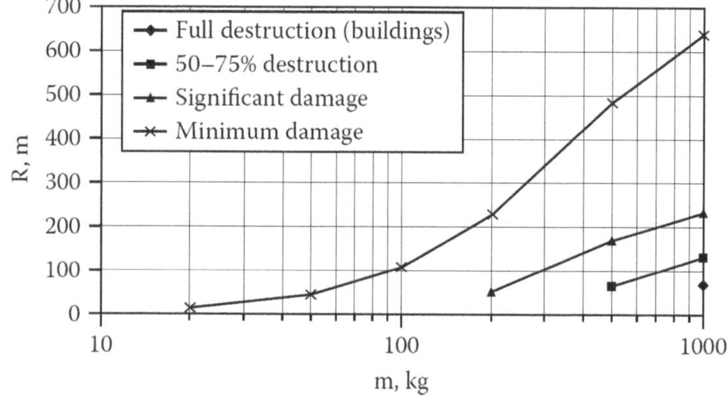

FIGURE 7.6 Blast radii for various degrees of building damage as a function of mass of hydrogen released in the case of low congestion. With permission.

Source: [36].

7.7 MEASUREMENT OF INHERENT SAFETY

As the previous sections have demonstrated, the principles of inherently safer design are widely applicable to the production, distribution, and utilization of hydrogen for a variety of purposes. This section addresses the issue of how to decide on the most appropriate application of these principles drawn from a range of alternatives.

In general, a relatively large number of ISD assessment tools exist for attempting to measure the degree of inherent safety of a given process or processing methodology. These techniques have been reviewed by, among others, Khan and Amyotte [38], Khan, Sadiq, and Amyotte [39], and Kletz and Amyotte [3]. Khan et al. [78] have reviewed and emphasized past progress concerning the development of methods and models for process safety and risk management.

Kidam et al. [79] have updated and improved the existing inherently safer design review (ISDR) practices during the design of chemical process plants by exploiting major accident cases from the US CSB and Failure Knowledge Database (FKD).

Athar et al. [80] performed an extensive survey regarding the bibliometrics and the features of the state-of-the-art inherent assessment methods. Gao et al. [81] developed a Systematic Inherent Safety Metric (SISM) for measuring the inherently safer modifications and applied SISM in a chlorine liquefaction process.

Amyotte and Khan [82] have investigated the integration of ISD into various process safety systems, activities, and applications and propose a conceptual framework of the Integrated Inherent Safety Index (I2SI) as described by Khan and Amyotte [44]. The conceptual framework of the I2SI index is shown in Figure 7.7.

Many researchers focus on the perspective of the inclusion of ISD into early design stages and conceptual process design such as Park et al. [83], Athar et al. [84], Zainal Abidin et al. [85], Gangadharan et al. [86], Gao et al. [87], Sultana and Haugen [88], Pasha et al. [89], Crivellari et al. [90] and Zhu et al. [91].

Other authors propose the co-consideration of various aspects including occupational health and biological hazards, process safety, and economic aspects to generate a holistic safer chemical plant via an inherent safety-oriented approach [92–95]. Bolla et al. [96] implemented specific inherent safety indices to compare spent coffee grounds treatment alternatives and design intrinsically safer plants, thus contributing to the transition to green processes.

Rayner Brown et al. [97] developed a protocol to identify ISD-based barriers in bow tie diagrams, while Sun et al. [98] proposed a new hazard index for reactivity management (HIRM) in chemical processes. The HIRM method is able to assess various runaway scenarios and quantify the reactivity hazards based on the consequence and probability.

Concerning various unit operations, Athar et al. [99] developed a new technique to improve the safety level of process piping from potential fire risk using the inherent safety concept. This inherently safer process piping (ISPP) technique utilizes the fundamentals of fluid flow to predict the potential damage from a major fire accident. Ebadollahi et al. [100] present a dual-loop bi-evaporator combined cooling and power (CCP) system to simultaneously supply electricity, air-conditioning, and freezing demands. The system is investigated from energy, exergy, exergo-economic,

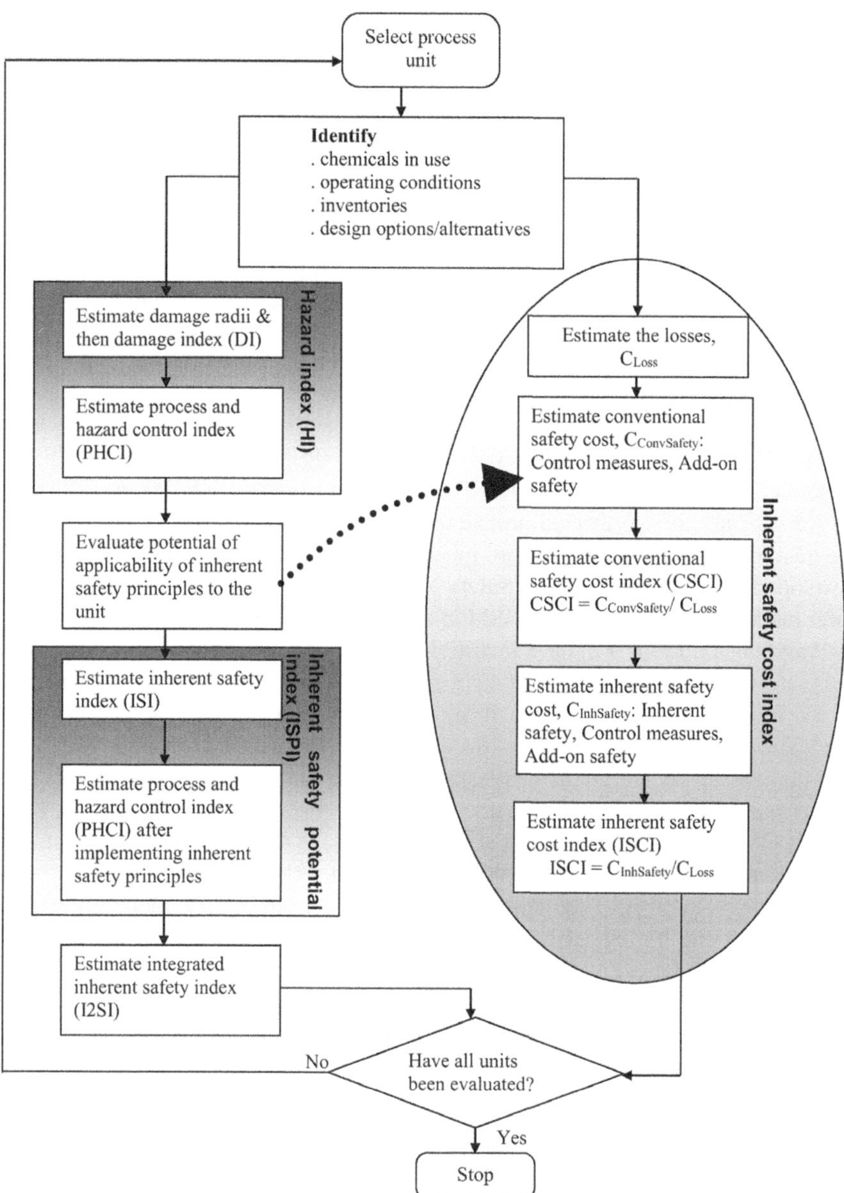

FIGURE 7.7 Conceptual framework of the Integrated Inherent Safety Index (I2SI). With permission from the Canadian Journal of Chemical Engineering. Open Access, public domain.

Source: [82].

exergo-environmental, and inherent safety perspectives. Ye et al. [101] propose an inherently safer synthesis approach for thermally coupled distillation sequences with structure constraints. The framework can effectively quantify and optimize the inherent safety index of each column configuration in the sequence, based on a State-Task Network and Dow's Fire and Explosion Index (F&EI).

Regarding development of new inherent safety techniques, Ahmad et al. [102] propose a framework of inherent safety assessments called the Numerical Descriptive Inherent Safety Technique (NuDIST). Song et al. [103] developed a methodology based on new fuzzy logic inherent safety performance indices which focus on chemical hazard indices and process hazard indices. It also takes into account other hazard factors such as process safety, process complexity, and operability. Ortiz-Espinoza et al. [104] compared different safety metrics to evaluate the risk associated with a given process design. The indices under consideration were Dow's F&EI, the fire and explosion damage index (FEDI), the process route index (PRI), and the process stream index (PSI).

Janošovský et al. [105] introduced a decision-making concept that is based on inherently safer design principles. The method was applied to a case study of the novel process of the conversion of refinery waste into the valuable product ammonium thiosulphate. As a second case study, the production of ethyl acetate by esterification was selected, in which the concept of process intensification was applied.

Graphical inherent safety assessment techniques have also been proposed. Ahmad et al. [106] developed the Graphical Descriptive Technique for Inherent Safety Assessment (GRAND) for inherent safety evaluation during the research and development (R&D) stage of process design. These authors have also developed the Graphical Inherent Safety Assessment Technique (GISAT) for a preliminary stage of engineering design which takes into account additional information obtained from the process flow diagram (PFD) in the assessment of inherent safety characteristics of a process [107]. Finally, Ahmad et al. [108] introduced the Inherent Safety and Economic Graphical Rating (InSafE), which is a graphical inherent safety and economic evaluation method suitable for R&D designs. InSafE takes into account its inherent safety assessment, both gross profit and net profit margin for its economic assessment, and eight inherent safety parameters taken from the GRAND method.

One of the earliest ISD assessment tools in the 1990s was the INSIDE (*IN*herent *S*HE [Safety, Health, Environment] *In DE*sign) project conducted in Europe, and which resulted in the INSET (*IN*herent *S*HE *E*valuation *T*ool) toolkit to identify inherently safer design options throughout the life of a process and to evaluate the options [3]. There are also well-known process safety methodologies such as checklist and what-if analyses that have been adapted for ISD consideration, largely in terms of identifying hazards that can be addressed by the principles of inherent safety.

Well-established indexing methods such as the Dow F&EI and Dow Chemical Exposure Index (CEI) have numerous inherent safety aspects associated with their calculation procedures. Etowa et al. [40] quantified the ISD features of both Dow indices (F&EI and CEI) and demonstrated the beneficial impact of the principles of minimization, substitution, and moderation. The work of Edwards and coworkers at

Loughborough University in the United Kingdom resulted in one of the first indices designed specifically to address inherently safer processing opportunities, the *prototype index of inherent safety*, or PIIS [41, 42].

Over the past decade or so there has been a proliferation of ISD indexing procedures and assessment methodologies appearing in the process safety literature. Table 7.2 gives a summary of such techniques developed over the period 2002–2010. Observations on the entries in Table 7.2 include the following [3]:

- Many of the methods deal specifically with the early concept and route-selection stages of the design process.
- Some of the approaches use sophisticated mathematical and problem-solving techniques such as fuzzy logic.
- There has been a growing trend to link inherent safety with environmental and health issues in an effort to achieve an integrated approach.
- There have been attempts to incorporate inherent safety assessment into process design simulators.
- Some of the indexing methods have been in existence long enough for comparative evaluations to be made among them.
- There is a trend to apply ISD to various old and new technological and environmental issues such as occupational health assessment, green processes, and energy transition.

When commenting in 2005 on potential barriers to wider adoption of inherently safer design principles in the process industries, Edwards [109] noted that the issue may not be the availability of inherent safety assessment tools but rather the limited use of these tools by industry. Reasons might include the subjective judgment required by some of these tools and also their attendant complexity [3]. In 2011, it seems that the same availability of tools, yet limited uptake by industry in general, exists. It is too early to comment in this regard with specific reference to the hydrogen industry, as hydrogen-specific ISD assessment methodologies have only recently appeared in the process safety literature.

While some of the more general entries in Table 7.2 could undoubtedly be applied to ISD assessments involving hydrogen, it does appear that the most concerted effort concerning hydrogen ISD evaluation has come from the team with members based in Pisa and Bologna, Italy [55–59]. This series of publications provides an interesting and valuable contribution to the field of hydrogen safety. Beginning in the early design stages and with the basic method shown in Figure 7.8, ISD assessments of storage [55] and production [56] alternatives were extended to include distribution and utilization considerations generally [58] and specifically with respect to the automotive sector [59]. *Key performance indicators* (KPIs) were introduced to account for unit, overall, and domino hazard indices [57].

Several features of this body of work [55–59] are noteworthy in terms of potential adoption of the developed methodologies by industrial designers and other practitioners. The first is the previously mentioned emphasis on early design stages when ISD considerations typically have the greatest impact. Second, there is a clear recognition that the preferred term is inherently *safer* design, not inherently *safe*

TABLE 7.2

Examples of Development of Inherent Safety Assessment Methodologies and Elucidation of Assessment Considerations

Reference	Contribution
Khan and Amyotte [43]	Integrated Inherent Safety Index (I2SI).
Khan and Amyotte [44]	Further development of I2SI to include cost model.
Carvalho et al. [45]	Method for identifying retrofit design alternatives of chemical processes. Uses Inherent Safety Index (ISI) developed by other researchers.
Hurme and Rahman [46]	Discussion of implementation of inherent safety throughout process life-cycle phases. Use of ISI developed earlier.
Rahman et al. [47]	Comparative evaluation of three inherent safety indices with expert judgment at process concept phase.
Hassim and Hurme [48]	Inherent Occupational Health Index (OHI) developed to assess the health risk of process routes during the process research and development stage. The index can be used to compare process routes or to determine the level of inherent occupational health hazards.
Hassim and Hurme [49]	OHI developed for assessment during the basic engineering stage. This method relies on the information available in piping and instrumentation diagrams and the plot plan. The health aspects considered are chronic and acute inhalation risks, and dermal/eye risk.
Hassim and Hurme [50]	Health Quotient Index (HQI) developed for assessment during the preliminary process design phase. This index quantifies a worker's health risk from exposure to fugitive emissions by using data from process flow diagrams. This method can be used to quantify the level of risk from a process or to compare alternative processes.
Hassim and Hurme [51]	Method to estimate inhalation exposures and risks; for use early in the design stages by utilizing process flow diagrams (PFDs). The risk of chemical exposure can be evaluated either through the hazard quotient method or by calculating the carcinogenic intake and resulting cancer risk.
Hassim and Edwards [52]	Process Route Healthiness Index (PRHI) for quantification of health hazards arising from alternative chemical process routes. Application is in early stages of chemical plant design.
Gupta and Edwards [53]	Graphical approach for evaluating inherent safety based on the earlier-developed Loughborough prototype index of inherent safety (PIIS).
Cozzani et al. [54]	Procedure for assessment of hazards arising from decomposition products formed due to loss of chemical process control. Applicable to consideration of substitution principle.
Landucci et al. [55]	Procedure and indices for evaluating inherent safety at preliminary PFD stage for hydrogen storage options.
Landucci et al. [56]	Consequence-based method for identification and assessment of inherently safer plant design alternatives in developing steam reforming processes for hydrogen production.

(continued)

TABLE 7.2 (Continued)
Examples of Development of Inherent Safety Assessment Methodologies and Elucidation of Assessment Considerations

Reference	Contribution
Landucci et al. [57]	Further development of PFD method [55] by use of quantitative key performance indicators (KPIs) to remove subjective judgment.
Tugnoli et al. [58]	Quantitative inherent safety assessment method utilizing PFDs in early design stages for hydrogen supply chain options. The result of the assessment is a quantification of the inherent safety of the process scheme by a set of KPIs.
Landucci et al. [59]	Consequence-based approach for inherent safety assessment of production, distribution, and utilization systems with respect to hydrogen vehicles.
Cordella et al. [60]	Further development of procedure for decomposition product analysis [54] to account for acute and long-term harm to human health, ecosystem damage, and environmental media contamination.
Shariff et al. [61]	Integrated Risk Estimation Tool (iRET) for inherent safety application at preliminary design stage. iRET links the design simulation software HYSYS with an explosion consequence model.
Leong and Shariff [62]	Further development of iRET [61] to incorporate a quantitative Inherent Safety Level (ISL), thus enabling integration of design simulation software with an Inherent Safety Index Module (ISIM). Application is again at the preliminary design stage.
Leong and Shariff [63]	Evolution of ISIM [62] to a Process Route Index (PRI) for comparison and ranking of different routes to manufacture the same product based on hazard potential of routes.
Shariff and Leong [64]	Method of evaluating inherent risk within a process as a result of the chemicals used and the process conditions. Through integration with HYSYS, the method can be used as early as the initial design stages to determine the probability and consequence of possible risk due to major accidents.
Shariff and Zaini [65]	Development of toxic release consequence analysis tool (TORCAT), a tool for consequence analysis and design improvement through use of inherent safety principles. The method utilizes an integrated process design simulator and a toxic release consequence analysis model.
Rusli and Shariff [66]	Qualitative assessment for inherently safer design (QAISD) method for application during preliminary design. This qualitative method combines hazard review techniques with inherently safer design concepts to generate inherently safer plant options/proactive measures.
Kossoy and Akhmetshin [67]	Use of nonlinear optimization method to select inherently safer operational parameters for given configuration of reactor equipment and materials. Primary concern is cooling failure.
Shah et al. [68]	Substance, reactivity, equipment, and safety technology (SREST) layer assessment method for environment, health, and safety (EHS) aspects in early phases of chemical process design.

TABLE 7.2 (Continued)
Examples of Development of Inherent Safety Assessment Methodologies and Elucidation of Assessment Considerations

Reference	Contribution
Adu et al. [69]	Comparative evaluation of various methods for assessing EHS hazards in early phases of chemical process design.
Palaniappan et al. [70]	Methodology for integrated inherent safety and waste minimization analysis during process design.
Palaniappan et al. [71]	Indexing procedure for inherent safety analysis at process route selection stage.
Palaniappan et al. [72]	Indexing procedure for inherent safety analysis at process flowsheet development stage. Discussion of *i*Safe, an expert system for automating procedures developed by Palaniappan et al. [71, 72].
Srinivasan and Nhan [73]	Inherent benign-ness indicator (IBI), a statistical analysis-based method for comparing alternative chemical process routes.
Srinivasan and Kraslawski [74]	Application of Theory of Inventive Problem Solving (TRIZ) methodology for creative problem solving to design of inherently safer chemical processes.
Meel and Seider [75]	Use of game theory to achieve inherently safer operation of chemical reactors.
Gentile et al. [76]	Fuzzy logic-based index for evaluation of inherently safer process alternatives with the aim of linking to process simulation.
Al-Mutairi et al. [77]	Linking of inherent safety and environmental concerns with optimization of process scheduling.
Khan et al. [78]	Development of methods and models for process safety and risk management and highlights of the research trends up to 2015. Opinions of the authors regarding the future research direction in the field.
Kidam et al. [79]	Update and improvement of the existing inherently safer design review (ISDR) practices during design of chemical process plant by exploiting major accident cases from the US Chemical Safety Board (CSB) and Failure Knowledge Database (FKD).
Athar et al. (2019a) [80]	Extensive survey regarding the bibliometrics and the features of the state-of-the-art inherent assessment methods.
Gao et al. (2020a) [81]	Development of a Systematic Inherent Safety Metric (SISM) for measuring the inherently safer modifications.
Amyotte and Khan [82]	Collaborative research efforts aimed at integrating ISD into various process safety systems, activities, and applications.
Park et al. [83]	Suggestions to process engineers on how to analyze ISD and inherent safety assessment tools (ISATs) from the perspective of inclusion in conceptual process design.
Athar et al. (2019b) [84]	New technique that integrates the mutually shared attributes of process equipment, in order to offer inherently safer process design at the preliminary design stage.
Zainal Abidin et al. [85]	Methodology for evaluating inherently safer design alternatives and identifying the benefits using an index-based approach at preliminary design stage.

(continued)

TABLE 7.2 (Continued)
Examples of Development of Inherent Safety Assessment Methodologies and Elucidation of Assessment Considerations

Reference	Contribution
Gangadharan et al. [86]	New comprehensive inherent safety index (CISI) for use in the early process design stage. The CISI assigns equipment safety scores to individual units in the process based on chemical, process, and connectivity scores.
Gao et al. (2020b) [87]	Systematic review on the implementation methodologies of inherent safety.
Sultana and Haugen [88]	Establishment of an inherent system safety index (ISSI) to evaluate inherently safer design during the concept development stage.
Pasha et al. [89]	Coherent framework to outline an inherently safer heat exchanger network in the preliminary design stage. Newly developed safety indices are introduced to estimate the inherent safety level of a single heat exchanger and the heat exchanger network.
Crivellari et al. [90]	Approach for the early design of offshore oil and gas facilities, based on multi-criteria KPIs, which addresses different targets (people, assets, environment) and provides a quantitative assessment of the safety score, accounting for both the possible accident consequences and their credibility.
Zhu et al. [91]	Development of a general inherent safety assessment tool at early design stage of chemical process.
Ee et al. [92]	Consideration of the occupational health hazards and biological hazards in a new developed safety index.
Ying et al. [93]	Creation of an assessment method that integrates the existing Inherent Occupational Health Index (IOHI) with suitable Layers of Protection (LOP) strategies for each of the enlisted hazard parameters.
Pu et al. [94]	Development of a tool called Safety Metric for Plant Design (SMPD), integrating occupational safety and process safety aspects, to generate a holistic safer chemical plant via inherent safety-oriented modifications.
Eini et al. [95]	Proposition for a novel framework in which the safety criteria are quantified and aggregated with the economic performance using multi-objective optimization programming.
Bolla et al. [96]	Implementation of specific inherent safety indices to compare spent coffee grounds treatment alternatives and design intrinsically safer plants, contributing to the transition to green processes.
Rayner Brown et al. [97]	Development of a protocol to identify ISD-based barriers in bow tie diagrams.
Sun et al. [98]	Development of a hazard index for reactivity management (HIRM) in chemical processes. The HIRM method is able to assess various runaway scenarios and quantify the reactivity hazards based on the consequence and probability.
Athar et al. 2019c [99]	New technique to improve the safety level of process piping from potential fire risk using the inherent safety concept.

TABLE 7.2 (Continued)
Examples of Development of Inherent Safety Assessment Methodologies and Elucidation of Assessment Considerations

Reference	Contribution
Ebadollahi et al. [100]	Presentation of a dual-loop bi-evaporator combined cooling and power (CCP) system to simultaneously supply electricity, air-conditioning, and freezing demands.
Ye et al. [101]	Presentation of an inherently safer synthesis framework for thermally coupled distillation sequences with structure constraints.
Ahmad et al. 2014 [102]	Development of a new approach of inherent safety assessments called the Numerical Descriptive Inherent Safety Technique (NuDIST).
Song et al [103]	Methodology based on new fuzzy logic inherent safety performance indices which focus not only on chemical hazard indices and process hazard indices but also incorporate more hazard factors such as process safety, process complexity, and operability.
Ortiz-Espinoza et al. [104]	Comparison of different safety metrics to evaluate the risk associated with a given process design. The indices selected for consideration are the Dow Fire and Explosion Index (F&EI), the fire and explosion damage index (FEDI), the process route index (PRI), and the process stream index (PSI).
Janošovský et al. [105]	Introduction of a new decision-making concept which is based on inherently safer design principles.
Ahmad et al. (2016) [106]	Graphical Descriptive Technique for Inherent Safety Assessment (GRAND) for inherent safety evaluation during the research and development (R&D) stage of process design.
Ahmad et al. (2019) [107]	Graphical Inherent Safety Assessment Technique (GISAT) for the preliminary engineering design stage which incorporates information obtainable from the PFD in the assessment of inherent safety features of a process.
Ahmad et al. (2021) [108]	Introduction of the Inherent Safety and Economic Graphical Rating (InSafE), a graphical inherent safety and economic evaluation for R&D design stage application.

design. This is demonstrated in Figure 7.9 which facilitates, for example, comparisons between cryogenic storage of liquefied hydrogen (schemes (a) and (b)) and bulk storage of metal hydrides (scheme (c)). A third important feature of this work is the explicit and unambiguous use of inherent safety terminology, for example, *substitution* and *moderation* when examining metal hydride options [58], as well as a clear placement of ISD technologies within the hierarchy of controls. All of these points are key to the advancement of the inherent safety concept within the various sectors of the hydrogen industry.

The energy industry is currently facing the challenge of replacing fossil fuels with renewable alternatives. Oliyide [110] claims hydrogen is the alternative fuel that is gaining increasing popularity, as evidenced by the UK government's intention to increase the capacity of hydrogen production to 10 GW by 2030. However, to combat

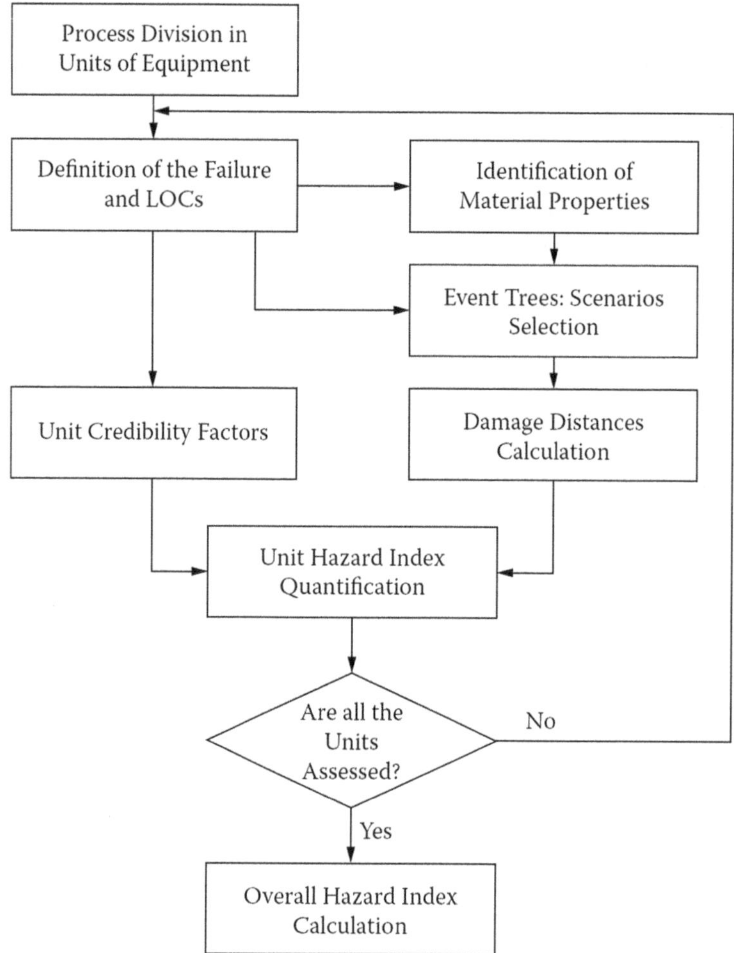

FIGURE 7.8 Flowchart of a hydrogen-specific ISD assessment method, where LOC stands for *loss of containment*. With permission.

Source: [56].

climate change by keeping the net-zero goals safe, hydrogen must be green. Green hydrogen production requires the expertise of safety engineers to apply the principles of inherently safer design to achieve safe and reliable solutions for operators and the public. It is a thrilling area of engineering that will provide new growth opportunities as the green hydrogen era seems to emerge.

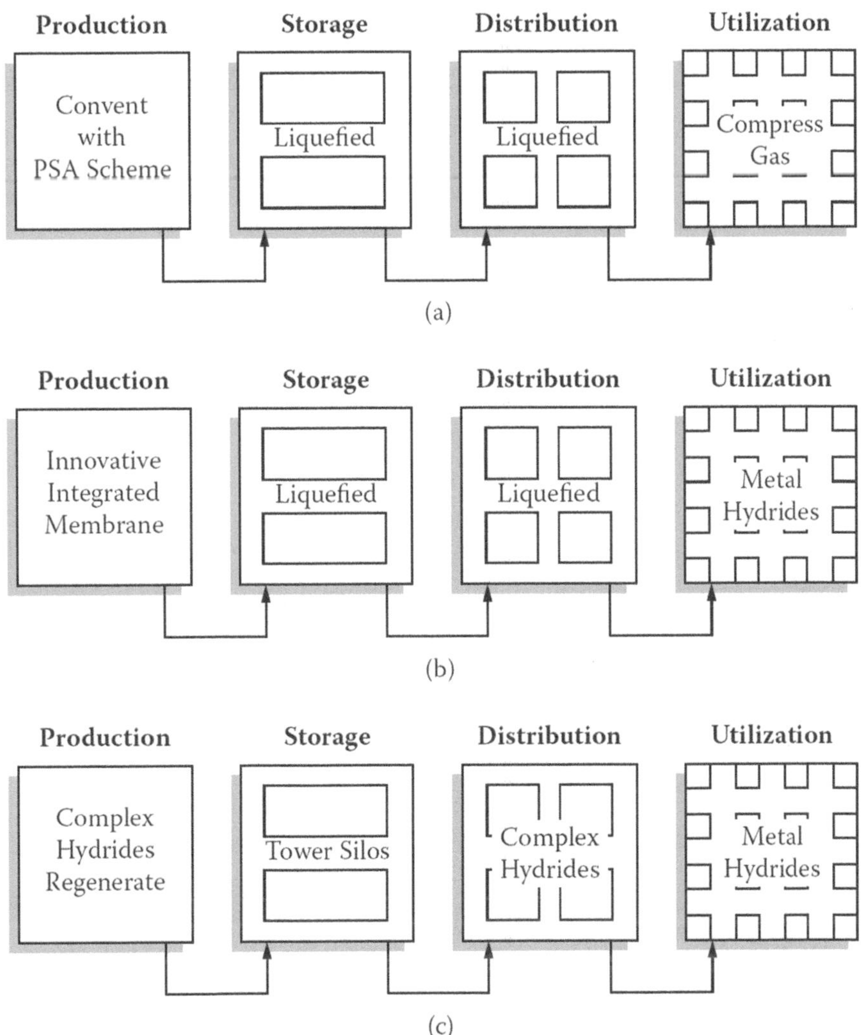

FIGURE 7.9 Alternative hydrogen supply chains for ISD assessment. With permission.

Source: [58].

REFERENCES

1. Amyotte, P.R., MacDonald, D.K., and Khan, F.I., An analysis of CSB investigation reports for inherent safety learnings, in *Paper No. 44a, Proceedings of 13th Process Plant Safety Symposium, 7th Global Congress on Process Safety (AIChE 2011 Spring National Meeting)*, Chicago, IL, March 13–16, 2011.
2. CCPS, *Inherently Safer Chemical Processes. A Life Cycle Approach*, 2nd ed., John Wiley & Sons, Hoboken, NJ, 2009.
3. Kletz, T. and Amyotte, P., *Process Plants: A Handbook for Inherently Safer Design*, 2nd ed., CRC Press/Taylor & Francis Group, Boca Raton, FL, 2010.

4. Khan, F.I. and Amyotte, P.R., How to make inherent safety practice a reality, *Canadian Journal of Chemical Engineering*, 81 (1), 2–16, 2003.

5. Hendershot, D.C., A summary of inherently safer technology, *Process Safety Progress*, 29 (4), 389–392, 2010.

6. Crowl, D.A. and Jo, Y.-D., The hazards and risks of hydrogen, *Journal of Loss Prevention in the Process Industries*, 20 (2), 158–164, 2007.

7. Markert, F., Nielsen, S.K., Paulsen, J.L., and Andersen, V., Safety aspects of future infrastructure scenarios with hydrogen refuelling stations, *International Journal of Hydrogen Energy*, 32 (13), 2227–2234, 2007.

8. General Hydrogen Corporation, *Material Safety Data Sheet: Compressed Hydrogen*, Washington, PA (undated).

9. Hydrogen Association, *Hydrogen Safety*, Fact Sheet Series (undated).

10. Molkov, V., Hydrogen safety research: State-of-the-art, in *Proceedings of the 5th International Seminar on Fire and Explosion Hazards*, Edinburgh, UK, April 23–27, 2007.

11. Miller, M., *Hydrogen Fueling Stations*, Institute of Transportation Studies, November 15, 2004.

12. Pasman, H.J. and Rogers, W.J., Safety challenges in view of the upcoming hydrogen economy: An overview, *Journal of Loss Prevention in the Process Industries*, 23 (6), 697–704, 2010.

13. Roads2HyCom, *Hydrogen Safety Measures*, Document Tracking ID 5031, March 30, 2011. www.roads2hy.com

14. HySafe, *Chapter V: Hydrogen Safety Barriers and Safety Measures*, Biennial Report on Hydrogen Safety, Version 1.0, May 2006.

15. Groethe, M., Merilo, E., Colton, J., Chiba, S., Sato, Y., and Iwabuchi, H., Large-scale hydrogen deflagrations and detonations, *International Journal of Hydrogen Energy*, 32 (13), 2125–2133, 2007.

16. Benard, P., Mustafa, V., and Hay, D.R., Safety assessment of hydrogen disposal on vents and flare stacks at high flow rates, *International Journal of Hydrogen Energy*, 24 (5), 489–495, 1999.

17. Astbury, G.R., Venting of low pressure hydrogen gas: A critique of the literature, *Process Safety and Environmental Protection*, 85 (4), 289–304, 2007.

18. Wald, M.L. and Pollack, A., Core of stricken reactor probably leaked, U.S. says, *New York Times*, April 6, 2011.

19. Tabuchi, H., Bradsher, K., and Wald, M.L., In Japan reactor failings, danger signs for the U.S., *New York Times*, May 17, 2011.

20. Alsheyab, M., Jiang, J.-Q., and Stanford, C., Risk assessment of hydrogen gas production in the laboratory scale electrochemical generation of ferrate (VI), *Journal of Chemical Health & Safety*, 15 (5), 16–20, 2008.

21. CSB, *Urgent Recommendations from Kleen Energy Investigation*, U.S. Chemical Safety and Hazard Investigation Board, Washington, DC, 2010.

22. Bradsher, K., Pressing ahead where others have failed, *New York Times,* March 24, 2011.

23. Sherman, D., At milepost 1 on the hydrogen highway, *New York Times,* April 29, 2007.

24. Guy, K.W., The hydrogen economy, *Process Safety and Environmental Protection*, 78 (4), 324–327, 2000.

25. Motavalli, J., A universe of promise (and a tankful of caveats), *New York Times*, April 29, 2007.

26. Leary, W.E., Use of hydrogen as fuel is moving closer to reality, *New York Times*, April 16, 1995.

27. Leary, W.E., With shuttle back in space, NASA returns to leak problem, *New York Times*, October 9, 1990.

28. Rainer, D., Hydrogen, *Journal of Chemical Health and Safety*, 15 (4), 49–50, 2008.

29. Middha, P., Engel, D., and Hansen, O.R., Can the addition of hydrogen to natural gas reduce the explosion risk? *International Journal of Hydrogen Energy,* 36 (3), 2628–2636, 2011.

30. Yang, J.C., Material-based hydrogen storage, *International Journal of Hydrogen Energy*, 33 (16), 4424–4426, 2008.

31. Xu, P., Zheng, J., Liu, P., Chen, R., Kai, F., and Li, L., Risk identification and control of stationary high-pressure hydrogen storage vessels, *Journal of Loss Prevention in the Process Industries*, 22 (6), 950–953, 2009.

32. Janssen, H., Bringmann, J.C., Emonts, B., and Schroeder, V., Safety-related studies on hydrogen production in high-pressure electrolysers, *International Journal of Hydrogen Energy*, 29 (7), 759–770, 2004.

33. CSB, *Positive Material Verification: Prevent Errors During Alloy Steel Systems Maintenance*, Safety Bulletin, No. 2005-04-B, U.S. Chemical Safety and Hazard Investigation Board, Washington, DC, 2006.

34. Segal, L., Wallace, J.S., and Keffer, J.F., Safety considerations in the design of a gaseous hydrogen fuel supply for engine testing, *International Journal of Hydrogen Energy*, 11 (11), 737–743, 1986.

35. Matthijsen, A.J.C.M., and Kooi, E.S., Safety distances for hydrogen filling stations, *Fuel Cells Bulletin*, 2006 (11), 12–16, 2006.

36. Dorofeev, S.B., Evaluation of safety distances related to unconfined hydrogen explosions, *International Journal of Hydrogen Energy*, 32 (13), 2118–2124, 2007.

37. Marangon, A., Carcassi, M., Engebo, A., and Nilsen, S., Safety distances: Definition and values, *International Journal of Hydrogen Energy*, 32 (13), 2192–2197, 2007.

38. Khan, F.I. and Amyotte, P.R., How to make inherent safety practice a reality, *Canadian Journal of Chemical Engineering*, 81 (1), 2–16, 2003.

39. Khan, F.I., Sadiq, R., and Amyotte, P.R., Evaluation of available indices for inherently safer design options, *Process Safety Progress,* 22 (2), 83–97, 2003.

40. Etowa, C.B., Amyotte, P.R., Pegg, M.J., and Khan, F.I., Quantification of inherent safety aspects of the Dow indices, *Journal of Loss Prevention in the Process Industries*, 15 (6), 477–487, 2002.

41. Edwards, D.W., and Lawrence, D., Assessing the inherent safety of chemical process routes, *Process Safety and Environmental Protection*, 71 (B4), 252–258, 1993.

42. Edwards, D.W., Rushton, A.G., and Lawrence, D., Quantifying the inherent safety of chemical process routes, in *Paper presented at the 5th World Congress of Chemical Engineering*, San Diego, CA, July 14–18, 1996.

43. Khan, F.I. and Amyotte, P.R., Integrated Inherent Safety Index (I2SI): A tool for inherent safety evaluation, *Process Safety Progress*, 23 (2), 136–148, 2004.

44. Khan, F.I. and Amyotte, P.R., I2SI: A comprehensive quantitative tool for inherent safety and cost evaluation, *Journal of Loss Prevention in the Process Industries*, 18 (4–6), 310–326, 2005.

45. Carvalho, A., Gani, R., and Matos, H., Design of sustainable chemical processes: Systematic retrofit analysis generation and evaluation of alternatives, *Process Safety and Environmental Protection*, 86 (5), 328–346, 2008.

46. Hurme, M. and Rahman, M., Implementing inherent safety throughout process lifecycle, *Journal of Loss Prevention in the Process Industries*, 18 (4–6), 238–244, 2005.

47. Rahman, M., Heikkila, A.-M., and Hurme, M., Comparison of inherent safety indices in process concept evaluation, *Journal of Loss Prevention in the Process Industries*, 18 (4–6), 327–334, 2005.

48. Hassim, M.H. and Hurme, M., Inherent occupational health assessment during process research and development stage, *Journal of Loss Prevention in the Process Industries*, 23 (1), 127–138, 2010.

49. Hassim, M.H. and Hurme, M., Inherent occupational health assessment during basic engineering stage, *Journal of Loss Prevention in the Process Industries*, 23 (2), 260–268, 2010.

50. Hassim, M.H. and Hurme, M., Inherent occupational health assessment during preliminary design stage, *Journal of Loss Prevention in the Process Industries*, 23 (3), 476–482, 2010.

51. Hassim, M.H. and Hurme, M., Occupational chemical exposure and risk estimation in process development and design, *Process Safety and Environmental Protection*, 88 (4), 225–235, 2010.

52. Hassim, M.H. and Edwards, D.W., Development of a methodology for assessing inherent occupational health hazards, *Process Safety and Environmental Protection*, 84 (5), 378–390, 2006.

53. Gupta, J.P. and Edwards, D.W., A simple graphical method for measuring inherent safety, *Journal of Hazardous Materials*, 104 (1–3), 15–30, 2003.

54. Cozzani, V., Barontini, F., and Zanelli, S., Assessing the inherent safety of substances: Precursors of hazardous products in the loss of control of chemical systems, in *Proceedings of American Institute of Chemical Engineers Spring National Meeting*, Orlando, FL, April 23–27, 2006.

55. Landucci, G., Tugnoli, A., Nicolella, C., and Cozzani, V., Assessment of inherently safer technologies for hydrogen storage, in *IChemE Symposium Series No. 153, in 12th International Symposium on Loss Prevention and Safety Promotion in the Process Industries*, Edinburgh, UK, 2007a.

56. Landucci, G., Tugnoli, A., Nicolella, C., and Cozzani, V., Assessment of inherently safer technologies for hydrogen production, in *Proceedings of the 5th International Seminar on Fire and Explosion Hazards*, Edinburgh, UK, April 23–27, 2007b.

57. Landucci, G., Tugnoli, A., and Cozzani, V., Inherent safety key performance indicators for hydrogen storage systems, *Journal of Hazardous Materials*, 159 (2–3), 554–566, 2008.

58. Tugnoli, A., Landucci, G., and Cozzani, V., Key performance indicators for inherent safety: Application to the hydrogen supply chain, *Process Safety Progress*, 28 (2), 156–170, 2009.

59. Landucci, G., Tugnoli, A., and Cozzani, V., Safety assessment of envisaged systems for automotive hydrogen supply and utilization, *International Journal of Hydrogen Energy*, 35 (3), 1493–1505, 2010.

60. Cordella, M., Tugnoli, A., Barontini, F., Spadoni, G., and Cozzani, V., Inherent safety of substances: Identification of accidental scenarios due to decomposition products, *Journal of Loss Prevention in the Process Industries*, 22 (4), 455–462, 2009.

61. Shariff, A.M., Rusli, R., Leong, C.T., Radhakrishnan, V.R., and Buang, A., Inherent safety tool for explosion consequences study, *Journal of Loss Prevention in the Process Industries*, 19 (5), 409–418, 2006.

62. Leong, C.T. and Shariff, A.M., Inherent Safety Index Module (ISIM) to assess inherent safety level during preliminary design stage, *Process Safety and Environmental Protection*, 86 (2), 113–119, 2008.

63. Leong, C.T. and Shariff, A.M., Process Route Index (PRI) to assess level of explosiveness for inherent safety quantification, *Journal of Loss Prevention in the Process Industries*, 22 (2), 216–221, 2009.
64. Shariff, A.M. and Leong, C.T., Inherent risk assessment: A new concept to evaluate risk in preliminary design stage, *Process Safety and Environmental Protection,* 87 (6), 371–376, 2009.
65. Shariff, A.M. and Zaini, D., Toxic release consequence analysis tool (TORCAT) for inherently safer design plant, *Journal of Hazardous Materials*, 182 (1–3), 394–402, 2010.
66. Rusli, R. and Shariff, A.M., Qualitative assessment for inherently safer design (QAISD) at preliminary design stage, *Journal of Loss Prevention in the Process Industries*, 23 (1), 157–165, 2010.
67. Kossoy, A. and Akhmetshin, Yu., Simulation-based approach to design of inherently safer processes, in *IChemE Symposium Series No. 153, in 12th International Symposium on Loss Prevention and Safety Promotion in the Process Industries*, Edinburgh, UK, 2007.
68. Shah, S., Fischer, U., and Hungerbuhler, K., A hierarchical approach for the evaluation of chemical process aspects from the perspective of inherent safety, *Process Safety and Environmental Protection*, 81 (6), 430–443, 2003.
69. Adu, I.K., Sugiyama, H., Fischer, U., and Hungerbuhler, K., Comparison of methods for assessing environmental, health and safety (EHS) hazards in early phases of chemical process design, *Process Safety and Environmental Protection,* 86 (3), 77–93, 2008.
70. Palaniappan, C., Srinivasan, R., and Halim, I., A material-centric methodology for developing inherently safer environmentally benign processes, *Computers and Chemical Engineering*, 26 (4–5), 757–774, 2002.
71. Palaniappan, C., Srinivasan, R., and Tan, R., Expert system for the design of inherently safer processes, Part 1: Route selection stage, *Industrial Engineering Chemistry Research*, 41 (26), 6698–6710, 2002.
72. Palaniappan, C., Srinivasan, R., and Tan, R., Expert system for the design of inherently safer processes, Part 2: Flowsheet development stage, *Industrial Engineering Chemistry Research*, 41 (26), 6711–6722, 2002.
73. Srinivasan, R. and Nhan, N.T., A statistical approach for evaluating inherent benign-ness of chemical process routes in early design stages, *Process Safety and Environmental Protection*, 86 (3), 163–174, 2008.
74. Srinivasan, R. and Kraslawski, A., Application of the TRIZ creativity enhancement approach to design of inherently safer chemical processes, *Chemical Engineering and Processing*, 45 (6), 507–514, 2006.
75. Meel, A. and Seider, W.D., Dynamic risk assessment of inherently safe chemical processes: Accident precursor approach, in *Presented at American Institute of Chemical Engineers Spring National Meeting*, Atlanta, GA, 2005.
76. Gentile, M., Rogers, W.J., and Mannan, M.S., Development of a fuzzy logic-based inherent safety index, *Process Safety and Environmental Protection*, 81 (6), 444–456, 2003.
77. Al-Mutairi, E.M., Suardin, J.A., Mannan, S.M., and El-Halwagi, M.M., An optimization approach to the integration of inherently safer design and process scheduling, *Journal of Loss Prevention in the Process Industries*, 21 (5), 543–549, 2008.
78. Khan, F., Rathnayaka, S., and Ahmed, S., Methods and models in process safety and risk management: Past, present and future. *Process Safety and Environmental Protection*, 98, 116–147, 2015.

79. Kidam, K., Sahak, H.A., Hassim, M.H., Shahlan, S.S., and Hurme, M., Inherently safer design review and their timing during chemical process development and design. *Journal of Loss Prevention in the Process Industries*, 42, 47–58, 2016.

80. Athar, M., Shariff, A.M., and Buang, A., A review of inherent assessment for sustainable process design. *Journal of the Cleaner Production*, 233, 242–263, 2019a.

81. Gao, X., Abdul Raman, A.A., Hizaddin, H.F., Buthiyappan, A., and Bello, M.M., Systematic inherent safety and its implementation in chlorine liquefaction process. *Journal of Loss Prevention in the Process Industries*, 65, 104133, 2020.

82. Amyotte, P.R. and Khan, F.I., The role of inherently safer design in process safety. *The Canadian Journal of Chemcial Engineering*, 99, 853–871, 2021.

83. Park, S., Xu, S., Rogers, W., Pasman, H., and El-Halwagi, M.M., Incorporating inherent safety during the conceptual process design stage: A literature review. *Journal of Loss Prevention in the Process Industries*, 63, 104040, 2020.

84. Athar, M., Shariff, A.M., Buang, A., Nazir, S., Hermansyah, H., and See, T.L., Process equipment common attributes for inherently safer process design at preliminary design stage. *Process Safety and Environmental Protection*, 128, 14–29, 2019b.

85. Zainal Abidin, M., Rusli, R., Khan, F., and Mohd Shariff, A., Development of inherent safety benefits index to analyse the impact of inherent safety implementation. *Process Safety and Environmental Protection*, 117, 454–472, 2018.

86. Gangadharan, P., Singh, R., Cheng, F., and Lou, H.H., Novel methodology for inherent safety assessment in the process design stage. *Industrial & Engineering Chemistry Research*, 52, 5921–5933, 2013.

87. Gao, X., Raman, Abdul, Hizaddin, A.A., Bello, H.F., and Buthiyappan, A.M.M., Review on the inherently safer design for chemical processes: Past, present and future. *Journal of Cleaner Production*, 305, 127154, 2021.

88. Sultana, S. and Haugen, S., Development of an inherent system safety index (ISSI) for ranking of chemical processes at the concept development stage. *Journal of Hazardous Materials*, 421, 126590, 2022.

89. Pasha, M., Zaini, D., and Mohd Shariff, A. Inherently safer design for heat exchanger network. *Journal of Loss Prevention in the Process Industries*, 48, 55–70, 2017.

90. Crivellari, A., Bonvicini, S., Tugnoli, A., and Cozzani, V. Multi-target inherent safety indices for the early design of offshore oil & gas facilities. *Process Safety and Environmental. Protection.* 148, 256–272, 2021.

91. Zhu J., Liu, Z., Cao, Z., Han, X., and Hao, L., Development of a general inherent safety assessment tool at early design stage of chemical process, *Process Safety and Environmental Protection*, 167, 356–367, 2022.

92. Ee, A.W.L., Kuznetsova, E., Lee, T.E.J., and Ng, A.T.S., Extended inherent safety index -analysis of chemical, physical and biological inherent safety. *Journal of Cleaner Production*, 248, 119258, 2020.

93. Ying, So, W., Hassim, M.H., Ahmad, S.I., and Rashid, R., Inherent occupational health assessment index for research and development stage of process design. *Process Safety and Environmental Protection*, 147, 103–114, 2021.

94. Pu, W., Abdul Raman, A.A., Hamid, M.D., Gao, X., Lin, S., Buthiyappan, A., Process & occupational safety integrated inherently safer chemical plant design: Framework development and validation. *Journal of Loss Prevention in the Process Industries*, 87, 105204, 2024.

95. Eini, S., Shahhosseini, H.R., Javidi, M., Sharifzadeh, M., and Rashtchian, D., Inherently safe and economically optimal design using multi-objective optimization: the case of a refrigeration cycle. *Process Safety and Environmental Protection*, 104, 254–267, 2016.

96. Bolla, M., Pettinato, M., and Fabiano, B., Inherent safety development applied to green processes and energy transition. *Chemical Engineering Transactions*, 104, 151–156, 2023.

97. Rayner Brown, K., Hastie, M., Khan, F.I., and Amyotte, P.R., Inherently safer design protocol for process hazard analysis. *Process Safety and Environmental Protection*, 149, 199–211, 2021.

98. Sun, F., Xu, W., Wang, G., Shi, N., and Yang, Z., Development of a hazard index for reactivity management (HIRM) in chemical process. *Process Safety and Environmental Protection*, 149, 22–34, 2021.

99. Athar, M., Shariff, A.M., Buang, A., Shaikh, M.S., and See, T.L., Inherent safety for sustainable process design of process piping at the preliminary design stage. *Journal of Cleaner Production*, 209, 1307–1318, 2019c.

100. Ebadollahi, M., Rostamzadeh, H., Pourali, O., Ghaebi, H., and Amidpour, M., Inherently safety design of a dual-loop bi-evaporator combined cooling and power system: 4E and safety based optimization approach. *Process Safety and Environmental Protection*, 154, 393–409, 2021.

101. Ye, H., Gao, W., Dong, H.-g, Bi, M., An inherently safer development approach for thermally coupled distillation sequences: Application in hazardous chemical separation. *Process Safety and Environmental Protection*, 160, 786–802, 2022.

102. Ahmad, S.I., Hashim, H., and Hassim, M.H., Numerical descriptive inherent safety technique (NuDIST) for inherent safety assessment in petrochemical industry. *Process Safety and Environmental Protection*, 92, 379–389, 2014.

103. Song, D., Yoon, E.S., and Jang, N., A framework and method for the assessment of inherent safety to enhance sustainability in conceptual chemical process design. *Journal of Loss Prevention in the Process Industries*, 54, 10–17, 2018.

104. Ortiz-Espinoza, A.P., Jim´enez-Guti´errez, A., El-Halwagi, M.M., Kazantzis, N.K., and Kazantzi, V., Comparison of safety indexes for chemical processes under uncertainty. *Process Safety and Environmental Protection*, 148, 225–236, 2021.

105. Janoˇsovský, J., Rosa, I., Vincent, G., ˇSulgan, B., Variny, M., Labovsk´a, Z., Labovský, J., and Jelemenský, Ľ., Methodology for selection of inherently safer process design alternatives based on safety indices. *Process Safety and Environmental Protection*, 160, 513–526, 2022.

106. Ahmad, S.I., Hashim, H., and Hassim, M.H., A graphical method for assessing inherent safety during research and development phase of process design. *Journal of Loss Prevention in the Process Industries*, 42, 59–69, 2016.

107. Ahmad, S.I., Hashim, H., Hassim, M.H., and Rashid, R., A graphical inherent safety assessment technique for preliminary design stage. *Process Safety and Environmental Protection*, 130, 275–287, 2019.

108. Ahmad, S.I., Hashim, H., Hassim, M.H., and Rashid, R., Inherent safety and economic graphical rating (InSafE) method for inherent safety and economic assessment. *Process Safety and Environmental Protection*, 149, 602–609, 2021.

109. Edwards, D.W., Are we too risk-averse for inherent safety? An examination of current status and barriers to adoption, *Process Safety and Environmental Protection*, 83 (2), 90–100, 2005.

110. Oliyide, T., *Why safety engineering is key for green hydrogen and net zero*, Ingenia Org., February 2023. www.ingenia.org.uk/articles/why-safety-engineering-is-key-for-green-hydrogen-and-net-zero/#:~:text=While%20designing%20a%20green%20hydrogen,and%20pressures%2C%20and%20process%20simplification

8 Safety Management Systems

A key engineering tool for industrial practice is a management system appropriate for the risks being addressed (process safety, occupational safety, health, environment, asset integrity, etc.) [1]. Such safety management systems are recognized and accepted worldwide as best-practice methods for managing risk. They typically consist of ten to 20 program elements that must be effectively carried out to manage risks in an acceptable way. This need is based on the understanding that once a risk is identified, it does not go away; there is always a possibility that the adverse event will occur unless the management system is actively monitoring company operations for concerns and taking proactive actions to correct potential problems [2]. Safety management system (SMS) implementation may differ across industries, but the basic principles remain constant: proactive hazard identification, risk assessment, continuous monitoring, and a commitment to continuous improvement in safety performance. Applying a reliable SMS assists organizations in systematically managing safety and reducing the likelihood of accidents and incidents.

In this chapter, safety management systems are first discussed from a general perspective. This is followed by an examination of the elements that make up a typical management system for specifically dealing with issues related to process safety. The important and timely topic of safety culture is then briefly addressed. The chapter is structured in a manner similar to the discussion in Kletz and Amyotte [3] (with relevant excerpts) on the relationship between process safety management and inherently safer design; here, of course, the emphasis is on the importance of adopting a management approach to ensuring hydrogen safety.

8.1 INTRODUCTION TO SAFETY MANAGEMENT SYSTEMS

As discussed in Chapter 2, root causes identified for the 1989 vapor cloud explosion in Pasadena, Texas, were a lack of both a hazard assessment study and an effective permit system for the control of maintenance activities. These factors are clearly related to the safety management system in place at the facility as they are effectively beyond the control of individual workers. Although it is incumbent upon plant operators to follow safe work procedures for maintenance and other activities, it is

DOI: 10.1201/9781003313007-8

a management responsibility to develop those procedures in the first place and to ensure they are implemented and revised as needed. It is well established by modern incident causation theory that management system deficiencies represent the ultimate cause of industrial accidents [4].

There is thus a need for a safety management system approach in the hydrogen industry, a fact that has been recognized by various organizations. For example, the US Department of Energy (DOE) has commented in its technical planning document for hydrogen safety that (with italics added here for emphasis) [5], "The continued safe operation, handling and use of hydrogen and related systems require *comprehensive safety management*."

What form, then, should such comprehensive safety management take? As noted in the introductory sentence to this chapter, safety management systems are most effective when they are tailored to the specific hazards and risks of concern. Regardless of whether the management approach is aimed at preventing occupational incidents related to individuals or process incidents that are more systemwide, certain features will (or should) be universal. Stelmakowich [6] describes a set of framework components for Occupational Health and Safety Assessment Series (OHSAS) 18000 that are equally applicable to other areas within the safety field:

- Continuous improvement.
- Policy development and support by senior management personnel.
- Planning (e.g., hazard identification, risk assessment, risk control).
- Implementation and operation (e.g., assignment of responsibilities, training, emergency preparedness and response).
- Checks and corrective actions (e.g., incident investigation, auditing).
- Management review.

Management system requirements such as those given in this list are widely known by the generic titles of *plan, do, check, act*. These four essential management functions apply to the safety management system as a whole and also to each of the individual elements comprising the management system. This point is arguably the most important feature of the management system approach to safety (hydrogen or otherwise).

The provision of additional focus for a safety management system is often addressed by reference to the familiar *incident pyramid* shown in Figure 8.1. Various studies over the years, covering a broad range of industries and typically from an occupational health and safety (OH&S) perspective, have resulted in different category totals for the pyramid levels but generally the same ratios between levels (see, for example, [7]). As explained by Creedy [8]:

> The idea of the pyramid is that really serious incidents such as fatalities occur so rarely in most organizations these days that it's not practical to use them as a measure for monitoring and improving an organization's safety effectiveness – there simply aren't enough such incidents to know whether things are getting better or worse. However, it is practical to track the larger number of less serious incidents and use that as a performance measure.

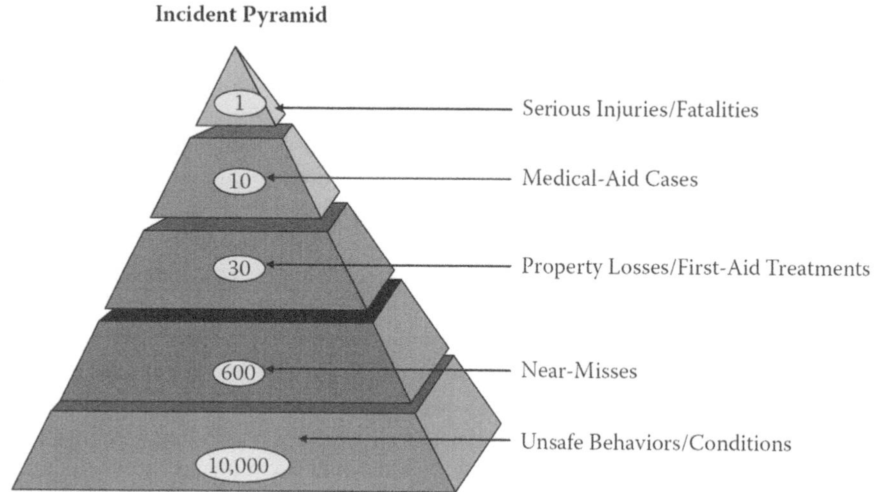

FIGURE 8.1   The incident pyramid. With permission.

Source: [8].

Hence, many safety management systems – particularly those dealing with OH&S and avoidance of lost-time injuries (LTIs) – rightfully emphasize the prevention, control, and investigation of near-misses and unsafe (or at-risk) acts and conditions. Creedy [8] does, however, provide a cautionary note:

> The pyramid is certainly a useful concept. Nevertheless, it does have a serious problem that is often unrecognized, even by workplace health and safety professionals. The problem is that the conditions that can lead to really serious incidents – those which could kill or seriously injure a large number of people – may not be identified by focusing on the bottom end of the pyramid, which could thus become a distraction rather than a help.

The thinking expressed in this passage allows us to make a further distinction between *occupational* safety and *process* safety. As previously mentioned, occupational or "traditional" safety largely aims to control individual exposures – often referred to as slips, trips, and falls. Process safety, when properly addressed with a systems approach, aims to prevent and mitigate process-related injuries and incidents. The scope of process safety largely deals with fires, explosions, and toxic releases. Returning to the incident pyramid (Figure 8.1), if a management system is designed to facilitate effective process safety efforts, then investigation of near-misses must involve process-related near-misses such as temperature excursions and overpressurizations inside process vessels. Investigation of OH&S near-misses involving working at height, for example, are ineffective and may be counterproductive if the desired focus is process safety. (These measures are, of course, entirely appropriate if the focus is on OH&S.)

This distinction between occupational and process safety is far from a simple academic exercise. As noted in Chapter 1, hydrogen is widely produced and used in the chemical and oil industries (broadly speaking, the process industries) with growing use in the public realm [9]. While the former area of focus would obviously necessitate process safety considerations, the latter would involve transportation uses in which occupational safety issues for individuals are paramount.

In its document on safety plan development for DOE-funded hydrogen and fuel cell projects, the DOE addresses this point by providing the following guidance on the first element of such a plan – the scope of work [10]:

> The plan should briefly describe the specific nature of the work being performed to set the context for the safety plan. It should distinguish between laboratory-scale research, bench-scale testing, engineering development, and prototype operation. All intended project phases should be described.

Implicit in this guidance is the need to identify the hazards of concern and the affected personnel when establishing the scope of work to be addressed by the safety plan (which is in essence a description of a safety management system). Interestingly, the section on scope of work [10] goes on to describe the value of quantifying the amounts of hazardous materials (including hydrogen) generated, used, and stored, thus emphasizing the linkage between inherently safer design (Chapter 7) and effective safety management. The complete set of DOE safety plan elements is given in Table 8.1; further discussion of this framework is undertaken in the next section.

OHSAS 18000 was replaced by OHSAS 18001, which was an international standard for occupational health and safety management systems. Subsequently, OHSAS 18001 was canceled to adopt ISO 45001, which was published by the International Organization for Standardization in March 2018. Compliance with it requires organizations to demonstrate that they have established a reliable system for occupational health and safety. Organizations that were certified to OHSAS 18001 were able to migrate to ISO 45001 by March 2021 to retain a recognized certification (Wikipedia OHSAS 18001) [11].

8.2 PROCESS SAFETY MANAGEMENT

Process safety management aims to prevent major accidents, like explosions, while occupational safety management aims to prevent more individual-level safety incidents, like falls. Process safety management is a regulatory standard published by the Occupational Safety and Health Administration (OSHA) concerning processes that use highly hazardous chemicals (HHCs). The OSHA process safety management standard includes guidelines for preventing catastrophic releases of toxic, reactive, flammable, or explosive chemicals.

The standard might also be useful for chemical safety managers to understand the basic definitions of toxic, reactive, flammable, and explosive. These are properties or characteristics of HHCs. According to the Center for Chemical Process Safety (CCPS) [12]:

TABLE 8.1
**Safety Plan Elements for US DOE-Funded
Hydrogen and Fuel Cell Projects**

No.	Element
1	Scope of work
2	Organizational safety information
	Organizational policies and procedures
	Hydrogen and fuel cell experience
3	Project safety
	Identification of safety vulnerabilities
	Risk reduction plan
	Operating procedures
	Equipment and mechanical integrity
	Management of change procedures
	Project safety documentation
4	Communication plan
	Employee training
	Safety reviews
	Safety events and lessons learned
	Emergency response
	Self-audits
5	Safety plan approval
6	Other comments or concerns

Source: [10].

- *Toxic Material* – an airborne agent that could result in acute adverse human health effects.
- *Reactive Material* – a substance which enters into a chemical reaction with other stable or unstable material.
- *Flammable* – a gas that can burn with a flame if mixed with a gaseous oxidizer such as air and then ignited.
- *Explosive* – a chemical that causes a sudden release of pressure, gas, and heat when subjected to sudden shock, pressure, or high temperature.

Having established the prevalence of hydrogen use in the process industries, as well as the clear need to prevent and mitigate process-related incidents, it is appropriate to now turn solely to the matter of process safety management. To reiterate, process safety management deals with the identification, understanding, and control of process hazards to prevent process-related injuries and incidents (fires, explosions, toxic releases). As previously mentioned, the discussion here is based on Kletz and Amyotte [3], with specific reference to hydrogen safety.

In his recounting of the history of safety philosophy in the process industries, Creedy [8] describes four phases of development.

1. The late nineteenth and early twentieth century: Here the objective was primarily the protection of capital assets, providing the origins of basic safety thinking in the explosives industry.
2. World War II through the 1950s and 1960s: During this period the objectives were greater efficiency and the creation of a better society. The concepts of loss prevention and investing in people were introduced. Safety measures were largely rule based.
3. The 1970s and 1980s: The objectives here were the same as the previous phase. However, recognition of consequence seriousness and causation mechanisms led to a focus on the process rather than the individual worker, and hence the development of a management approach to process safety (*process* being defined by the US Occupational Health and Safety Administration as any activity involving a highly hazardous chemical, including any use, storage, manufacturing, handling, or the on-site movement of such chemicals, or combination of these activities).
4. The 1990s and beyond: Here we see a realization of the significance of sociocultural factors in human thought processes and behaviors at the individual and organizational levels. This has led to increased understanding of the importance of such concepts as human factors and safety culture.

At present, we appear to be in a period that merges the last two of the above phases – process safety management coupled with a recognition that without a strong safety culture (to be defined later in this chapter), even the best management system on paper can become dysfunctional. An approach widely used in Canada is known simply as PSM (Process Safety Management). The complete suite of PSM elements is shown in Table 8.2, taken from the Process Safety Management Guide of the Canadian Society for Chemical Engineering (CSChE) [13].

This guide was prepared by the Process Safety Working Group of the former Major Industrial Accidents Council of Canada (MIACC) in conjunction with the Process Safety Management Committee of the Canadian Chemical Producers' Association (CCPA), now known as the Chemistry Industry Association of Canada. With the dissolution of MIACC in 1999, rights to the guide were transferred to the CSChE. The material in the CSChE PSM guide [13] is based on that developed by the CCPS of the American Institute of Chemical Engineers (e.g., [14]). This route was adopted because the CCPS approach to process safety management was determined to be comprehensive, well-supported by reference materials, tools, and an organizational structure, and based on a benchmark of leading or good industry practice rather than on minimum standards [13].

Table 8.2 will therefore likely be familiar to PSM practitioners throughout North America. It will also be relevant to those engaged in process safety efforts in other parts of the world given the incorporation of best practices in the elements and components shown in Table 8.2. Other systems may have more or fewer elements, the terminology may be somewhat different, or a specific management system may be mandated by regulation, but the underlying concepts outlined in the previous section are the same.

TABLE 8.2
Elements and Components of Process Safety Management

No.	Element	Component
1	Accountability: objectives and goals	1.1 Continuity of operations; 1.2 Continuity of systems; 1.3 Continuity of organization; 1.4 Quality process; 1.5 Control of exceptions; 1.6 Alternative methods; 1.7 Management accessibility; 1.8 Communications; 1.9 Company expectations.
2	Process knowledge and documentation	2.1 Chemical and occupational health hazards; 2.2 Process definition/design criteria; 2.3 Process and equipment design; 2.4 Protective systems; 2.5 Normal and upset conditions (operating procedures); 2.6 Process risk management decisions; 2.7 Company memory (management of information).
3	Capital project review and design	3.1 Appropriation request procedures; 3.2 Hazard reviews; 3.3 Siting; 3.4 Plot plan; 3.5 Process design and review procedures; 3.6 Project management procedures and controls.
4	Process risk management	4.1 Hazard identification; 4.2 Risk analysis of operations; 4.3 Reduction of risk; 4.4 Residual risk management; 4.5 Process management during emergencies; 4.6 Encouraging client and supplier companies to adopt similar risk management practices; 4.7 Selection of businesses with acceptable risk.
5	Management of change	5.1 Change of process technology; 5.2 Change of facility; 5.3 Organizational changes; 5.4 Variance procedures; 5.5 Permanent changes; 5.6 Temporary changes.
6	Process and equipment integrity	6.1 Reliability engineering; 6.2 Materials of construction; 6.3 Fabrication and inspection procedures; 6.4 Installation procedures; 6.5 Preventative maintenance; 6.6 Process, hardware and systems inspection and testing; 6.7 Maintenance procedures; 6.8 Alarm and instrument management; 6.9 Decommissioning and demolition procedures.
7	Human factors	7.1 Operator–process/equipment interface; 7.2 Administrative control versus hardware; 7.3 Human error assessment.
8	Training and performance	8.1 Definition of skills and knowledge; 8.2 Design of operating and maintenance procedures; 8.3 Initial qualifications assessment; 8.4 Selection and development of training programs; 8.5 Measuring performance and effectiveness; 8.6 Instructor program; 8.7 Records management; 8.8 Ongoing performance and refresher training.
9	Incident investigation	9.1 Major incidents; 9.2 Third party participation; 9.3 Follow-up and resolution; 9.4 Communication; 9.5 Incident recording, reporting, and analysis; 9.6 Near-miss reporting.
10	Standards, codes, and laws	10.1 External codes/regulations; 10.2 Internal standards.
11	Audits and corrective actions	11.1 Process safety management systems audits; 11.2 Process safety audits; 11.3 Compliance reviews; 11.4 Internal/external auditors.
12	Enhancement of process safety knowledge	12.1 Quality control programs and process safety; 12.2 Professional trade and association programs; 12.3 CCPS program; 12.4 Research, development, documentation, and implementation; 12.5 Improved predictive system; 12.6 Process safety resource centre and reference library.

Source: [13]. With permission.

TABLE 8.3
Risk-Based Process Safety Management System

Accident Prevention Pillar	Risk-Based Process Safety Element
Commit to process safety	Process safety culture
	Compliance with standards
	Process safety competency
	Workforce involvement
	Stakeholder outreach
Understand hazards and risk	Process knowledge management
	Asset integrity and reliability
	Contractor management
	Training and performance assurance
	Management of change
	Operational readiness
	Conduct of operations
	Emergency management
	Hazard identification and risk analysis
Manage risk	Operating procedures
	Safe work practices
Learn from experience	Incident investigation
	Measurement and metrics auditing
	Management review and continuous improvement

Source: [12]. With permission.

As previously discussed, continuous improvement is a key feature of safety management systems; in this regard it is important to note the recent CCPS work on developing a framework for risk-based process safety (RBPS) management [15]. This 20-element, risk-based system is shown in Table 8.3. As an evolution of the PSM approach given in Table 8.2, the RBPS management system makes an explicit link between process safety *management* and process safety *culture.*

The generic safety management systems shown in Tables 8.2 and 8.3 share many commonalities with the hydrogen-specific safety plan given in Table 8.1. The implication here is that classical process safety concepts and methodologies are entirely applicable to the production, storage, and use of hydrogen. This is especially the case for *process risk management* (element 4 in Table 8.2) as illustrated later in this section. A specific example involving element 6 in Table 8.2 will serve to further illustrate the validity of this claim. This element, expressed in slightly different terms meaning the same thing, is as follows:

- Table 8.1, equipment and mechanical integrity.
- Table 8.2, process and equipment integrity.
- Table 8.3, asset integrity and reliability.

TABLE 8.4
Incident Causation (2004 PRIM Data) According to PSM
Elements Given in Table 8.2

No.	Element	Percent of Incidents
6	Process and equipment integrity	23.8
2	Process knowledge and documentation	21.2
4	Process risk management	16.8
7	Human factors	8.9
5	Management of change	7.3
3	Capital project review and design procedures	6.5

Source: [16]. With permission.

The former CCPA Process Safety Management Committee has collected and analyzed data on an annual basis for process-related incidents reported by then CCPA member companies using a procedure known as PRIM (Process-Related Incidents Measure). The 2004 PRIM analysis of 89 reported incidents demonstrated that six of the PSM elements in Table 8.2 contributed to 85% of the total incidents [16]. As shown in Table 8.4, deficiencies in *process and equipment integrity* (element 6 in Table 8.2) contributed to approximately 24% of the total reported incidents. Figure 8.2 shows a breakdown of this PSM element into its components (or sub-elements) for 2004 as well as the five reporting periods prior to 2004. Here we see the predominant role of *preventative maintenance* (component 6.5) and *maintenance procedures* (component 6.7). This observation is consistent with the comment made earlier in this chapter with respect to inadequate permit controls for maintenance activities being one of the root causes of the 1989 Pasadena, Texas, vapor cloud explosion involving hydrogen (and other hazardous materials).

The PRIM methodology uses multiple analysts (i.e., self-reporting by companies) with overall review by a team of process safety experts. It is therefore best to draw only broad conclusions as to the relative importance of particular PSM elements, especially with respect to trend analysis from year to year. For example, it seems reasonable to conclude from Figure 8.3 that for the seven-year reporting period, deficiencies in PSM elements 2 to 7 inclusive (Table 8.2) have been viewed as key contributors to process-related incidents in Canadian process industry companies. The importance attached to each of these six elements has varied over this period, but each has crossed an arbitrary threshold of having contributed to at least 10% of the total incidents during a given year (at least once during the seven-year period).

Further validation of the conclusions reached in the preceding paragraph is provided by Figure 8.4. This comes from the work of Amyotte, MacDonald, and Khan [17], who reviewed approximately 60 investigation reports produced by the US Chemical Safety Board (CSB); the review was conducted from a PSM perspective, looking for examples of application (or lack thereof) of safety measures categorized

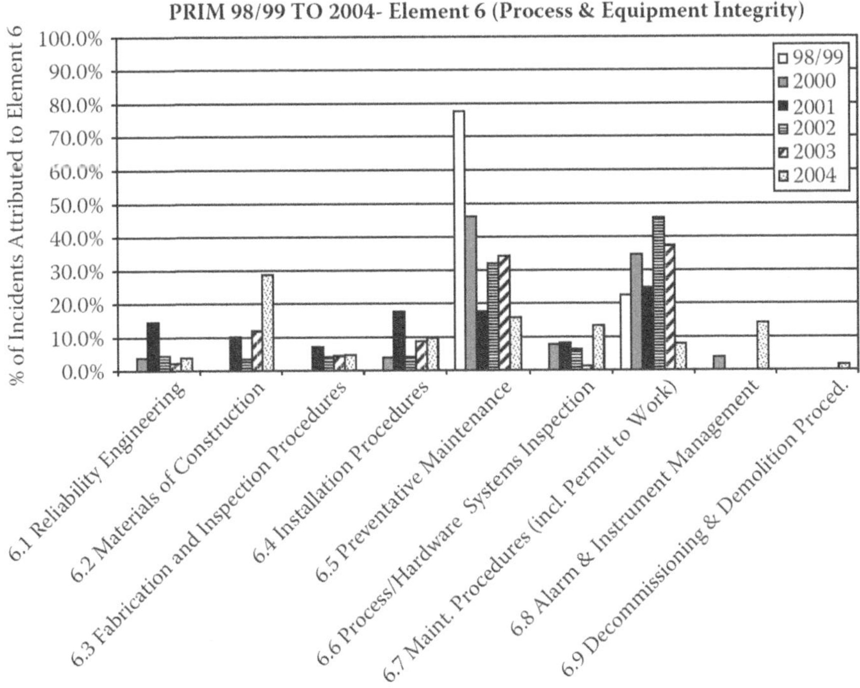

FIGURE 8.2 Incident causation (PRIM data) according to PSM element 6 given in Table 8.2. With permission.

Source: [16].

by the hierarchy of controls – inherent, passive engineered, active engineered, and procedural (see Chapter 7). Here again we see the importance of *process and equipment integrity* (element 6 in Table 8.2, or *asset integrity and reliability* in the US DOE hydrogen safety plan [10] elements shown in Table 8.1). Included in the bar for PSM element 6 in Figure 8.4 is the inherently safer design (ISD) simplification example [18] described in Chapter 7 involving a hydrogen release and subsequent fire due to incompatible materials of construction. Figure 8.4 also shows that PSM elements 2 to 6 inclusive (Table 8.2) are again dominant, as in Figure 8.3. The role of element 8 (*training and performance*) in Figure 8.4 is heightened over that in Figure 8.3 because of the large number of procedural safety measures identified in the CSB reports reviewed.

While the PRIM type of analysis does offer a method for prioritized resource allocation aimed at improvement, opportunities for risk reduction will be missed if all management system elements are not examined. For example, it is not surprising that *accountability: objectives and goals* (element 1 in Figure 8.3) is rated lower than other elements as an incident causation factor in a system of self-reporting by engineers who may be predisposed to finding technical solutions. *Incident investigation*

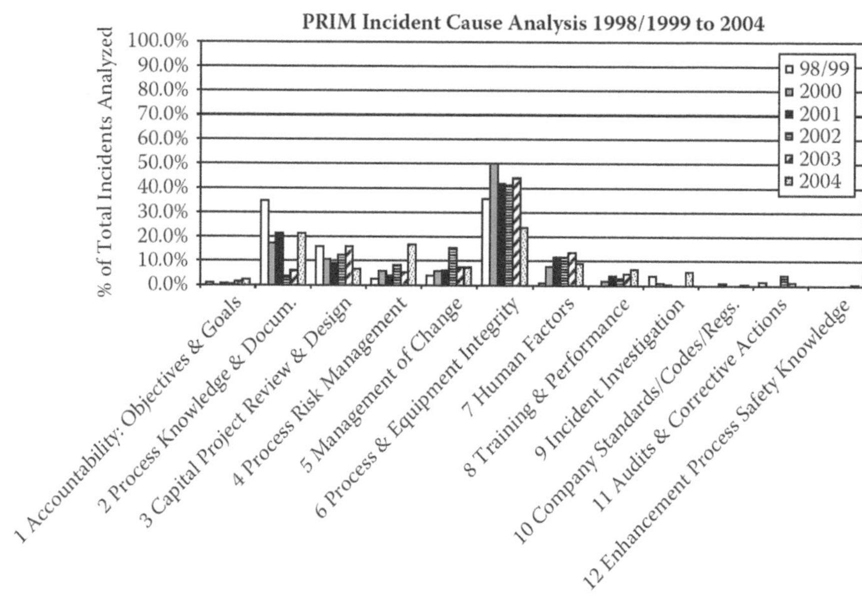

FIGURE 8.3 Incident causation (PRIM data) according to PSM elements given in Table 8.2. With permission.

Source: [16].

(element 9 in Figure 8.3) might be thought of as being reactive and therefore of limited use as a preventive measure; this is an erroneous conclusion. With respect to *enhancement of process safety knowledge* (element 12 in Figure 8.3), it can be difficult to know what you do not know.

In Figure 8.4 a breakdown of hierarchy of risk control examples is depicted according to applicable PSM elements given in Table 8.2 [17].

With these points in mind, the suite of PSM elements referenced in Table 8.2 and Figure 8.3 is now examined for its application to hydrogen safety via the system elements given in Table 8.1 and other relevant examples.

8.2.1 PSM Element 1 – Accountability: Objectives and Goals

Management commitment at all levels is necessary for PSM to be effective. The objectives for establishing accountability are to demonstrate the status of process safety compared to other business objectives (e.g. production and cost), to set objectives for safe process operation and to set specific process safety goals. These objectives should be internally consistent.

[13]

The fundamental underpinning of an effective process safety management system is the belief that safety is a corporate value and that performance improvement requires

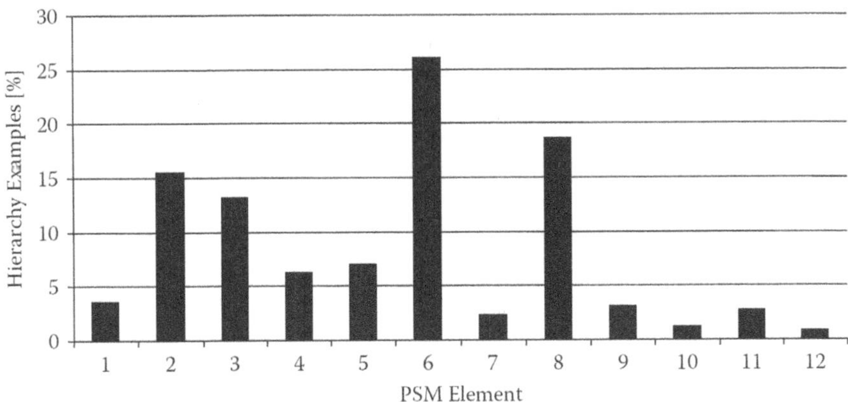

FIGURE 8.4 Breakdown of hierarchy of risk control examples according to applicable PSM elements given in Table 8.2. With permission.

Source: [17].

leadership for breakthrough results [19]. This issue of whether a company believes it is possible to achieve a higher standard of safety has been addressed in a recent book by sociologist Andrew Hopkins of the Australian National University. Hopkins [20] describes three concepts that address a company's cultural approach to safety, and makes the argument that the three are essentially alternative ways of talking about the same phenomenon.

1. *Safety culture.* The concept of a safety culture embodies the following subcultures [20]:
   - A *reporting culture* in which people report errors, near-misses, substandard conditions, inappropriate procedures, etc.
   - A *just culture* in which blame and punishment are reserved for behavior involving defiance, recklessness, or malice, such that incident reporting is not discouraged.
   - A *learning culture* in which a company learns from its reported incidents, processes information in a conscientious manner, and makes changes accordingly.
   - A *flexible culture* in which decision-making processes are not so rigid that they cannot be varied according to the urgency of the decision and the expertise of the people involved.
2. *Collective mindfulness.* The concept of collective mindfulness embodies the principle of *mindful organizing,* which incorporates the following processes [20]:
   - A *preoccupation with failure* so that a company is not lulled into a false sense of security by periods of success. A company that is preoccupied with failure would have a well-developed reporting culture.
   - A *reluctance to simplify* data that may at face value seem unimportant or irrelevant, but which may in fact contain the information needed to reduce

the likelihood of a future surprise. (Note that simplification here is not a desirable goal, unlike the ISD principle of simplification.)
- A *sensitivity to operations* in which frontline operators and managers strive to remain as aware as possible of the current state of operations, and to understand the implications of the present situation for future functioning of the company.
- A *commitment to resilience* in which companies respond to errors or crises in a manner appropriate to dealing with the difficulty, and a *deference to expertise* in which decisions are made by the people in the company hierarchy who have the most appropriate knowledge and ability to deal with the difficulty.

3. *Risk awareness.* Hopkins [20] states that risk awareness is synonymous with collective mindfulness (which is obviously closely related to the concept of a safety culture). He also describes a culture of *risk denial* in which it is not simply a matter of individuals and companies being unaware of risks, but rather that there exist mechanisms that deny the existence of risk.

From an overall perspective, element 2 (*organizational safety information*) in the DOE hydrogen safety plan (Table 8.1) pertains to this PSM element, particularly the setting of *organizational policies and procedures*. Management accountability must also permeate the entire safety management system. This is demonstrated by the relevance of Hopkin's points on safety culture [20] to the element 4 component *safety events and lessons learned* (Table 8.1). If a just culture does not exist, incident reporting will be sporadic at best and there will be no lessons that can be learned from nonexistent investigation reports. The concepts of collective mindfulness and risk awareness [20] can be seen to relate directly to several of the element 3 components in Table 8.1, for example, *identification of safety vulnerabilities* and *risk reduction plan.*

8.2.2 PSM ELEMENT 2 – PROCESS KNOWLEDGE AND DOCUMENTATION

Information necessary for the safe design, operation and maintenance of any facility should be written, reliable, current, and easily accessible by people who need to use it.

[13]

There is a close correspondence between this PSM element and the element 3 component *project safety documentation* (as well as *operating procedures*) in Table 8.1. The DOE hydrogen safety plan [10] contains the following requirements for *project safety documentation*, all of which relate directly to the PSM element 2 components in Table 8.2:

- Information pertaining to the technology of the project.
- Information pertaining to the equipment or apparatus.
- Safety systems (e.g., alarms, interlocks, detection or suppression systems).
- Safety review documentation, including identification of safety vulnerabilities.

- Operating procedures (including response to deviation during operation).
- Material safety data sheets.
- References such as handbooks and standards.

Also relevant here is PSM component 2.7, *company memory (management of information)*. The intention of this component is to ensure that knowledge and information gained from plant experience, and which is likely to be important for the future safety of a facility, is well documented so it is not forgotten or overlooked as personnel and organizational changes occur [13]. Mannan, Prem, and Ng [21] comment that organizations lose valuable information after about ten years because of such changes. Given the relative newness of some features of the hydrogen industry, as well as the sensitivity of public opinion to some aspects of hydrogen usage, it would seem opportune to heed the lessons learned – sometimes with significant hardship – by more-established process industry sectors.

8.2.3 PSM Element 3 – Capital Project Review and Design Procedures

Many industrial practitioners hold the opinion that careful attention to this element can have the greatest impact on the effectiveness of process safety management [22]. The key here is to conduct *hazard reviews* (PSM component 3.2 in Table 8.2) early in the design sequence by employing a preliminary hazard analysis. This is essentially the advice given in the DOE hydrogen safety plan [10] under the element 4 component *safety reviews*, which extends the review concept throughout the life-cycle of a project. Life-cycle considerations must, by definition, include the front end of the project.

PSM components 3.3, *siting*, and 3.4, *plot plan*, therefore take on notable significance. In siting a proposed expansion or new plant, the exposure hazard to and from adjacent plants or facilities is a critical consideration; similarly, the location of control rooms, offices, and other buildings should be carefully considered in conducting a plot plan review [13]. This is in accordance with the safety discussion in Section 7.6 on avoiding knock-on or domino effects, and the use of unit segregation in the hierarchy of controls (Figure 7.1 in Chapter 7).

Well-known examples where greater attention to facility siting and plot plan review (temporary as well as capital) would have assisted with consequence mitigation include the administration and control buildings at Flixborough [23] and the contractor trailers at the BP Texas City refinery [24]. Several hydrogen-specific examples were given in Section 7.6; for example, the work of Matthijsen and Kooi [25] in determining safety distances for hydrogen filling stations. The motivation for this work was to minimize external or third-party risk (i.e., the risk exposure of people living or working in the vicinity of facilities handling large amounts of hazardous substances) [25].

8.2.4 PSM Element 4 – Process Risk Management

The PSM guide comments that component 4.1, *hazard identification*, is the most important step in process risk management: If hazards are not identified, they cannot be considered in implementing a risk reduction program, nor addressed by emergency response plans [13].

This is similar to the distinction between hazard and risk made by Crowl and Jo [26]; as illustrated in Figure 8.5, effective assessment of the risk components of incident probability (or likelihood) and severity of consequences can only be carried out after thorough identification of the relevant hazards. The important aspects of system description (i.e., establishing the physical and analytical scopes) and scenario identification (i.e., identifying credible scenarios for study) have been addressed

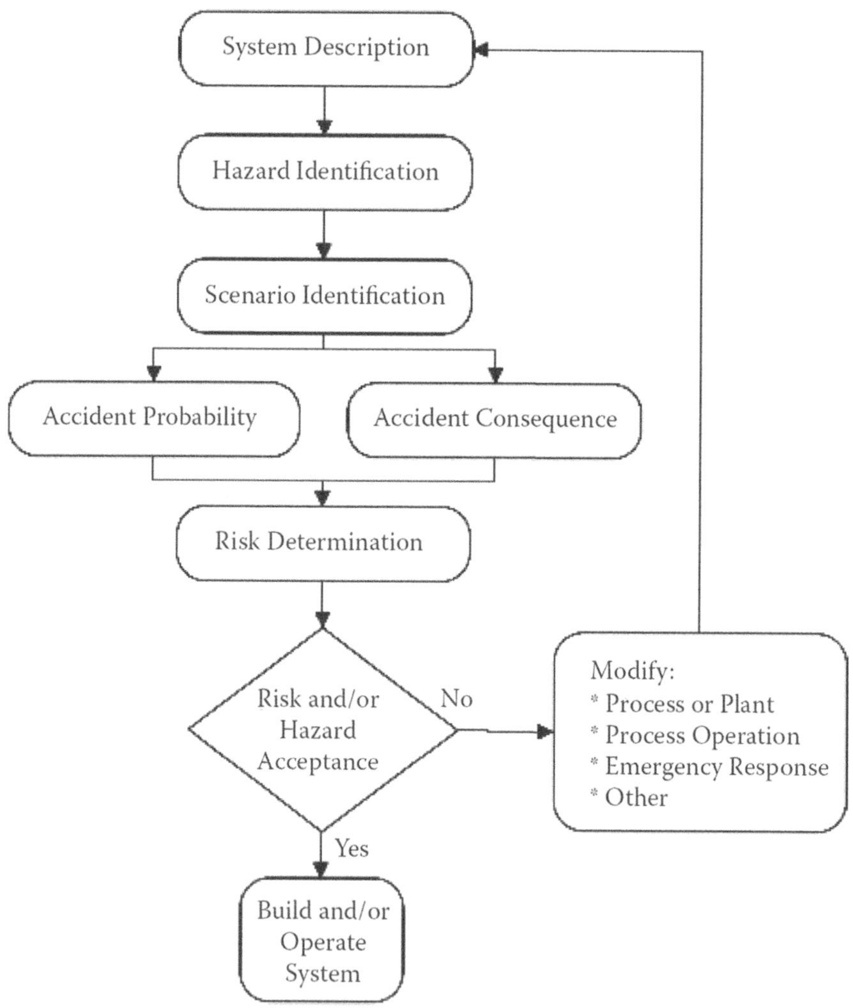

FIGURE 8.5   Risk management process involving hazard identification and risk assessment. With permission.

Source: [26].

by Takeno et al. [27] and Gerboni and Salvador [28]. Pasman and Rogers [29] further comment on the need to focus on prevention and mitigation measures aimed at *both* risk components so as to facilitate the smooth introduction of large-scale use of hydrogen as a transportation fuel.

Several techniques are referenced in the PSM guide [13] for identifying and assessing hazards, including what-if (WI) analysis, checklist (CL) analysis, hazard and operability (HAZOP) study, failure modes and effects analysis (FMEA), fault tree analysis (FTA), and the Dow Fire and Explosion Index (F&EI) and Chemical Exposure Index (CEI). These and a number of other hazard identification/risk assessment methodologies have been well described by the CCPS of the American Institute of Chemical Engineers [28, 30]. The DOE hydrogen safety plan [10] provides similar guidance under the element 3 component *identification of safety vulnerabilities* (ISV), which lists the following ISV methods also given in the PSM guide [13]: FMEA, WI, HAZOP, CL, and FTA. Additional methods referenced include event tree analysis (ETA) and probabilistic (or quantitative) risk assessment (PRA or QRA) [10]. These techniques find application in both nonemergency situations and those requiring an *emergency response* (element 4 component in Table 8.1).

A review of the hydrogen safety literature reveals that essentially all of the typical process safety hazard identification techniques have been successfully applied to various sectors of the hydrogen industry. Knowlton [31] used HAZOP to study the safety aspects of hydrogen as a ground transportation fuel almost 30 years ago. More recently, HAZOP and FMEA were used by Kikukawa, Mitsuhashi, and Miyake [32] to identify hazards and assess the ensuing risks for liquid hydrogen fueling stations (Figure 8.6). Their risk assessment process is shown in Figure 8.7; this flowchart is similar to the general one given in Figure 8.5, with the understanding that any risk reduction measures recommended through use of Figure 8.7 should be thoroughly examined for the introduction of new hazards.

The risk matrix used by Kikukawa, Mitsuhashi, and Miyake [32] to discern whether a given risk is tolerable is shown in Figure 8.8. Risk matrices are a commonly used decision tool in the process industries and are referenced in the DOE hydrogen safety plan [10] under the element 3 component *risk reduction plan* by means of the term "risk binning matrix". Additional information on risk matrices and an associated concept, the ALARP (as low as reasonably practicable) principle, has been given for hydrogen refueling stations by Norsk Hydro ASA and DNV [33].

FTA, along with HAZOP and PHA (preliminary hazard analysis), were used by Brown and Buchier [34] in their study of a hydrogen gaseous effluent treatment and purification plant. FTA is a logic diagram with "AND" and "OR" gates. In the context of FTA, events and conditions are categorized as either basic events or intermediate events. Basic events are events that are considered indivisible or are not further analyzed within the fault tree. "Q" typically stands for "Basic Event Probability" or "Probability of the Top Event."

Correa-Jullian and Groth [35] performed a quantitative safety assessment for a hydrogen station leakage. Figure 8.9 shows the FTA developed by Seong et al. for the top event for internal factors of hydrogen leakage. In Figure 8.9 an AND gate

FIGURE 8.6   Liquid hydrogen fueling station. With permission.

Source: [32].

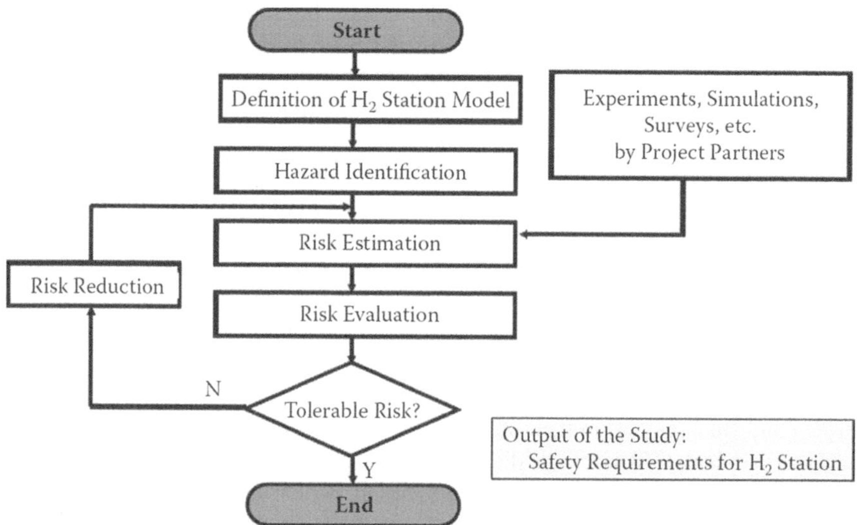

FIGURE 8.7   Risk assessment process for liquid hydrogen fueling station. With permission.

Source: [32].

		Probability Level			
		A	B	C	D
Consequence Severity Level		Improbable	Remote	Occasional	Probable
1	Extremely Severe Damage	H	H	H	H
2	Severe Damage	M	H	H	H
3	Damage	M	M	H	H
4	Limited Damage	L	L	M	H
5	Minor Damage	L	L	L	M

FIGURE 8.8 Risk matrix for liquid hydrogen fueling station. H = High risk; M = medium risk; L = low risk. With permission.

Source: [32].

beneath the Hydrogen Leakage connects internal and external factors. Figure 8.9 demonstrates the usefulness of such graphical techniques for identifying the potential hazards leading to a hazardous event.

FTA and ETA were used by Roysid, Jablonski, and Hauptmanns [36] and Rodionov, Wilkening, and Moretto [37] in their safety studies of life-cycle-related hydrogen usage and private vehicles with a hydrogen-driven engine, respectively. The use of event trees figured prominently in the work of Gerboni and Salvador [28] on hydrogen transportation systems and that of Rigas and Sklavounos [38] on hydrogen storage facilities.

Figure 5.4 shows an event tree for the case of a hydrogen release, drawn from the study of Rigas and Sklavounos [38]. The usefulness of graphical hazard/risk techniques is again demonstrated in Figure 5.4 by the clear elucidation of the ultimate result of various mitigating factors following an undesired event. For example, a hydrogen release followed by immediate ignition in a region of significant confinement is seen to lead to either a deflagration or a detonation. On the other hand, the same release without subsequent ignition and with no confinement results in dissolution of the hydrogen plume. Risk reduction measures can then be decided upon with the aid of Figure 5.4.

The Dow F&EI, often considered as a form of relative risk ranking (RRR), was used by Bernatik and Libisova [39] in conjunction with HAZOP, FTA, and ETA to conduct a risk assessment study of large gasholders in municipal areas. The F&EI is ideally suited for use with hydrogen given its high value of US National Fire Protection Association (NFPA) flammability number ($N_F = 4$); the Dow CEI, however, would be of limited use with hydrogen because of the low NFPA health hazard number ($N_H = 0$). The use of RRR was also adopted by Kim, Lee, and Moon [40] by means of FMEA and FTA coupled with a relative risk index for comparison of hydrogen production, storage, and transportation activities.

There are also numerous reports in the hydrogen safety literature describing knowledge acquisition to better understand specific industrial hazards; again, this is a key step in the risk assessment process which first relies upon effective hazard identification. For example, Petukhov, Naboko, and Fortov [41] investigated the

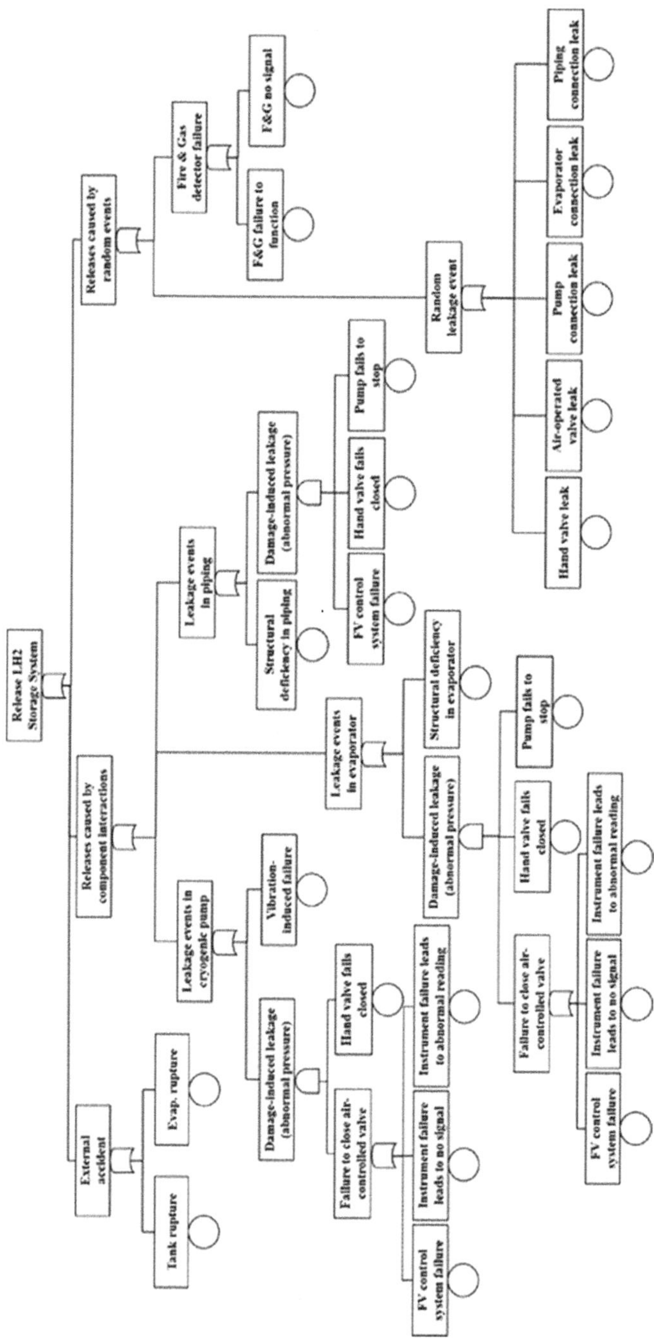

FIGURE 8.9 An FTA for the top event "Major Liquid Hydrogen Leakage" with the purpose of determining a general initiating event frequency for LH_2 release in a storage system. With permission from Carbon Brief (www.carbonbrief.org).

Source: [35].

FIGURE 8.10 Ventilation duct outlet at hydrogen fueling station in Japan. With permission.

Source: [43].

deflagration/detonation relationship for hydrogen–air mixtures in large volumes by conducting explosion tests in a 12 m diameter chamber. Sommersel et al. [42] undertook experiments in a 3 m long channel to gain a better understanding of hydrogen dispersion, ignition, and explosion development in such configurations.

Imamura, Mogi, and Wada [43] obtained empirical data on hydrogen ignition by electrostatic discharge at the outlet of a ventilation duct. The practical concern in this case was the possibility of ignition during routine and emergency release of hydrogen to the atmosphere through ventilation ducts such as the one shown in Figure 8.10. These researchers determined that when the ventilation duct itself was not grounded, electrostatic discharge between grounded conductors and the duct outlet, with subsequent hydrogen ignition, was a distinct possibility. Grounding of the ventilation duct outlet effectively eliminated this problem, thus demonstrating the efficacy of grounding as a risk reduction measure in this particular application [43].

Quantification of one or both of the components of risk (incident likelihood and consequence severity) has become increasingly popular in the process industries, especially as more reliable failure frequency databases and computational tools for consequence analysis have become available. The same statement can be made with confidence for the hydrogen industry as evidenced by the proliferation of articles describing the use of: (1) the "Dutch Purple Book" for determining risk scenarios and failure frequencies [25, 39], (2) semiquantitative risk assessment methodologies [44], (3) probit functions for translating radiant heat flux calculations to probabilities of first- and second-degree burns as well as fatalities [45], (4) Gaussian dispersion modeling [46], (5) Bayesian networks as a supplement to FTA and ETA [47], and (6) QRA as the foundation for making risk-based management decisions [48].

It can be argued that in this regard, some of the most important developments in recent years have been in the field of computational fluid dynamics (CFD). The

GexCon CFD-tool FLACS (FLame ACceleration Simulator), for example, has been adapted for use in the hydrogen industry to simulate accidental releases and dispersion of hydrogen followed by ignition and flame and overpressure development [49, 50]. Further investigation with FLACS has been reported for a wide range of hydrogen applications in various natural and manufactured environments [51–58]; these studies have been carried out within the framework of the European Commission-funded Network of Excellence HySafe (Hydrogen Safety as an Energy Carrier), which is briefly described in Chapter 9.

Concluding comments can be made for this PSM element with respect to component 4.6, *encouraging client and supplier companies to adopt similar risk management practices* (Table 8.2). This component is particularly important because of the common practice of outsourcing or contracting of engineering services. This is often the case on large projects where partnerships and joint ventures are formed but may also apply to smaller projects through the awarding of subcontracts. Success on these projects is in large measure determined by the degree of commonality in risk management practices among the different parties. Not the least of the concerns is whether there is a common set of expectations for safety performance and risk awareness. The DOE hydrogen safety plan [10] addresses these points under element 1, *scope of work* (Table 8.1), with reference to the need for such safety plans to cover the work of all subcontractors.

8.2.5   PSM Element 5 – Management of Change

A system to manage change is critical to the operation of any facility. A written procedure should be required for all changes except replacement-in-kind. The system should address: a clear definition of change (scope of application); a description and technical basis for the proposed change; potential impact of the proposed change in health, safety and environment; authorization requirements to make the change; training requirements for employees or contractors following the change; updating of documentation including process safety information, operating procedures, maintenance procedures, alarm and interlock settings, fire protection systems, etc.; and contingencies for "emergency" changes.

[11]

There is clearly a direct correspondence between this PSM element and the DOE hydrogen safety plan [10] element 3 component *management of change procedures* (Table 8.1). The authors of this plan [10] have seemingly heeded the advice of Hansen and Gammel [59], who state that while management of change (MOC) is critical to process safety, it is a concept that, if well implemented, could help prevent accidents in many other industries as well.

MOC is of particular importance when employing the inherently safer design principles described in Chapter 7. Simply put, inherent safety involves change, and change in any industrial sector must be managed. Potential hazards brought about by inherent safety changes must therefore be identified and the ensuing risk reduced to an acceptable level. This is as important as the concept of looking for inherent safety opportunities when making a process change. For example, the development of nano-porous carbon tubes and their substitution for liquid hydrogen storage [9] would necessitate careful consideration of new hazards that might be introduced through the use of nanomaterials.

8.2.6   PSM Element 6 – Process and Equipment Integrity

Procedures for fabricating, inspecting and maintaining equipment are vital to process safety. Written procedures should be used to maintain ongoing integrity of process equipment such as: pressure vessels and storage tanks; piping, instrumentation and electrical systems; process control software; relief and vent systems and devices; emergency and fire protection systems; controls including monitoring devices and sensors, alarms and interlocks; and rotating equipment. A documented file should be maintained for each piece of equipment.

[13]

The relationship between this PSM element and the DOE hydrogen safety plan has been discussed extensively in the introductory material to this section.

An additional comment concerning *maintenance procedures* (PSM component 6.7 in Table 8.2), follows from a safety alert issued by the US Environmental Protection Agency [60], which offered the advice that facilities with storage tanks containing flammable vapors (such as hydrogen) should review their equipment and operations in the following areas:

- Design of atmospheric storage tanks.
- Inspection and maintenance of storage tanks.
- Hot-work safety.
- Ignition source reduction.

While the middle two items in this list are largely procedural in nature, the first and last items have inherent safety overtones (simplification and minimization, respectively). This again demonstrates the linkage between the subject matter of the current chapter (safety management systems) and Chapter 7 (inherently safer design).

8.2.7   PSM Element 7 – Human Factors

Human factors are a significant contributor to many process accidents. Three key areas are operator–process/equipment interface, administrative controls and human error assessment.

[13]

Human error and the underlying human factors have been recognized as causation factors for several industrial accidents involving hydrogen [61]. Xu et al. [59] present data indicating that for a subset of such accidents occurring over the period 1972 to 2005, human error was a root cause in 26% of the total number of incidents (with management system deficiencies accounting for a further 27%). It is therefore surprising that the DOE hydrogen safety plan [10] does not contain a separate element or component to explicitly address human factors. This is clearly one area where hydrogen safety management can benefit from the lessons of process safety management in general.

The PSM element on human factors has a strong relationship with the principles of ISD, particularly simplification (Chapter 7). Component 7.1, *operator–process/ equipment interface* (Table 8.2), refers to issues such as [13]:

- The design of equipment increasing the potential for error (e.g., confusing equipment, positioning of dials, color coding, different directions for on/ off, etc.).
- The need for a task analysis (a step-by-step approach to examine how a job will be done) to determine what can go wrong during the task and how potential problem areas can be controlled.

These issues are applicable to equipment and procedures involved with hydrogen usage as well as other hazardous materials.

The description of component 7.2 of this PSM element, *administrative control versus hardware control*, (Table 8.2) includes:

Hazards may be controlled by the use of procedures or by the addition of protective equipment. This balance is often a matter of company culture and economics. If procedures are well understood, kept current and are used, then they are likely to be effective. Similarly protective systems need regular testing and maintenance to be effective. The problem of administrative versus hardware controls should be considered and a balance selected by conscious choice rather than allowing it to happen by default.

[13]

This description may leave some readers with the unfortunate impression that only procedural (administrative) and engineered (add-on) measures are available, or are effective, for hazard control. As demonstrated in Chapter 7, these categories in the hierarchy of controls are indeed helpful in facilitating hydrogen safety; the predominant effectiveness of ISD should not, however, be ignored when attempting to combat human error.

The third and final component of human factors, *7.3 human error assessment* (Table 8.2), is perhaps one of the more challenging areas of process safety management. Human error assessment is becoming increasingly important in industry and is a growing area of concern for the public and for regulators. This is especially the case for the hydrogen industry given the findings previously described for human error as a key incident causation factor [61, 62].

8.2.8 PSM Element 8 – Training and Performance

People need to be trained in the right skills and to have ongoing retraining to maintain these skills.

[11]

As with MOC (Section 8.2.5), there is a clear and direct correspondence between this PSM element and the DOE hydrogen safety plan [10] – here, with the element 4 component *employee training* (Table 8.1). This helps to explain, in part, the rationale behind initiatives such as the development of a curriculum for an online hydrogen fuel training program [63].

Embodied within the components of this PSM element (Table 8.2) is the management cycle comprised of *plan, do, check, act* (Section 8.1). The activities in the training protocol given by DiBerardinis [64] are particularly appropriate in this regard:

- Conducting a needs analysis.
- Setting learning objectives.
- Deciding on the method of presentation and delivery of training.
  - Styles: lectures, group discussion, hands-on exercises, self-learning, etc.
  - Materials: slides, videos, workbooks, etc.
- Evaluating the instruction.
- Providing feedback on the instruction.

Excellent advice is given by Felder and Brent [65] concerning the setting of instructional (learning) objectives. They comment that in setting such objectives, there are four leading verbs that should be avoided: *know, learn, appreciate*, and *understand*. Thus, while it would be desirable for plant employees to know, learn, appreciate, and understand various issues related to hydrogen safety, these are not valid instructional objectives because it is not possible to directly see whether they have been done. It is necessary to consider what trainees should be asked to *do* to demonstrate their knowledge, learning, appreciation, and understanding of hydrogen safety, and then make those activities the instructional objectives [65].

Felder and Brent [65] further describe the concept of using action verbs to set instructional objectives. Using their breakdown according to *Bloom's Taxonomy of Educational Objectives*, examples of instructional objectives for hydrogen safety training would be:

- Knowledge: *List* the key material hazards of hydrogen.
- Comprehension: *Explain* in your own words the concept of moderation as applied to inherently safer design issues for hydrogen.
- Application: *Calculate* the Dow Fire and Explosion Index for several hydrogen inventory scenarios.
- Analysis: *Identify* the safety features in a given design for hydrogen storage using the hierarchy of risk control measures as a guide.
- Synthesis: *Develop* an original case study involving some aspect of hydrogen safety.
- Evaluation: *Select* from available techniques for assessing the risk of a given design for a hydrogen infrastructure project and justify your choice.

(*NOTE:* Although *Bloom's Taxonomy* has been revised somewhat in recent years, the original form has been retained here for consistency with the work of Felder and Brent [63].)

8.2.9 PSM Element 9 – Incident Investigation

The element 4 component *safety events and lessons learned* (Table 8.1) is the DOE hydrogen safety plan [10] counterpart to this PSM element. As discussed in Section 8.2.1, three of the safety subcultures (just, reporting, and learning) identified by Hopkins [20] relate directly to the issue of incident investigation. This illustrates another key point about safety management systems: the various elements and components comprising these systems are not stand-alone modules having little interaction among them.

The process safety literature contains numerous descriptions of how to effectively investigate industrial accidents and near-misses. For example, Goraya, Amyotte, and Khan [66] developed an inherent safety-based incident investigation methodology that is easily adaptable as a basic protocol for investigating hydrogen-related incidents. Key features of their work include the use of [66]:

- An overall framework having best-practice industry consensus.
- An integrated approach in considering all potential categories of loss (people, assets, business operation, and the environment).
- Evidence classification into data categories (position, people, parts, and paper).
- An established loss causation model for identification of causal factors.
- A layered investigation approach for making recommendations of an immediate technical nature, to remove the underlying hazards, and to improve the safety management system.

8.2.10 PSM Element 10 – Standards, Codes, and Relevant Laws

A management system is needed to ensure that the various internal and external published guidelines, standards and regulations are current, disseminated to appropriate people and departments, and applied throughout the plant.

[13]

As noted above, the PSM guide [11] breaks this element into two components: external to the company and internal to its operations. External codes and regulations include legislated items; internal standards are comprised of items such as design principles and standard operating procedures. The DOE hydrogen safety plan [10] addresses these points, in part, under the element 3 component *project safety documentation* (Table 8.1) with reference to documenting the design codes and standards employed for equipment or apparatus.

From a process industry perspective, codes and standards for assuring hydrogen safety would be addressed in the first instance by the relevant process safety management regulatory regime (e.g., the highly regulated and prescriptive requirements promulgated in the United States). In other sectors of the broader hydrogen industry (e.g., transportation), considerable efforts are underway worldwide to develop risk-informed codes and standards with appropriate stakeholder input [29, 48].

8.2.11 PSM Element 11 – Audits and Corrective Actions

> The purpose of safety audits is to determine the status and effectiveness of safety
> management efforts versus goals and also the progress toward those goals.
>
> [13]

The element 4 component *self-audits* (Table 8.1) is the DOE hydrogen safety plan [10]
counterpart to this PSM element (which is arguably one of the more self-explanatory
elements given in Table 8.2). Although the term "self-audit" is used in the DOE plan
[10], the supporting documentation does indicate the need for verification of the audit
findings by a third party external to the project.

8.2.12 PSM Element 12 – Enhancement of Process Safety Knowledge

> A management system for process safety should be designed for continuous
> improvement. Safety requirements are becoming more stringent, while know-
> ledge of systems and technology is growing, e.g. consequence modelling
> techniques. Safe operation of a process plant calls for personnel to stay abreast
> of current developments, and for safety information to be readily accessible.
>
> [13]

This PSM element is partially addressed in the DOE hydrogen safety plan [10] by the
element 2 component *hydrogen and fuel cell experience* (Table 8.1). Enhancement of
knowledge, perhaps more than any other element, is where the continuous improvement aspect of a safety management system should be highly evident. The following
example illustrates this point.

Frank [67] asks the following question of urban fire department personnel who,
although well aware of the hazards in tenement building fires, may go many years
without responding to a fire in a power plant: *How do you find out that power plants
use hydrogen?* His answers are to: (1) tour the plant regularly, (2) ask where hydrogen
is stored in bulk, (3) ask to be shown the location of the hydrogen lines and the
hydrogen-cooled generator, and (4) think about the required actions should an incident occur [64]. While this excellent advice relates to many PSM elements, for
example those dealing with hazard identification, emergency response, training, etc.,
it also aligns well with the intent of the quote from the PSM guide [13] given at the
start of this section.

8.3 SAFETY CULTURE

Having established the link between PSM and hydrogen safety, some thoughts are
now given on the role of safety culture in today's industrial world. As identified
earlier for previous sections of the current chapter, the discussion on safety culture
follows that given by Kletz and Amyotte [3].

There can be no doubt that safety culture is an important topic, especially since
the BP Texas City incident in 2005 [68]. Recent emphasis has also been placed on
safety culture in other fields (e.g., occupational health and safety [69]) and in other
applications (e.g., offshore safety [70]). Section 8.2.1 has addressed safety culture

from the perspective of management leadership and accountability within a process safety management system. Table 8.3 concerning risk-based process safety (RBPS) management [15] lists the first of its four RBPS pillars as committing to process safety, and emphasizes the need to develop and sustain a culture that embraces process safety.

It is beyond the scope of this book to provide an extensive review of the literature on process safety culture. Here it is sufficient to say that two key points have emerged:

- Typical occupational safety indicators such as lost-time injuries (LTIs) are inappropriate as a primary indicator with respect to process safety (as previously discussed in Section 8.1).
- Leading indicators are generally viewed as being more useful than lagging indicators.

Similar to the development of methods to measure the inherent "safeness" of a process (Section 7.7), safety culture metrics are currently an area of significant interest in both academia and industry. According to Hopkins [71], perhaps the most important consideration for process safety indicators is that they measure the effectiveness of the various controls comprising the risk control system. This affords an opportunity to link back to the elements making up the process safety management system. Little work appears to have been done in this area, which has been identified by Glendon [72] as a major challenge for industry (i.e., the linking of safety culture methodologies with process safety approaches and broader systems safety and risk management concepts).

The coming years will undoubtedly see advances in the area of metrics for process safety culture; this will be a welcome development, especially where the measurement tools specifically address hydrogen safety. Perhaps, however, there is something to be gained by looking to the past as well as to the future. It would seem that while it has not necessarily been named as such, safety culture has been a subject of consideration for centuries.

One of the first recorded accounts of a dust explosion was written by Count Morozzo, who gave a detailed account of an explosion in a flour warehouse in Turin, Italy [73]. In the final paragraph of his report, the Count writes:

Ignorance of the fore-mentioned circumstances, and a culpable negligence of those precautions which ought to be taken, have often caused more misfortunes and loss than the most contriving malice. It is therefore of great importance that these facts should be universally known, that public utility may reap from them every possible advantage.

[74]

The above passage makes an eloquent case for the importance of incident investigation and the sharing of lessons learned (Section 8.2.9), and for a strong safety culture. It is instructive to also note that it was written over 200 years ago.

Additional insight into the importance of safety culture for the hydrogen industry can be gained by looking at other areas of application. In addressing the role of safety culture in the nanotechnology field, Amyotte [75] made the following comments:

The nanotechnology world does not want, and neither should it need, a Bhopal, Buncefield or Gulf oil leak (all major and/or recent process/environmental incidents) to drive its safety culture. Simply put, nanotechnology industries cannot afford to ignore the hard safety lessons that have been learned and at times ignored by the chemical process industries.

[75]

The safety culture is addressed to some extent in the "attitudes" descriptor, but the safety culture as a whole is rather a reflection of the safety culture in the entire organization than just the attitudes of the workers (see Figure 8.11).

The *consequences of a poor safety culture* within an organization which neglects safety practices can lead to serious repercussions [76]:

1. *Increased Injuries and Deaths*:
   - A poor safety culture significantly raises the risk of workplace injuries and fatalities among employees.
   - Approximately 7 to 9% of the US workforce sustains work-related injuries annually.
   - Improving safety practices can reduce the number of injuries and contribute to overall well-being [77].

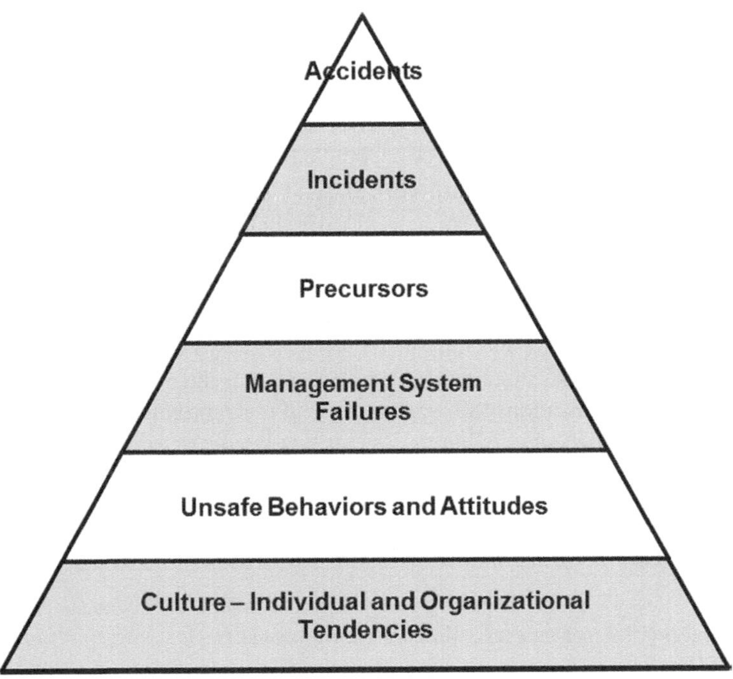

FIGURE 8.11   The Modified Safety Triangle. With permission from Wiley.

Source: [12].

2. *Higher Costs*:
   - Inadequate safety management leads to increased costs:
     - *Employee Absence*: Workplace injuries result in more sick days and absenteeism.
     - *Healthcare Expenses*: Higher healthcare costs due to injuries.
     - *Workers' Compensation Payments*: Financial burden for the company.
     - *Lost Productivity*: Reduced efficiency and disrupted operations [77].
3. *Business Impact*:
   - Disengaged employees, due to poor safety management, can cost businesses up to USD550 billion annually in lost productivity [78].
4. *Insurance Premiums*:
   - A company with a weak safety culture faces higher insurance premiums.
   - More accidents and health issues lead to increased claims, prompting insurers to offset risks by raising premiums [79].
5. *Loss of Reputation*
   - The legal responsibility of a company is to make the workplace a safe environment for its employees. If the company neglects its legal responsibility, it may face high legal costs, hefty fines, and the possibility of a jail sentence [79].

In conclusion, a positive safety culture is essential for employee well-being, productivity, and overall business success. Companies that prioritize safety benefit from reduced injuries, lower costs, and a healthier work environment.

These comments apply equally to industries involved in the production, distribution, storage, and use of hydrogen. As described in Chapter 10 and elsewhere in this book, there have been numerous industrial accidents involving hydrogen. The avoidance of further incidents will be accomplished by successful implementation of many factors – chief among which must be acknowledgment of the importance of, and adoption of, specific measures to ensure a sound safety culture.

The hydrogen industry would be well advised to look to the social and management sciences for advice on how to avoid complacency and how best to heed the warning signs that always precede industrial accidents. These signs may appear to be vague and ambiguous, and may invite what is known as *normalization of deviance* (wherein abnormal situations become accepted as the norm) [80]. The study of high-reliability organizations [80] appears to hold much promise in providing widely applicable lessons on safety culture – lessons that would be beneficial to the various sectors of the hydrogen industry.

REFERENCES

1. New integrated management system attempts to link environment, health, safety and process management, *Workplace Environment Health & Safety Reporter*, 7 (1), 1166, 2001.
2. Amyotte, P.R. and McCutcheon, D.J., *Risk Management: An Area of Knowledge for all Engineers*, Discussion paper prepared for Canadian Council of Professional Engineers, Ottawa, Ontario, 2006.

3. Kletz, T. and Amyotte, P., *Process Plants: A Handbook for Inherently Safer Design*, 2nd ed., CRC Press/Taylor & Francis Group, Boca Raton, FL, 2010.
4. Amyotte, P.R. and Oehmen, A.M., Application of a loss causation model to the Westray mine explosion, *Process Safety and Environmental Protection*, 80 (1), 55–59, 2002.
5. DOE, *Hydrogen safety*, Technical Plan – Safety; Multi-Year Research, Development and Demonstration Plan, U.S. Department of Energy, 3.8-1–3.8-12, 2007. www1.eere.energy.gov/hydrogenandfuelcells/mypp/pdfs/safety.pdf
6. Stelmakowich, A., Continuous improvement, *OHS Canada*, 19 (7), 38–39, 2003.
7. Bird, F.E. and Germain, G.L., *Practical Loss Control Leadership*, DNV, Loganville, GA, 1996.
8. Creedy, G., *Process safety management*, PowerPoint presentation prepared for Process Safety Management Division, Chemical Institute of Canada, Ottawa, Ontario, 2004.
9. Guy, K.W.A., The hydrogen economy, *Process Safety and Environmental Protection*, 78 (4), 324–327, 2000.
10. DOE, *Safety planning guidance for hydrogen and fuel cell projects*, Fuel Cell Technologies Program, U.S. Department of Energy, Washington, DC, 2010.
11. Wikipedia OHSAS 18001. https://en.wikipedia.org/wiki/OHSAS_18001
12. CCPS, *Guidelines for Implementing Process Safety Management*, 2nd ed., John Wiley & Sons, Hoboken, NJ, 2016.
13. Canadian Society for Chemical Engineering, *Process Safety Management*, 3rd ed., Canadian Society for Chemical Engineering, Ottawa, Ontario, 2002.
14. CCPS, *Guidelines for technical management of chemical process safety*, Center for Chemical Process Safety, American Institute of Chemical Engineers, New York, 1989.
15. CCPS, *Guidelines for Risk Based Process Safety*, John Wiley & Sons, Hoboken, NJ, 2007.
16. Amyotte, P.R., Goraya, A.U., Hendershot, D.C., and Khan, F.I., Incorporation of inherent safety principles in process safety management, *Process Safety Progress*, 26 (4), 333–346, 2007.
17. Amyotte, P.R., MacDonald, D.K., and Khan, F.I., An analysis of CSB investigation reports concerning the hierarchy of controls, *Process Safety Progress*, 30, 261–265, 2011.
18. CSB, *Positive material verification: Prevent errors during alloy steel systems maintenance*, Safety Bulletin, No. 2005-04-B, U.S. Chemical Safety and Hazard Investigation Board, Washington, DC, 2006.
19. Griffiths, S., Leadership, Commitment & Accountability—The driver of safety performance, in *Process Safety and Loss Management Symposium, 55th Canadian Chemical Engineering Conference*, Canadian Society for Chemical Engineering, Toronto, Ontario, 2005.
20. Hopkins, A., *Safety, Culture and Risk: The Organizational Causes of Disasters*, CCH Australia Limited, Sydney, Australia, 2005.
21. Mannan, M.S., Prem, K.P., and Ng, D., Challenges and needs for process safety in the new millennium, in *Proceedings of 13th International Symposium on Loss Prevention and Safety Promotion in the Process Industries*, Vol. 1, Bruges, Belgium, June 6–9, 2010, 5–13. DOI: 10.1016/j.psep.2011.08.002
22. Creedy, G., private communication, 2005.
23. Sanders, R.E., Designs that lacked inherent safety: Case studies, *Journal of Hazardous Materials*, 104 (1–3), 149–161, 2003.
24. CSB, *Refinery explosion and fire*, Investigation Report, No. 2005-04-I-TX, U.S. Chemical Safety and Hazard Investigation Board, Washington, DC, 2007.

25. Matthijsen, A.J.C.M. and Kooi, E.S., Safety distances for hydrogen filling stations, *Fuel Cells Bulletin*, (11), 12–16, 2006.

26. Crowl, D.A. and Jo, Y.-D., The hazards and risks of hydrogen, *Journal of Loss Prevention in the Process Industries*, 20 (2), 158–164, 2007.

27. Takeno, K., Okabayashi, K., Kouchi, A., Nonake, T., Hashiguchi, K., and Chitose, K., Dispersion and explosion field tests for 40 MPa pressurized hydrogen, *International Journal of Hydrogen Energy*, 32 (13), 2144–2153, 2007.

28. Gerboni, R. and Salvador, E., Hydrogen transportation systems: Elements of risk analysis, *Energy*, 34 (12), 2223–2229, 2009.

29. Pasman, H.J. and Rogers, W.J., Safety challenges in view of the upcoming hydrogen economy: An overview, *Journal of Loss Prevention in the Process Industries*, 23 (6), 697–704, 2010.

30. CCPS, *Guidelines for Hazard Evaluation Procedures*, 2nd ed., Center for Chemical Process Safety, American Institute of Chemical Engineers, New York, 1992.

31. Knowlton, R.E., An investigation of the safety aspects in the use of hydrogen as a ground transportation fuel, *International Journal of Hydrogen Energy*, 9 (1–2), 129–136, 1984.

32. Kikukawa, S., Mitsuhashi, H., and Miyake, A., Risk assessment for liquid hydrogen fueling stations, *International Journal of Hydrogen Energy*, 34 (2), 1135–1141, 2009.

33. Norsk Hydro ASA, and DNV, *Risk Acceptance Criteria for Hydrogen Refuelling Stations*, European Integrated Hydrogen Project (EIHP2), Contract: ENK6-CT2000-00442, February 2003. www.eihp.org/public/documents/acceptance_criteria_for_H2-refuelling_stations_FEB2003.pdf

34. Brown, A.E.P. and Buchier, P.M., Hazard identification analysis of a hydrogen plant, *Process Safety Progress*, 18 (3), 166–169, 1999.

35. Correa-Jullian, C. and Groth, K.M., Data requirements for improving the quantitative risk assessment of liquid storage systems, *International Journal of Hydrogen Energy*, 47, 4222–4235, 2022.

36. Rosyid, O.A., Jablonski, D., and Hauptmanns, U., Risk analysis for the infrastructure of a hydrogen economy, *International Journal of Hydrogen Energy*, 32 (15), 3194–3200, 2007.

37. Rodionov, A., Wilkening, H., and Moretto, P., Risk assessment of hydrogen explosion for private car with hydrogen-driven engine, *International Journal of Hydrogen Energy*, 36 (3), 2398–2406, 2011.

38. Rigas, F. and Sklavounos, S., Evaluation of hazards associated with hydrogen storage facilities, *International Journal of Hydrogen Energy*, 30 (13–14), 1501–1510, 2005.

39. Bernatik, A. and Libisova, M., Loss prevention in heavy industry: Risk assessment of large gasholders, *Journal of Loss Prevention in the Process Industries*, 17 (4), 271–278, 2004.

40. Kim, J., Lee, Y., and Moon, I., An index-based risk assessment model for hydrogen infrastructure, *International Journal of Hydrogen Energy*, 36 (11), 6387–6398, 2011.

41. Petukhov, V.A., Naboko, I.M., and Fortov, V.E., Explosion hazard of hydrogen-air mixtures in the large volumes, *International Journal of Hydrogen Energy*, 34 (14), 5924–5931, 2009.

42. Sommersel, O.K., Bjerketvedt, D., Vaagsaether, K., and Fannelop, T.K., Experiments with release and ignition of hydrogen gas in a 3 m long channel, *International Journal of Hydrogen Energy*, 34 (14), 5869–5874, 2009.

43. Imamura, T., Mogi, T., and Wada, Y., Control of the ignition possibility of hydrogen by electrostatic discharge at a ventilation duct outlet, *International Journal of Hydrogen Energy*, 34 (6), 2815–2823, 2009.

44. Moonis, M., Wilday, A.J., and Wardman, M.J., Semi-quantitative risk assessment of commercial scale supply chain of hydrogen fuel and implications for industry and society, *Process Safety and Environmental Protection*, 88 (2), 97–108, 2010.

45. LaChance, J., Tchouvelev, A., and Engebo, A., Development of uniform harm criteria for use in quantitative risk analysis of the hydrogen infrastructure, *International Journal of Hydrogen Energy*, 36 (3), 2381–2388, 2011.

46. Ramamurthi, K., Bhadraiah, K., and Murthy, S.S., Formation of flammable hydrogen-air clouds from hydrogen leakage, *International Journal of Hydrogen Energy*, 34 (19), 8428–8437, 2009.

47. Haugom, G.P. and Friis-Hansen, P., Risk modelling of a hydrogen refuelling station using Bayesian network, *International Journal of Hydrogen Energy*, 36 (3), 2389–2397, 2011.

48. MacIntyre, I., Tchouvelev, A.V., Hay, D.R., Wong, J., Grant, J., and Benard, P., Canadian hydrogen safety program, *International Journal of Hydrogen Energy*, 32 (13), 2134–2143, 2007.

49. Middha, P. and Hansen, O.R., Using computational fluid dynamics as a tool for hydrogen safety studies, *Journal of Loss Prevention in the Process Industries*, 22 (3), 295–302, 2009.

50. Middha, P., Hansen, O.R., and Storvik, I.E., Validation of CFD-model for hydrogen dispersion, *Journal of Loss Prevention in the Process Industries*, 22 (6), 1034–1038, 2009.

51. Makarov, D., Verbecke, F., Molkov, V., Roe, O., Skotenne, M., Kotchourko, A., Lelyakin, A., Yanez, J., Hansen, O., Middha, P., Ledin, S., Baraldi, D., Heitsch, M., Efimenko, A., and Gavrikov, A., An inter-comparison exercise on CFD model capabilities to predict a hydrogen explosion in a simulated vehicle refuelling environment, *International Journal of Hydrogen Energy*, 34 (6), 2800–2814, 2009.

52. Middha, P. and Hansen, O.R., CFD simulation study to investigate the risk from hydrogen vehicles in tunnels, *International Journal of Hydrogen Energy*, 34 (14), 5875–5886, 2009.

53. Venetsanos, A.G., Papanikolaou, E., Delichatsios, M., Garcis, J., Hansen, O.R., Heitsch, M., Huser, A., Jahn, W., Jordan, T., Lacome, J.-M., Ledin, H.S., Makarov, D., Middha, P., Studer, E., Tchouvelev, A.V., Teodorczyk, A., Verbecke, F., and Van der Voort, M.M., An inter-comparison exercise on the capabilities of CFD models to predict the short and long term distribution and mixing of hydrogen in a garage, *International Journal of Hydrogen Energy*, 34 (14), 5912–5923, 2009.

54. Baraldi, D., Kotchourko, A., Lelyakin, A., Yanez, J., Middha, P., Hansen, O.R., Gavrikov, A., Efimenko, A., Verbecke, F., Makarov, D., and Molkov, V., An inter-comparison exercise on CFD model capabilities to simulate hydrogen deflagrations in a tunnel, *International Journal of Hydrogen Energy*, 34 (18), 7862–7872, 2009.

55. Venetsanos, A.G., Papanikolaou, E., Hansen, O.R., Middha, P., Garcia, J., Heitsch, M., Baraldi, D., and Adams, P., HySafe standard benchmark problem SBEP-V11: Predictions of hydrogen release and dispersion from a CGH2 bus in an underpass, *International Journal of Hydrogen Energy*, 35 (8), 3857–3867, 2010.

56. Garcia, J., Baraldi, D., Gallego, E., Beccantini, A., Crespo, A., Hansen, O.R., Hoiset, S., Kotchourko, A., Makarov, D., Migoya, E., Molkov, V., Voort, M.M., and Yanez, J., An intercomparison exercise on the capabilities of CFD models to reproduce a large-scale hydrogen deflagration in open atmosphere, *International Journal of Hydrogen Energy*, 35 (9), 4435–4444, 2010.

57. Ham, K., Marangon, A., Middha, P., Versloot, N., Rosmuller, N., Carcassi, M., Hansen, O.R., Schiavetti, M., Papanikolaou, E., Venetsanos, A., Engebo, A., Saw, J.L. Saffers,

J.-B., Flores, A., and Serbanescu, D., Benchmark exercise on risk assessment methods applied to a virtual hydrogen refuelling station, *International Journal of Hydrogen Energy,* 36 (3), 2666–2677. 2011.

58. Venetsanos, A.G., Adams, P., Azkarate, I., Bengaouer, A., Brett, L., Carcassi, M.N., Engebo, A., Gallego, E., Gavrikov, A.I., Hansen, O.R., Hawksworth, S., Jordan, T., Kessler, A., Kumar, S., Molkov, V., Nilsen, S., Reinecke, E., Stocklin, M., Schnidtchen, U., Teodorczyk, A., Tigreat, D., and Versloot, N.H.A., On the use of hydrogen in confined spaces: Results from the internal project InsHyde, *International Journal of Hydrogen Energy,* 36 (3), 2693–2699, 2011.

59. Hansen, M.D., and Gammel, G.W., Management of change: A key to safety—not just process safety, *Professional Safety,* 53 (10), 41–50, 2008.

60. EPA, *Catastrophic failure of storage tanks,* Chemical Safety Alert, U.S. Environmental Protection Agency, Washington, DC, 1997.

61. Alsheyab, M., Jiang, J.-Q., and Stanford, C., Risk assessment of hydrogen gas production in the laboratory scale electrochemical generation of ferrate (VI), *Journal of Chemical Health & Safety,* 15 (5), 16–20, 2008.

62. Xu, P., Zheng, J., Liu, P., Chen, R., Kai, F., and Li, L., Risk identification and control of stationary high-pressure hydrogen storage vessels, *Journal of Loss Prevention in the Process Industries,* 22 (6), 950–953, 2009.

63. *TEEX Developing Hydrogen Fuel Training Program, Occupational Health & Safety,* April 24, 2011. https://ohsonline.com/Articles/2011/04/24/TEEX-Developing-Hydro gen-Fuel-Training-Program.aspx?admgarea=ht.Training

64. DiBerardinis, L.J. (Ed.), *Handbook of Occupational Safety and Health,* 2nd ed., John Wiley & Sons, New York, 1999.

65. Felder, R.M. and Brent, R., Objectively speaking, *Chemical Engineering Education,* 31 (3), 178–179, 1997.

66. Goraya, A., Amyotte, P.R., and Khan, F.I., An inherent safety-based incident investigation methodology, *Process Safety Progress,* 23 (3), 197–205, 2004.

67. Frank, J., Observations on pre-emergency planning, *Industrial Fire World,* June 2007.

68. Hendershot, D.C., Process safety culture, *Journal of Chemical Health and Safety,* 14 (3), 39–40, 2007.

69. Erickson, J.A., Corporate culture. Examining its effects on safety performance, *Professional Safety,* 53 (11), 35–38, 2008.

70. Antonsen, S., The relationship between safety culture and safety on offshore supply vessels, *Safety Science,* 47 (8), 1118–1128, 2009.

71. Hopkins, A., *Thinking about process safety indicators,* Working Paper 53 (Paper Prepared for Presentation at the Oil and Gas Industry Conference), National Research Centre for OHS Regulation, Australian National University Manchester, UK, November 2007.

72. Glendon, I., Safety culture and safety climate: How far have we come and where should we be heading? *Journal of Occupational Health and Safety – Australia and New Zealand,* 24 (3), 249–271, 2008.

73. Amyotte, P.R. and Eckhoff, R.K., Dust explosion causation, prevention and mitigation: An overview, *Journal of Chemical Health and Safety,* 17 (1), 15–28, 2010.

74. Morozzo, C., *Account of a violent explosion which happened in the flour-Warehouse, at Turin, December the 14th, 1785; To which are added some observations on spontaneous inflammations,* From the Memoirs of the Academy of Sciences of Turin, The Repertory of Arts and Manufactures, London, 1795.

75. Amyotte, P.R., Are classical process safety concepts relevant to nanotechnology applications? Nanosafe2010: International Conference on safe production and use of nanomaterials, *Journal of Physics: Conference Series*, 304, 2011.
76. Javed B., *40 negative indicators of health and safety culture*, hseblog, June 17, 2023. www.hseblog.com/negative-indicators-of-health-and-safety-culture-at-workplace/
77. Ross Technology, The Risks of Poor Safety Culture, January 2024. www.rosstechnol ogy.com/news/poor-safety-culture-risks/
78. Safetybank, The Business Impact of Poor Health and Safety Management, October 15, 2019. www.safetybank.co.uk/blog/the-business-impact-of-poor-health-and-saf ety-management
79. Martinelli, K., High speed training, what are the consequences of poor health & safety procedures? September 23, 2017. www.highspeedtraining.co.uk/hub/consequences-of-poor-health-safety-procedures/
80. Hopkins, A. (Ed.), *Learning from High Reliability Organisations*, CCH Australia Limited, Sydney, Australia, 2009.

9 HySafe
Safety of Hydrogen as an Energy Carrier

This chapter provides a brief look at one of the more comprehensive hydrogen safety initiatives undertaken worldwide, HySafe (Safety of Hydrogen as an Energy Carrier). The intent here is not to give an exhaustive account of all HySafe activities and accomplishments, but rather to encourage readers to explore further the various research and educational packages that may be of interest. The chapter is therefore arranged in the following manner: an overview of the Network of Excellence HySafe, followed by a description of important research packages and projects, and an educational curriculum for teaching hydrogen safety.

Two primary resources used in writing this chapter (and hence recommended for further reading) are the HySafe website [1] and the recent (2011) article by Jordan et al. [2]. Jordan [3] and HySafe [4] provide perspectives on the network as of 2006 and 2007, respectively. Molkov [5], in his preface to the hydrogen safety special issue of the *Journal of Loss Prevention in the Process Industries* published in 2008, provides additional commentary on HySafe activities designed to contribute to safer use of hydrogen and fuel cell technologies.

In his presentation to the first European Hydrogen Energy Conference in 2003 (approximately six months before the formal start of the HySafe network), Dorofeev [6] identified the following HySafe objectives:

- To contribute to common understanding and approaches for addressing hydrogen safety issues.
- To integrate experience and knowledge on hydrogen safety in Europe.
- To integrate and harmonize the fragmented research base.
- To provide contributions to EU (European Union) safety requirements, standards and codes of practice.
- To contribute to an improved technical culture on handling hydrogen as an energy carrier.
- To promote public acceptance of hydrogen technologies.

DOI: 10.1201/9781003313007-9

In 2011, these same objectives have been concisely expressed by the following [2]:

- To strengthen, focus, and integrate the fragmented research on hydrogen safety.
- To form a self-sustained competitive scientific and industrial community.
- To promote public awareness and trust in hydrogen technologies.
- To develop an excellent safety culture.

The reason behind presenting these two listings of the same objectives is to demonstrate the evolution of hydrogen safety thinking, even over this relatively short eight-year period. This does not constitute a redefinement of the HySafe objectives, but rather their re-expression in a manner commensurate with recent developments in the safety field. One clearly sees an emphasis in the 2011 formulation of objectives [2] on issues of public *awareness and trust* (compared with *acceptance*) and an excellent *safety culture* (compared with an improved *technical culture*). These are certainly welcome developments that are consistent with arguments previously advanced in this book (e.g., Section 8.3).

9.1 OVERVIEW OF EC NETWORK OF EXCELLENCE FOR HYDROGEN SAFETY

With these objectives in mind, the Network of Excellence (NoE) HySafe (Safety of Hydrogen as an Energy Carrier) was established and supported by the European Commission (EC), with a start date of March 1, 2004 [2]. A follow-up organization, the International Association for Hydrogen Safety (HySafe (IA)), was founded by most of the HySafe consortium members on February 26, 2009 [2]. Official logos for both HySafe and HySafe (IA) are given in Figure 9.1.

The HySafe consortium itself was co-ordinated by the Forschungszentrum Karlsruhe in Germany [2], now merged with Universität Karlsruhe since October 1, 2009 to form the Karlsruhe Institute of Technology (KIT) [7]. HySafe participants included approximately 120 scientific researchers from 25 institutions located in 13 countries (12 European countries and Canada) [2]. The organizations involved consisted of 12 public research institutions, seven industrial partners, five universities, and one governmental body [2]; details are given in Table 9.1. Further thoughts on the origin and objectives of the HySafe network have been given by Molkov [8].

FIGURE 9.1 Logos for the Network of Excellence HySafe and the International Association for Hydrogen Safety, HySafe (IA). With permission.

Source: [2].

TABLE 9.1
Network of Excellence HySafe Consortium Members

Organization	Abbreviation	Country
Forschungszentrum Karlsruhe GmbH	FZK	Germany
L'Air Liquide	AL	France
Federal Institute for Materials Research and Testing	BAM	Germany
BMW Forschung und Technik GmbH	BMW	Germany
Building Research Establishment Ltd	BRE	United Kingdom
Commissariat a l'Energie Atomique	CEA	France
Det Norske Veritas AS	DNV	Norway
Fraunhofer-Gesellschaft ICT	Fh-ICT	Germany
Forschungszentrum Jülich GmbH	FZJ	Germany
GexCon AS	GexCon	Norway
The United Kingdom's Health and Safety Laboratory	HSE/HSL	United Kingdom
Foundation INASMET	INASMET	Spain
Institut National de l'Environnement Industriel et des Risques	INERIS	France
European Commission – JRC – Institute for Energy	JRC	Netherlands
National Center for Scientific Research Demokritos	NCSRD	Greece
StatoilHydro ASA	SH	Norway
DTU/Risø National Laboratory	DTU/Riso	Denmark
TNO	TNO	Netherlands
University of Calgary	UC	Canada
University of Pisa	UNIPI	Italy
Universidad Politécnica de Madrid	UPM	Spain
University of Ulster	UU	United Kingdom
Volvo Technology Corporation	Volvo	Sweden
Warsaw University of Technology	WUT	Poland
Russian Research Center Kurchatov Institute	KI	Russia

Source: [2]. With permission.

9.2   HYSAFE WORK PACKAGES AND PROJECTS

As described (and illustrated graphically) by Jordan et al. [2], HySafe activities consisted of 15 work packages and three internal projects accommodated within four "activity clusters":

- Basic research.
- Risk management.
- Dissemination.
- Management.

These clusters indicate a primary focus on knowledge acquisition through fundamental and applied research, coupled with effective overall management and publication and presentation of findings.

The 15 work packages as given on the HySafe website [1] are as follows:

- Biennial report on hydrogen safety.
- Integration of experimental facilities.
- Scenario and phenomenon ranking.
- Hydrogen incident and accident database.
- Principal computational fluid dynamics (CFD) exercises and guidelines.
- Mapping priorities and assessment.
- Hydrogen release, distribution, and mixing.
- Hydrogen ignition and jet fires.
- Hydrogen explosions.
- Mitigation.
- Risk assessment methodologies.
- International conference on hydrogen safety.
- e-Academy of hydrogen safety (discussed in Section 9.3).
- Contribution to standards and legal requirements.
- Material compatibility and structural integrity.

Not surprisingly, one sees in these work packages a clear linkage to the activity clusters focused on basic and applied research, network management, and dissemination of both research and educational products.

Also identified on the HySafe website [1] are the three internal projects:

- InsHyde
  - This project had its origins in the understanding that hydrogen releases in confined and partially confined environments – even seemingly small releases with low flow rates – could pose significant problems because of the potential for ignition of flammable gas accumulations leading to deflagration and possibly detonation [2]. Issues related to the indoor use of hydrogen were investigated under the InsHyde project by means of a combination of theoretical and experimental studies on the release, mixing, and combustion of hydrogen [2]. Sensor evaluations were also conducted [2], thus addressing the active engineered category of the hierarchy of safety controls (Section 7.1).
- HyTunnel
  - The HyTunnel project also dealt with confined environments – as the name itself illustrates – the specific geometry being a tunnel. Motivation for this work came from the need for management of the hazards and risks related to tunnel fires, particularly in light of the relevant EU regulations concerning road-tunnel safety [2]. In addition to the performance of fire and explosion simulations, numerical modeling was used extensively in an attempt to better understand the complex interaction among the following parameters affecting dispersion: (1) hydrogen release, (2) tunnel geometry, and (3) ventilation system design [2]. Buoyancy, one of the inherent safety aspects of hydrogen discussed in Chapter 7, was observed to play a key role in these dispersion studies wherein increased tunnel height yielded a lower risk profile [2].

- HyQRA
  - This project served as a vehicle for linking fundamental scientific research to industrially relevant applications [2]. The primary HyQRA objective, again as the name implies, was to develop quantitative risk assessment (QRA) tools for hydrogen technologies that were appropriate for the level of detail required and/or feasible [2]. Improvements were sought for various aspects of the QRA process: (1) screening models, (2) scenario selection, (3) ignition probability, (4) fire modeling, (5) structural response, and (6) acceptance criteria [2]. Several of these points were previously discussed in Section 8.2.4 under the Process Safety Management (PSM) element *process risk management*.

The Hysafe website [1] and Jordan et al. [2] provide information on the products (i.e., deliverables) arising from the various work packages and internal projects listed earlier. Several research findings have also been published in the archival literature; selected illustrative examples from the *International Journal of Hydrogen Energy* are now presented to close the current section. The emphasis in these examples is clearly on both process safety and occupational safety (as defined and discussed in Section 8.1), although not necessarily in what might be termed a "pure" process environment such as the hydrogen upgrader section of an oil refinery. Industrial workers must obviously be protected from hydrogen hazards but so too must the public. Hence one sees in these HySafe deliverables a focus on application to garages, transportation tunnels, and refueling stations.

One of the rigorous scientific research features of the HySafe work was the use of standard benchmark exercise problems (SBEPs). Venetsanos et al. [9] provide details on one of the InsHyde SBEPs aimed at predicting distribution and mixing patterns of hydrogen in a garage. Ten different CFD codes employing eight different turbulence submodels were used in this intercomparison exercise in an attempt to reduce the variation in predicted results from the numerical simulators [9]. In a subsequent publication, Venetsanos et al. [10] summarize the results from the InsHyde project in terms of, among other features, experimental data obtained from dispersion and combustion experiments, performance evaluation of commercial hydrogen detectors, and further CFD code intercomparisons.

In a similar vein, Baraldi et al. [11] describe a HyTunnel SBEP in which five CFD codes, again having different turbulence and combustion submodels, were used to simulate explosions of stoichiometric hydrogen–air mixtures in a 78.5 m long tunnel. The numerical simulation results were compared against one another and also with experimental data obtained for the same geometry; good agreement was determined to exist with respect to experimental and predicted maximum overpressures [11]. Middha and Hansen [12] used the GexCon code FLACS (Section 8.2.4) to conduct hydrogen-vehicle risk assessments for different tunnel configurations and both worst-case and more realistic (i.e., credible) scenarios. Venetsanos et al. [13] investigated, again using various CFD codes, hydrogen release and dispersion from a hydrogen-fueled bus in an underpass; that is, a semiconfined environment as in the previously referenced tunnel studies, but also one having roof obstructions as would be present in roof-and-slab style construction.

There are two fundamental aspects to QRA: quantification of probability (or likelihood) and quantification of consequence severity. Garcia et al. [14] conducted an intercomparison exercise using CFD codes to predict the latter of these two risk parameters – in this case for the explosion of a 2094 m^3 stoichiometric hydrogen–air mixture in an unconfined environment. The practical application of this work was envisaged as an accidental leak and subsequent explosion of hydrogen at a facility such as a refueling station [14]. This scenario was considered by Ham et al. [15] in their application of various QRA methodologies involving both analytical and CFD consequence modeling to a "virtual" hydrogen refueling station complete with vulnerable surroundings. A key finding of this research was that the numerical models provided more realistic results in the near-field and/or confined areas, while there was no discernible advantage of numerical over analytical processing in the open air [15]. Clearly there is a need for further work in this area which has implications for reducing risk to both on-site workers and off-site members of the general public [15].

9.3   E-ACADEMY OF HYDROGEN SAFETY

The previous section has highlighted some of the research achievements of the HySafe Network of Excellence – in particular the issues of CFD code validation with experimental data and the assessment of underlying code assumptions and submodels. To close this chapter, we turn to the matter of education and training in hydrogen safety.

Dahoe and Molkov [16] describe the establishment of an "e-academy of hydrogen safety" led by the University of Ulster in collaboration with other HySafe partners. They give details on a postgraduate certificate course in hydrogen safety delivered in the distance-learning mode, as well as an annual European summer school on hydrogen safety. Both of these offerings are aimed at helping to achieve the following objectives of the e-academy of hydrogen safety [1]:

- Integration of academic and other institutions through the development and implementation of an international curriculum on hydrogen safety engineering.
- Development of a database of organizations working in the hydrogen industry to form a market of potential trainees and to disseminate results of mutual activities of the network.
- Creation of a pool of specialists for the delivery of joint training/teaching on hydrogen safety and joint supervision of research students.

The key element in the first of these objectives is the international curriculum on hydrogen safety engineering (the "backbone" of the e-academy [16]). The current version of this curriculum posted on the HySafe website [1] is shown in Table 9.2. Topics are arranged in 15 modules organized into three thematic areas – basic, fundamental, and applied – with the majority of the basic modules intended for the undergraduate level and the majority of the fundamental and applied modules designed for the graduate level [1]. This thinking is consistent with the graphic displayed in Figure 9.2, which underpins the developers' philosophy of the undergraduate program being grounded with a strong engineering science core and augmented with educational material focused on hydrogen safety [1]. The graduate program then covers

TABLE 9.2
Structure of the International Curriculum on Hydrogen Safety Engineering

Level	Module	Contents
Basic	Thermodynamics	Fundamental concepts and first principles; volumetric properties of a pure substance; first law of thermodynamics; first law of thermodynamics and flow processes; second law of thermodynamics; second law of thermodynamics and flow processes; first and second laws of thermodynamics, and chemically reacting systems; phase equilibrium; thermodynamics and electrochemistry.
	Chemical kinetics	Rates of chemical reactions; kinetics of complex reactions; surface reactions; application of sensitivity analysis to reaction mechanisms; reduction of complex reaction systems to simpler reaction mechanisms; chemical kinetics and the detonation of combustible mixtures.
	Fluid dynamics	Fluid statics; kinematics of the flow field; kinematics of incompressible potential flow; kinematics of compressible flow; incompressible laminar viscous flow; mathematical models of fluid motion; dimensional analysis and similitude; incompressible turbulent flow; waves in fluids and the stability of fluid flow; compressible turbulent flow.
	Heat and mass transfer	Basic modes of heat transfer and particular laws; isothermal mass transfer; heat conduction; convection heat transfer; forced convection; natural convection; heat transfer with phase change; radiation heat transfer; simultaneous heat and mass transfer.
	Solid mechanics	Analysis of stress; deformation and strain; tension and compression; statically indeterminate force systems; thin-walled pressure vessels; direct shear stresses; torsion; shearing force and bending moment; centroids, moments of inertia, and products of inertia of plane areas; stresses in beams; elastic deformation of beams: double integration method; statically indeterminate elastic beams; special topics in elastic beam theory; plastic deformation of beams; columns; strain energy methods; combined stresses; members subject to combined loadings.
Fundamental	Hydrogen as an energy carrier	Introduction to hydrogen as an energy carrier; introduction to hydrogen applications and case studies; equipment for hydrogen applications; possible accident scenarios; definitions and overview of phenomena and methodologies related to hydrogen safety.

TABLE 9.2 (Continued)
Structure of the International Curriculum on Hydrogen Safety Engineering

Level	Module	Contents
	Fundamentals of hydrogen safety	Hydrogen properties, compatibility of metallic materials with hydrogen; hydrogen thermochemistry; governing equations of multicomponent reacting flows; premixed flames; diffusion flames; partially premixed flames; turbulent premixed combustion; turbulent non-premixed combustion; ignition and burning of liquids and solids; fire through porous media.
	Release, mixing, and distribution	Fundamentals of hydrogen release and mixing; handling hydrogen releases.
	Hydrogen ignition	Hydrogen ignition properties and ignition sources; prevention of hydrogen ignition.
	Hydrogen fires	Fundamentals of hydrogen fires.
	Explosions: deflagrations and detonations	Deflagrations; detonation; transitional hydrogen explosion phenomena.
Applied	Fire and explosion effects on people, structures, and the environment	Thermal effects of hydrogen combustion; blast waves; calculation of pressure effects of explosions; structural response, fragmentation and missile effects; fracture mechanics.
	Accident prevention and mitigation	Prevention, protection and mitigation; basic phenomena underpinning mitigation technologies; standards, regulations, and good practices related to hydrogen safety; inertization; containment; explosion venting; flame arresters and detonation arresters.
	Computational hydrogen safety engineering	Introduction to CFD; introduction to thermodynamic and kinetic modeling; mathematical models in fluid dynamics; finite difference method; solution of the generic transport equation; solution of weakly compressible Navier–Stokes equations; solution of compressible Navier–Stokes equations; turbulent flow modeling; combustion modeling; high-speed reactive flows; modeling of hydrogen–air diffusion flames and turbulence–radiation interactions; modeling of liquid hydrogen pool fires; multiphase flows; special topics.
	Risk assessment	General risk assessment and protective measures for hazardous materials processing and handling; regulations, codes, and standards; risk assessment methodologies; hazard identification and scenario development; effect analysis of hydrogen accidents; vulnerability analysis; risk reduction and control in the hydrogen economy.

Source: [1].

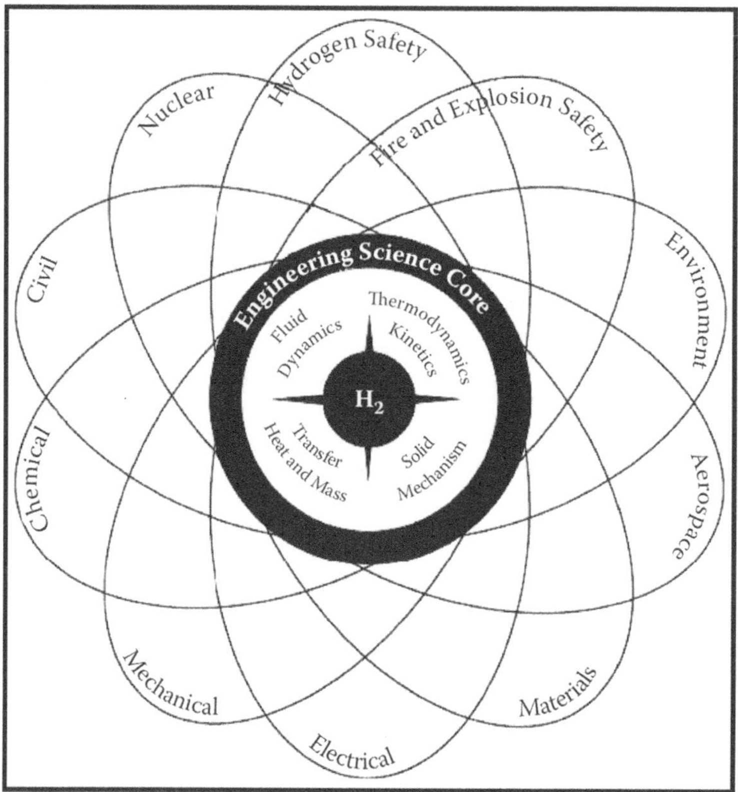

FIGURE 9.2 Hydrogen safety in relation to other branches of engineering science.

Source: [1].

more advanced hydrogen safety topics consistent with the overall HySafe research activities [1].

What one sees in this approach is a deliberate attempt to integrate hydrogen safety material into the undergraduate curriculum while providing opportunities for acquisition of specialized knowledge at the graduate (e.g., master's) level. This is an approach the current authors wholeheartedly endorse and which is consistent with arguments previously advanced for the topic of risk management [17]. Further, the international curriculum shown in Table 9.2 provides numerous examples of possible training topics on hydrogen safety according to the hierarchical learning continuum known as *Bloom's Taxonomy of Educational Objectives* (described in Section 8.2.8). The fundamental and applied modular topics, although more extensive in depth of coverage, are also generally consistent with the scope of the current book.

In keeping with the introductory remarks to this chapter, our intention here has been to highlight key research and educational achievements of the HySafe Network of Excellence. The HySafe website [1] and the summary paper by Jordan et al. [2] are recommended as starting points for further study.

9.4 HYDROGEN SAFETY FOR ENERGY APPLICATIONS HANDBOOK

HySafe published on March 25, 2022 the first edition of the handbook "Hydrogen Safety for Energy Applications" with Alexei Kotchourko and Thomas Jordan as editors [18]. The handbook describes among other things: engineering design, risk assessment, codes and standards, risks, and safety connected with hydrogen systems. It mainly presents hydrogen technologies and investigates possible sources and consequences of accidental hydrogen events, including its vehicular applications. Risk assessment and safety measures to prevent and/or mitigate hydrogen incidents are also considered. A survey concerning legal aspects is also presented.

The key features of the book are the following [18]:

- Provides useful information, such as models, tables, nomograms and formulas for the safe use of hydrogen.
- Presents reference data and detailed descriptions and guidelines to be used in risk assessments.
- Covers accident phenomena and consequences of accidents stemming from hydrogen.

9.5 INTERNATIONAL CONFERENCES ON HYDROGEN SAFETY HYDROGEN SAFETY

HySafe organizes the worldwide biennial International Conference on Hydrogen Safety, ICHS. The ninth International Conference on Hydrogen Safety (ICHS 2021) was held in Edinburgh, Scotland on 21–24 September, 2021 [19]. The all-encompassing theme for ICHS 2021 was "Safe Hydrogen for Net Zero" and addressed most of the hydrogen safety topics. The tenth International Conference on Hydrogen Safety (ICHS 2023) was held in Québec City, Québec, Canada on September 19–21, 2023.

In the first nine ICHS biennial conferences from 2005 to 2021 experts from all over the world participated and contributed with presentations and discussions on hydrogen safety, from basic to applied research and from good practice to regulatory issues.

By 2023, hydrogen started to be dominant in the transition to clean, safe, and sustainable energy issues. Among others, ICHS 2023 addressed industrial feedstocks, commercial and domestic heat, energy storage, and energy transportation across continents.

The overarching theme for ICHS 2023 addressed most hydrogen safety issues including:

- Hydrogen production and supply infrastructure safety.
- Hydrogen and hydrogen carrier functioning.
- Physical effects.
- Consequence and risk assessment.
- Incidents, accidents, and near-misses.

- Hydrogen impact on materials.
- Safety of energy storage.
- Risk management.
- Best practices, regulations, codes, and standards.
- Targeted approach for public awareness and acceptance of hydrogen.

The next International Conference on Hydrogen Safety (ICHS 2025) will be held in the Republic of Korea under the auspices of the International Association for Hydrogen Safety (HySafe (IA)) in September, 2025.

9.6　RESEARCH PRIORITIES WORKSHOPS

Every even year, the HySafe standing research committee organizes a Research Priorities Workshop (RPW). The last workshop was held in Québec City, Québec on November 21–23, 2022 [20].

Speakers at the meetings discuss their research topics and perspectives on the present state of the art. Their research proposals have usually a foresight of five to ten years. Extended abstracts submitted by the attendees are included in the final reports.

Analysis of accident phenomena takes priority at the meetings, followed by reports on progress in mitigation sensors, hazard prevention, and risk assessments. Although there is still a knowledge gap on hydrogen's impact on materials, the committee ranked this as the less interesting issue. Accident physics for the gas phase is currently ranked relatively low since this is a topic on which significant effort has been spent already. These workshops proved to be very efficient in addressing the knowledge gaps in hydrogen safety [20].

9.7　HYDROGEN INCIDENT AND ACCIDENT DATABASE

In Europe, the Hydrogen Incident and Accident Database (HIAD) [21] was created by the Joint Research Center (JRC), DNV, and HySafe project partners, and launched in July 2006. It is an open communication platform for recording and categorizing accidental events related to hydrogen and a database for risk assessment, acting as a repository [22]. Some documents based on studies performed in European countries on hydrogen safety are available in HIAD [23–24].

Research priorities in hydrogen safety were considered in a relevant gap analysis performed by the JRC and HySafe network, and the results were presented in several reports and in the HIAD database [25–26].

REFERENCES

1. The International Association for Hydrogen Safety, www.hysafe.org/IAHySafe
2. Jordan, T., Adams, P., Azkarate, I., Baraldi, D., Barthelemy, H., Bauwens, L., Bengaouer, A., Brennan, S., Carcassi, M., Dahoe, A., Eisenrich, N., Engebo, A., Funnemark, E., Gallego, E., Gavrikov, A., Haland, E., Hansen, A.M., Haugom, G.P., Hawksworth, S., Jedicke, O., Kessler, A., Kotchourko, A., Kumar, S., Langer, G.,

Stefan, L., Lelyakin, A., Makarov, D., Marangon, A., Markert, F., Middha, P., Molkov, V., Nilsen, S., Papanikolaou, E., Perrette, L., Reinecke, E.-A., Schmidtchen, U., Serre-Combe, P., Stocklin, M., Sully, A., Teodorczyk, A., Tigreat, D., Venetsanos, A., Verfondern, K., Versloot, N., Vetere, A., Wilms, M., and Zaretskiy, N., Achievements of the EC network of excellence HySafe, *International Journal of Hydrogen Energy*, 36 (3), 2656–2665, 2011.

3. Jordan, T., *HySafe – The Network of Excellence for Hydrogen Safety*, WHEC 16, Lyon, France, June 13–16, 2006.

4. HySafe, 2004 – 2009: A Network of Excellence in the 6th Framework Programme of the European Commission, with 25 European Partners, HYSAFE – Safe Use of Hydrogen as an Energy Carrier, 2007. www.hysafe.org

5. Molkov, V., Preface: Special issue on hydrogen safety, *Journal of Loss Prevention in the Process Industries*, 21 (2), 129–130, 2008.

6. Dorofeev, S., Safety aspects of hydrogen as an energy carrier, in *Presented at 1st European Hydrogen Energy Conference*, Grenoble, France, September 2–5, 2003.

7. KIT – Karlsruhle Institute of Technology. www.kit.edu/kit/english/index.php (accessed August 5, 2011).

8. Molkov, V., Hydrogen safety research: State-of-the-art, in *Proceedings of the 5th International Seminar on Fire and Explosion Hazards*, Edinburgh, UK, April 23–27, 2007. www.academia.edu/74112001/Hydrogen_Safety_Research_State_Of_The_Art

9. Venetsanos, A.G., Papanikolaou, E., Delichatsios, M., Garcis, J., Hansen, O.R., Heitsch, M., Huser, A., Jahn, W., Jordan, T., Lacome, J.-M., Ledin, H.S., Makarov, D., Middha, P., Studer, E., Tchouvelev, A.V., Teodorczyk, A., Verbecke, F., and Van der Voort, M.M., An inter-comparison exercise on the capabilities of CFD models to predict the short and long term distribution and mixing of hydrogen in a garage, *International Journal of Hydrogen Energy*, 34 (14), 5912–5923, 2009.

10. Venetsanos, A.G., Adams, P., Azkarate, I., Bengaouer, A., Brett, L., Carcassi, M.N., Engebo, A., Gallego, E., Gavrikov, A.I., Hansen, O.R., Hawksworth, S., Jordan, T., Kessler, A., Kumar, S., Molkov, V., Nilsen, S., Reinecke, E., Stocklin, M., Schnidtchen, U., Teodorczyk, A., Tigreat, D., and Versloot, N.H.A., On the use of hydrogen in confined spaces: Results from the internal project InsHyde, *International Journal of Hydrogen Energy*, 36 (3), 2693–2699, 2011.

11. Baraldi, D., Kotchourko, A., Lelyakin, A., Yanez, J., Middha, P., Hansen, O.R., Gavrikov, A., Efimenko, A., Verbecke, F., Makarov, D., and Molkov, V., An intercomparison exercise on CFD model capabilities to simulate hydrogen deflagrations in a tunnel, *International Journal of Hydrogen Energy*, 34 (18), 7862–7872, 2009.

12. Middha, P. and Hansen, O.R., CFD simulation study to investigate the risk from hydrogen vehicles in tunnels, *International Journal of Hydrogen Energy*, 34 (14), 5875–5886, 2009.

13. Venetsanos, A.G., Papanikolaou, E., Hansen, O.R., Middha, P., Garcia, J., Heitsch, M., Baraldi, D., and Adams, P., HySafe standard benchmark problem SBEP-V11: Predictions of hydrogen release and dispersion from a CGH2 bus in an underpass, *International Journal of Hydrogen Energy*, 35 (8), 3857–3867, 2010.

14. Garcia, J., Baraldi, D., Gallego, E., Beccantini, A., Crespo, A., Hansen, O.R., Hoiset, S., Kotchourko, A., Makarov, D., Migoya, E., Molkov, V., Voort, M.M., and Yanez, J., An intercomparison exercise on the capabilities of CFD models to reproduce a large-scale hydrogen deflagration in open atmosphere, *International Journal of Hydrogen Energy*, 35 (9), 4435–4444, 2010.

15. Ham, K., Marangon, A., Middha, P., Versloot, N., Rosmuller, N., Carcassi, M., Hansen, O.R., Schiavetti, M., Papanikolaou, E., Venetsanos, A., Engebo, A., Saw, J.L. Saffers, J.-B., Flores, A., and Serbanescu, D., Benchmark exercise on risk assessment methods applied to a virtual hydrogen refuelling station, *International Journal of Hydrogen Energy*, 36 (3), 2666–2677, 2011.

16. Dahoe, A.E., and Molkov, V.V., On the implementation of an international curriculum on hydrogen safety engineering into higher education, *Journal of Loss Prevention in the Process Industries*, 21 (2), 222–224, 2008.

17. Amyotte, P.R. and McCutcheon, D.J., *Risk Management: An Area of Knowledge for All Engineers*, Discussion paper prepared for Canadian Council of Professional Engineers, Ottawa, Ontario, 2006. www.academia.edu/74112001/Hydrogen_Safety_Research_State_Of_The_Art

18. Kotchourko, A. and Jordan, Th. (Eds.), *Hydrogen Safety for Energy Applications*. 1st ed, Elsevier, 2022. www.sciencedirect.com/book/9780128204924/hydrogen-safety-for-energy-applications

19. HySafe, in *International Conference on Hydrogen Safety Hydrogen Safety*, Québec, Canada, September 19–21, 2023. https://hysafe.info/ichs2023/

20. Research Priorities Workshops (RPW), November 21–23, 2022. https://hysafe.info/activities/research-priorities-workshops/rpw-2023-quebec-city/

21. HIAD 2.0 – Free access to the renewed hydrogen incident and accident database, June 28, 2018. https://hysafe.info/hiad-2-0-free-access-to-the-renewed-hydrogen-incident-and-accident-database/

22. Ruiz, P., Vega, L.F., Mar Arxer, M., Jimenez, C., and Rausa, A., *Hydrogen: Applications and Safety Considerations*, 1st ed., MATGAS 2000 AIE, Barcelona, Spain, 2015.

23. O'Garra, T., *AcceptH2 full analysis report-comparative analysis of the impact of the hydrogen bus trials on public awareness, attitudes and preferences: A comparative study on four cities*, Work Package 6 – Deliverable 9, 2005. www.accepth2.com/results/docs/AcceptH2_D9_Full-Analysis-Report_050804.pdf

24. Hickson, A., Philips, A., and Morales, G., Public perception related to a hydrogen hybrid internal combustion engine transit bus demonstration and hydrogen fuel, *Energy Policy*, 35 (4), 2249–2255, 2007.

25. Baraldi D., Papanikolaou E., Heitsch M., Moretto P., Cant R.S., Roekaerts D., Dorofeev S., Kotchourko A., Middha P., Tchouvelev A.V., Ledin S., Wen J., Venetsanos A., and Molkov V.V., *Prioritisation of research and development for modelling the safe production, storage, delivery and use of hydrogen*. EUR – Scientific and Technical Research series – ISSN 1831-9424 (online), ISSN 1018-5593 (print) ISBN 978-92-79-21602-2 (PDF) ISBN 978-92-79-21601-5, , JRC European Commission publication, 2011. (print). doi:10.2790/36578

26. Kotchourko, A., Baraldi, D., Bénard P., Eisenreich N., Jordan T., Keller J., Kessler A., LaChance J., Molkov V., Steen M., Tchouvelev A., and Wen J. *State of the Art and Research Priorities in Hydrogen Safety*, European Comission JRC Sience and Policy Report Publication, Luxembourg, 2014. ISBN 978-92-79-34719-1 (PDF) ISSN 1831-9424 (online). doi: 10.2790/99638.

10 Case Studies

The purpose of this chapter is to present guidance on the development and use of case studies (or as they are sometimes called, case histories). The role of incident investigation reports in this regard is highlighted, and several examples drawn from a range of industrial applications involving hydrogen are given.

To answer the question of whether case studies are helpful in enhancing the overall safety initiative, one need only consider the following quote from Crowl and Louvar [1]: "To paraphrase G. Santayana, one learns from history or one is doomed to repeat it". Case studies therefore incorporate the concept of *lessons learned* as the primary motivation for their use. This usually means that the most effective case studies are those giving details of a failure or shortcoming of some sort; it is human nature to pay attention when a story is being told and a loss – whether catastrophic or not – is involved [2]. The quote cited earlier illustrates the importance of learning from case histories and avoiding hazardous situations; the alternative is to ignore the mistakes of others and be involved in potentially life-threatening incidents [1, 2].

10.1 CASE STUDY DEVELOPMENT

The development of a case study involves first the selection of an appropriate incident (accident or near-miss); this is followed by an exposition of the lessons that can be learned from the incident and that ideally have universal or at least industry-specific application. For example, Chapter 2 describes several incidents and lessons learned that are drawn from the archival literature and various databases, the latter relying on incident investigation reports as discussed here, in Section 10.3.

With respect to inherently safer design (ISD, as presented in Chapter 7), Kletz and Amyotte [2] have presented thoughts on the following 12 avenues for developing case studies:

- Everyday life experiences.
- Newspapers and magazines.
- Topical conferences.
- Technical papers.
- Process safety books.

DOI: 10.1201/9781003313007-10

- Case study books.
- Other inherent safety books.
- Books from other industries and applications.
- Training packages.
- Trade literature.
- *Loss Prevention Bulletin.*
- Chemical Safety Board reports.

Because ISD is concept based, some of these suggestions may not translate literally to the world of hydrogen safety, which is focused on the nature and use of a particular substance. Few people would have first-hand experience in their daily lives with hydrogen, while many more would be able to relate to the notion that a tricycle (having three wheels) is inherently safer (more stable) than a bicycle (having two wheels).

The popular media (newspapers and magazines), on the other hand, seem to be a ready source of hydrogen-related case information, at least with respect to incident descriptions if not lessons learned beyond the impact of such incidents on public perception. Witness the numerous articles published in the *New York Times* on, for example, hydrogen leaking from a road tanker involved in an accident while carrying 11,000 kg of compressed hydrogen [3], liquid hydrogen fuel leaks during operation of the US space shuttle program [4–6], and gaseous hydrogen leaks from the non-nuclear operations (electrical generator cooling system) of a nuclear power plant [7].

Topical conferences such as the International Conference on Hydrogen Safety (ICHS) may prove to be quite helpful with respect to case study development. The fourth ICHS (September 12–14, 2011, in San Francisco, California) incorporated the topic of case studies under the theme of risk management of hydrogen technologies [8].

Similarly, technical papers cited throughout this book have a key role to play in developing hydrogen safety case studies. General papers such as Sanders [9] contain information on lessons related to items including human error and human factors, which can be easily tailored to applications involving hydrogen. Hydrogen-specific technical case study papers abound in the archival literature (see Chapter 2). As an additional example also referenced later in Section 10.3, Weiner, Fassbender, and Quick [10] present an interesting combination of *Hydrogen Safety Best Practices* (www.h2bestpractices.org) and *Hydrogen Incident Reporting and Lessons Learned* (www.h2incidents.org).

Process safety books (e.g., [1]) and case study books (e.g., [11]) provide both general cases and widely applicable lessons, as well as specific information on hydrogen-related incidents such as the 1989 Pasadena, Texas, vapor cloud explosion (see Chapters 2 and 8). There are, of course other books on hydrogen usage and safety (e.g., [12]); these may also provide relevant case study details. Books from other industries and applications would seem at first glance to be of limited use in hydrogen case study development. These texts have proven, however, to be quite useful in demonstrating the general lessons on safety culture (Section 8.3) to be learned from the railway and aircraft maintenance industries [13] and air traffic control activities [14].

Even only a quick Internet search for hydrogen training packages yields multiple hits (for example, reference [15]). Some Internet sources provide helpful case studies on topics such as hydrogen-induced crack-resistant (HIC) steel [16] and hydrogen fueling stations [17].

Trade (or industry sector) literature sources often provide incident descriptions that can be followed up to develop case studies, such as the explosion and fire at an ammonia plant involving a mixture of gases including hydrogen, nitrogen, and ammonia [18]. The same publication also describes problems associated with the degradation of munitions in stockpiles maintained by the US Army [19]. One potential issue is the explosion risk from hydrogen forming as stockpiled mustard gas degrades; risk reduction recommendations given include regular testing of all stored chemical agents, establishing databases for accurate information recording, and using statistical analysis for early identification of trends [19]. Interestingly, the inherently safer approach of swiftly destroying all the munitions is described as "ultimately the only effective way to reduce risks to the public" [19].

From the world of occupational health and safety comes a general (non-incident-specific) case study on battery fire and explosion hazards [20]. In the final stages of battery recharging, hydrogen can be produced by a process known as *gassing*; recommended risk reduction measures from the hierarchy of controls include air movement and exchange, hydrogen gas monitoring, and appropriate signage aimed at avoiding ignition sources [20]. In accordance with the discussion in Section 10.3, post-incident investigations of battery fires and explosions often indicate causes related to operational errors and safety shortcuts [20]. To further develop the lessons learned from these incidents, one must ask whether root causes at the management system level existed (as per Sections 10.2.3 and 10.3), and whether inherently safer prevention and mitigation measures were available.

The *Loss Prevention Bulletin* (LPB) is published by the Institution of Chemical Engineers (IChemE) in the United Kingdom. LPB issues represent a good source of case studies arising from process accident and near-miss reports [2]. For example, Carson and Mumford [21] illustrate various concerns related to the hazards and control of batch processing. In a listing of potentially hazardous exothermic processes, they include both *hydrogenation* (addition of hydrogen atoms to both sides of a double- or triple-bond) and *halogenation* (substitution of atoms such as hydrogen in organic molecules by halogens). In keeping with the case study philosophy of describing incidents and elucidating the accompanying lessons learned, an online search of the *Loss Prevention Bulletin* with the keyword *hydrogen* will reveal several LPB issues [22] that provide such information. Examples are given in Table 10.1.

As noted on its website [23], the US Chemical Safety and Hazard Investigation Board (Chemical Safety Board or CSB) is an independent, nonregulatory federal agency that conducts root cause investigations of chemical accidents at fixed industrial facilities. The reports of its investigations are available on the CSB website [23] for downloading and are often accompanied by video footage and animation of the incident sequence and root causes/lessons learned [2]. The safety bulletin [24, 25] on the ISD practice of *making incorrect assembly impossible* (as previously described in Section 7.6) is an excellent hydrogen-specific example of the significant case study

TABLE 10.1

Hydrogen Case Study Examples Found in the *Loss Prevention Bulletin*

Issue	Year	Case Study
015	1977	Fires involving hydrogen in a naphtha cracker and in a hydrogenation reaction
068	1986	Vapor cloud explosions involving hydrogen-rich gases
083	1988	Failure of a woven-steel braided flexible hose on a hydrogen installation area
156	2000	Explosion of hydrogen in a pipeline intended for transfer of CO_2 gas from an ammonia plant
207	2009	Hydrogen explosions from charging batteries; see also Ramsey [20]
		Unexpected generation of hydrogen in a new process for production of aluminum chloride
		Hydrogen generation inside sealed components at a refinery storage terminal
215	2010	Hydrogen explosion in an electrolyter plant

Source: Institute of Chemical Engineers, www.icheme.org.

value of CSB reports. This point is reiterated in Section 10.3, and other CSB examples are given in Section 10.4.

In short, there is an abundance of mechanisms for developing case studies relating to the lessons that can be learned from occupational and process incidents involving hydrogen. The next section suggests ways to use these case studies once they have been sourced and/or prepared.

10.2　USE OF CASE STUDIES

Case studies are particularly useful in conducting training exercises as part of an overall process safety management (PSM) scheme (see Section 8.2.6). They have also been shown to have value in addressing other PSM components such as hazard identification and risk analysis [26] (see Table 8.2). Further, educational efforts within undergraduate and graduate science and engineering programs can be significantly enhanced by the use of case studies to emphasize the practical implications of various safety concepts and methodologies.

As discussed in the introduction to the current chapter, it is the concept of *lessons learned* that makes case studies so effective regardless of their ultimate end use. The lessons that can be learned by the study of previous incidents fall within one or more of the following general categories: (1) legacy lessons, (2) engineering lessons, and (3) management lessons. The meaning of each of these categories can be best understood by briefly examining the Bhopal, India, incident – without doubt, the most significant process safety event that has occurred worldwide to the present time.

As excerpted from Khan and Amyotte [27], the Bhopal plant was a pesticide-producing facility owned by the Union Carbide Corporation of the United States and local interests. The plant was located in the town of Bhopal, the capital of Madhya Pradesh state in central India. At 12:45 am on December 3, 1984, several thousand kilograms of a toxic gas, rich in methyl isocyanate (MIC), were released into the atmosphere from the plant, causing over 2000 civilian casualties [28].

The physical properties of MIC make it highly hazardous. It has a flash point of –18.1°C and boils at 39.1°C, reacts exothermically with water, and has a maximum allowable 8 h exposure concentration of as low as 0.02 ppm. MIC was the intermediate product of carbaryl (a pesticide) manufacturing. It was stored in three horizontal stainless steel vessels, referred to as tanks 610, 611, and 619. Tanks 610 and 611 were used in normal operation, and the third tank, 619, was for emergency use. The MIC was usually stored under refrigerated conditions, but in the days before the incident, the refrigeration unit had been shut down.

On the day of the incident, storage tank 610 became contaminated with some other substance (most probably water). An exothermic reaction heated the MIC to a temperature beyond its boiling point. This led to the concrete mounds above the tank cracking and MIC vapor being released through a relief valve. The scrubber and flare systems designed to remove and destroy the MIC were not in operation on the day of the incident. After an hour and a quarter, the relief valve on tank 610 repositioned and the MIC release was stopped. By that time about 36,000 kg of material had escaped from tank 610, of which about 25,000 kg were MIC vapor [1, 28].

10.2.1 Case Study Legacy Lessons

The legacy of Bhopal is its long-lasting impact on the chemical process industries and numerous other areas of industrial endeavor [29–31]. As described by process safety expert Dennis Hendershot [29]: "A significant impact of Bhopal was to make everybody – corporate management, government, communities – aware of the potential magnitude of a chemical accident". The results of this heightened awareness have been manifested positively in new process safety regulatory regimes, worldwide best-practice initiatives such as Responsible Care®, and a greater sense of the need for corporations to export technology – not unacceptable risk – to the developing world.

Most importantly, however, coupled forever with Bhopal are the chemical MIC and the immeasurable loss caused to thousands of human beings in 1984 – loss that continues to this day in the form of ongoing suffering from respiratory distress caused by exposure to MIC and other chemicals remaining on the plant site, which have the potential for further health and environmental impacts [30]. As illustrated in Figure 10.1, the legacy of a process incident can involve negative and positive aspects.

Arguably, the closest the hydrogen industry has come to having a case study with Bhopal-type legacy lessons is the *Hindenburg* disaster (and perhaps also the *Challenger* explosion) described in Chapter 2. Whether warranted or not, most people who are aware of the *Hindenburg* will automatically associate that word with another – *hydrogen*. This may not be especially problematic with respect to air travel given that few people will experience a trip on a buoyancy-driven airship during their lifetime. It is imperative, however, that the legacy lessons from incidents such as the *Hindenburg* not be forgotten as public travel via hydrogen-fueled vehicles grows in feasibility and popularity.

FIGURE 10.1   Memorial commemorating the Bhopal disaster.

Source: [32].

10.2.2 Case Study Engineering Lessons

The engineering lessons of a case study relate to the hierarchy of safety controls as discussed in Section 7.1. With respect to Bhopal, it is evident that safety at the facility was heavily dependent on engineered and procedural safeguards, the effectiveness and operational state of which were questionable. This catastrophe might have been prevented, or the consequences mitigated, through appropriate inherent safety considerations [27].

The relevance of the previous paragraph to the hydrogen industry can be found in Chapter 7. Following a specific incident involving hydrogen, the engineering lessons that are transferable to other hydrogen applications would include the use and effectiveness of passive engineered, active engineered, and procedural measures, as well as the potential for implementation of the principles of inherently safer design (ISD). For example, the case of a gaseous hydrogen release necessitates an examination of the detection and mitigation measures in place and should also be followed by investigation of hydrogen storage possibilities in a liquid state or as metal hydrides.

10.2.3 Case Study Management Lessons

Root cause analysis of an incident must result in identification of the lessons to be learned by management – both the organizational personnel in management positions and the safety management system itself. As described in the previous section, many of the engineered and procedural safeguards at the Bhopal plant were non-functional, and there was a lack of commitment to inherently safer design. Where then does responsibility for these shortcomings exist? Squarely on the shoulders of management; no modern root cause analysis methodology could conclude otherwise. (See Section 8.1.)

From a hydrogen safety perspective, the issues of process safety management and safety culture are described in Chapter 8. Deficiencies in these areas constitute key case study lessons that are critical to the prevention of future incidents. This point is expanded upon in the following section.

10.3 INCIDENT INVESTIGATION REPORTS

Management deficiencies identified via incident investigation (company-sponsored or third-party) represent universal findings that can be generalized across industry sectors. Lack of adequate loss prevention and management is generally due to deficiencies in one or more of three areas: the safety management system itself, the standards identified and set for the safety management system, and the degree of compliance with such standards [33]. A particular management system element may have missing components or may be entirely absent; alternatively, the management system element may be present to some degree, but could have inappropriate standards, or perhaps standards with which there is little or no compliance [33].

These points are (unfortunately) well illustrated in the Westray case study, as demonstrated by the following excerpt from Amyotte and Eckhoff [33].

The Westray coal mine explosion occurred in Plymouth, Nova Scotia, Canada, on May 9, 1992, killing 26 miners. An indication of the destructive overpressures generated underground can be seen in Figure 10.2, which shows surface damage at the mine site. The methane levels in the mine were consistently higher than

FIGURE 10.2 Westray No. 1 main portal post-explosion.

Source: [34].

regulations permitted, which was caused by inadequate ventilation in the mine. Dust accumulations also exceeded permissible levels due to inadequate cleanup of coal dust; additionally, there was no crew in charge of rock dusting (inerting the coal dust with limestone or dolomite). These and many other factors contributed to the poor work conditions that continually existed in the Westray mine and made it the site of an incident waiting to happen. All of these substandard conditions and practices could be attributed to the lack of concern that management had toward safety issues in the mine, which was one of the primary root causes of the problem at Westray.

Management system elements (see Chapter 8) that contributed to the Westray explosion include inadequate:

- Management commitment and accountability to safety matters (which is a key element in establishing an effective company safety culture).
- Management of change procedures.
- Incident investigation (including near-miss reporting and investigation).
- Training (orientation, safety, task-related, etc.).
- Task definition and safe work practices and procedures.
- Workplace inspections and more proactive hazard identification methodologies.
- Program evaluation and audits.

System standards that contributed to the Westray explosion include standards (i.e., levels of performance) relating to virtually all the system elements just listed, including inadequate:

- Concern expressed by management toward safety matters (in terms of the standard of care one would reasonably expect and which, from a legal perspective, should be considered mandatory).
- Follow-through on inspections for substandard practices and conditions.
- Action on hazard reports submitted by employees.
- Job instructions for employees.
- Equipment maintenance.
- Scheduling of management/employee meetings to discuss safety concerns.

Compliance factors that contributed to the Westray explosion include:

- Poor correlation between management actions and official company policy concerning the relationship between safety and production (as evidenced by the same management personnel holding responsibility for both production and underground safety).
- Inadequate compliance with industry practice and legislated standards concerning numerous aspects of coal mining, such as methane concentrations, rock dusting, control of ignition sources underground, etc.

This type of analysis, that is, critical assessment of the management-level root causes of an incident, is routinely conducted by the US CSB as described in Section 10.1 for a case study involving hydrogen. Other CSB examples for hydrogen-related incidents are given in Section 10.4 with an emphasis on management-level issues. These additional case study illustrations are consistent with the discussion in Chapter 2 and Section 8.1 concerning the 1989 Pasadena, Texas, vapor cloud explosion in which the management system concepts of hazard assessment and maintenance control were determined to be inadequate.

As also described in Chapter 2, several accident databases have been used to collect and classify past accidents related to hydrogen applications. By definition, such databases are the result of incident descriptions and/or investigation reports, the quality of which can be quite variable (ranging from newspaper articles to public inquiry findings). Database analysis can nevertheless be quite valuable as a learning method; in the words of Sepeda [35]:

> Learning from the experiences of others … is an essential tool since industry has neither the time and resources nor the willingness to experience an incident before taking corrective or preventive steps.

Useful databases include the general Major Hazard Incident Database Service (MHIDAS) [36–38], the material-specific Hydrogen Incident Reporting and Lessons Learned (H2 Incidents)/Hydrogen Safety Best Practices (H2 Best Practices) [10], and the HySafe work package (Chapter 9) Hydrogen Incidents and Accidents Database (HIAD) [39].

Figure 10.3 gives an example of the lessons that can be learned from such databases, in this case, MHIDAS. Here we see the results of the analysis conducted by Gerboni and Salvador [36], who demonstrated the importance of minimizing human error and mechanical failure to prevent accidents involving hydrogen. These findings also illustrate the need for effective process safety management according to the corresponding PSM elements of *human factors* (Section 8.2.7) and *process and equipment integrity* (Section 8.2.6), respectively.

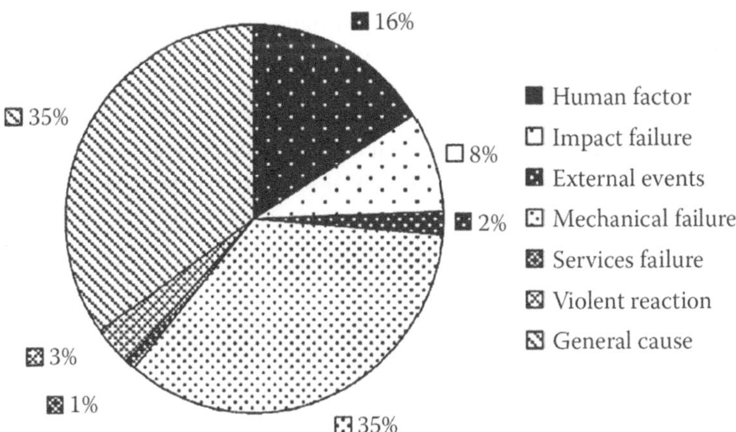

FIGURE 10.3 Causation factors for hydrogen accidents drawn from analysis of MHIDAS. With permission.

Source: [36].

10.4 OTHER EXAMPLES

To conclude this chapter on case studies, examples are now given of incidents involving hydrogen in conjunction with other hazardous materials. We first examine the issue of hydrogen as a flammable gas coexisting with an explosible dust; this is followed by a brief look at the hazards associated with hydrogen being chemically bound with sulfur in the form of hydrogen sulfide.

10.4.1 Hydrogen and Explosible Dusts

A dust explosion can occur when particulate solid material is suspended in air and a sufficiently energetic ignition source is present within a confined or semiconfined environment [33]. In other words, dust explosions occur when the components of the explosion pentagon are all present: fuel, oxidant, ignition source, mixing, and confinement. The following two paragraphs, adapted from Amyotte and Eckhoff [33], provide additional background for the discussion here on explosions of hydrogen and combustible dusts.

Dust explosions usually occur in industry inside process vessels and units such as mills, grinders, and driers, that is, inside equipment where the conditions of the explosion pentagon are satisfied. Such occurrences are often called *primary* explosions, especially if they result in *secondary* explosions external to the process unit (as described in the next paragraph). The reason for the majority of dust explosions being initiated in this manner is relatively straightforward; the range of explosible dust concentrations in air at normal temperature and pressure is orders of magnitude greater than airborne dust concentrations permitted in areas inhabited by workers (i.e., in the context of industrial hygiene).

Notwithstanding the discussion in the previous paragraph, dust explosions do occur in process areas, not just inside process units. A *secondary* explosion can be initiated due to entrainment of dust layers by the blast waves arising from a *primary* explosion. The primary event might be a dust explosion originating in a process unit, or could be any disturbance energetic enough to disperse explosible dust layered on the floor and various work surfaces. An example of such an energetic disturbance (other than a primary dust explosion) would be a gas explosion leading to a dust explosion. This is a well-documented phenomenon in the coal mining industry, where devastating effects can result from the overpressures and rates of pressure rise generated in a coal dust explosion that has been triggered by a methane explosion. (See Section 10.3 and the description of the Westray coal mine explosion.)

10.4.2 Three Accidents at the Hoeganaes Iron Powder Facility

The scenario described in Section 10.4.1 appears to have been realized recently at the Hoeganaes facility in Gallatin, Tennessee. On May 27, 2011, an explosion and ensuing fire resulted in the death of two workers and significant injuries to a third [40]. Two previous incidents involving flash fires (solely iron dust) had occurred at this facility during 2011, resulting in the death of two workers on January 31 and injuries to another on March 29 [40]. As described in the CSB news conference statement [40]

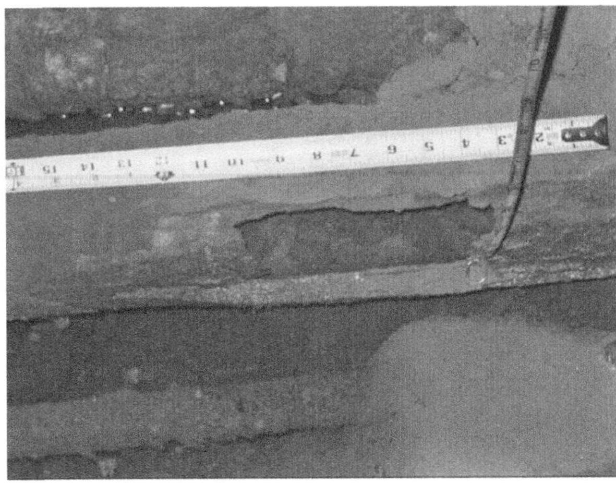

FIGURE 10.4 Photograph of a hole in hydrogen piping at Hoeganaes. Public domain.

Source: [41].

prepared following the May 27 incident, the Gallatin Hoeganaes facility manufactures atomized iron powder for the production of metal parts in the automotive and other industries. Hydrogen is used in the continuous annealing furnaces to prevent oxidation of the iron powder.

The initial explosion in the May 27 incident was caused by the ignition of hydrogen leaking into a process-pipe trench from a 7.6 cm by 17.8 cm hole in a vent pipe [40] (see Figure 10.4). This primary explosion disturbed layered iron dust (Figure 10.5), which subsequently burned in suspension as a flash fire – the secondary event.

On January 5, 2012, the US CSB released its final report on the Hoeganaes Corporation Case Study (No. 2011-4-I-TN) [41]. The report investigated thoroughly all three accidents on January 31, March 29, and May 27, 2011. The partially destroyed factory after the last explosion is shown in Figure 10.6.

In the trench where the hydrogen leak was initiated on the May 27, 2011 incident some more pipes were housed, including nitrogen and hydrogen supply and vent pipes for the band furnaces. In addition to these pipes, the trench also served as a drain for the cooling water used in the band furnaces. At the time of the incident the hot cooling water found its way out of the furnaces and drained onto the pipes and into the trench (Figure 10.7).

Concerning ignition sources, witnesses indicated that the May 27, 2011, hydrogen explosion was ignited by sparks generated during the lifting of the trench cover. This is reasonable considering that the minimum ignition energy (MIE) of hydrogen is 0.02 mJ, and the energy of mechanical sparks from metal-to-metal contact can be several mJ [42].

The testing performed on a contract by Hoeganaes in 2010 found that the minimum ignition energy for similar iron dust samples was greater than 500 mJ. Consequently, the arc generated during the accident was sufficient to ignite the disturbed layered iron dust. As witnessed, at ground level a noise was heard at the time of the incident,

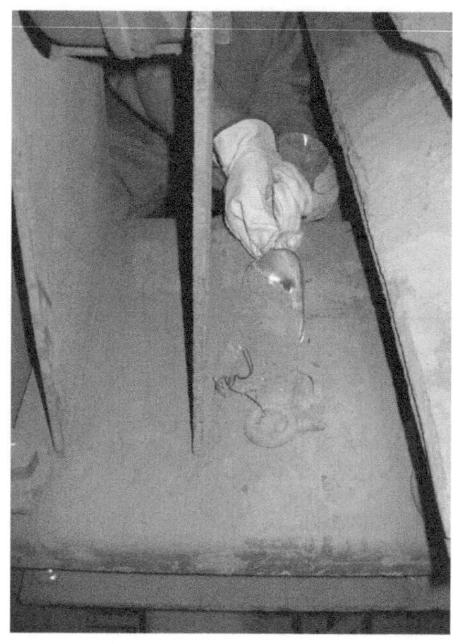

FIGURE 10.5   Photograph of layered iron dust deposits at Hoeganaes. Public domain. Source: [41].

FIGURE 10.6   The partially destroyed factory after the last gas and dust explosion on May 27, 2011. Public domain.

Source: [41]

FIGURE 10.7 Water flow and corroded pipes in the trench involved in the May 27, 2011, incident. Public domain.

Source: [41].

characterized as "electric sound". The motor operating bucket elevator #12 found nearby in the area is thought to have been the source of ignition. This is quite probable because the motor had exposed wiring, was insufficiently grounded, and was near the dust cloud created. Moreover, the electrical conduit supplying power to this motor was not firmly connected to the motor's junction box (Figure 10.8).

The discussion to this point has focused on series-type events, that is, a hydrogen explosion followed by a dust explosion (or fire). What happens when these events occur in parallel, that is, when hydrogen and dust ignition occur simultaneously? This scenario of admixture of a flammable gas and an atmosphere sustaining combustion of a dust is known as the creation of a hybrid mixture. Each component of the mixture (flammable gas and combustible dust) may be present in an amount less than its lower limit of reactivity (lower flammable limit (LFL) for the gas, and minimum explosible concentration (MEC) for the dust), yet still give rise to an explosible mixture [33].

The CSB's final report on the Hoeganaes Corporation Case Study after investigating the process and the incidents made known the key findings, and suggested recommendations.

The *Key Findings* in the CSB's report on the accidents in the Hoeganaes Corporation are the following:

1. *Accumulations of combustible iron powder at the Hoeganaes facility fueled fatal flash fires.*

FIGURE 10.8   Exposed electrical wiring on an elevator motor near the January 2011 incident site. Public domain.

Source: [41].

2. *Hoeganaes facility management were aware of the iron powder combustibility hazard two years prior to the fatal flash fire incidents but did not take necessary measures.*

3. *Hoeganaes did not establish safe procedures or training for employees to avoid flammable gas fires and explosions.*

4. *OSHA did not include iron and steel mills in its Combustible Dust National Emphasis Program when it was first issued in 2007 or when it was re-issued in 2008.*

5. *The 2006 International Fire Code for Combustible Dusts does not enforce the more demanding NFPA standards for the prevention of dust fires and explosions.*

6. *The Tennessee Fire Code and the City of Gallatin do not enforce recommended standards or practices of the IFC.*

7. *The Gallatin Fire Department inspected the Hoeganaes facility after the first two iron powder flash fires but did not refer to combustible dust hazards to avoid the third fatal hydrogen accident.*

8. *The flame-resistant clothing supplied by the company to its employees did not provide substantial protection against dust flash fires and hydrogen explosions.*

9. *GKN and Hoeganaes did not provide sufficient corporate supervision to the Hoeganaes Gallatin facility.*

Based on these findings, the CSB suggested specific, well-defined measures for the company, the city of Gallatin, the Gallatin Fire Department, the International Code Council, and the Metal Powder Producers Association to take.

10.4.3 TESORO REFINING AND MARKETING COMPANY ACCIDENT

According to the CSB Report 2010-08-I-WA [43] released in May 2014, on April 2, 2010, the Tesoro Refining and Marketing Company petroleum refinery in Anacortes, Washington experienced a catastrophic rupture of a heat exchanger in the catalytic reformer/naphtha hydrotreater (NHT) unit. The cause of this accident was the rupture of the heat exchanger following a *high-temperature hydrogen attack (HTHA)*. HTHAs are the damaging effects on carbon steel of hydrogen at high temperatures and pressures. In this accident, hydrogen and naphtha at more than 260°C (500°F) were released from the split-open heat exchanger, ignited, and exploded. A fierce fire followed that burned for more than three hours.

The Material Safety Data Sheet (MSDS) for the naphtha used by Tesoro gives its autoignition temperature as 437°F (225°C). Since the naphtha's temperature inside the heat exchanger exceeded 500°F, autoignition was inevitable. Seven Tesoro employees working near the heat exchanger at the time of the incident were killed. This is the worst fatal accident at a US petroleum refinery after the BP Texas City accident in March 2005 which had 15 fatalities and 180 injuries. When the incident occurred, the process was found in the start-up activity of putting all the heat exchangers back in service after cleaning them. Taking into account the refinery's history of frequent leaks (sometimes followed by fires) during this start-up activity, the CSB was persuaded that this work was nonroutine and rather hazardous. While putting the heat exchangers back in service, the corroded heat exchanger catastrophically ruptured as a result of the carbon steel weakening by the HTHA effect [43].

The HTHA damage mechanism has been proved to occur when carbon steel equipment is attacked by hydrogen at high temperatures and pressures resulting in fissures and cracking that deteriorate the strength of materials. The degradation effect is even more severe in areas where intense stress concentration occurs, for instance at non-post-weld heat-treated welds. This is what happened in the ruptured heat exchanger since HTHA damage was verified in their welds. Furthermore, the selection of the construction material was not the proper one as carbon steel was used instead of the more resistant and inherently safer chromium steels suggested by the American Petroleum Institute (API). The initial constructor and owner of the installations of Tesoro Anacortes Refinery was Shell Oil Company whose design and process operating conditions anticipated fouling inside the heat exchangers during operation. This refers to the development of a buildup of process contaminant byproducts, both inside and outside of the heat exchanger tubes, as shown in Figure 10.9. This fouling hinders heat transfer between the two sides of the tubes thus lowering the heat transfer efficiency [43].

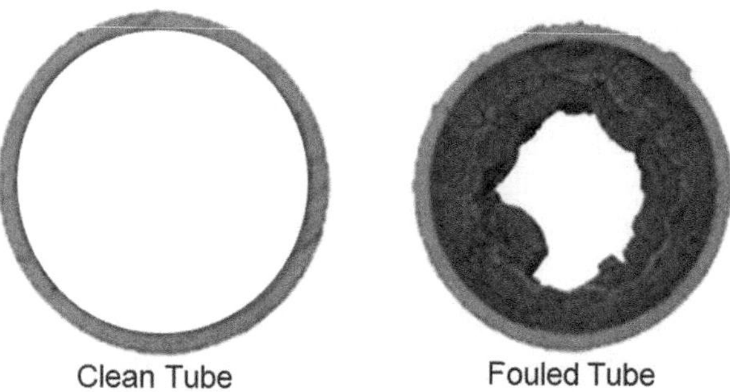

FIGURE 10.9 Example of fouling deposits on the inside of heat exchanger tubes. Fouling greatly reduces heat transfer between the shell-side and tube-side process fluids. Public domain.

Source: [43].

Thus, periodic cleaning of the tubes was necessary to maintain the efficient operation of the distillery. To do this, one bank of heat exchangers would be taken out of service while the spare bank would continue operating. This was a nonroutine alternate operation, which both Shell Oil and Tesoro followed to avoid a total shutdown of the NHT unit. Yet the procedure of putting the heat exchangers back into service has been proven to be hazardous as there was a history of frequent leaks and occasional fires occurring in the company [43].

On the evening of April 1, 2010, Tesoro operators started the alternate operation of the two heat exchanger banks as planned while two leaks from the heat exchangers were reported during the start-up. Despite the leaks, the operators did not stop operations because they were used to such frequent leaks during the start-up of these heat exchangers. It seems that the leaks had become a "normal" part of the start-up. On the other hand, based on their experience, these leaks usually stopped after a while when the heat exchangers reached the typical operating temperature. Finally, at 12:30am on April 2, while the heat exchanger bank start-up operations were in progress, one heat exchanger in proximity to the in-service bank catastrophically ruptured. Both circumferential and longitudinal weld seams of the exchanger failed. The pressure-containing "shell" of the heat exchanger split up at the weld seams, as shown in Figure 10.10, spewing large quantities of hot hydrogen and naphtha. Upon release into the atmosphere, autoignition of naphtha and hydrogen took place, creating a large fireball. All seven outside operators were severely burned, and within 22 days of the incident, all succumbed to their injuries [43].

In accordance with its authority, and towards fostering safer operations at petroleum refineries and protecting workers and communities from future severe accidents both in the state of Washington and nationally, the CSB made in their report extensive safety recommendations to:

- The US Environmental Protection Agency.
- Washington State Legislature, Governor of Washington.

FIGURE 10.10   Post-incident view of D/E/F NHT heat exchanger bank. Public domain.

Source: [43].

- Washington State Department of Labor and Industries – Division of Occupational Safety and Health.
- American Petroleum Institute.
- Tesoro Refining & Marketing Company LLC.
- Tesoro Anacortes Refinery.
- United Steelworkers Local 12-591.

10.4.4 NUCLEAR REACTORS' SAFETY, SECURITY, AND ENVIRONMENTAL CHALLENGES

10.4.4.1 Direct Safety Challenges

With the unfolding of the Fukushima disaster, the world was made aware of the potential for hydrogen explosions in nuclear facilities when exposed zircaloy fuel cladding comes in contact with steam. In a somewhat similar vein, a series of recent papers [44–46] has addressed the issue of combined hydrogen and graphite/tungsten dust explosions in a next-generation fusion machine known as the International Thermonuclear Experimental Reactor (ITER). As described by Chuyanov and Topilski [44], interaction of reactor components over time can be expected to liberate beryllium, graphite, and tungsten dust having characteristic diameters of less than 100 μm. In the event of a water leak leading to contact between water and hot beryllium particles and the subsequent generation of hydrogen, the possibility exists for hydrogen/graphite/tungsten explosions. Thus,

FIGURE 10.11 Transmission electron microscope (TEM) photograph of 4 μm graphite dust.

Source: [45].

although this situation has not occurred, such an incident could further intensify the spotlight on hydrogen safety.

With respect to ITER-related dusts, Figure 10.11 illustrates the lower boundary of graphite particle size (4 μm) examined by Denkevits and Dorofeev [45] in their research on graphite/tungsten explosibility. This work enabled a baseline dataset to be established for graphite dust alone (4–45 μm), tungsten dust alone (1 μm), and graphite (4 μm)/tungsten (1 μm) mixtures. A follow-up study by Denkevits [46] investigated the explosibility of hybrid mixtures of the 4 μm graphite and lean hydrogen–air mixtures having hydrogen concentrations of 8 to 18 vol%. While valuable technical information was gained on the explosion behavior of the various fuel/air systems studied, the broadly applicable lesson to be learned from this case is the importance of credible scenario selection during the performance of risk assessments. Inadequate explosion prevention and mitigation measures could well be the result of neglecting a significant component of the total fuel loading.

Regarding *fusion reactors* in general, there is an almost universal belief that they are the perfect solution to energy issues offering abundant availability of "fuel source" combined with little radioactive waste, no plutonium byproducts, and no runaway chain reactions that can lead to a meltdown. All these constitute the main disadvantages of nuclear fission. In addition, no carbon dioxide emissions will be put into the atmosphere and, consequently, there will be no effect on climate change.

Furthermore, the amount of energy produced from fusion is enormous (four times as much as nuclear fission reactions).

It is well known that, in our sun, two ordinary hydrogen nuclei fuse to create a helium nucleus at tremendous density (due to the gigantic mass of the star) and enormous temperature (due to the huge energy produced). Nevertheless, to accomplish these conditions on Earth demands completely different means since density is millions of times lower than on the Sun. To overcome this problem deuterium and tritium are employed, which are 24 orders of magnitude more reactive than ordinary hydrogen [47].

Although the reactivity of tritium is 20 times higher than that of deuterium-fueled reactions, tritium fuel cannot be fully replenished in fusion reactors and, moreover, it is extremely scarce, as a consequence of its half-life being only 12.3 years. In fact, the main source of tritium is fission nuclear reactors, while deuterium is readily available in water. To overcome this disadvantage a lithium-containing blanket is placed in fusion reactors around the reacting fully ionized hot gas (plasma). This blanket aims at utilizing the neutrons produced by the fusion reaction to produce tritium from the lithium blanket. Theoretically, the tritium consumed in fusion could be fully regenerated to sustain the nuclear reactions, yet this is not possible in real reactors with only 15% overall efficiency due to technical difficulties.

The blanket is usually a liquid lithium-fluoride and beryllium-fluoride (FLiBe) compound in its eutectic composition ($2LiF\text{-}BeF_2$). In addition, due to increased radiation fluxes on the magnets, a highly efficient radiation shield is necessary to properly shield the magnets and the surrounding components. Thus, the lithium-beryllium blanket is beneficial since it acts both as a tritium regenerator and radiation shield [48]. Additional shields have been proposed to reduce the radiation levels in the reactor building. For instance, Kuang et al. [49] proposed a 20 cm thick ZrH_2 blanket specifically for reducing the neutron flux on the mentioned coils.

However, the high neutron energy involved suggests that gamma radiation is also produced, which can become harmful to sensitive functional materials such as superconductive magnets and thermal insulators. It is therefore necessary to evaluate additional backup materials for effectively shielding these two forms of radiation. In addition, ZrH_2 and the more common B_4C tungsten-based ceramics have been proposed and investigated, as well as WC and W_xB_y compounds, which have recently attracted interest in the radiation shielding field [48].

Besides performance, the use of cWCs (cemented tungsten carbide alloys) as shielding rather than metal-based shields is considered safer because of higher resistance to oxidation in the worst-case scenario of exposure of hot reactor surfaces to atmospheric oxygen [50].

Many factors influence the selection of a safe structural material to be used as a breeding blanket. These materials should be compatible with other components and meet blanket thermal, mechanical, and safety performance requirements. Structural integrity of the blanket under both normal and abnormal load conditions is a prerequisite [51].

The *damage in the vessel wall* by the neutron radiation in fusion reactors is expected to be worse than in fission reactors because of the higher neutron energies, thus raising increased safety concerns. The more energetic neutrons knock atoms out of their lattice positions, causing swelling and fracturing of the structure. Furthermore, neutron-induced reactions create large interstitial helium and hydrogen gas pockets that result in swelling, embrittlement, and material fatigue. These phenomena endanger the integrity of the reaction vessel, possibly leading to a catastrophic loss of containment.

The extreme conditions on the first wall inside a nuclear fusion reactor demand carefully selected materials that can resist these high temperatures and pressures, as well as neutron bombardment, without becoming eroded and brittle, which would result in probable loss of containment. Reliable experimental and computational data are valuable to determine the properties of candidate materials to be used. These data are evaluated by International Atomic Energy Agency (IAEA) experts through Coordinated Research Projects and Technical Meetings and are distributed through its free online databases [52].

The problem of neutron-deteriorated materials may be mitigated in fusion reactors by encapsulating the fusion fuel core in a 1 m thick liquid lithium sphere or cylinder. However, this will transform the fuel assemblies into tons of highly radioactive waste that need to be removed and properly disposed of annually from each reactor, which is a severe environmental issue. In fact, the radioactivity per kilogram of waste from a fusion reactor would be much smaller than for fission reactor wastes, but the volume and mass of waste would be many times larger in fusion rectors. New low-activation structural alloys are under development that would allow discarded reactor materials to be considered as low-level radioactive waste to be disposed of by shallow land burial [52]. Yet even if these alloys become available on a commercial scale, very few municipalities would gladly accept landfills for low-level radioactive waste.

Regarding the *radiation exposure of fusion plant workers* to neutron radiation, sufficient biological shielding will be needed even when the reactor is off. Remote handling equipment and robots will be required for all maintenance work in the intensely radioactive areas of the installation.

Molten *lithium itself also is a fire and explosion-hazardous material*, a drawback that is common to liquid-metal-cooled fission reactors as well [47]. Furthermore, the extreme operating conditions of fusion reactors are a severe challenge. In the core of the Sun, enormous gravitational pressures enable this to occur at temperatures of around 10 million degrees Celsius. However, possible pressures on Earth are much lower, so the temperatures needed to initiate fusion should be more than 100 million degrees Celsius. For the time being, the energy input required to reach the temperatures and pressures that make possible a significant fusion reaction utilizing hydrogen isotopes far exceeds the fusion energy generated.

From the safety viewpoint, since fusion reactions can only take place under such extreme conditions, a "runaway" chain reaction is impossible. Furthermore, according to the IAEA [52], fusion reactions are self-limiting processes and depend on the continuous input of fuel; if you cannot control the reaction, it stops by itself. Given that a fusion reaction could stop within seconds, *the process is considered inherently safe*. Safety in fusion reactors is ensured by applying the existing safety requirements for fission. Nevertheless, taking into account the fundamental differences in probable accidents, a targeted approach is needed for fusion to avoid overregulation. The standards to be established for fusion should be proportionate with specific risks to make sure that the new technology will be feasible and safe as well.

The First Joint IAEA–ITER Technical Meeting on Safety and Radiation Protection for Fusion, at Treviso, Italy in November 2020, concentrated on methods under development and aimed to determine and propose potential types and amounts of radioactive or hazardous material derived from fusion installations that would be acceptable to dispose of into the environment, drawing at the same time suitable safety documents for fusion [52].

10.4.4.2 Future Prospects of Fusion Technology

Nevertheless, using magnetic confinement and laser-based inertial confinement, we are addressing the problem and aiming to get more energy from a fusion reactor than the amount of energy put in to obtain these extreme conditions, thus producing net energy. A collaborative, multinational project on this topic is the *International Thermonuclear Experimental Reactor (ITER)*, which was initiated in 2010 in France with the first experiments on its fusion machine, called a Tokamak, expected to begin in 2025. ITER is the biggest fusion experimental installation worldwide and has collected specialists from 35 countries to try to make the fusion project workable, facing effectively any safety and security issues [52]. When completed, the ITER fusion reactor will contain the world's largest magnet and the chamber will weigh more than 25,000 tons, at nearly 30 meters tall, withstanding temperatures of up to 302 million degrees Fahrenheit (168 million degrees Celsius).

Another experimental project running is the *Joint European Torus (JET)* in Oxfordshire, England (see Figure 10.12), which co-operates also with the European

FIGURE 10.12 Inside the Joint European Torus fusion reactor in Oxfordshire, England. Public domain. (From EUROfusion, https://euro-fusion.org.)

Community. In the USA, the best known fusion project is a laser-based *Inertial Confinement Fusion (ICF)* research device, at Lawrence Livermore National Laboratory in Livermore, California [47]. SPARC is also a tokamak under development in an experimental project conducted by Commonwealth Fusion Systems (CFS) collaborating with the Massachusetts Institute of Technology (MIT) Plasma Science and Fusion Center (PSFC) [53].

The fusion reaction with a deuterium–tritium mixture is more favorable because its reactivity is 20 times higher than a deuterium–deuterium fueled reaction, and furthermore the former reaction is obtained at one-third of the temperature required for deuterium-only fusion. Nevertheless, since external tritium production is incredibly expensive, it is likely that only fusion reactors fueled solely with deuterium would be feasible from the viewpoint of fuel supply, taking into account also that tritium is produced in fission reactors thus aggravating nuclear proliferation issues.

Another problem with fusion reactors is that a huge electricity-generating industry is needed to support the system and the facilities. About 75 to 100 MWe (megawatts electric) are consumed continuously by parasitic power drains, such as liquid-helium refrigerators, water pumping, vacuum pumping, heating, ventilating, and air-conditioning for numerous buildings, tritium processing, and so forth, as verified by the ITER fusion project in France. Moreover, when the fusion output is stopped for any reason, this energy must be supplied and purchased from the regional grid at retail prices, additionally destabilizing the grid. Thus, to run a fusion power plant economically its size should be more than 1000 MWe to offset the parasitic power drain [47].

On December 13, 2022, physicists at the National Ignition Facility (NIF) at Lawrence Livermore National Laboratory in California announced that they had succeeded in irradiating a tiny fuel pellet containing two hydrogen isotopes with a laser of around 2 MJ of energy, turning the atoms into plasma and producing 3 MJ of energy; that is a 50% increase. Although the scientists responsible are very enthusiastic about the results, others remain skeptical. In fact, there is not really a net gain of energy because parasitic power drain as described by [47] was not taken into account. As Turner [54] put it, this nuclear fusion reactor "breakthrough" is significant, but lightyears away from being useful, adding that *"Useful, cost-effective nuclear fusion remains a distant dream, despite a small step in the right direction from the government's NIF reactor"*.

10.4.4.3 Global Security and Environmental Challenges

With regard to *global security*, there is an issue with nuclear weapons proliferation with the probable utilization of fusion technology. This is because plutonium 239 can be produced in a fusion reactor when natural or depleted uranium oxide is radiated with neutrons emitted from the fusion reactor. Consequently, a reactor operating with deuterium–tritium or deuterium-only will have the capacity to produce many kilograms of tritium, offering chances to produce plutonium for nuclear weapons. Thus, the International Atomic Energy Agency would release safeguards to prevent nuclear weapons proliferation using plutonium produced in future fusion reactors, as it did with fission reactors [55].

Probable accidental dispersion of tritium derived from fusion plants poses *environmental concerns*. Dispersion of radioactive tritium into the atmosphere or local water resources could be the result of corrosion in the heat exchangers or release in the reactor's vacuum ducts. The dispersed tritium in the environment will then create hazardous tritiated water. On the other hand, fission reactors also produce tritium, yet in insignificant quantities compared with fusion reactors. It is remarkable that in Fukushima the release of even very small amounts of tritium resulting from the cooling water of the damaged fission reactors into the sea caused public fright in 2023.

Tritium permeation rates through certain solids have not yet been sufficiently studied. In the USA, the National Nuclear Security Administration has been producing tritium in fission power reactors to be used in nuclear weapons by absorbing neutrons in lithium-containing substitute control rods. It was found that leakage of tritium occurred from the rods into the reactor cooling water which was eventually released to the environment, to the degree that annual tritium production in the USA was drastically reduced, thus causing a decrease in the tritium stockpile for nuclear weapons [56].

Concerning the environment, the water demand of a fusion reactor is higher than any type of thermal power plant, either fossil or nuclear. Thus, considering climate change and the drought conditions already occurring in some regions of the world, many countries would not physically support large fusion reactors.

Ball [57] has also questioned the future of fusion concluding that *"Nuclear fusion won't arrive in time to fix climate change, but it could be essential for our future energy needs"*.

To sum up, the issues of fusion technology mentioned here demonstrate that the claims of its proponents of "unlimited, clean, safe, and cheap energy" are not backed up by any evidence. Thus, according to Jassby [47]: *"Terrestrial fusion energy is not the ideal energy source extolled by its boosters, but to the contrary: It's something to be shunned"*.

REFERENCES

1. Crowl, D.A. and Louvar, J.F., *Chemical Process Safety: Fundamentals with Applications*, 3rd ed., Prentice Hall PTR, Upper Saddle River, NJ, 2011.
2. Kletz, T. and Amyotte, P., *Process Plants: A Handbook for Inherently Safer Design*, 2nd ed., CRC Press/Taylor & Francis Group, Boca Raton, FL, 2010.
3. Stuart, C., Hydrogen leak in Connecticut forces dozens from homes, *New York Times*, February 13, 2008.
4. Leary, W.E., Hopes rise on fixing shuttle leaks so the fleet can resume flying, *New York Times*, July 14, 1990.
5. Broad, W.J., Shuttle astronomy mission is postponed by fuel leak, *New York Times*, September 6, 1990.
6. Leary, W.E., Fuel leak delays launching of space shuttle, *New York Times*, April 5, 2002.
7. Hu, W., Leak at Indian Point 2 plant leads it to curtail operations, *New York Times*, September 12, 2002.
8. International Conference on Hydrogen Safety (ICHS). www.ichs2011.com/themestop ics.htm (accessed August 17, 2011).

9. Sanders, R.E., Designs that lacked inherent safety: Case histories, *Journal of Hazardous Materials,* 104 (1–3), 149–161, 2003.

10. Weiner, S.C., Fassbender, L.L., and Quick, K.A., Using hydrogen safety best practices and learning from safety events, *International Journal of Hydrogen Energy,* 36 (4), 2729–2735, 2011.

11. Kletz, T., *What Went Wrong? Case Histories of Process Plant Disasters and How They Could Have Been Avoided,* 5th ed., Gulf Professional Publishing, Oxford, UK, 2009.

12. Gupta, R.B. (Ed.), *Hydrogen Fuel: Production, Transport, and Storage,* CRC Press/Taylor & Francis Group, Boca Raton, FL, 2009.

13. Hopkins, A., *Safety, Culture and Risk: The Organizational Causes of Disasters,* CCH Australia Limited, Sydney, Australia, 2005.

14. Hopkins, A. (Ed.), *Learning from High Reliability Organisations,* CCH Australia Limited, Sydney, Australia, 2009.

15. Hydrogen Safety for Employees. www.safetyinstruction.com/hydrogen_safety_for_employees_power_point.htm (accessed August 18, 2011).

16. Brown McFarlane Case Studies. www.brownmac.com/media-down-loads/case-studies/?ID=18 (accessed August 18, 2011).

17. U.S. Department of Energy Hydrogen Program. www.hydrogen.energy.gov/permitting/fueling_case_studies_washington.cfm (accessed August 18, 2011).

18. Sanderson, K., Explosion at ammonia plant, *Chemistry World,* June 1, 2006. www.chemistryworld.com/news/explosion-at-ammonia-plant/3001584.article

19. Evans, J., Weapons of mass degradation, *Chemistry World,* 2011.

20. Ramsey, B., Big batteries? Big electric shock potential! *Occupational Health & Safety,* February 1, 2004. https://ohsonline.com/Articles/2004/02/Big-Batteries-Big-Electric-Shock-Potential.aspx?m=1

21. Carson, P. and Mumford, C., Batch reactor hazards and their control, *Loss Prevention Bulletin,* 171, 13–24, 2003.

22. Institute of Chemical Engineers (IChemE). www.icheme.org/sitecore/content/icheme_home/shop/searchresults.aspx?keywords=hydrogen&product=&CurrentPage=1&SortBy=Relevance&OrderBy=Asce (accessed August 18, 2011).

23. U.S. Chemical Safety Board (CSB). www.csb.gov/ (accessed August 18, 2011).

24. U.S. Chemical Safety Board (CSB), *Positive material verification: Prevent errors during alloy steel systems maintenance,* Safety Bulletin, No. 2005-04-B, U.S. Chemical Safety and Hazard Investigation Board, Washington, DC, 2006.

25. Litterick, D., Texas refinery fire inquiry critical of BP, *The Telegraph,* October 17, 2006.

26. Mahnken, G.E., Use case histories to energize your HAZOP, *Chemical Engineering Progress,* 97, 73–78, 2001.

27. Khan, F.I. and Amyotte, P.R., How to make inherent safety practice a reality, *Canadian Journal of Chemical Engineering,* 81 (1), 2–16, 2003.

28. Etowa, C.B., Amyotte, P.R., Pegg, M.J., and Khan, F.I., Quantification of inherent safety aspects of the Dow indices, *Journal of Loss Prevention in the Process Industries,* 15 (6), 477–487, 2002.

29. West, A.S., Hendershot, D., Murphy, J.F., and Willey, R., Bhopal's impact on the chemical industry, *Process Safety Progress,* 23 (4), 229–230, 2004.

30. Willey, R., Hendershot, D., and Berger, S., The accident in Bhopal: Observations 20 years later, *Process Safety Progress,* 26 (3), 180–184, 2007.

31. Louvar, J.F., Editorial. Bhopal, CCPS and 25 years, *Process Safety Progress,* 28 (4), 299, 2009.

32. Bhopal Memorial. http://upload.wikimedia.org/wikipedia/commons/d/d8/Bhopal-Union_Carbide_1.jpg (accessed September 8, 2011).

33. Amyotte, P.R., and Eckhoff, R.K., Dust explosion causation, prevention and mitigation: An overview, *Journal of Chemical Health and Safety*, 17 (1), 15–28, 2010.

34. Richard, K.P., *The Westray story: A predictable path to disaster*, Report of the Westray Mine Public Inquiry, Province of Nova Scotia, Canada, 1997.

35. Sepeda, A.L., Lessons learned from process incident databases and the Process Safety Incident Database (PSID) approach sponsored by the Center for Chemical Process Safety, *Journal of Hazardous Materials*, 130 (1–2), 9–14, 2006.

36. Gerboni, R., and Salvador, E., Hydrogen transportation systems: Elements of risk analysis, *Energy*, 34 (12), 2223–2229, 2009.

37. Imamura, T., Mogi, T., and Wada, Y., Control of the ignition possibility of hydrogen by electrostatic discharge at a ventilation duct outlet, *International Journal of Hydrogen Energy*, 34 (6), 2815–2823, 2009.

38. Astbury, G.R. and Hawksworth, S.J., Spontaneous ignition of hydrogen leaks: A review of postulated mechanisms, *International Journal of Hydrogen Energy*, 32 (13), 2178–2185, 2007.

39. Kirchsteiger, C., Vetere Arellano, A.L., and Funnemark, E., Towards establishing an International Hydrogen Incidents and Accidents Database (HIAD), *Journal of Loss Prevention in the Process Industries*, 20 (1), 98–107, 2007.

40. U.S. Chemical Safety Board (CSB). www.csb.gov/assets/news/document/Final_Statement_6_3_2011.pdf (accessed September 8, 2011).

41. CSB, Chemical Safety and Hazard Investigation Board, *Final Report on the Hoeganaes Corporation Case Study* (No. 2011-4-I-TN), CSB_Case_Study_Hoeganaes_Feb3_300-1, www.csb.gov/hoeganaes-corporation-fatal-flash-fires/ (accessed January, 2024).

42. Babrauskas, V., *Ignition Handbook*, Fire Science Publishers, Issaquah, WA, 2003.

43. CSB News Release, 2010 Tesoro Refinery Fatal Explosion (No. 2010-08-I-WA), Seattle, Washington, January 29, 2014. www.csb.gov/csb-investigation-finds-2010-tesoro-refinery-fatal-explosion-resulted-from-high-temperature-hydrogen-attack-damage-to-heat-exchanger/ (accessed January, 2024).

44. Chuyanov, V. and Topilski, L., Prevention of hydrogen and dust explosion in ITER, *Fusion Engineering and Design*, 81 (8–14), 1313–1319, 2006.

45. Denkevits, A. and Dorofeev, S., Explosibility of fine graphite and tungsten dusts and their mixtures, *Journal of Loss Prevention in the Process Industries*, 19 (2–3), 174–180, 2006.

46. Denkevits, A., Explosibility of hydrogen-graphite dust hybrid mixtures, *Journal of Loss Prevention in the Process Industries*, 20 (4–6), 698–707, 2007.

47. Jassby, D., Fusion reactors: Not what they're cracked up to be, *Bulletin of the Atomic Scientists*, April 19, 2017. https://thebulletin.org/2017/04/fusion-reactors-not-what-theyre-cracked-up-to-be/

48. Segantin, S., Meschini, S., Testoni, R., and Zucchetti, M., Preliminary investigation of neutron shielding compounds for ARC-class tokamaks, *Fusion Engineering and Design*, 185, 113335, 2022.

49. Kuang, A.Q., Cao, N.M., Creely, A.J., et al., Conceptual design study for heat exhaust management in the ARC fusion pilot plant, *Fusion Engineering and Design*, 137, 221–242, 2018.

50. Windsor, C.G., Marshall, J.M., Morgan, J.G., et al., Design of cemented tungsten carbide and boride-containing shields for a fusion power plant, *Nuclear Fusion*, 58, 076014, 2018.

51. Sawan, M.E. and Abdou, M.A., Physics and technology conditions for attaining tritium self-sufficiency for the DT fuel cycle, *Fusion Engineering and Design*, 81, 1131–1144, 2006.

52. IAEA Bulletin, *Fusion Energy*, International Atomic Energy Agency, May 2021. www.iaea.org/bulletin

53. Creely, A.J., Greenwald, M.J., Ballinger, S.B., Brunner, D., Canik, J., Doody, J., Fülöp, T., Garnier, D. T., Granetz, R., Gray, T.K., and Holland, C., Overview of the SPARC tokamak. *Journal of Plasma Physics*, 86, 865860502, 2020.

54. Turner, B., Nuclear fusion reactor 'breakthrough' is significant, but light-years away from being useful, *LiveScience*, December 2022. www.livescience.com/fusion-ignit ion-scientists-skeptical-explained?fbclid=IwAR3WxtRu701BGIDFYsDz6Mcke9Uea bFlg_MvocY_wRU7Jb0HhdTsgsLj0Ys

55. Kalinowski, M.B. and Colschen, L.C., International control of tritium to prevent horizontal proliferation and to foster nuclear disarmament, *Science & Global Security*, 5, 131–203, 1995.

56. GAO (U.S Government Accountability Office), *Nuclear weapons: National Nuclear Security Administration needs to ensure continued availability of tritium for the weapons stockpile*, GAO-11-100, Released: October 07, 2010. www.gao.gov/produ cts/gao-11-100

57. Ball, Ph., *What is the future of fusion energy?* Scientific American, June 1, 2023. www. scientificamerican.com/article/what-is-the-future-of-fusion-energy/

11 Effects of Hydrogen on Materials of Construction

Chapter 4 presented in detail the primary physiological, physical, and chemical hazards of hydrogen. In this chapter we return briefly to the issue of the physical hazards of hydrogen from the perspective of the subsequent effects on materials of construction. To recapitulate, Chapter 4 categorized these physical hazards as being related to either hydrogen's small molecular size or its storage at generally low temperatures. Two key issues arise from these factors: degradation of structural materials because of hydrogen embrittlement, and loss of thermal stability, respectively.

From a functional point of view, broad advice on the effects of hydrogen on materials of construction is given in a number of references. A self-study training guide on hydrogen gas safety prepared by Los Alamos National Laboratory [1] reiterates the point that due to its small molecular size, hydrogen is able to readily pass through porous materials and be absorbed by some containment materials (the end result being loss of ductility, or embrittlement, at a rate that is temperature dependent). The guide therefore recommends the selection of materials for use with hydrogen that are not subject to degradation by hydrogen embrittlement (such as 300-series stainless steels, copper, and brass) [1]. Similar guidance is given by Molkov [2], who cautions on the decarburization (decrease in carbon content) and embrittlement of mild steels caused by exposure to hydrogen at elevated temperature and pressure. Selection of appropriate materials such as special alloy steels is critical to address the issue of hydrogen embrittlement [2].

Metals are not the only structural materials for which hydrogen-related issues may arise. As noted in the HySafe work package on material compatibility and structural integrity (see Section 9.2 and reference [3]), hydrogen components and hydrogen systems also involve nonmetallic materials such as polymers. HySafe researchers have recommended that materials intended for hydrogen service be carefully evaluated for use under the design, operating, and emergency conditions to which they will be exposed [3]. This includes storage conditions such as solid hydrides based on nanostructured carbonaceous materials [3] for which hazard uncertainties exist [3, 4].

DOI: 10.1201/9781003313007-11

All of the points mentioned earlier in this introductory section are consistent with the main materials-related findings given in a key report produced by the Health and Safety Laboratory in the United Kingdom [5]:

- Primary material considerations with hydrogen usage are permeability, hydrogen embrittlement, and properties of materials at low (cryogenic) temperatures typically involved in the storage and use of liquid hydrogen.
- Increased hydrogen prominence will likely result in the use of new materials, for example, high-performance composites for pressurized hydrogen cylinders, combinations of materials for pipelines, and materials such as metal hydrides and carbon nanotubes for hydrogen storage.
- Materials proposed for use with hydrogen must be tested in the environment in which they will actually be employed. (Reference [5] notes that this requirement may necessitate the use of specialist equipment and techniques.)

This report by Hobbs [5] is also of interest because it extends the discussion on materials-related effects to considerations beyond the physical hazards of hydrogen. By drawing on a NASA (US National Aeronautics and Space Administration) document [6] (also reference [6] in Chapter 4), Hobbs [5] lists 12 aspects to be considered when selecting materials for use with hydrogen. The following excerpted points are as relevant today as they were in 2005 [5] and also in 1997 [6]; account should be taken of:

- Properties suitable for the design and operation conditions.
- Compatibility with the operating environment.
- Availability of selected materials and appropriate test data.
- Corrosion resistance.
- Ease of fabrication, assembly, and inspection.
- Consequences of failure.
- Toxicity.
- Hydrogen embrittlement.
- Potential for exposure to high temperature from a hydrogen fire.
- Cold embrittlement.
- Thermal contraction.
- Property changes that occur at cryogenic temperatures.

In this list we see not only the familiar topics of hydrogen embrittlement and cryogenic-temperature property changes, but also considerations related to material availability and failure consequence analysis. These are important issues with respect to inherently safer design substitutions (Section 7.3) and process risk management (Section 8.2.4). From a general process safety management perspective, and again in accordance with the discussion in Chapter 4, Hobbs [5] comments that these points apply to all hydrogen-service components, not just the more obvious ones such as containment vessels. Such components include valve bodies and seats, electrical systems, insulation, gaskets, seals, tubing, lubricants, and adhesives [5].

11.1 HYDROGEN EMBRITTLEMENT

Chapter 4 covered in some detail various elements of hydrogen embrittlement, including types, causes, and rate-influencing parameters. Here we present information of an applied, example-based nature aimed at a general understanding of hydrogen embrittlement impacts on constructional materials.

The article by Still [7] is helpful in this regard, with a particular emphasis on the offshore oil and gas industry; the author uses the example of a floating, production, storage, and offloading (FPSO) vessel consisting of piping systems connecting subsea reservoirs to onboard processing equipment. Two main sources of hydrogen ingress are described as being problematic for carbon and carbon-manganese steel pipelines and weld metals: (1) atomic hydrogen ingress resulting from the processing of reservoir fluids, and (2) hydrogen introduction into the weld pool during welding of process piping because of the presence of moisture [7].

Still [7] describes in detail the various types of hydrogen embrittlement shown in Table 4.1, including *environmental hydrogen embrittlement* (termed HSC or hydrogen stress cracking). Also elucidated is a form of *internal hydrogen embrittlement* (Table 4.1) known as a *fisheye* (Figure 11.1). Fisheyes are described as a type of hydrogen embrittlement occurring in as-welded structures, wherein hydrogen is retained within a weld defect such as a pore [7]. Still [7] emphasizes the need to adhere to standardized welding controls to avoid this form of hydrogen failure.

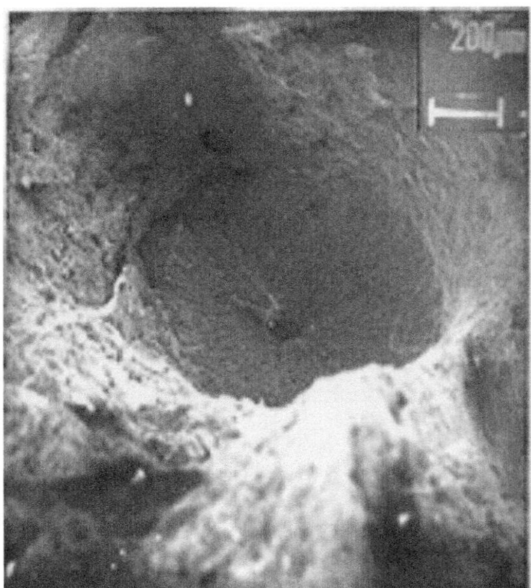

FIGURE 11.1 Example of a fisheye, a form of hydrogen embrittlement. With permission. Source: [7].

An example of *hydrogen reaction embrittlement* (Table 4.1) is given in the CSB safety bulletin [8] describing an incident at the BP Texas City refinery. This incident was previously discussed in Sections 7.6, 8.2, and 10.1 from the perspectives of inherently safer design principles, safety management systems, and case studies, respectively. Here the focus is on the material damage caused by high-temperature hydrogen attack (HTHA).

The following excerpt from reference [8] provides incident background and consequence details:

> On July 28, 2005, 4 months after a devastating incident in the Isomerization (Isom) unit that killed 15 workers and injured 180, the BP Texas City refinery experienced a major fire in the Resid Hydrotreater Unit (RHU) that caused a reported $30 million in property damage. One employee sustained a minor injury during the emergency unit shutdown and there were no fatalities.
>
> The RHU incident investigation determined that an 8-inch diameter carbon steel elbow inadvertently installed in a high-pressure, high-temperature hydrogen line ruptured after operating for only 3 months. The escaping hydrogen gas from the ruptured elbow quickly ignited.
>
> This incident occurred after a maintenance contractor accidentally switched a carbon steel elbow with an alloy steel elbow during a scheduled heat exchanger overhaul in February 2005. The alloy steel elbow was resistant to high-temperature hydrogen attack (HTHA) but the carbon steel elbow was not. Metallurgical analyses of the failed elbow concluded that HTHA severely weakened the carbon elbow steel.

Figure 11.2 illustrates the potential that existed at this facility for human error to occur with respect to elbow replacement. As discussed in Section 8.2.7, this

Upper Left & Top Arrow – Alloy steel elbows 2 and 3
Lower Left Arrow – Carbon steel elbow 1

FIGURE 11.2 Heat-exchanger piping elbows with identical dimensions. With permission. **Source:** [8].

potential is most effectively countered by human factor considerations during process design – in this case by a design that avoids configurations that allow critical alloy piping components to be interchanged with incompatible piping components [8].

The CSB report [8] remarks that the occurrence of incidents involving HTHA has been documented since the 1940s. It is known that carbon steel is vulnerable to HTHA at pressures above 100 psia (7 bar(a)) and temperatures above about 450°F (230°C); these conditions promote the permeation of hydrogen through the steel and the reaction of hydrogen with carbon and carbides to produce methane gas [8]. This loss of carbon is known as decarburization (as described earlier [2]) and results in steel degradation and eventually, in the case of this particular incident, pipeline rupture [8]. These effects of hydrogen reaction embrittlement (in particular HTHA) are shown in Figure 11.3.

Thus, the HTHA damage mechanism occurs when steel equipment is exposed to hydrogen at high temperatures and partial pressures as described in the CSB final investigation report [9] on the 2010 Tesoro Refinery fatal explosion in Seattle, Washington (see Section 10.4.3). The resulting damage severely degrades the mechanical properties of the carbon steel [10]. HTHA occurs when atomic hydrogen diffuses into the steel walls of process equipment and reacts with carbon in the steel, thus removing carbon (decarburization) and rendering the wall more fragile through the following mechanism [11, 12].

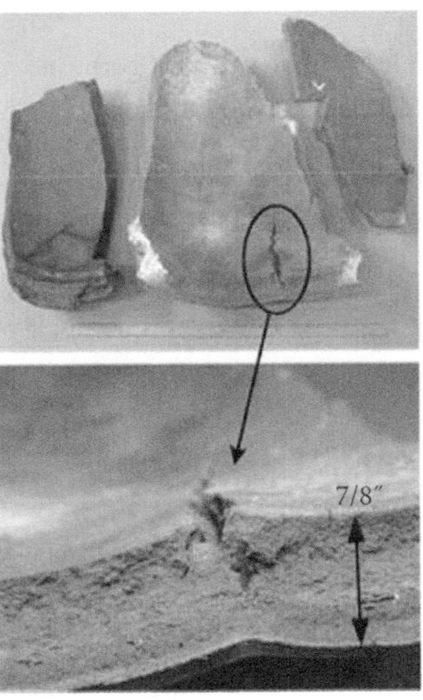

FIGURE 11.3 Ruptured carbon steel pipe elbow subjected to high-temperature hydrogen attack.

Source: [8].

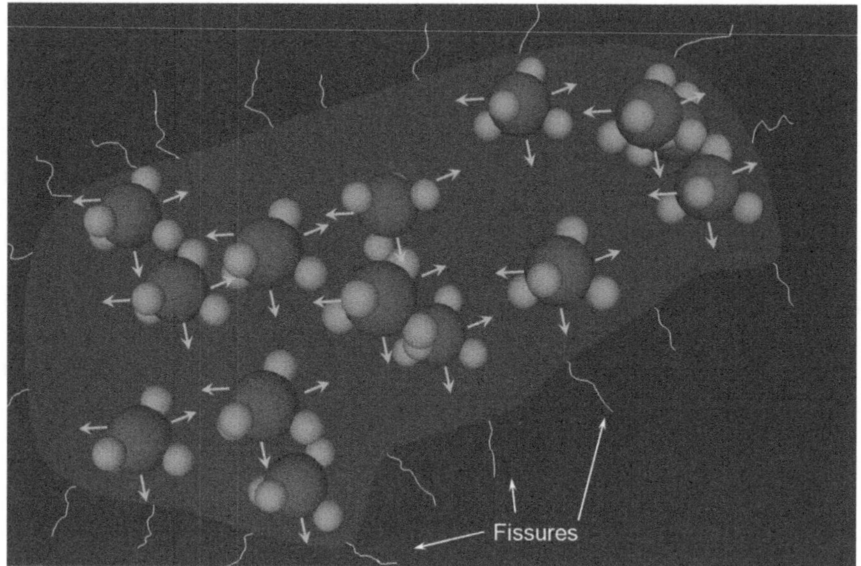

FIGURE 11.4 Methane fissures. When methane molecules cannot diffuse out of the steel, they accumulate inside of the steel, exerting high pressure that forms fissures in steel. Public domain.

Source: [9].

Methane, a much larger molecule than atomic hydrogen, cannot permeate and diffuse out of the steel plate. Thus, it accumulates inside the vessel walls [10] pressing the surrounding steel. The more methane gas is formed, the more methane pressure increases. The high pressure inside the steel due to methane gas can form fissures, as illustrated in Figure 11.4, or blisters in the steel, as shown in Figure 11.5 [13].

As more fissures are formed, they can link, forming microcracks in the steel [14]. The microcracks can then link to form larger cracks (Figure 11.6). These cracks greatly weaken the steel leading to rupture of the vessel. This process occurred in one heat exchanger at the Tesoro Anacortes Refinery.

11.2 UNDERSTANDING HYDROGEN EMBRITTLEMENT OF STEELS

In a review of experimental, modeling, and design progress, Barrera et al. [15] make an effort to understand and mitigate hydrogen embrittlement of steels. Despite the significant research effort to understand the mechanisms of hydrogen embrittlement and to develop mitigating measures, the relevant mechanisms are still not well comprehended. In the literature, there are debatable theories regarding the in-depth mechanisms. There is also debate concerning experimental studies supporting each of these theories. Barrera et al. [15] provided a detailed review of the current state of the art of the impact of hydrogen on metals, particularly on steels. Their study extends

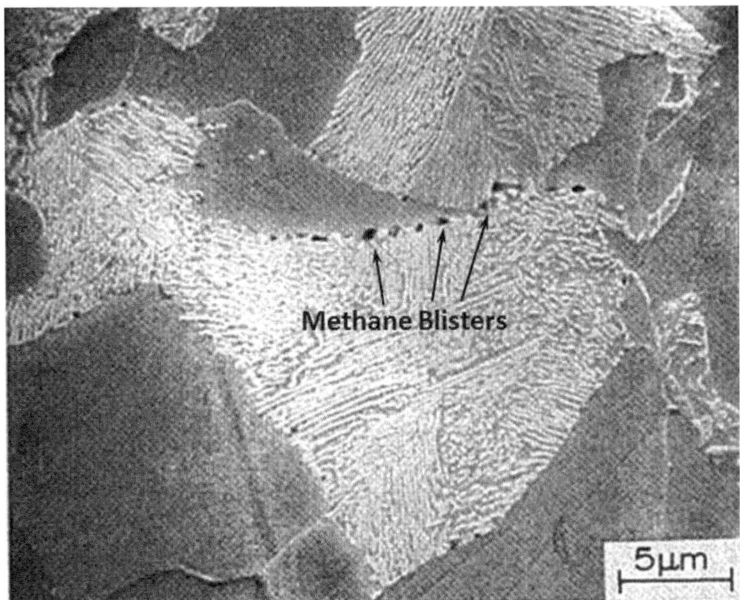

FIGURE 11.5 Methane blisters. Accumulation of methane in steel can also form blisters in the metal. Public domain.

Source: [9].

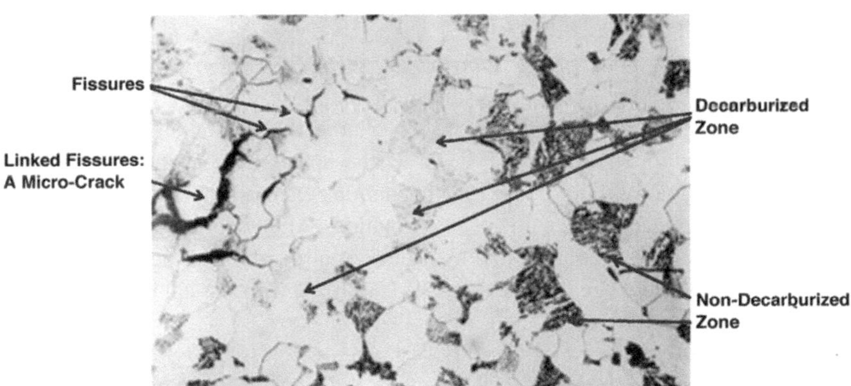

FIGURE 11.6 Microcrack resulting from linked HTHA fissures. This image from API RP 941 (Steels for Hydrogen Service at Elevated Temperatures and Pressures in Petroleum Refineries and Petrochemical Plants) shows fissures formed as a result of HTHA linked together to form a microcrack. Decarburized regions appear lighter in color (because of an absence of carbon) than unaffected regions. Public domain.

Source: [9].

from the atomistic to the continuum scale citing theoretical works based on quantum calculations, experimental methods, and macroscopic phenomena that have an impact on the mechanical properties of steels, thus establishing damaging mechanisms for the embrittlement of steels. They finally describe current approaches and novel mitigation strategies utilized to design new steels resistant to hydrogen embrittlement.

In this detailed review, Barrera et al. [15] describe the following mechanisms: hydrogen-induced decohesion (HID), hydrogen-enhanced localized plasticity (HELP), hydrogen-induced phase transformation (HIPT), and hydrogen-enhanced strain-induced vacancy (HESIV) formation. A brief description of the adsorption-induced dislocation emission (AIDE) is also included in the paper.

The *hydrogen-induced decohesion (HID) mechanism* was initially proposed by Gerberich et al. [16] in the 1970s in an attempt to explain the increase in crack-tip-opening angle, as a result of reduced cleavage toughness with increasing hydrogen content in steels. The HID mechanism proved successful in explaining the brittle intergranular fracture surface observed in high-strength steels. Nevertheless, direct experimental methods cannot explain the observed hydrogen-induced decohesion.

The *hydrogen-enhanced localized plasticity (HELP) theory* [17] states that hydrogen embrittlement results from the increased mobility of dislocations due to the presence of hydrogen. Yet this phenomenon is not in accordance with the well-established observation of ductility enhancement in metals with increased dislocation mobility. The HELP theory claims that hydrogen has a protective effect on the elastic stress field of dislocations, thus enhancing dislocation mobility and slip localization [17, 18]. The mechanism of failure by HELP has been verified in many steels by macroscopic flow stress measurements, fractographic evidence, in situ transmission electron microscopy (TEM) studies, measurements of dislocation motion, and theoretical treatments [19–25].

The *hydrogen-induced phase transformation (HIPT) theory* claims that face-centered cubic (FCC) alloys such as austenitic stainless steels are characterized by a hydrogen supersaturation under hydrogen charge that can alter to an important degree the structure of the surface layer of the specimens. In this phenomenon, known as HIPT, two types of phase transformations exist: (a) hydride formation and (b) hydrogen-induced martensitic transformation. Microscopic observations and thermodynamic analysis confirm that hydride formation and cleavage fracture occur at sites where hydrides are stable or can be stabilized with the application of stress [24, 26].

The *hydrogen-enhanced strain-induced vacancy (HESIV) formation theory* proposes that hydrogen enhances the density and clustering of vacancies. Then, vacancies tend to merge to form microvoids, which in turn may conjoin to form larger voids resulting in lower ductile crack growth resistance [27]. Although HESIV alone does not present a convincing mechanism of embrittlement, Neeraj et al. [28] have proposed a combination mechanism of plasticity-generation (HELP), hydrogen-enhanced vacancy formation (HESIV), and nanovoid coalescence to explain the fracture mechanism of X65 and X80 steels which failed by quasi-brittle fracture.

Adsorption-induced dislocation emission (AIDE) mechanism. Based on the remarkable similarities between hydrogen-assisted cracking, stress corrosion cracking, and adsorption-induced liquid-metal embrittlement, Lynch [29] proposed the theory of hydrogen-assisted cracking caused by the adsorption of hydrogen at the crack tip.

The widely cited belief that high-strength steels are more susceptible to hydrogen embrittlement than simpler steels is opposed by an interesting work which demonstrates that hydrogen embrittlement also has an impact on low-strength pure iron[30].

11.3 HYDROGEN EMBRITTLEMENT MITIGATION IN STEELS

To mitigate hydrogen embrittlement of metals and alloys it is imperative to limit the quantity of hydrogen in the material. This can be accomplished by either preventing hydrogen ingress or by enhancing the resistance within the metallic microstructure [15]:

a) *The prevention or limitation of hydrogen ingress* in the material can be obtained by applying a hydrogen-resistant coating, with methods such as electroplating or hot-dip galvanizing processes. Various efficient coatings have been proposed in the literature, such as cadmium or Zn–Ni coating, or the on-site formation of resistant oxides, such as "black oxide" magnetite (Fe_3O_4) [31], TiO_2 [32], Al_2O_3, and Si_3N_4, [33–35]. Nitriding of the metal surface layer has also been successfully used in austenitic stainless steels [36, 37]. The electrolytic plating process produces hydrogen, so an annealing process is required to remove hydrogen from the surface of the austenitic stainless steels, with annealing times of 8–24 h, at 200°C [38]. Other electroplating materials proposed in the literature to provide a barrier for hydrogen ingress are Zn and Zn–Ni. However, there have been contradictory findings about their effectiveness [39].

b) *The introduction of an intrinsic resistance within the microstructure* is needed when the application of a coating is not effective. In this case, an intrinsic resistance may be considered during the stage of designing a material's microstructure. Degassing treatment to release hydrogen from the material is also an option [40–42]. Takasawa et al. [43] and Hejazi et al. [44] have also experimentally found that a reduction in dislocation density and grain refinement is effective in reducing hydrogen embrittlement vulnerability.

Despite the recent significant progress by both experimental works and theoretical modeling, there are many hard questions still open. The crucial issue is that there are arguments within the scientific community concerning knowledge gaps with regard to mechanisms that are responsible for hydrogen embrittlement [15].

In a recent review, Poorhaydari [45] discusses the current state of knowledge in the areas which are still in the development stage. Significant uncertainty remains around such questions as whether decarburization is a form of HTHA, or whether the incubation period is composed of detectable and non-detectable stages of attack, and these are discussed in American Petroleum Institute, Recommended Practice (API RP) 941 (Steels for Hydrogen Service at Elevated Temperatures and Pressures in Petroleum Refineries and Petrochemical Plants).

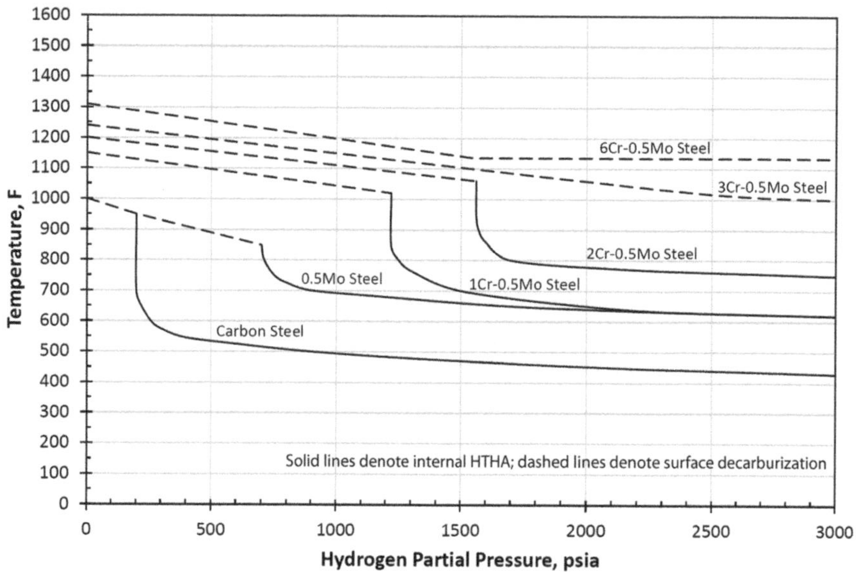

FIGURE 11.7   The first published operating limit curves by Nelson, redrawn by Poorhaydari from the original graph. With permission from Springer Nature BV.

Source: [45].

The least resistant materials to HTHA are carbon steel as well as the low-alloy steels and certain medium-alloy and high-alloy steels. Austenitic stainless steels are the most resistant. Some non-ferrous alloys, such as nickel alloys (e.g., K-Monel, Inconel, and Hastalloy B) and copper alloys, are also vulnerable to high temperatures and high hydrogen pressure[45].

Assessment of the susceptibility of a metal to HTHA is possible using the *Nelson curves*. These curves were first published in 1949 by Nelson [46] and were adopted in 1970 by the American Petroleum Institute as API publication 941 [11]. The first published operating limit curves by Nelson were redrawn from the original graph by Poorhaydari [45] and are depicted in Figure 11.7. Nelson commented that the limits in his curves are to be used for design purposes.

The terminology related to HTHA which is adopted by Poorhaydari [45] is as follows:

- *Hydrogen Damage: This is the broadest term that refers to mechanical property deterioration of metals by hydrogen through various forms and mechanisms. It includes (low-temperature) hydrogen embrittlement, hydrogen-induced blistering/cracking, HTHA, and hydride formation (in certain alloys).*
- *Hydrogen Embrittlement (HE): This refers to low-temperature (i.e., typically less than 100°C and mostly at ambient temperature) loss of ductility and/or cracking in metals/alloys by the actions of dissolved hydrogen in the atomic form (i.e., excluding embrittlement/damage upon reaction and*

combination of a dissolved hydrogen atom with itself, with impurities, or with metal atoms). It covers reversible embrittlement and cracking that are induced during manufacturing and develop prior to service as well as those that develop in service.

- *Hydrogen Attack (HA): This was the most prevalent term in the twentieth century for irreversible high-temperature hydrogen damage. However, the term was also used occasionally for low-temperature HE in aqueous corrosive environments, which was confusing.*
- *High-Temperature, High-Pressure Hydrogen Attack: This was a transitional term between HA and HTHA that was used to exclude the low-temperature HE.*
- *Hot Hydrogen Attack (HHA): Another transitional term that was used towards the end of the twentieth century to exclude the low-temperature HE.*
- *High-Temperature Hydrogen Attack (HTHA): This has been the most prevalent term in the twenty-first century for irreversible high-temperature hydrogen damage.*

11.4 LOSS OF THERMAL STABILITY

The discussion in Chapter 4 on loss of thermal stability focused on low-temperature embrittlement and thermal contraction at cryogenic temperatures. Hobbs [5] reiterates the familiar concept that because of the low temperatures at which liquid hydrogen is stored, special attention must be paid to the properties of the materials used for such storage. A primary concern is the possibility of transition from ductility to brittleness as temperature decreases. This was addressed quantitatively in Chapter 4 by the Charpy impact data shown in Figure 4.1; these data clearly demonstrate the preferential use of 304 stainless steel over carbon steel and 201 stainless steel for low-temperature hydrogen service.

The importance of proper material selection for operations involving low temperatures is amply demonstrated by the Longford gas plant explosion in Australia. Although not a liquid hydrogen storage incident, as an additional case study of lessons to be learned from process incidents the treatment of Longford by sociologist Andrew Hopkins [47] is a key resource for understanding the potential consequences of a loss of thermal stability (from the perspectives of both low-temperature embrittlement and thermal contraction).

It should be noted that the focus of Hopkins's analysis is on incident causation (September, 1998, Esso Australia's gas plant at Longford in Victoria, Australia) due to organizational failures such as the safety management system and safety culture deficiencies [9]. Readers are directed to reference [47] for a full exposition of this analysis and the resulting generalized lessons. For our purposes here, the following excerpt provides sufficient detail on loss of thermal stability [47]:

Because the circulation of warm lean oil had stopped, two of the heat exchangers became abnormally cold and a thick layer of frost formed on their exterior pipework. The temperature dropped below the design limit and *the metal in one exchanger contracted to the point that it began to leak oil onto the ground.*

Unsuccessful attempts were made to fix this leak by tightening certain bolts. Operators decided to stop the flow into GP 1 [Gas Plant 1] at this point to try to deal with the situation. This stopped any further flow of cold condensate within the plant. But operators did not depressurize the plant. Rather they decided to try again to restart the pumps to rewarm the heat exchanger. This was a critical error. *The metal in the vessel was by this time so cold that it was brittle* and it needed time to thaw out before being rewarmed. Operators succeeded in restarting the pumps and the reintroduction of warm liquid caused fracturing and catastrophic failure of one of the heat exchangers. A large quantity of volatile liquid and gas escaped and was ignited by a nearby ignition source.

The passages emphasized with italics (author added) in this excerpt give practical examples of, respectively, thermal contraction and low-temperature embrittlement (cold-temperature or cold-metal embrittlement as it is termed in reference [47]). The ultimate consequences of these physical phenomena occurring at Longford include two fatalities, extensive asset damage, and significant business interruption for the plant owner and the surrounding communities.

11.5 ONGOING RESEARCH

While practical information such as that derived from case studies is important in reducing hydrogen storage risks, fundamental research is essential to elucidate corrosion mechanisms and embrittlement reaction pathways. Obtaining a better understanding of such features is the first step in designing enhanced prevention and mitigation techniques.

To conclude, Table 11.1 was developed from a search of the archival publications *Corrosion Science* and the *International Journal of Hydrogen Energy*. Table 11.1 is

TABLE 11.1

Examples from the Literature of Research on the Effects of Hydrogen on Material Properties

Reference	Research Focus
Smiyan et al. [48]	Causes and mechanism of initiation and growth of corrosion sites on the internal surface of the shell of high-pressure steam boilers.
Torres-Islas et al. [49]	Effects of hydrogen on the mechanical properties of X70 pipeline steel under different heat treatments.
Rogante et al. [50]	Intercrystalline and intergranular crack detection in narrow welded zones of pipelines.
Nikiforov et al. [51]	Corrosion resistance of commercially available stainless steels, nickel-based alloys, titanium, and tantalum under conditions corresponding to those in high-temperature proton exchange membrane (PEM) steam electrolyzers.
Figueroa and Robinson (2008b) [52]	Feasibility of zinc/nickel and aluminum-based coatings as replacements for cadmium in steel-coating processes.

TABLE 11.1 (Continued)
Examples from the Literature of Research on the Effects of Hydrogen on Material Properties

Reference	Research Focus
Michler et al. (2008b) [53]	Measurement of environmental hydrogen embrittlement of several heats of austentitic stainless steels by slow strain rate tensile testing at various combinations of temperature and pressure.
Kittel et al. [54]	Investigation of hydrogen-induced cracking of pipeline steel by means of immersion testing and hydrogen permeation measurements.
Capelle et al. [55]	Assessment of hydrogen absorption abilities of X52, X70, and X100 pipeline steels through consideration of cathodic hydrogen charging, time of exposure, and applied stress.
Yao et al. [56]	Effects of hydrogen on the fracture stress of passive films formed on 316L stainless steel.
Venegas et al. [57]	Resistance to hydrogen-induced cracking of low-carbon pipeline steel samples by control of crystallographic texture and grain-boundary distribution.
Li et al. [58]	Effect of pre-strain on hydrogen embrittlement of high-strength steels was investigated by slow strain-rate tensile (SSRT) testing.
Alvaro et al. [59]	Hydrogen embrittlement susceptibility of a weld-simulated X70 heat-affected zone under hydrogen pressure.
Laliberté-Riverin et al. [60]	Internal hydrogen embrittlement of pre-cracked, cadmium-plated AISI 4340 high-strength steel with sustained load tests and incremental step-loading tests.
Hong et al. [61]	Effect of hydrogen and strain rate on nanoindentation creep of austenitic stainless steel.
Kim et al. [62]	Carbon effect on hydrogen diffusivity and embrittlement in austenitic stainless steels.
Zhang et al. [63]	Twinning behavior and hydrogen embrittlement of a pre-strained twinning-induced plasticity (TWIP) steel.
Luo et al. [64]	Interactions of nanoprecipitates, metastable austenite, and Lüders banding in an ultra-low carbon medium manganese steel with abnormal strength dependence of hydrogen embrittlement.
Wang [65]	Influence of microstructure on hydrogen embrittlement in hot-rolled medium manganese steels.

not intended as a comprehensive listing, but rather as a snapshot of some of the recent and ongoing research related to properties of materials for use with hydrogen. This is clearly an important area of hydrogen safety that will benefit from continued effort as new opportunities for hydrogen usage arise and new materials for hydrogen storage emerge.

REFERENCES

1. Basquin, S., and Smith, K., *Hydrogen gas safety. Self-study*, Document No. ESH13-401-sb-8/00, Los Alamos National Laboratory, Los Alamos, NM, 2000.
2. Molkov, V., Hydrogen safety research: State-of-the-art, in *Proceedings of the 5th International Seminar on Fire and Explosion Hazards*, Edinburgh, UK, April 23–27, 2007. www.academia.edu/74112001/Hydrogen_Safety_Research_State_Of_The_Art
3. *HySafe – Safety of Hydrogen as an Energy Carrier*, www.hysafe.net/ (accessed September 22, 2011).
4. Amyotte, P.R., Are classical process safety concepts relevant to nanotechnology applications? *Journal of Physics: Conference Series,* 304 (Nanosafe2010: International Conference on Safe Production and Use of Nanomaterials), 012071, 2011.
5. Hobbs, J., *The hydrogen economy: Evaluation of the materials science and engineering issues*, HSL/2006/59, Health and Safety Laboratory, Buxton, UK, 2005.
6. NASA, *Safety standard for hydrogen and hydrogen systems*, National Aeronautics and Space Administration, Report NSS 1740.16, Office of Safety and Mission Assurance, Washington, DC, 1997.
7. Still, J.R., Understanding hydrogen failures, *Welding Journal*, American Welding Society, January 2004. www.aws.org.wj/jan04/still_feature.html (accessed September 22, 2011).
8. U.S. Chemical Safety Board (CSB), *Positive material verification: Prevent errors during alloy steel systems maintenance*, Safety Bulletin, No. 2005-04-B, U.S. Chemical Safety and Hazard Investigation Board, Washington, DC, 2006.
9. *CSB Final Report on 2010 Tesoro Refinery Fatal Explosion and Fire*, Seattle, Washington, January 29, 2014. www.csb.gov/tesoro-anacortes-refinery-fatal-explosion-and-fire-/ (accessed January, 2024).
10. Shih, H.M. and Johnson, H.H. A model calculation of the Nelson curves for hydrogen attack, *Acta Metallurgica*, 30, 537–545, 1982.
11. API Technical Report 941, *Steels for Hydrogen Service at Elevated Temperatures and Pressures in Petroleum Refineries and Petrochemical Plants,* 8th ed., API Publishing Services, Washington, DC, February 2016.
12. Weiner, L.C., Kinetics and mechanism of hydrogen attack of steel, *Corrosion*, 17, 109–115, 1961.
13. Allen, R.E., Jansen, R.J., Rosenthal, P.C., and Vitovec, F.H., The rate of irreversible hydrogen attack of steel at elevated temperatures, in *26th Midyear Meeting of AIChE*, May 9, 1961.
14. Lai, G., *High Temperature Corrosion and Materials Applications*, ASM International, Materials Park, 2007.
15. Barrera, O., Bombac, D., Chen, Y., Daff, T.D., Galindo-Nava, E., Gong, P., Haley, D., Horton, R., Katzarov, I., Kermode, J.R., Liverani, C., Stopher, M., and Sweeney, F., Understanding and mitigating hydrogen embrittlement of steels: A review of experimental, modelling and design progress from atomistic to continuum, *Journal of Materials Science*, 53, 6251–6290, 2018.
16. Gerberich, W.W., Oriani, R.A., Lji, M.J., Chen, X., and Foecke, T., Necessity of both plasticity and brittleness in the fracture thresholds of iron, *Philosophical Magazine. A, Physics of Condensed Matter, Defects and Mechanical Properties*, 63 (2), 363–376, 1991.
17. Birnbaum, H.K. and Sofronis, P., Hydrogen-enhanced localized plasticity—a mechanism for hydrogen-related fracture, *Materials Science Engineering A,* 176 (1),

191–202, 1994. https://doi.org/10.1016/0921-5093(94)90975-X. www.sciencedirect.com/science/article/pii/092150939490975X

18. Robertson, I., The effect of hydrogen on dislocation dynamics, *Engineering Fracture Mechanics*, 68 (6), 671–692, 2001.

19. Han, J., Nam, J.H., and Lee, Y.K., The mechanism of hydrogen embrittlement in intercritically annealed medium Mn TRIP steel, *Acta Materialia*, 113, 1–10, 2016.

20. Hatano, M., Fujinamo, M., Arai, K., Fuji, H., and Nagumo, M., Hydrogen embrittlement of austenitic stainless steels revealed by deformation microstructures and strain-induced creation of vacancies, *Acta Materialia*, 67, 342–353, 2014.

21. Koyama, M., Tasan, C.C., Akiyama, E., Tsuzaki, K., and Raabe, D., Hydrogen-assisted decohesion and localized plasticity in dual-phase steel, *Acta Materialia*, 70, 174–187, 2014.

22. Laureys, A., Depover, T., Petrov, R., and Verbeken, K., Microstructural characterization of hydrogen induced cracking in trip-assisted steel by EBSD, *Materials Characterization*, 112, 169–179, 2016.

23. Robertson, I.M. and Birnbaum, H.K., An HVEM study of hydrogen effects on the deformation and fracture of nickel, *Acta Metallurgica*, 34 (3), 353–366, 1986. https://doi.org/10.1016/0001-6160(86)90071-4

24. Shih, D.S., Robertson, I.M., and Birnbaum, H.K., Hydrogen embrittlement of a titanium: in situ TEM studies, *Acta Metallurgica*, 36 (1), 111–124, 1988. https://doi.org/10.1016/0001-6160(88)90032-6

25. You, Y., Teng, Q., Zhang, Z., and Zhong, Q., The effect of hydrogen on the deformation mechanisms of 2.25Cr–1Mo low alloy steel revealed by acoustic emission, *Materials Science Engineering A*, 655, 277–282, 2016.

26. Flanagan, T., Mason, N., and Birnbaum, H., The effect of stress on hydride precipitation, *Scripta Metallurgica*, 15(1):109–112, 1981. https://doi.org/10.1016/0036-9748(81)90148-

27. Nagumo, M., Hydrogen related failure of steels—A new aspect, *Materials Science Technology*, 20(8), 940–950, 2004.

28. Neeraj, T., Srinivasan, R., and Li, J., Hydrogen embrittlement of ferritic steels: Observations on deformation microstructure nanoscale dimples and failure by nanovoiding, *Acta Materialia*, 60, 5160–5171, 2012.

29. Lynch, S.P., A fractographic study of hydrogen-assisted cracking and liquid–metal embrittlement in nickel, *Journal of Materials Science*, 21(2), 692–704, 1986. https://doi.org/10.1007/BF01145543

30. Pfeil, L.B., The effect of occluded hydrogen on the tensile strength of iron, *Proceedings of the Royal Society A: Mathematical, Physical and Engineering Sciences*, 112(760), 182–195, 1926. https://doi.org/10.1098/rspa.1926.0103

31. Hurd, R. and Hackerman, N., Kinetic studies on formation of black-oxide coatings on mild steel in alkaline nitrite solutions, *Journal of the Electrochemical Society*, 104(8), 482–485, 1957. https://doi.org/10.1149/1.2428631

32. Bates, J.B., Wang, J.C., and Perkins, R.A., Mechanisms for hydrogen diffusion in Tio2, *Physical Review B*, 19, 4130–4139, 1979. https://doi.org/10.1103/PhysRevB.19.4130

33. Forcey, K., Ross, D., and Wu, C., The formation of hydrogen permeation barriers on steels by aluminizing, *Journal of Nuclear Materials*, 182, 36–51, 1991. https://doi.org/10.1016/0022-3115(91)90413-2

34. Tamura, M. and Eguchi, T., Nanostructured thin films for hydrogen-permeation barrier. *Journal of Vacuum Science & Technology A Vacuum Surface and Films*, 33(4), 041503, 2015. https://doi.org/10.1116/1.4919736

35. Yu, G.C. and Yen, S., Hydrogen diffusion coefficient of silicon nitride thin films, *Applied Surface Science*, 201 (1-4), 204–207, 2002. https://doi.org/10.1016/S0169-4332(02)00934-0

36. Michler, T., Influence of plasma nitriding on hydrogen environment embrittlement of 1.4301 austenitic stainless steel, *Surface Coatings Technology*, 202 (9), 1688–1695, 2008a. https://doi.org/10.1016/j.surfcoat.2007.07.036

37. Murray, G.T., Bouffard, J.P., and Briggs, D., Retardation of hydrogen embrittlement of 17–4ph stainless steels by nonmetallic surface layers, *Metallurgical and Materials Transactions A*, 18(1), 162–164, 1987. https://doi.org/10.1007/BF02646236

38. Zamanzadeh, M., Allam, A., Kato, C., Ateya, B., and Pickering, H., Hydrogen absorption during electrodeposition and hydrogen charging of sn and cd coatings on iron, *Journal of the Electrochemical Society*, 129(2), 284–289, 1982. https://doi.org/10.1149/1.2123813

39. Figueroa, D. and Robinson, M., The effects of sacrificial coatings on hydrogen embrittlement and re-embrittlement of ultra high strength steels, *Corrosion Science*, 50(4), 1066–1079, 2008a. https://doi.org/10.1016/j.corsci.2007.11.023

40. Pressouyre, G.M., Trap theory of hydrogen embrittlement, *Acta Metallurgica*, 28(7), 895–911, 1980. https://doi.org/10.1016/0001-6160(80)90106-6

41. Pressouyre, G.M. and Bernstein, I.M., A quantitative analysis of hydrogen trapping, *MTA*, 9(11), 1571–1580, 1978. https://doi.org/10.1007/bf02661939

42. Yamasaki, S. and Bhadeshia, H., M 4 C 3 precipitation in Fe–C–Mo–V steels and relationship to hydrogen trapping, *Proceedings of the Royal Society A: Mathematical, Physical and Engineering Sciences*, 462 (2072), 2315–2330, 2006. https://doi.org/10.1098/rspa.2006.1688

43. Takasawa, K., Ikeda, R., Ishikawa, N., and Ishigaki, R., Effects of grain size and dislocation density on the susceptibility to high-pressure hydrogen environment embrittlement of high-strength low-alloy steels. *International Journal of the Hydrogen Energy*, 37(3), 2669–2675, 2012. https://doi.org/10.1016/j.ijhydene.2011.10.099

44. Hejazi, D., Haq, A., Yazdipour, N., Dunne, D., Calka, A., Barbaro, F., and Pereloma, E., Effect of manganese content and microstructure on the susceptibility of X70 pipeline steel to hydrogen cracking, *Materials Science and Engineering: A*, 551:40–49, 2012. https://doi.org/10.1016/j.msea.2012.04.076

45. Poorhaydari, K., A comprehensive examination of high-temperature hydrogen attack— A review of over a Century of Investigations, *Journal of Materials Engineering and Performance*, 30, 7875–7908, 2021.

46. Nelson, Hydrogenation plant steels, *American Petroleum Institute—Proceedings*, 29 (Sec 3), 163–172, 1949.

47. Hopkins, A., *Lessons from Longford. The Esso Gas Plant Explosion*, CCH Australia Limited, Sydney, Australia, 2000.

48. Smiyan, O.D., Grigorenko, G.M., and Vainman, A.B., Effect of hydrogen on corrosion damage of metal of the high-pressure energetic boiler drum, *International Journal of Hydrogen Energy*, 27 (7–8), 801–812, 2002.

49. Torres-Islas, A., Salinas-Bravo, V.M., Albarran, J.L., and Gonzalez-Rodriguez, J.G., Effect of hydrogen on the mechanical properties of X-70 pipeline steel in diluted $NaHCO_3$ solutions at different heat treatments, *International Journal of Hydrogen Energy*, 30 (12), 1317–1322, 2005.

50. Rogante, M., Battistella, P., and Cesari, F., Hydrogen interaction and stress-corrosion in hydrocarbon storage vessel and pipeline weldings, *International Journal of Hydrogen Energy*, 31 (5), 597–601, 2006.

51. Nikiforov, A.V., Petrushina, I.M., Christensen, E., Tomas-Garcia, A.L., and Bjerrum, N.J., Corrosion behaviour of construction materials for high temperature steam electrolysers, *International Journal of Hydrogen Energy*, 36 (1), 111–119, 2011.

52. Figueroa, D., and Robinson, M.J., The effects of sacrificial coatings on hydrogen embrittlement and re-embrittlement of ultra high strength steels, *Corrosion Science*, 50 (4), 1066–1079, 2008h.

53. Michler, T., Yukhimchuk, A.A., and Naumann, J., Hydrogen environment embrittlement testing at low temperatures and high pressures, *Corrosion Science*, 50 (12), 3519–3526, 2008b.

54. Kittel, J., Smanio, V., Fregonese, M., Garnier, L., and Lefebvre, X., Hydrogen induced cracking (HIC) testing of low alloy steel in sour environment: Impact of time of exposure on the extent of damage, *Corrosion Science*, 52 (4), 1386–1392, 2010.

55. Capelle, J., Dmytrakh, I., and Pluvinage, G., Comparative assessment of electrochemical hydrogen absorption by pipeline steels with different strength, *Corrosion Science*, 52 (5), 1554–1559, 2010.

56. Yao, Y., Qiao, L.J., and Volinsky, A.A., Hydrogen effects on stainless steel passive film fracture studied by nanoindentation, *Corrosion Science*, 53 (9), 2679–2683, 2011.

57. Venegas, V., Caleyo, F., Baudin, T., Espina-Hernandez, J.H., and Hallen, J.M., On the role of crystallographic texture in mitigating hydrogen-induced cracking in pipeline steels, *Corrosion Science*, In press, 2011. doi:10.1016/j.corsci.2011.08.031.

58. Li, X., Wang, Y., Zhang, P., Li, B., Song, X., and Chen, J., Effect of pre-strain on hydrogen embrittlement of high strength steels, *Materials Science & Engineering A*, 616, 116–122, 2014.

59. Alvaro, A., Olden, V., Macadre, A., and Akselsen, O.M., Hydrogen embrittlement susceptibility of a weld simulated X70 heat affected zone under H2 pressure, *Materials Science & Engineering A*, 597, 29–36, 2014.

60. Laliberté-Riverin, S., Bellemare, J., Sirois, F., and Brochu, M., Internal hydrogen embrittlement of pre-cracked, cadmium-plated AISI 4340 high strength steel with sustained load tests and incremental step-loading tests, *Engineering Fracture Mechanics*, 223, 106773, 2020.

61. Hong, Y., Zhou, C., Zheng, Y., Zhang, L., Zheng, J., and Chen, X., Effect of hydrogen and strain rate on nanoindentation creep of austenitic stainless steel, *International Journal of Hydrogen Energy*, 44, 1253–1262, 2019.

62. Kim, K.-S., Kang, J.H., and Kim, S.J., Carbon effect on hydrogen diffusivity and embrittlement in austenitic stainless steels, *Corrosion Science*, 180, 109226, 2021.

63. Zhang, C., Yu, H., Zhi, H., Antonov, S., and Su, Y., Twinning behavior and hydrogen embrittlement of a pre-strained twinning-induced plasticity (TWIP) steel, *Corrosion Science*, 192, 109791, 2021.

64. Luo, H., Li, Y., Zhong, S., Xu, C., and Jia, X., Interactions of nanoprecipitates, metastable austenite, and Lóders banding in an ultra-low carbon medium Mn steel with abnormal strength dependence of hydrogen embrittlement, *Materialia*, 22, 101380, 2022.

65. Wang, Z., Xu, J., and Li, J., Influence of microstructure on hydrogen embrittlement in hot-rolled medium Mn steels, *Materials Science & Engineering A*, 780, 139147, 2020.

12 Future Requirements for Hydrogen Safety

In this chapter we present some thoughts on areas requiring further research to more fully realize the anticipated benefits of the hydrogen economy. In this regard, Guy [1] notes that there are several issues concerning safe handling that must be successfully addressed before hydrogen usage gains wide public acceptance. These constraints have been described by Dahoe and Molkov [2] as involving both technical and non-technical barriers. Nontechnical issues include the perception that hydrogen is a dangerous substance with a propensity for creating fire and explosion hazards; technical issues relate to the requirement to ensure hydrogen technologies have the same level of safety as those based on fossil fuels [2].

Salvi [3] extends the latter point in his description of the European Union's "Grand Challenges", one of which concerns sustainable energy and the greening of transport. According to Salvi [3], with italics added for emphasis:

> The safe development of clean energy alternatives [requires] that several problems have to be solved regarding the technology itself but also the integration of the new technology in the present systems; the greening of transport has to be treated with the same systemic view, e.g. the existing underground infrastructures have to be adapted to the new energy vehicles using *hydrogen*, batteries or compressed natural gas.

The following sections outline hydrogen research needs from several different perspectives: public safety, occupational safety, and process safety. Section 12.1 is a brief review of the literature, followed, in Section 12.2, by a discussion of research needs that we have identified.

12.1 HYDROGEN SAFETY RESEARCH GAPS IDENTIFIED IN THE LITERATURE

In their commentary on hydrogen knowledge gaps, Pasman and Rogers [4] consider two primary sources of information: (1) the Hydrogen Research Advisory Council working under the umbrella of the Fire Protection Research Foundation (linked to the US National Fire Protection Association, or NFPA), and (2) the International Energy Agency (IEA) under the Hydrogen Implementing Agreement.

DOI: 10.1201/9781003313007-12

The NFPA-linked initiative identified 27 items, with the following nine topics being the most urgent, in order of priority (as described by Pasman and Rogers [4]):

- Refinement of explosion models for flame speeds, blast waves, etc.
- Development and evaluation of wide-area sensing technology.
- Effects on materials of construction (in particular, fatigue loading).
- Hydrogen gas cabinets (for storage of pressurized-gas cylinders).
- Deflagrations with partial confinement.
- Reliability of pressure-relief devices.
- Strategies for mitigation of confined releases.
- Design, installation, testing, and maintenance of hydrogen detection systems.
- Ignition limits and criteria for large-leak, dynamic scenarios.

Pasman and Rogers [4] further explain that much is unknown about hydrogen ignition probability and its dependence on release conditions and leak characteristics. This is a particularly important concern in risk assessment studies [4], specifically infrastructure safety studies and effectiveness of fire barriers.

The white paper produced under the IEA initiative categorized hydrogen safety requirements into three groups of knowledge gaps described by Pasman and Rogers [4] as related to:

- Existing codes and standards, as well as the ongoing development of such codes and standards.
- Existing risk assessment methods and their application to hydrogen systems.
- Fundamental knowledge (with a link to modeling approaches, including those employing computational fluid dynamics, or CFD).

Additional thoughts on unresolved items related to hydrogen safety can be found in the description of the achievements of the HySafe Network of Excellence (see Chapter 9) by Jordan et al. [5]. The following issues have been identified as requiring further research from a technical/scientific perspective [5]:

- Properties and behavior of cold, liquid hydrogen releases.
- Mitigation strategies in response to accidental release scenarios (e.g., optimization of pressure-relief device placement and operation).
- Ignition phenomena (particularly in relation to modeling ignition probability).
- Impinging and wall-attached jets and jet fires (in relation to safe blowdown conditions).
- Sensor technologies.
- Transitional combustion phenomena under realistic scenarios (e.g., low-temperature, congested environments with nonuniform mixing) and the associated impact on mitigation measures to deal with, for example, the processes of flame acceleration and deflagration-to-detonation transition (DDT) with water sprays.

- Permit requirements for hydrogen vehicle use in confined spaces.
- Fundamental understanding of hydrogen behavior in confined spaces.
- Reference quantitative risk assessment (QRA) methodology for application to scenarios involving garages and tunnels.
- Reference simulation tool for combustion that is widely available to researchers.
- Composite storage and vehicle safety testing strategies.
- Pipeline field tests.

One sees in this list several items in common with the previous two from Pasman and Rogers [4], for example, ignition phenomena and probability modeling, risk assessment methodologies, and sensor/detection systems. Also evident in the list from Jordan et al. [5], as would be expected, is an emphasis on specific HySafe focus areas such as garages, tunnels, and confined spaces in general.

Two recent contributions from Vladimir Molkov of the University of Ulster, United Kingdom, provide further insight into hydrogen safety research needs from a technical perspective. The first is a review of the state of the art of hydrogen safety research [6], which identifies the following broad areas having several unresolved issues:

- Releases.
- Ignitions.
- Jet flames.
- Premixed and partially premixed combustion.
- Deflagration-to-detonation transition.
- CFD for hazard and risk assessment.

Reference [6] provides a detailed breakdown of research needs within each area, and there is considerable agreement between their list of research topics and that given by Jordan et al. [5]. One area of particular importance is the spontaneous ignition of pressurized hydrogen releases [6]. This is also evident in the nature of the submissions to a 2008 special issue on hydrogen safety of the *Journal of Loss Prevention in the Process Industries*; in his preface to the issue, Molkov [7] comments that four of the ten papers are related to the practical problem of spontaneous ignition of sudden high-pressure releases of hydrogen. The remaining papers deal with the familiar topics of hydrogen–air detonations, metal hydride fire incidents, CFD modeling, and hydrogen safety training and education [7] – all issues that have been identified in this section.

The needs identified by other researchers and practitioners that have been summarized here can serve as an entry point to hydrogen safety research for those interested in the field. Although other aspects of hydrogen research dealing with technical feasibility issues are of obvious importance, the general theme of this book has been to illustrate the overarching need for complementary work on the safety of hydrogen as an energy carrier.

12.2 HYDROGEN SAFETY NEEDS: SOME THOUGHTS ON RESEARCH

The concluding chapter of this book presents a brief recapitulation of our thoughts on the need for continued research on hydrogen safety. The process industries have a decades-old track record of handling hydrogen in a manner that attempts to reduce the risk to an acceptable level. While process incidents involving hydrogen have indeed occurred (see Chapter 10), it would seem helpful to look to this industry sector for guidance on future research needs.

Reference [8], on process safety research needs from an industrial perspective, was written to stimulate discussion at a workshop, "Frontiers of Research", organized by Professor Sam Mannan of the Mary Kay O'Connor Process Safety Center at Texas A&M University. The remainder of this chapter is adapted from Amyotte [8], taking hydrogen-specific issues into consideration.

12.2.1 GENERAL PROCESS SAFETY RESEARCH NEEDS

Assuming that one's own process safety research topics represent true industrial needs may seem somewhat self-centered. Perhaps this is a valid assumption, however, when the research is directly supported by industry and funded by granting agencies through strategic, targeted programs. With this reasoning, current research on dust explosions by Amyotte and Khan [9] can be viewed as industrially relevant.

This work is being conducted under four themes: (1) experimentation, (2) modeling, (3) development of risk management protocols, and (4) communication to industry, academia, and the general public. The fuel/air systems being studied experimentally include selected nanomaterials, flocculent (fibrous) materials, and hybrid mixtures (mixtures of a combustible dust and a flammable gas). The modeling work includes phenomenological, thermokinetic, and CFD approaches. The risk management component is an attempt to help advance the field of dust explosion prevention and mitigation from an emphasis on *hazards* (with an accompanying reliance on primarily engineered or add-on safety features) to a focus on *risk* (with an accompanying reliance on hierarchical, risk-based decision-making tools). The final theme is a recognition of the serious problem that exists with respect to effectively communicating the results of dust explosion research in a meaningful manner to various stakeholders.

Content-wise, what likely facilitated the funding of the research described are the following points:

- A choice of fuel/air systems for which there is industry consensus on the need for additional data.
- The coupling of experimental work with modeling research (so as to eventually lessen the very empirical nature of this field).
- A move away from absolute safety measures to relative measures selected on a risk basis.
- An expressed desire to communicate in ways other than journal papers and conference presentations (as important as these dissemination avenues are).

Again, it seems self-centered to promote these points as universal factors for judging what constitutes research having industrial relevance. But it also seems difficult to argue that these points are pertinent only to the case presented here [9] for illustrative purposes.

To broaden the search for industrial process safety research needs, a number of thematic papers on the topic [10–14] were consulted to identify recurrent keywords. Notice was also taken of the session titles from recent editions of the *Global Congress on Process Safety* sponsored by the Center for Chemical Process Safety (CCPS) and the Safety and Health Division of the American Institute of Chemical Engineers (AIChE). This event now consists of the *Loss Prevention Symposium, Process Plant Safety Symposium*, and *CCPS International Conference*. Also helpful in this regard were the *Hazards Symposia* of the Institution of Chemical Engineers and the *Loss Prevention and Safety Promotion in the Process Industries Symposia* of the European Federation of Chemical Engineering.

This procedure resulted in the general keywords shown in the first column of Table 12.1. Papers incorporating these keywords were then identified by searching three archival journals whose scope is primarily related to process safety: (1) the *Journal of Loss Prevention in the Process Industries* (JLPPI, over the period 1988–2011), (2) *Process Safety and Environmental Protection* (PSEP, over the period 1996–2011), and (3) *Process Safety Progress* (PSP, over the period 1993–2011). The entries in the column for each journal in Table 12.1 give the number of papers for which the authors self-reported the given keyword. Table 12.2 repeats this process for a set of more specific keywords that were identified in the thematic papers referenced in the previous paragraph. (Note that discussion of the final column in Tables 12.1

TABLE 12.1

General Process Safety Keywords and Corresponding Journal Papers

Keywords	JLPPI	PSEP	PSP	IJHE
Risk assessment	110	39	21	13
Quantitative risk assessment	20	5	6	7
Safety management *or* Safety management system	41	10	29	1
Accident investigation *or* Incident investigation	7	4	7	0
Human error *or* Human factors	23	12	13	0
Safety culture	6	7	11	0
Performance indicator *or* Key performance indicator	3	2	7	2
Inherent safety *or* Inherently safer design	21	20	14	2
Security *or* Plant security	9	6	1	4
Offshore Safety	12	3	1	0
Reactive chemistry *or* Reactivity hazards	3	1	2	0
Dust explosion	113	10	9	1
Gas explosion	64	12	9	1

Note: JLPPI, *Journal of Loss Prevention in the Process Industries* (1988–2011); PSEP, *Process Safety and Environmental Protection* (1996–2011); PSP, *Process Safety Progress* (1993–2011); IJHE, *International Journal of Hydrogen Energy* (1976–2011), as discussed in Section 12.2.2.

TABLE 12.2
Specific Process Safety Keywords and Corresponding Journal Papers

Keywords	JLPPI	PSEP	PSP	IJHE
HAZOP	18	13	8	1
FMEA	1	0	0	2
Fault tree	13	6	2	3
Event tree	5	4	0	1
Bow-tie	3	4	2	0
Bayesian	9[a]	3	0	1

Note: JLPPI, *Journal of Loss Prevention in the Process Industries* (1988–2011); PSEP, *Process Safety and Environmental Protection* (1996–2011); PSP, *Process Safety Progress* (1993–2011); IJHE, *International Journal of Hydrogen Energy* (1976–2011), as discussed in Section 12.2.2.

[a] All of these papers were published in 2006 or later, with the exception of Kirchsteiger [15].

and 12.2, IJHE or the *International Journal of Hydrogen Energy,* is deferred until Section 12.2.2.)

Comparisons among the data shown in Tables 12.1 and 12.2, as well as any conclusions drawn from the data, are fraught with potential pitfalls, such as those related to the following questions:

- Because a topic is the subject of extensive research, does that necessarily mean the topic is reflective of an actual industry need? While some of the papers appearing in JLPPI, PSEP, and PSP are written by industrial practitioners, many submissions to these journals come from researchers based in academia and government centers.
- What is the impact of special issues that are devoted to a particular topic? For example, JLPPI has published special issues consisting of papers presented at the *International Symposia on Hazards, Prevention, and Mitigation of Industrial Explosions* (ISHPMIE); these papers deal almost exclusively with gas and dust explosions. This undoubtedly helps to explain the high number of JLPPI papers listed in the last two rows of Table 12.1 and limits the usefulness of any attempts to normalize the data in Tables 12.1 and 12.2.
- Should one conclude that a topic showing low activity is not as applicable to industry as one having more papers that cite the topic as a keyword? As one example, is reactive chemistry/reactivity hazards research less important to industry than human error/human factors research? Such a conclusion drawn solely from the data in Table 12.1 would be speculative at best. At worst, such a conclusion could direct research effort in a manner that is counterproductive.

In spite of the difficulties just articulated, two key industry needs can be postulated and put forward for discussion based on Tables 12.1 and 12.2. First, continued research is required in the following five broad areas: (1) risk assessment (both qualitative and quantitative), (2) safety management systems (including various elements such as incident investigation and human factors), (3) safety culture (including the concept of key performance indicators or KPIs), (4) inherently safer design (including security as well as safety issues, and offshore as well as land-based facilities), and (5) material hazards (as determined by chemical reactivity, flammability, and explosibility). Second, both long-standing methodologies (e.g., fault trees) and newer techniques (e.g., Bayesian networks; see footnote to Table 12.2) have a role to play in advancing the body of knowledge within these five broad research areas.

The following additional points are offered concerning industry needs with respect to process safety research:

- Industry needs (or rather, *wants*) process safety analysis tools that are simple to use. The phrase *simple to use* should not be confused with *simplistic*. Scientific and engineering rigor are expected in all process safety research efforts. This may be an obvious statement, but it is one that bears repeating.
- In determining industry needs, we should consider case histories of process-related incidents. Of significant value in this regard are investigation reports produced by the United States Chemical Safety and Hazard Investigation Board (US Chemical Safety Board, or CSB). A recent analysis of these reports by Amyotte, MacDonald, and Khan [16] focused on the lessons to be learned with respect to inherently safer design and safety management systems.
- Interdisciplinary considerations are key to future successes in meeting industry's process safety research needs. The process industries have been moderately successful with the incorporation of chemistry as practiced by industrial chemists into the domain of chemical/process engineering. (Much, however, remains to be done in this regard; witness the continuing need for the greater participation of chemistry researchers in the field of inherently safer design.) Elements of other disciplines such as game theory and multi-attribute decision making have also been adopted by process safety researchers. What has not been as widely adopted in process safety research are some of the research findings from the social sciences. Significant in this regard is the work of sociologist Andrew Hopkins (e.g., [17–20]).

This last bullet point touches on an issue that is arguably at the heart of process safety research needs: *Are technical solutions enough, or are more people-centered considerations required?* Consider two of the slides (adapted) from the presentation by Hendershot et al. [11]:

- What next? Future challenges.
- Globalization of the chemical industry.
  - Establishing safety culture in developing countries.
- Continued economic pressures – the need to do more with fewer resources.
- Maintaining a good process safety culture and management system at a time of frequent mergers, acquisitions, divestitures, and other business environment changes.

- Chemical process safety in industries other than the traditional chemical process industries (biotechnology, electronics, food, pharmaceuticals, etc.).
- Complacency.
  - Do some people think process safety is a problem that has been solved?
  - Will good experience threaten the programs that are responsible for that good experience?
- Education and awareness in the broader industry community.
- Examples are reactivity hazards, dust explosions.

The terms *globalization, culture, economics, management, complacency, education, awareness* all represent concepts that are clearly important to industry. But are these concepts firmly captured within the skill set of process safety researchers with a primary background in science and engineering? This is suggested as food for thought in contemplating the process safety research needs of the twenty-first century.

12.2.2 HYDROGEN-SPECIFIC PROCESS SAFETY RESEARCH NEEDS

Is there reason to expect that process safety research needs specific to hydrogen are different from those in general? We would argue that the answer is no – at least at the macroscopic level. There will always be material hazard issues, such as those identified in Section 12.1, that a generic review (Section 12.2.1) cannot capture. The points raised in Section 12.2.1 are, however, consistent with the extensive analysis in previous chapters (e.g., Chapter 7 on inherently safer design and Chapter 8 on safety management systems).

The last columns in Tables 12.1 and 12.2 show the results of process safety keyword searches in the *International Journal of Hydrogen Energy*. It is encouraging to see some level of activity in most of the areas surveyed, albeit typically at a lower level than in the three process safety journals. Two thoughts emerge:

- Should the non-process-oriented hydrogen industries attempt to learn more from the process industries in terms of transferable research knowledge? (Witness, for example, the comment that hydrogen distribution systems including refueling stations should be built with *inherently safer* features [4].)
- Is it time for an archival, peer-reviewed journal devoted solely to hydrogen safety?

12.3 RECENT RESEARCH ON HYDROGEN SAFETY

12.3.1 BIBLIOMETRIC APPROACH

Zhang et al. [21] performed an advanced bibliometric analysis extending to future research progress on the safety of hydrogen energy. Bibliometrics is a statistical approach that applies statistical tools to investigate literature in a specific field [22–24]. Databases such as Web of Science (WoS), Scopus, the Lens, and others are widely used by researchers. Every database has its pros and cons but WoS is advantageous mainly when searching for recent literature [25–27]. Concerning the accuracy of journal categorization, it was found that WoS is more precise in comparison to Scopus [28]. In addition, when compared to Scopus, WoS has a 7% lower error rate in document type tagging [29]. Taking into account all these criteria, Zhang et al. [21] chose the WoS database for their bibliometric analysis. Their paper deals with hydrogen safety and the authors have investigated 8283 research papers in this field, analyzing the 30 years from 1992 to 2022. The aspects covered by their analysis include countries of publication, affiliations of authors, frequently cited journals, research hot spots, keywords, and future trends.

12.3.2 ACCIDENT DATABASES REGULATIONS AND STANDARDS

Many national and international organizations have created hydrogen accident databases, simultaneously conducting analyses of hydrogen accidents to promote hydrogen safety. For instance, according to hydrogen accident statistics provided by the US Department of Energy, hydrogen accidents occur quite often in laboratories. Equipment malfunctions, such as pipeline ruptures and components/valves, are the main causes of hydrogen accidents. Considering fatality rates, hydrogen accident frequency is double that of natural gas accidents according to Yang et al. [30]. In Europe, databases serve as a reference for governments to enhance safety, emergency preparedness, and land planning. Moreover, Europe's databases aid stakeholders in analyzing hydrogen accidents, thus designing preventive measures, mitigating risks, and promoting safe market growth [31].

In addition, regulations and standards on hydrogen safety are created in various countries, aiming to enhance safe hydrogen utilization and promoting the entry of hydrogen technologies into the market[32, 33].

12.3.3 ANALYSIS OF PUBLICATIONS

Zhang et al. [21] found in their bibliometric analysis a limited number of papers related to hydrogen safety published before the year 2000, as shown in Figure 12.1. The annual global circulation of relevant papers during that period was less than 50, while the number of authors dealing with hydrogen safety annually was less than ten. The total citations these papers received per year were below 1000. From 2000 to 2008, the number of publications increased slightly, and only in 2006 did the number of published papers exceed 100 for the first time. This increase was probably caused by the recognition that the contribution of the transportation industry was important to energy consumption and environmental pollution [34]. Moreover, there were in that period extensive discussions worldwide regarding hydrogen as the new renewable and clean energy source [35]. Consequently, governments as well as businesses invested considerable resources in the research and promotion of hydrogen technologies [35, 36].

Nevertheless, hydrogen promotion began to decline by 2010 due to challenges such as the cost of hydrogen production, hydrogen storage problems, and lack of infrastructure. This resulted in a decreasing interest in hydrogen utilization, and thus the number of researchers in 2011 was significantly lower compared to 2010, as shown in Figure 12.1. On the other hand, due to frequent accidents in hydrogen

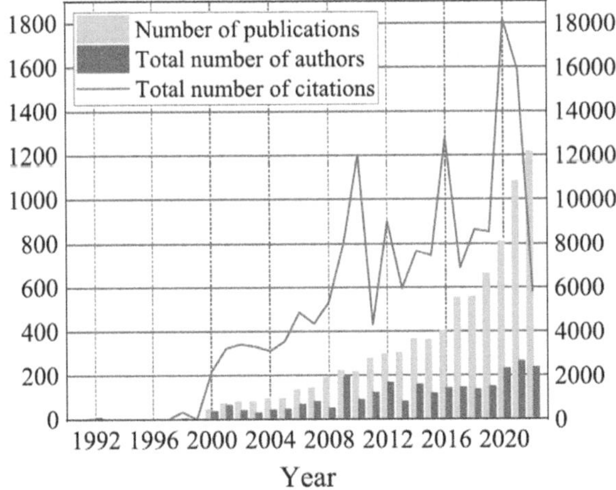

FIGURE 12.1 Chart of the total number of papers, number of authors, and citation counts on the topic of hydrogen safety from 1992 to 2022. With permission from Elsevier.

Source: [21].

energy installations, some countries increased their research funding on hydrogen safety [37–39]. Yet, from 2017 to 2022, there was an important increase in research activity with the average annual number of publications surpassing 500 research papers. The study performed by Zhang et al. [21], revealed that 4876 research papers were published during this period, accounting for 58.8% of the total 8283 papers from 1992 to 2022.

12.3.4 TOP TEN COUNTRIES RANKED BY NUMBER OF PUBLICATIONS

China ranks first with 2414 published papers on the topic of hydrogen safety, which is 29.14% of the total 8232 published papers. The United States is second with 1493 published papers (18.02% of the total 8283 papers). Germany ranks third with 634 papers and (7.65% of the total 8283 papers) followed by Japan, France, the United Kingdom, South Korea, Italy, India, and Canada. Analyzing the nationality distribution of the top ten researchers in terms of publication output, they come from China, Northern Ireland, Germany, the Netherlands, Italy, and Greece[21].

12.3.5 ANALYSIS OF PUBLISHING JOURNALS AND RESEARCH AREAS

The *International Journal of Hydrogen Energy* is the leading scientific journal with 931 published articles, possessing an admirable impact factor of 7.139. *Nuclear Engineering and Design* ranks second with 149 published papers and an impact factor of 1.900. The *Journal of Loss Prevention in the Process Industries* is third with 139 articles and an impact factor of 3.916. It is notable that the *Journal of Power Sources* and the *Applied Energy* journal, although they have not published many papers on the topic of hydrogen safety, are very influential in hydrogen safety research. These journals have published important articles with many citations. The *Journal of Power Sources* has an impact factor of 9.794, while *Applied Energy* is proud of an impact factor of 11.446. Both impact factors demonstrate the significant influence of these journals in the ongoing scientific research on hydrogen safety [21].

12.4 CHARACTERISTICS OF RESEARCH ON HYDROGEN SAFETY

12.4.1 DISTINGUISHING FEATURES OF RESEARCH ON HYDROGEN SAFETY

Zhang et al. [21] identified the top keywords in research articles on hydrogen safety from 1992 to 2022. The most frequent keywords were "combustion", "mechanism", "storage", "explosion", "kinetics", "dispersion", "ignition", "risk assessment", and "fuel cell", signifying that the research on hydrogen safety principally focuses on hydrogen combustion phenomena, hydrogen storage, and hydrogen leakage safety, as well as risk assessment in hydrogen utilization.

Regarding hydrogen combustion phenomena, there are currently three principal research approaches. The *first approach* deals with the study of the hydrogen self-ignition mechanism [40–45]. The crucial factors affecting the spontaneous ignition of high-pressure hydrogen gas leaks that have attracted interest recently are the leakage pipeline length [46], inner diameter [47], shape [48], rupture disks [49], and the physicochemical properties of hydrogen [50]. Great interest has been manifested recently from researchers from Germany, the United Kingdom, China, and Russia, among others, in the phenomenon of hydrogen autoignition [21].

The *second approach* focuses on hydrogen jet flames regarding the flame shape (length, width), heat flux, and other critical aspects. This is of great importance due to the hazardous range of jet flames, the delineation of safety zones, and the definition of emergency measures. To this end, experimental and numerical simulation methods are often applied by researchers from the United States, the United Kingdom, China, France, Japan, and other countries to study hydrogen jet fires. For instance, Schefer et al. [51] have offered valuable experimental data for jet fire research and contributed to effective risk management. Proust et al. [52] studied the influence of nozzle pressure on jet fires, while Delichatsios [53] proposed a calculation method for defining the flame length in under-expanded jet fires. Moreover, more studies are needed to improve engineering practices for successful suppression practices of jet fires. These challenges regarding the mitigation of jet flames demand the study of the impact of key factors, such as the mist spray (type and jet angle) and application location, on the suppression of hydrogen jet flames.

The *third approach* is related to the study of flammable hydrogen clouds and the phenomenon of deflagration-to-detonation transition (DDT). This research area concerns the study of flame acceleration (FA) and the factors influencing DDT [54–56]. High-speed Schlieren systems have been used to study the damping effect of different materials on deflagration waves to prevent their transition to detonation waves (DDT effect). Concerning hydrogen storage and leakage safety, various mitigation measures have been proposed. To limit liquid hydrogen leaks, recent research methods suggest the validation of liquid hydrogen jet leak models using experimental data obtained from liquid nitrogen leaks [21].

12.4.2 RESEARCH ON HYDROGEN EMBRITTLEMENT

Hydrogen embrittlement of materials, and especially various types of steel, is certainly a critical challenge as regards hydrogen safety. Thus, research is directed to the development of embrittlement-resistant materials suitable for the construction of hydrogen installations. Many countries have dealt with the problem and among them is the United States, where researchers at Sandia National Laboratories have created a huge database of tests conducted to investigate hydrogen embrittlement phenomena along with proposals for proactive suggestions to address hydrogen embrittlement [57]. Many unsolved issues still have to be addressed concerning the hydrogen embrittlement crack propagation mechanisms. Thus, recent efforts focus on international co-operation to improve the existing theoretical models and the mechanisms of hydrogen explosion. Luo et al. [58] found that some oxides can act as barriers that efficiently reduce the hydrogen absorption rate at metal surfaces, thus hindering hydrogen embrittlement.

12.4.3 OTHER RESEARCH TOPICS ON HYDROGEN SAFETY

Many papers focus on research on the topic of risk assessment of liquid hydrogen refueling stations. Risk matrices have been proposed by Kikukawa et al. [59] to investigate accident scenarios, thus providing valuable knowledge for the safe functioning of fuel cells.

Significant progress has been achieved in the sensor technology for hydrogen detection, thus addressing current technical challenges. The experience gained shows that a single sensor type is not always impeccable, so using a combination of different sensors can compensate for the deficiencies of individual sensors [60].

Taylor et al. [61] conducted an important study on the chemical kinetics of hydrogen detonation, while Kolla et al. [62] applied numerical simulation methods to model the characteristics of hydrogen jet flames.

Taking into account that the safety of fuel cells is crucial for the public embrace of fuel cell cars, special attention should be paid to critical safety properties of hydrogen itself, such as detonation [63]. On the other hand, Zhang et al. [64] proposed protection strategies of fast-filling hydrogen storage systems for fuel cell vehicle applications and solutions for the transportation safety of fuel cell vehicles.

12.5 FUTURE RESEARCH ON HYDROGEN SAFETY

Additional research on hydrogen gas leakage and ventilation control
Available hydrogen diffusion experiments and simulation studies come mainly from onshore studies regarding garages [65], refueling stations [66], tunnels [67], and hydrogen-powered vehicles [68]. Nevertheless, there is a research gap for offshore studies on the leakage diffusion models on hydrogen-powered ships [69].

Further research on the liquid hydrogen leakage mechanism and emergency measures to address liquid hydrogen accidents
The effect of multiple factors on the diffusion of liquid hydrogen accompanied by hydrogen cloud dispersion needs elucidation [70]. There is a knowledge gap in risk assessment studies and emergency protection countermeasures to address large liquid hydrogen leakage, and also a lack of models to estimate safety distances in case of a severe liquid hydrogen release [71, 72].

More research on spontaneous combustion of hydrogen release
More large-scale research on the spontaneous combustion of hydrogen release is needed since most of the existing studies were performed on small-scale tube experiments [73, 74]. These phenomena are complicated, engaging many mechanisms of gas leakage and spontaneous ignition. Consequently, the coupling effects of release and combustion should be thoroughly elucidated[75].

Elucidation of hydrogen flame propagation characteristics
Most flame propagation experiments are based on macroscopic measurements of overall parameters such as flame propagation velocity [76] and explosion overpressure [77]. Thus, more in-depth research is needed to clarify flame and pressure dynamics, both related to wall friction, equivalence ratio, and opening ratio.

Elucidation of hydrogen deflagration dynamics in real terrain conditions
Many studies on hydrogen deflagration dynamics are performed currently inside obstructed pipelines [78]. Since in real industrial accidents irregular obstacles play a crucial role in the propagation of a deflagration wave, further research should be performed to take into account existing structures that contribute to the propagation mechanisms of hydrogen deflagrations [79].

Further research on hydrogen embrittlement of materials
Despite the remarkable progress in elucidating hydrogen embrittlement phenomena, much more is needed to complete the current knowledge gaps, especially the lack of data on the charge of materials with hydrogen under high temperature and pressure that create brittle fracture and brittle crack propagation [80].

Lack of quantitative risk assessment studies
Although quite a few qualitative studies are found in the literature on the estimation of safety distances from hydrogen installations, these are of limited value because they are mainly based on experience from past accidents [81, 82].

To fill the knowledge gap, quantitative risk assessment (QRA) studies are needed to more accurately investigate real hydrogen-caused accidents onshore and offshore, thus enabling the determination of convincing safety distances from hydrogen facilities [83].

REFERENCES

1. Guy, K.W.A., The hydrogen economy, *Process Safety and Environmental Protection*, 78 (4), 324–327, 2000.
2. Dahoe, A.E., and Molkov, V.V., On the implementation of an international curriculum on hydrogen safety engineering into higher education, *Journal of Loss Prevention in the Process Industries*, 21 (2), 222–224, 2008.
3. Salvi, O., Process Safety Research and its Impact on Sustainability and Resilience of the Society, Plenary Paper, A Frontiers of Research Workshop, Mary Kay O'Connor Process Safety Center, Texas A&M University, College Station, TX (October 21-22, 2011).
4. Pasman, H.J., and Rogers, W.J., Safety challenges in view of the upcoming hydrogen economy: An overview, *Journal of Loss Prevention in the Process Industries*, 23 (6), 697–704, 2010.
5. Jordan, T., Adams, P., Azkarate, I., Baraldi, D., Barthelemy, H., Bauwens, L., Bengaouer, A., Brennan, S., Carcassi, M., Dahoe, A., Eisenrich, N., Engebo, A., Funnemark, E., Gallego, E., Gavrikov, A., Haland, E., Hansen, A.M., Haugom, G.P., Hawksworth, S., Jedicke, O., Kessler, A., Kotchourko, A., Kumar, S., Langer, G., Stefan, L., Lelyakin, A., Makarov, D., Marangon, A., Markert, F., Middha, P., Molkov, V., Nilsen, S., Papanikolaou, E., Perrette, L., Reinecke, E.-A., Schmidtchen, U., Serre-Combe, P., Stocklin, M., Sully, A., Teodorczyk, A., Tigreat, D., Venetsanos, A., Verfondern, K., Versloot, N., Vetere, A., Wilms, M., and Zaretskiy, N., Achievements of the EC network of excellence hysafe, *International Journal of Hydrogen Energy*, 36 (3), 2656–2665, 2011.
6. Molkov, V., Hydrogen safety research: State-of-the-art, in *Proceedings of the 5th International Seminar on Fire and Explosion Hazards*, Edinburgh, UK (April 23-27, 2007).
7. Molkov, V., Preface. Special issue on hydrogen safety, *Journal of Loss Prevention in the Process Industries*, 21 (2), 129–130, 2008.
8. Amyotte, P.R., Process Safety Research Needs from the Industry Perspective, Plenary Paper, A Frontiers of Research Workshop, Mary Kay O'Connor Process Safety Center, Texas A&M University, College Station, TX (October 21-22, 2011).
9. Amyotte, P.R., and Khan, F.I., *An Inherently Safer Approach to Dust Explosion Risk Reduction*, Strategic Project Grant (Safety and Security) No. 396398, Natural Sciences and Engineering Research Council of Canada, 2010.
10. Amyotte, P.R., *Are Classical Process Safety Concepts Relevant to Nanotechnology Applications? Journal of Physics: Conference Series* 304 (Nanosafe2010: International Conference on Safe Production and Use of Nanomaterials), 012071, 2011.
11. Hendershot, D.C., Ventrone, T.A., Schwab, R.F., Ormsby, R.W., Davenport, J.A., and Bradford, W.J., *History of Process Safety and Loss Prevention in the American Institute of Chemical Engineers*, Presented at the American Chemical Society, National Meeting, Washington, D.C. (August 28-September 1, 2005).
12. Kletz, T.A., The origin and history of loss prevention, *Process Safety and Environmental Protection*, 77 (3), 109–116, 1999.
13. Knegtering, B., and Pasman, H.J., Safety of the process industries in the 21st century: A changing need of process safety management for a changing industry, *Journal of Loss Prevention in the Process Industries*, 22 (2), 162–168, 2009.
14. Qi, R., Prem, K., Ng, D., Ranes, M., Yun, G., and Mannan, M.S., Challenges and needs for process safety in the new millenium, *Process Safety and Environmental Protection*, 90 (2), 91–100, March 2012.

15. Kirchsteiger, C., Impact of accident precursors on risk estimates from accident databases, *Journal of Loss Prevention in the Process Industries*, 10 (3), 159–167, 1997.

16. Amyotte, P.R., MacDonald, D.K., and Khan, F.I., An analysis of CSB investigation reports concerning the hierarchy of controls, *Process Safety Progress*, 30 (3), 261–265, 2011.

17. Hopkins, A., *Lessons from Longford. The Esso Gas Plant Explosion*, CCH Australia Limited, Sydney, Australia, 2000.

18. Hopkins, A., *Safety, Culture and Risk. The Organisational Causes of Disasters*, CCH Australia Limited, Sydney, Australia, 2005.

19. Hopkins, A., *Failure to Learn. The BP Texas City Refinery Disaster*, CCH Australia Limited, Sydney, Australia, 2009.

20. Hopkins, A. (Editor), *Learning from High Reliability Organisations*, CCH Australia Limited, Sydney, Australia, 2009.

21. Zhang, D., Jiang, M., Li, G., and Tang, Y., An advanced bibliometric analysis and future research insights on safety of hydrogen energy, *Journal of Energy Storage*, 77, 109833, 2024.

22. Donthu, N., Kumar, S., Mukherjee, D., Pandey, N., and Lim, W.M., How to conduct a bibliometric analysis: an overview and guidelines, *Journal of Business Research*, 133, 285–296, 2021. https://doi.org/10.1016/j.jbusres.2021.04.070

23. Tan, H., Li, J., He, M., Li, J., Zhi, D., and Qin, F., et al., Global evolution of research on green energy and environmental technologies: a bibliometric study, *Journal of Environment Management*, 297, 113382, 2021. https://doi.org/10.1016/j.jenv man.2021.113382

24. Waltman, L., A review of the literature on citation impact indicators, *Journal of Informetrics*, 10 (2), 365–391, 2016. https://doi.org/10.1016/j.joi.2016.02.007

25. Falagas, M.E., Pitsouni, E.I., Malietzis, G.A., and Pappas, G., Comparison of PubMed, scopus, web of science, and google scholar: Strengths and weaknesses, *FASEB Journal*, 22 (2), 338–342, 2008. https://doi.org/10.1096/fj.07-9492LSF

26. Kulkarni, A.V., Aziz, B., Shams, I., and Busse, J.W., Comparisons of citations in web of science, scopus, and google scholar for articles published in general medical journals, *JAMA*, 302 (10), 1092–1096, 2009. https://doi.org/10.1001/jama.2009.1307

27. Bakkalbasi, N., Bauer, K., Glover, J., and Wang, L., Three options for citation tracking: Google scholar, scopus and web of science, *Biomedical Digital Libraries*, 3, 7, 2006. https://doi.org/10.1186/1742-5581-3-7

28. Wang, Q., and Waltman L., Large-scale analysis of the accuracy of the journal classification systems of Web of Science and Scopus, *Journal of Informetrics*, 10 (2), 347–364, 2016. https://doi.org/10.1016/j.joi.2016.02.003

29. Donner, P., Document type assignment accuracy in the journal citation index data of Web of Science, *Scientometrics*, 113 (1), 219–236, 2017. https://doi.org/10.1007/s11 192-017-2483-y

30. Yang, F., Wang, T., Deng, X., Dang, J., Huang, Z., and Hu, S., et al., Review on hydrogen safety issues: Incident statistics, hydrogen diffusion, and detonation process, *International Journal of Hydrogen Energy*, 46 (61), 31467–31488, 2021. https://doi. org/10.1016/j.ijhydene.2021.07.005

31. Kirchsteiger, C., Vetere Arellano A.L., and Funnemark, E., Towards establishing an International Hydrogen Incidents and Accidents Database (HIAD), *Journal of Loss Prevention in the Process Industries*, 20 (1), 98–107, 2007. https://doi.org/10.1016/j.jlp.2006.10.004

32. Dincer, I., and Acar, C., Innovation in hydrogen production, *International Journal of Hydrogen Energy*, 42 (22), 14843–14864, 2017. https://doi.org/10.1016/j.ijhyd ene.2017.04.107

33. Ambrose, A.F., Al-Amin, A.Q., Rasiah, R., Saidur, R., and Amin, N., Prospects for introducing hydrogen fuel cell vehicles in Malaysia, *International Journal of Hydrogen Energy*, 42 (14), 9125–9134, 2017. https://doi.org/10.1016/j.ijhydene.2016.05.122

34. El-Osta, W., and Zeghlam, J., Hydrogen as a fuel for the transportation sector, Appl. *Energy*, 65 (1–4), 165–171, 2000. https://doi.org/10.1016/s0306-2619(99)00092-6

35. Melton, N., Axsen, J., and Sperling, D., Moving beyond alternative fuel hype to decarbonize transportation, *Nature Energy*, 1 (3), 2016. https://doi.org/10.1038/nene rgy.2016.13

36. Verbong, G., Geels, F.W., and Raven, R., Multi-niche analysis of dynamics and policies in Dutch renewable energy innovation journeys (1970–2006): hype-cycles, closed networks and technology-focused learning, *Technology Analysis and Strategic Management*, 20 (5), 555–573, 2008. https://doi.org/10.1080/09537320802292719

37. Wen, J.X., Marono, M., Moretto, P., Reinecke, E.-A., Sathiah, P., and Studer, E., et al., Statistics, lessons learned and recommendations from analysis of HIAD 2.0 database, *International Journal of Hydrogen Energy*, 47 (38), 17082–17096, 2022. https://doi.org/10.1016/j.ijhydene.2022.03.170

38. Meng, X., Gu, A., Wu, X., Zhou, L., Zhou, J., and Liu, B., et al., Status quo of China hydrogen strategy in the field of transportation and international comparisons, *International Journal of Hydrogen Energy*, 46 (57), 28887–28899, 2021. https://doi.org/10.1016/j. ijhydene.2020.11.049

39. Li, Z., Zhang, W., Zhang, R., and Sun, H., Development of renewable energy multi-energy complementary hydrogen energy system (a case study in China): a review, *Energy Exploration & Exploitation*, 38 (6), 2099–2127, 2020. https://doi.org/10.1177/0144598720953512

40. Astbury, G., and Hawksworth, S., Spontaneous ignition of hydrogen leaks: A review of postulated mechanisms, *International Journal of Hydrogen Energy*, 32 (13), 2178–2185, 2007. https://doi.org/10.1016/j.ijhydene.2007.04.005

41. Dryer, F.L., Chaos, M., Zhao, Z., Stein, J.N., Alpert, J.Y., and Homer, C.J., Spontaneous ignition of pressurized releases of hydrogen and natural gas into air, *Combustion Science and Technology*, 179 (4), 663–694, 2007. https://doi.org/10.1080/0010220060 0713583

42. Michels, A., de Graaff, W., and Wolkers, G.J., Thermodynamic properties of hydrogen and deuterium at temperatures between 175°C and 150°C and at densities up to 840 amagat, *Physica*, 25 (7–12), 1097–1124, 1959. https://doi.org/10.1016/0031-8914(59)90029-1

43. Merilo, E.G., Groethe, M.A., Adamo, R.C., Schefer, R.W., Houf, W.G., and Dedrick, D.E., Self-ignition of hydrogen releases through electrostatic discharge induced by entrained particulates, *International Journal of Hydrogen Energy*, 37 (22), 17561–17570, 2012. https://doi.org/10.1016/j.ijhydene.2012.03.167

44. Maxwell, B.M., and Radulescu, M.I., Ignition limits of rapidly expanding diffusion layers: Application to unsteady hydrogen jets, Combust. *Flame*, 158 (10), 1946–1959, 2011. https://doi.org/10.1016/j.combustflame.2011.03.001

45. Maxwell, B.M., Tawagi, P., and Radulescu, M.I., The role of instabilities on ignition of unsteady hydrogen jets flowing into an oxidizer, *International Journal of Hydrogen Energy*, 38 (6), 2908–2918, 2013. https://doi.org/10.1016/j.ijhydene.2012.11.133

46. Wang, Z., Pan, X., Jiang, Y., Wang, Q., Yan, W., and Xiao, J., et al., Experiment study on the pressure and flame characteristics induced by high-pressure hydrogen spontaneous

ignition, *International Journal of Hydrogen Energy*, 45 (35), 18042–18056, 2020. https://doi.org/10.1016/j.ijhydene.2020.04.051

47. Oleszczak, P., and Wolanski, P., Ignition during hydrogen release from high pressure into the atmosphere, *Shock Waves* 20 (6), 539–550, 2010. https://doi.org/10.1007/s00 193-010-0291-x

48. Morii, Y., Terashima, H., Koshi, M., and Shimizu, T., Numerical study of the effect of obstacles on the spontaneous ignition of high-pressure hydrogen, *Journal of Loss Prevention in the Process Industries*, 34, 92–99, 2015. https://doi.org/10.1016/j.jlp.2015.01.020

49. Golovastov, S., and Bocharnikov, V., The influence of diaphragm rupture rate on spontaneous self-ignition of pressurized hydrogen: Experimental investigation, *International Journal of Hydrogen Energy*, 37 (14), 10956–10962, 2012. https://doi.org/10.1016/j.ijhydene.2012.04.070

50. Zeng, Q., Duan, Q., Sun, D., Li, P., Zhu, M., and Wang, Q., et al., Experimental study of methane addition effect on shock wave propagation, self-ignition and flame development during high-pressure hydrogen sudden discharge from a tube, *Fuel*, 277, 118217, 2020. https://doi.org/10.1016/j.fuel.2020.118217

51. Schefer, R.W., Merilo, E.G., Groethe, M.A., and Houf, W.G., Experimental investigation of hydrogen jet fire mitigation by barrier walls, *International Journal of Hydrogen Energy*, 36 (3), 2530–2537, 2011. https://doi.org/10.1016/j.ijhydene.2010.04.008

52. Proust, C., Jamois, D., and Studer, E., High pressure hydrogen fires, *International Journal of Hydrogen Energy*, 36 (3), 2367–2373, 2011. https://doi.org/10.1016/j.ijhyd ene.2010.04.055

53. Delichatsios, M.A., Transition from momentum to buoyancy-controlled turbulent jet diffusion flames and flame height relationships, *Combustion and Flame*, 92 (4), 349–364, 1993. https://doi.org/10.1016/0010-2180(93)90148-v

54. Grune, J., Sempert, K., Friedrich, A., Kuznetsov, M., and Jordan, T., Detonation wave propagation in semi-confined layers of hydrogen–air and hydrogen–oxygen mixtures, *International Journal of Hydrogen Energy*, 42 (11), 7589–7599, 2017. https://doi.org/10.1016/j.ijhydene.2016.06.055

55. Bauwens, C.R., Chaffee, J., and Dorofeev, S.B., Vented explosion overpressures from combustion of hydrogen and hydrocarbon mixtures, *International Journal of Hydrogen Energy*, 36 (3), 2329–2336, 2011. https://doi.org/10.1016/j.ijhydene.2010.04.005

56. Alekseev, V.I., Kuznetsov, M.S., Yankin, Y.G., and Dorofeev, S.B., Experimental study of flame acceleration and the deflagration-to-detonation transition under conditions of transverse venting, *Journal of Loss Prevention in the Process Industries*, 14 (6), 591–596, 2001. https://doi.org/10.1016/s0950-4230(01)00051-1

57. San Marchi, C., and Somerday, B.P., *Technical Reference for Hydrogen Compatibility of Material*, Sandia National Laboratories, California, 2012.

58. Luo, H., Sohn, S.S., Lu, W., Li, L., Li, X., and Soundararajan C.K. et al., A strong and ductile medium-entropy alloy resists hydrogen embrittlement and corrosion, *Nature Communications*, 11 (1), 3081, 2020. https://doi.org/10.1038/s41467-020-16791-8

59. Kikukawa, S., Mitsuhashi, H., and Miyake, A., Risk assessment for liquid hydrogen fueling stations, *International Journal of Hydrogen Energy*, 34 (2), 1135–1141, 2009. https://doi.org/10.1016/j.ijhydene.2008.10.093

60. Boon-Brett, L., Bousek, J., Black, G., Moretto, P., Castello, P., and Hübert, T. et al., Identifying performance gaps in hydrogen safety sensor technology for automotive and stationary applications, *International Journal of Hydrogen Energy*, 35 (1), 373–384, 2010. https://doi.org/10.1016/j.ijhydene.2009.10.064

61. Taylor, B.D., Kessler, D.A., Gamezo, V.N., and Oran, E.S., Numerical simulations of hydrogen detonations with detailed chemical kinetics, Proc. *Combustion Institute*, 34 (2), 2009–2016, 2013. https://doi.org/10.1016/j.proci.2012.05.045

62. Kolla, H., Grout, R.W., Gruber, A., and Chen, J.H., Mechanisms of flame stabilization and blowout in a reacting turbulent hydrogen jet in cross-flow, Combust. *Flame*, 159 (8), 2755–2766, 2012. https://doi.org/10.1016/j.combustflame.2012.01.012

63. Austin, J., Detonations in hydrocarbon fuel blends, *Combustion and Flame* 132 (1–2), 73–90, 2003. https://doi.org/10.1016/s0010-2180(02)00422-4

64. Zhang, C., Cao, X., Bujlo, P., Chen, B., Zhang, X., and Sheng, X., et al., Review on the safety analysis and protection strategies of fast filling hydrogen storage system for fuel cell vehicle application, *Journal of Energy Storage*, 45 2022. https://doi.org/10.1016/j.est.2021.103451

65. Huang, T., Zhao, M., Ba, Q., Christopher, D.M., and Li, X., Modeling of hydrogen dispersion from hydrogen fuel cell vehicles in an underground parking garage, *International Journal of Hydrogen Energy*, 47 (1), 686–696, 2022. https://doi.org/10.1016/j.ijhydene.2021.08.196

66. Li, Y., Wang, Z., Shi, X., and Fan, R., Safety analysis of hydrogen leakage accident with a mobile hydrogen refueling station, *Process Safety and Environmental Protection*, 171, 619–629, 2023. https://doi.org/10.1016/j.psep.2023.01.051

67. Shao, X., Shi, W., Li, P., Pu, L., Zheng, L., and Lu, C. et al., Explosion-prevention strategies of airflow controlling and closed-inerting for hydrogen dilution in utility tunnel, *International Journal of Hydrogen Energy*, 48 (37), 14095–14111, 2023. https://doi.org/10.1016/j.ijhydene.2022.12.221

68. Li, Y., Hou, X., Wang, C., Wang, Q., Qi, W., and Li, J. et al., Modeling and analysis of hydrogen diffusion in an enclosed fuel cell vehicle with obstacles, *International Journal of Hydrogen Energy*, 47 (9), 5745–5756, 2022. https://doi.org/10.1016/j.ijhydene.2021.11.205

69. Li, F., Yuan, Y., Yan, X., Malekian, R., and Li, Z., A study on a numerical simulation of the leakage and diffusion of hydrogen in a fuel cell ship, *Renewable and Sustainable Energy Reviews*, 97, 177–185, 2018. https://doi.org/10.1016/j.rser.2018.08.034

70. Pu, L., Shao, X., Zhang, S., Lei, G., and Li, Y., Plume dispersion behaviour and hazard identification for large quantities of liquid hydrogen leakage, *Asia Pacific Journal of Chemical Engineering*, 14 (2), 2019. https://doi.org/10.1002/apj.2299

71. Sun, R., Pu, L., Yu, H., Dai, M., Li, Y., Investigation of the hazardous area in a liquid hydrogen release with or without fence, *International Journal of Hydrogen Energy*, 46 (73), 36598–36609, 2021. https://doi.org/10.1016/j.ijhydene.2021.08.155

72. Li, Z., Pan, X., Meng, X., and Ma, J., Study on the harm effects of releases from liquid hydrogen tank by consequence modeling, *International Journal of Hydrogen Energy*, 37 (22), 17624–17629, 2012. https://doi.org/10.1016/j.ijhydene.2012.05.141

73. Jiang, Y., Pan, X., Cai, Q., Wang, Z., Klymenko, O.V., and Hua, M. et al., Physics and flame morphology of supersonic spontaneously combusting hydrogen spouting into air, *Renewable Energy*, 196, 959–972, 2022. https://doi.org/10.1016/j.renene.2022.06.153

74. Zhang, T., Jiang, Y., Pan, X., Wang, Z., Wang, Q., and Li, Y. et al., Effects of tubes with different inlet shapes on the shock wave and self-ignition induced by accidental release of pressurized hydrogen, *Fuel*, 317, 2022. https://doi.org/10.1016/j.fuel.2022.123554

75. Zhuo, X., and Gou, X., Kinetic analysis on spontaneous combustion of pressurized hydrogen in tubes, *ACS Omega*, 6 (40), 26509–26518, 2021. https://doi.org/10.1021/acsomega.1c03761

76. Shi, X., Chen, J.-Y., and Chen, Z., Numerical study of laminar flame speed of fuelstratified hydrogen/air flames, *Combustion and Flame*, 163, 394–405, 2016. https://doi.org/10.1016/j.combustflame.2015.10.014

77. Guo, L., Ba, Q., and Zhang, S., Study on hydrogen dynamic leakage and flame propagation at normal-temperature and high-pressure, *International Journal of Hydrogen Energy*, 48(70), 27416–27426, 2023. https://doi.org/10.1016/j.ijhyd ene.2023.03.445

78. Li, X., Dong, J., Jin, K., Duan, Q., Sun, J., and Li, M., et al., Flame acceleration and deflagration-to-detonation transition in a channel with continuous triangular obstacles: Effect of equivalence ratio, *Process Safety and Environmental Protection*, 167, 576–591, 2022. https://doi.org/10.1016/j.psep.2022.09.033

79. Xiao, H. Duan, Q., and Sun, J., Premixed flame propagation in hydrogen explosions, *Renewable and Sustainable Energy Reviews*, 81, 1988–2001, 2018. https://doi.org/ 10.1016/j.rser.2017.06.008

80. Barnoush, A., and Vehoff, H., Recent developments in the study of hydrogen embrittlement: Hydrogen effect on dislocation nucleation, *Acta Materialia*, 58 (16), 5274–5285, 2010. https://doi.org/10.1016/j.actamat.2010.05.057

81. Ade, N., Wilhite, B., and Goyette, H., An integrated approach for safer and economical design of hydrogen refueling stations, *International Journal of Hydrogen Energy*, 45 (56), 32713–32729, 2020. https://doi.org/10.1016/j.ijhydene.2020.08.232

82. Middha, P., and Hansen, O.R., CFD simulation study to investigate the risk from hydrogen vehicles in tunnels, *International Journal of Hydrogen Energy*, 34 (14), 5875–5886, 2009. https://doi.org/10.1016/j.ijhydene.2009.02.004

83. Seong, D.H., Kim, T.H., Oh, D.S., Oh, Y.D., Seo, D.H., Kim, Y.G., and Kim, E.J., Quantitative safety assessment for hydrogen station, *Journal of the Korean Society of Safety*, 111–116, January 2012.

13 Legal Requirements for Hydrogen Safety

13.1 GENERAL ASPECTS AND DEFINITIONS

Regulations, codes, and *directives* are legal requirements imposed by legislative bodies (parliaments, governments, the European Union) and are mandatory for everyone dealing with a certain activity. In contrast, *standards, guidelines,* and *codes of practice* are voluntary documents useful for interested industrial organizations engaged in the relevant activity.

Codes of practice refer usually to best practices with regard to safe handling, operation, and maintenance of a product. *Guidelines* or *guides* are documents written for a specific organization or specific users and provide guidance to ensure best practice and safety, or provide information and analysis of codes, standards, and regulations on the recommended way to meet these requirements. The *state of the art* is the most advanced technique available currently. *Best engineering practices* are the best available practices for design and construction, or for operation of machines and other devices in industry and other activities.

In the European Union, a *directive* is the legal requirement that fosters common ground in all European countries. Directives are compulsory only after a predetermined period of time after their adoption by the national legislation of every member country. Unfortunately, there is not yet a European directive for hydrogen. Matters concerning any substance not specifically covered by a directive are usually mentioned incidentally in other application-specific directives. As a result, only national legislation may define minimum levels for hazardous substances including hydrogen.

With reference to pressurized containers, hydrogen is covered by the European Pressure Equipment Directive (PED) and also by the Transportable Pressure Equipment Directive (TPED), which in turn refers to international agreements for safe transport of hazardous goods, such as ADR (road), RID (rail), IMO (sea), and ADNR (inland waterways). Permits to use motor vehicles internationally are obtained via the United Nations Economic Commission for Europe (UN ECE).

13.1.1 ATEX DIRECTIVE

With regard to regulation on the prevention of damage from the release of flammable gases (including hydrogen), the directives 94/9/EC (known as ATEX 100

DOI: 10.1201/9781003313007-13

Directive) and 99/92/EC (known as ATEX 118 Directive) are relevant. ATEX derives its name from the French title of the 94/9/EC Directive: *Appareils destinés à être utilisés en ATmosphères EXplosives*. Major accident prevention and mitigation of consequences to humans, installations, and the environment are covered by the Seveso III Directive.

ATEX Directive 1999/92/EC has been replaced by ATEX Directive 2014/34/EU [1] of the European Parliament and of the Council of 26 February 2014 on the harmonization of the laws of the Member States relating to equipment and protective systems intended for use in potentially explosive atmospheres (recast). Directive 2014/34/EU defines the minimum requirements for improving safety and health in a working environment by minimizing risk from a potentially explosive atmosphere. In this sense, hydrogen poses a risk of forming an explosive atmosphere when mixed with air.

This directive applies to the following, hereinafter referred to as "products":

a) *Equipment and protective systems intended for use in potentially explosive atmospheres.*

b) *Safety devices, controlling devices, and regulating devices intended for use outside potentially explosive atmospheres but required for or contributing to the safe functioning of equipment and protective systems for the risks of explosion.*

c) *Components intended to be incorporated into equipment and protective systems referred to in point (a).*

To prevent an explosive event, the employer is obliged to take a series of measures based on the following principles:

- Prevention of formation of an ATEX (explosive atmosphere) and, if this is not possible,
- Avoidance of ignition of ATEX, and
- Mitigation of explosion effects.

To accomplish this role, the employer has to perform risk assessment studies aimed at determining:

- The likelihood of an ATEX event.
- The likelihood of presence of active and effective ignition sources.
- The scale of anticipated effects in relation to installations, substances, and processes used.

The employer should also classify hazardous zones as follows:

- Zone 0: Places with continuous presence of ATEX, or for a long period of time, or with high frequency.
- Zone 1: Places with occasional risk of ATEX during normal operation.
- Zone 2: Places with unlikely risk of ATEX during normal operation and with short duration.

With regard to application of the ATEX Directive to hydrogen hazards, illustrative examples are the following:

- A process in which hydrogen is used (e.g., after ingress of air in a hydrogenation reactor).
- A closed space in which hydrogen may be produced via a metal powder reaction with water (especially when acidified).
- Hydrogen production during refueling of lead batteries.
- Release of hydrogen from pressurized facilities.

In addition to this ATEX Directive, many other directives are related to flammable gases including hydrogen, such as the Machinery Directive.

13.1.2 OTHER OFFICIAL DOCUMENTS

Considering the international standardization of hydrogen safety, the ISO Technical Committee 197 titled "Hydrogen Technologies" has published the following official documents:

- SO/AWI 13985:2006, Liquid hydrogen – Land vehicle fuel tanks. (Under development).
- ISO 12619-1:2014, Road vehicles – Compressed gaseous hydrogen and hydrogen/natural gas blend fuel system components – Part 1: General requirements and definitions.
- ISO/TS 19870:2023, Hydrogen technologies – Methodology for determining the greenhouse gas emissions associated with the production, conditioning and transport of hydrogen to consumption gate.
- ISO/TR 15916:2015, Basic considerations for the safety of hydrogen systems. (Provides guidelines for the use of hydrogen in its gaseous and liquid forms as well as its storage in either of these or other forms (hydrides).
- ISO 14687:2019, Hydrogen fuel quality – Product specification. (To be revised).
- ISO 6469-1:2019, Electrically propelled road vehicles – Safety specifications – Part 1: Rechargeable energy storage system (RESS)
- ISO 17268:2020, Gaseous hydrogen land vehicle refuelling connection devices. (To be revised).
- ISO/DIS 17268-1, Gaseous hydrogen land vehicle refuelling connection devices – Part 1: Flow capacities up to and including 120 g/s. (Defines the design, safety, and operation characteristics of gaseous hydrogen land vehicle (GHLV) refuelling connectors). (Under development).
- ISO 23273:2013, Fuel cell road vehicles – Safety specifications – Protection against hydrogen hazards for vehicles fuelled with compressed hydrogen. (Under review).

Some additional useful guidelines and other documents can be found in HySafe (Safety of Hydrogen as an Energy Carrier; see Chapter 9) [2] as listed here:

- Commission des Communautés Européennes, *Eléments pour un guide de sécurité hydrogène*, Expérimentations spécifiques, choix d'appareils et matériels adaptés, Volume 1, Rapport EUR 9689 FR, Luxembourg 1985.
- Commission des Communautés Européennes, *Eléments pour un guide de sécurité hydrogène, Aperçu d'ensemble*, Volume 2, Rapport EUR 9689 FR, Luxembourg 1985.
- FM Global, *"Hydrogen", Property Loss Prevention Data Sheets 7–91*, January 2024.
- IGC 15/96/E, *Gaseous Hydrogen Stations*, Industrial Gases Council, Brussels, Belgium.
- IGC 06/93/E, *Safety in Storage, Handling and Distribution of Liquid Hydrogen*, Industrial Gases Council, Brussels, Belgium.
- ISO/TR 15916, *Basic Considerations for the Safety of Hydrogen Systems (Considérations fondamentales pour la sécurité des systèmes à l'hydrogène)*, first edition, 2004-02-15.
- American National Standards Institute, *Guide to Safety of Hydrogen and Hydrogen Systems*, American Institute of Aeronautics and Astronautics, ANSI/AIAA G-095A-2017, Chap. 4, ANSI, Washington DC, 2017.
- NASA/TM – 2003-212059, *Guide for Hydrogen Hazards Analysis on Components and Systems*, Harold Beeson (Lyndon B. Johnson Space Center, White Sands Test Facility), Stephen Woods (Honeywell Technology Solutions Inc., White Sands Test Facility); Published as TP-WSTF-937, October 2003.
- NASA, *NASA Glenn Safety Manual*, Chap. 6 – Hydrogen, Revision Date: 9/03, Biannual Review.
- NFPA 50A, *Standard for Gaseous Hydrogen Systems at Consumer Sites*, National Fire Protection Association, Quincy, Massachusetts, 1999.
- NFPA 50B, *Standard for liquefied hydrogen systems at consumer sites*, National Fire Protection Association, Quincy, Massachusetts, 1999.
- NFPA 853, *Standard for the Installation of Stationary Fuel Cell Power Plants*, National Fire Protection Association, Quincy, Massachusetts, 2003.
- NRCC 27406, *Safety Guide for Hydrogen*, Hydrogen Safety Committee, National Research Council of Canada, Ottawa, 1987.
- Canadian Hydrogen Installation Code (CHIC), Bureau de Normalisation du Québec (BNQ), 2007.

HySafe has also released the first edition of the *Hydrogen Safety for Energy Applications* handbook (Editors: Alexei Kotchourko, Thomas Jordan) [3]. The book provides engineering design, risk assessment, and codes and standards, as well as discussing different aspects of contemporary knowledge regarding the hazards, risks, and safety connected with hydrogen systems.

Certain guidelines and codes of practice are more substance-oriented, among which are the American National Standard ANSI/AIAA G-095A-2017, *Guide to Safety of Hydrogen and Hydrogen Systems* [4] of the American Institute of Aeronautics and Astronautics, the Final Report on *Guidelines for Use of Hydrogen Fuel in Commercial Vehicles* of the US Department of Transportation (US DOT, [5]) and the Code of Practice (COP) on *Safety in Storage, Handling and Distribution of*

Liquid Hydrogen of the European Industrial Gases Association (EIGA) [6]. These are described in some detail in the following sections.

13.2 HYDROGEN FACILITIES

The following are general guidelines to ensure safety in hydrogen storage and transfer areas. Good illumination, lightning protection, alarm systems, and gas detection and sampling systems should be provided in such a facility. More detailed safety guidelines concerning safety policy, safety in construction, operation, maintenance, and final disposition of a hydrogen facility, as well as safety measures in buildings and test chambers used in hydrogen service and emergency procedures can be found in the American National Standard ANSI/AIAA G-095A-2017 [4], from which some synoptic information is given in this section, along with other references [7].

13.2.1 ELECTRICAL CONSIDERATIONS

Areas where flammable hydrogen mixtures are expected to occur or areas where hydrogen is stored, transferred, or used and where the hydrogen is normally contained are classified as highly hazardous zones according to internationally accepted regulations. All electrical sources of ignition should be prohibited in these areas, using approved explosion-proof equipment or selecting non-arcing approved equipment.

Explosion-proof equipment has an enclosure strong enough to contain the pressure produced by igniting a flammable mixture inside the enclosure. Since it is not gastight, the joints and threads must be tight enough and long enough to prevent issuance of flames or gases that would be hot enough to ignite a surrounding flammable mixture.

Another method to prevent a gas explosion is to locate the equipment in an enclosure purged and maintained above ambient pressure with an inert gas. Intrinsically safe installations used in the facility should be approved for hydrogen service.

It is acceptable to use general-purpose equipment in general-purpose housing, if it is continuously purged with clean air or nitrogen. In this case, no hydrogen sources must be plumbed into the equipment and positive indication of continued purge must be provided. By putting items that might become ignition sources outside the hazardous area, the cost of an installation will be reduced and safety increased. Equipment installed in a hazardous area, but not required during hazardous periods, may be built with general-purpose equipment provided it is disconnected before the hazardous period begins. The conduits for such systems must be sealed when they leave the hazardous area.

Electrical components for a hydrogen system should be installed as required by ATEX [1]. Equipment not classified for hazardous environments should be located outside the area rated as hazardous.

13.2.2 BONDING AND GROUNDING

Mobile hydrogen supply units must be electrically bonded to the system before discharging hydrogen according to regulations.

Liquefied hydrogen containers, static and mobile, and associated piping must be electrically bonded and grounded using proper sizes of grounding conductors and acceptable connections based on the expected amperage to ground.

All *off-loading facilities* must provide easily accessible grounding connections and be located outside the immediate transfer area. Facility grounding connections should have resistances less than 10 ohm. Transfer subsystem components should be grounded before subsystems are connected.

All hydrogen equipment should be electrically bonded and grounded as required by ATEX [1].

13.2.3 HYDROGEN TRANSMISSION LINES

Hydrogen transmission lines carrying hydrogen from trailers and storage vessels must be aboveground installations. Lines crossing roadways should be installed in concrete channels covered with an open grating. Hydrogen lines crossing roadways should be installed in concrete channels covered with an open grating and should not be located beneath electric power lines. Area surfaces located below liquefied hydrogen lines from which condensed liquid air may drop must be constructed of noncombustible materials such as concrete. Asphalt must not be used.

Piping leaks have been the initiating event in many hydrogen and other gaseous fuel accidents. A hydrogen leak, due to its very low density, can easily find its way to upper floors in a building, traveling through connecting ducts and other openings, and cause secondary explosions in other spaces.

13.2.4 ELIMINATION OF IGNITION SOURCES

Installations using hydrogen should be protected from *lightning* by lightning rods, aerial cable, and ground rods suitably connected. Lightning strikes may cause inducing sparks; therefore, all equipment in a building should be bonded and grounded to prevent sparks.

Static electricity may be generated in moving machinery belts or in flowing fluids containing solid or liquid particles. The measures taken to limit electrostatic charge generation and accumulation include:

- Bonding and grounding of all metal parts within a system.
- Use of conductive machinery belts.
- Personnel clothes made of antistatic fibers.
- Conductive and nonsparking floors.

Sparks may also be generated by other mechanisms such as *friction and impact*. Even spark-proof tools can cause ignitions because the energy required for ignition of flammable hydrogen–air mixtures is extremely small. So, spark-proof tools should be used and, in addition, they should be used with caution to prevent slipping, glancing blows, or dropping, all of which can cause sparks. Extra care should be taken if spark-proof tools are not available.

13.2.5 Hot Objects, Flames, and Flame Arrestors

The measures taken to eliminate ignition by flames and hot objects and prevent proliferation of a fire to other areas (domino effect) include the prohibition of open flames, welding, or cutting within the exclusion area around a hydrogen facility, and the use of internal combustion system equipment with exhaust system spark arrestors and carburetor flame arrestors. Only *flame arrestors* specifically designed for hydrogen applications must be used, taking into account the oxidant present. Flame arrestors can quench a flame on the basis of sufficient heat removal from the gas mixture. It is quite difficult to develop flame arrestors and explosion-proof equipment for hydrogen due to its small quenching distance (0.6 mm). Sintered-bronze flame arrestors may be effective in stopping hydrogen flames, whereas sintered stainless steel is not as effective. Arrestors should be well maintained to minimize accidental ignition because many accidents have been caused by poor maintenance of safety devices.

13.2.6 Design and Construction of Buildings

Buildings in which hydrogen is used must be constructed of light and noncombustible materials on a sufficiently strong frame. Windowpanes must be made of shatterproof glass or plastic. Floors, walls, and ceilings should be designed and installed to limit the generation and accumulation of static electricity and have a fire resistance rating of at least 2 h.

13.2.6.1 Explosion Venting

Explosion venting must be provided only in the exterior walls or the roof. The venting area should not be less than 0.11 m^2 per cubic meter of room volume. Vents, designed to relieve at a maximum internal pressure of 1.2 kPa, may consist of one or a combination of walls of light material, lightly fastened hatch covers, or outward-opening swinging doors in exterior walls or roof.

Doors should be hinged to swing outward in an explosion and must be readily accessible to personnel. *Walls* or partitions must be continuous from floor to ceiling and securely anchored. At least one wall must be an exterior wall, and the room must not be open to other parts of the building. Only indirect means, such as steam and hot water, should be used for *heating* in rooms containing hydrogen.

13.2.6.2 Test Facilities

Test facilities (chambers, cells, or stands) should be so constructed as to adhere to appropriate safety standards and guidelines. Test cells that cannot be ventilated sufficiently to prevent explosive hazards should be provided with an inert atmosphere of nitrogen, carbon dioxide, helium, steam, or other inert gas. The test cell pressure should be higher than atmospheric to avoid inflow of air. Nevertheless, the system design must prevent asphyxiation of personnel in adjacent areas. The system design must prevent the entrance of personnel into the cell unless confined space conditions are safe.

A partial vacuum may be used to restrict oxidants (e.g., oxygen) in a test chamber. In this case, the vacuum should be sufficient to limit the pressure of an explosion to a value that the system can withstand. Reflected shock waves should be taken into consideration in the design of heads, baffles, and other obstructions in a pipe run. Ultimate stress values should be used because of the violent nature of an explosion.

13.2.7 PLACARDING IN EXCLUSION AND CONTROL AREAS

Exclusion areas must have placards, postings, and labels displayed, so personnel are made aware of the potential hazard in the area.

Gaseous hydrogen systems must be permanently placarded as follows:

HYDROGEN – FLAMMABLE GAS – NO SMOKING – NO OPEN FLAMES.

Storage sites for liquefied hydrogen systems must be fenced and posted to prevent entrance by unauthorized personnel and placarded as follows:

LIQUEFIED HYDROGEN – FLAMMABLE GAS – NO SMOKING – NO OPEN FLAMES.

A sign must be placed on the container near the pressure-relief valve vent stack or on the vent stack warning against spraying water on or in the vent opening.

The maximum number of workers and visitors permitted at any time and the maximum amount of propellant materials and their Groups/Classes must be placarded and posted in a conspicuous place in all buildings, cells, rooms, and storage areas containing liquefied hydrogen. Safety showers must be placarded:

NOT TO BE USED FOR TREATMENT OF CRYOGENIC BURNS.

13.2.8 PROTECTION OF HYDROGEN SYSTEMS AND SURROUNDINGS

13.2.8.1 Barricades

Barricades are constructed to protect uncontrolled areas from the effects of a hydrogen system failure and to protect a hydrogen system from the hazards of adjacent or nearby operations. Pressure vessels, piping, and components are designed in a way that failure caused by overpressure or material deficiencies will not produce shrapnel. Barricades should be constructed adjacent to the expected fragment source and in a direct line of sight between it and the facility to be protected because barricades have been shown to be most effective against fragments and only marginally effective in reducing overpressures at extended distances from them [8].

The *housing of equipment* provides partial protection in many cases. Shrapnel protection can be achieved by blast curtains or blast mats placed adjacent to the equipment to be protected.

Barricades are commonly constructed as mounds and single-revetted barricades. A mound (earthworks) is an elevation of sloped dirt with a crest at least 0.91 m wide, whereas single-revetted barricades are mounds supported on a retaining wall on the side facing the hazard source.

Simulations [8] and experimental work [9] have shown that:

- Barricades should be designed to block the line of sight between equipment from which fragments can originate and the protected items.
- Barricades should be placed adjacent to the fragment source for maximum protection.
- The efficiency of barricades depends on the height above ground and the barricade location, dimensions, and configuration.
- Barricades can reduce peak overpressures and impulses behind the barricades, but reflection from obstacles may amplify blast waves behind barricades.
- Single-revetted barricades are more efficient than mounds.

13.2.8.2 Liquid Spills and Vapor Cloud Dispersion

A liquid H_2 (LH_2) spill from a storage vessel will result in a brief period of ground-level flammable cloud travel. The quick evaporation of the liquid causes the hydrogen vapors to mix quickly with air, dilute to nonflammable concentrations, warm up, and become positively buoyant. Accident prevention measures in such cases would be either natural dispersion or confinement of the spill [10].

If barricades are chosen, these should not excessively confine the vapor cloud formed because this may lead to detonation rather than simple burning of escaped hydrogen. LH_2 spills in an open-ended (U-shaped) bunker may produce detonation of the hydrogen–air mixture even without a roof [4].

The use of dikes and barricades around storage tanks is not recommended for LH_2, although this is required for liquid natural gas (LNG) because it may prolong evaporation and ground-level travel of the flammable cloud. Hydrogen detectors should be positioned to indicate the possible ground-level travel of flammable mixtures. Sewer drains must not be located in an area in which a liquefied hydrogen spill could occur.

13.2.8.3 Shields and Impoundment Areas

Impoundment areas and shields to control the extent of liquid and vapor travel caused by spills should be included in the design of a facility using hydrogen. The loading areas and the terrain below the transfer piping should be directed to a sump or impoundment area. Crushed stone should be used in the impoundment area to provide added surface area for liquefied hydrogen dissipation.

Ignition of hydrogen–air mixtures in free space usually results in combustion or deflagration, but with a certain degree of confinement a deflagration can evolve into a detonation (termed DDT). Consequently, careful design of installations should eliminate possible confinement caused by the equipment or buildings because initial combustion or deflagration of hydrogen–air mixtures may evolve to DDT.

13.2.9   QUANTITY–DISTANCE RELATIONSHIPS

In spite of the numerous efforts to determine quantity–distance relationships between LH_2 storage facilities and inhabited buildings in various countries, the results are largely uncertain because of the different assumptions made by each institution. The International Atomic Energy Agency (IAEA) [11] has taken up many of these quantity–distance relationships as specified in codes and standards of different countries and institutions. They are shown for comparison in Figure 13.1.

Many efforts to define a quantity–distance relationship have turned to the so-called *cube-root scaling law*. This law relates the mass of a flammable substance with distance to define safety distance and is expressed by the simple equation:

$$R = k * M^{1/3} \qquad (13.1)$$

where R is the distance (m) from a mass M (kg) of a flammable substance.

The factor k depends on the type of building to be protected. According to German recommendations, this factor takes the value 2.5 to 8 for working buildings, 22 for residential buildings, and 200 for supposed no damage. The factor k may be modified by damping parameters if protective measures such as earth coverage or protective walls are taken.

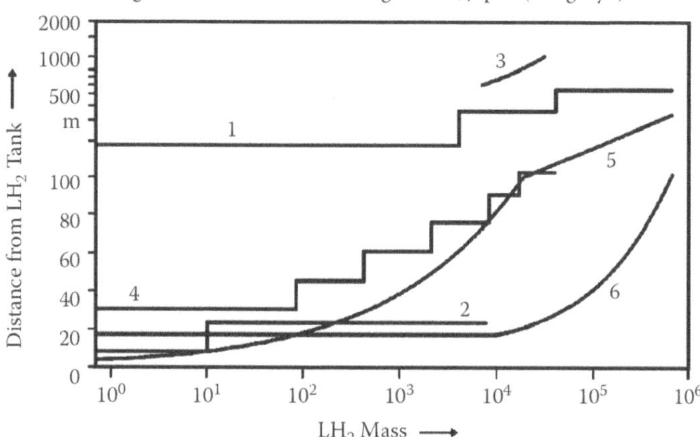

FIGURE 13.1   Safety distances of LH_2 storage tanks from inhabited buildings as a function of LH_2 mass in various countries and institutions (please note scale change on the ordinate).

Source: [11].

13.3 HYDROGEN FUELING OF VEHICLES

The Final Report on *Guidelines for Use of Hydrogen Fuel in Commercial Vehicles* of the US Department of Transportation [5] is a thorough text covering this issue in detail. Thus, the following sections make use of its suggestions in brief, and it is recommended that those dealing with hydrogen safety with regard to use as a vehicular fuel consult this valuable report.

13.3.1 GUIDELINES FOR HYDROGEN SYSTEMS ON VEHICLES

13.3.1.1 Compressed Hydrogen Systems

Vehicle design – For safety reasons hydrogen-fueled vehicles should be properly marked with diamond-shaped labels indicating "Compressed Hydrogen". The labels should be legible from 15 m in daylight. The gas pressure should be reduced gradually in three stages: fuel storage system (up to 345 bar), motive-pressure circuit (up to 12 bar), and low-pressure circuit (up to 1 bar). The whole system should be equipped with proper pressure-relief devices, isolation valves, and flow regulators.

A safety factor of three is recommended for the whole gas system according to NFPA 2005 [12], at least for vehicles circulating or constructed in the United States.

Furthermore, many other recommendations should be followed, some of which are listed here:

- The fuel cylinders should be permanently marked "Hydrogen" and securely mounted and protected from damage by road debris.
- All cylinders should be equipped with PRD (pressure-relief devices), TRD (thermal-relief devices) and manual shut-off valves.
- Isolation of the cylinders from the rest of the system should be done with electrically activated "fail safe" valves.
- Special care should be taken in the selection of construction materials in contact with hydrogen to ensure they are not subject to hydrogen embrittlement.
- Not only the fuel system but also the whole engine system should be electrically bonded and finally grounded to avoid the buildup of static electricity.
- All vehicle compartments should be efficiently ventilated to avoid accumulation of hydrogen in case of a leak exceeding a hydrogen concentration of 1 vol% (25% of the lower flammable limit).
- Hydrogen sensors should be installed on the vehicle and be properly connected to the vehicle control system to activate an alarm and automatic system shutdown when hydrogen concentration exceeds 1 to 2 vol%.
- An automatic system shutdown should be installed such that it is triggered by detection of leaked hydrogen, excess fuel flow, vehicle crash, or other system default.
- The vehicle operator should have easy accessibility to an on/off switch for shutting down the fuel cell system, disconnecting the traction power, de-energizing high-voltage equipment, and shutting off fuel supply.
- An interlock to the vehicle fueling port should prevent fueling unless the fuel cell system is shut down and the traction system is off.

- Electrical bonding of the fueling receptacle to the vehicle chassis is a must, accompanied by efficient grounding.
- When needed, hydrogen venting should be done by an outlet at the top of the vehicle.

Operation and maintenance – All *operators* of compressed-hydrogen-fueled vehicles should receive special training on hydrogen hazards and emergency response. Those dealing with *maintenance* should never use noncertified replacement parts. Parts intended for natural gas should not be used in hydrogen systems. Special caution is needed for checking any probable leak. Furthermore:

- The maintenance technicians should never loosen any joints under hydrogen pressure or over-tighten joints above the levels specified by manufacturers.
- If exposed to the atmosphere, purging of any component, and especially of hydrogen fuel cylinders, is only allowed with nitrogen before refilling.
- Thorough visual inspection of hydrogen cylinders for nicks, dents, and cuts must be conducted at least every 36 months or 58,000 km.
- Periodic checks and calibrations should always be performed according to the manufacturer's service manual.
- Warning lights and alarms should not be ignored and automatic system shutdown should not be overridden.
- No repairing of fuel lines is allowed, only replacement.
- Isolation of the fuel system, disconnection of the battery, and switching off of the main switch are mandatory before servicing the vehicle.
- No smoking or use of cell phones is allowed when servicing the vehicle.
- The internal pressure in hydrogen cylinders should always remain above atmospheric pressure; if not, purging with nitrogen is indispensable before refilling with hydrogen.

13.3.1.2 Liquid Hydrogen Systems

Experience of cryogenic systems onboard is not as common as for compressed gases. In addition, special care should be taken with respect to hazards stemming from the extremely low temperatures.

Vehicle design – Labels should be put on the exterior of the vehicle saying "Liquid Hydrogen" and should be legible from a distance of at least 15 m. The cryogenic tanks should also be marked with the word "Hydrogen", securely fastened, routed away from heat sources onboard, and protected from road debris. In the absence of certification standards for liquid hydrogen tanks, at least the existing standards for LNG should be applied for liquid hydrogen (including drop and flame tests).

Furthermore, many other recommendations should be followed, some of which are listed here:

- The cryotanks should be equipped with a pressure-relief valve, whose outlet should empty into a hydrogen diffuser. The latter should be able to dilute hydrogen with air to a concentration less than 1 vol%.

- A manual shutoff valve should be installed on the vehicle to isolate the hydrogen tank from the rest of the fuel system.
- A liquid level gauge should be installed on each cryotank, with indication in the driver's cab and a pressure gauge readable near the tank.
- Each cryotank should be equipped with at least one electrically activated valve isolating the tank from the rest of the fuel system. These valves should be of the "fail safe" type.
- Fuel lines should not pass through the passenger compartment.
- All materials coming into contact with liquid hydrogen should be chosen to withstand the liquid hydrogen temperatures, and the materials coming into contact with gaseous hydrogen should be resistant to hydrogen embrittlement.
- The fuel and engine system of a liquid hydrogen vehicle should be electrically bonded and finally grounded to prevent buildup of static electricity.
- A pressure-relief valve should be installed on every liquid hydrogen line isolated between two potentially closed valves to deal with venting hydrogen as it vaporizes because of line heat-up.
- Hydrogen sensors should be installed on the vehicle, providing an alarm and automatic system shutdown if the hydrogen concentration exceeds 1 to 2 vol%.
- Excess flow valves should also be installed on the vehicle capable of stopping fuel flow when the flow exceeds the set threshold.
- The liquid hydrogen vehicle should be equipped with an inertial crash sensor to automatically shut down the vehicle in the case of a crash. There should be the possibility to allow the vehicle to operate for a short time with the aid of a switch in an emergency, such as moving out of high-speed traffic or off a railroad track.
- The vehicle operator should have easy accessibility to an on/off switch for shutting down the fuel cell system, disconnecting the traction power, de-energizing high-voltage equipment, and shutting off the fuel supply.
- An interlock to the vehicle fueling port should prevent fueling unless the fuel cell system is shut down and the traction system is off.
- When needed, hydrogen venting should be done by an outlet at the top of the vehicle.

Operation and maintenance – All *operators* of LH_2-fueled vehicles should receive special training on hydrogen hazards and emergency response. Those dealing with *maintenance* should never use noncertified replacement parts. Parts intended for natural gas should not be used in hydrogen systems. Special caution is needed for checking any probable leak. Lines transmitting liquid hydrogen must be thermally insulated, with the outer layer of insulation vapor-sealed. All components coming into contact with liquid hydrogen must withstand the extremely low temperatures.

Furthermore, many other recommendations relating not only to high pressure but also to the extremely low temperatures encountered with liquid hydrogen should be followed, some of which are listed here:

- Maintenance technicians should never work on a liquid hydrogen system without wearing appropriate personal protective equipment (e.g., safety glasses or full-face shield, leather gloves and boots, long-sleeved shirt, long pants without cuffs. Pant legs should be worn outside of the boots).

- They should never loosen or over-tighten joints connected to hydrogen lines under pressure or containing liquid hydrogen.
- They should never disturb the insulation of liquid hydrogen lines or cryotanks still containing liquid hydrogen.
- They should never allow air to enter any component of the hydrogen system. If that happens accidentally (because of lower than ambient pressure in a cryotank), helium should be used to purge before refilling. Nitrogen is not allowed to be used in liquid hydrogen systems because it will liquefy and then solidify when in contact with liquid hydrogen, thus plugging lines and valves.
- Connections of the whole hydrogen system should be checked periodically according to the manufacturer's service manual.
- The outer surface of fuel lines and cylinders should be checked for any damage, as well as the exterior insulation layer of the cryotank for any damage causing loss of its function as a vapor barrier.
- The technicians should periodically check the hydrogen sensors and the hydrogen diffuser fan according to the manufacturer's service manual.
- Warning lights and alarms should not be ignored and the automatic system shutdown should not be overridden.
- No repairing of fuel lines is allowed, only replacement.
- Isolation of the fuel system, disconnection of the battery, and switching off of the main switch are mandatory before servicing the vehicle.
- No smoking or use of cell phones is allowed when servicing the vehicle.
- Before fueling, the vehicle should be bonded and grounded. The ullage space specified by the manufacturer should be maintained and no overfilling is allowed.

13.3.1.3 Liquid Fuel Reformers Onboard

In an effort to minimize the load of a hazardous material in a process according to the principles of inherent safety discussed in Chapter 7, *fuel reformers* have been developed, which reform a liquid fuel such as diesel fuel, gasoline, or methanol to a *reformate* containing 45 to 75% hydrogen. The other constituents are carbon dioxide, nitrogen, water, and sometimes carbon monoxide.

The reformate is produced on demand and can be used to feed a hydrogen fuel cell, thus eliminating the need to carry significant quantities of either gaseous or liquid hydrogen on the vehicle. After shutting down the engine, the remaining reformate is usually vented and no hydrogen is left onboard.

Vehicle design – Labels should be put on the exterior of the vehicle saying "Liquid Hydrogen", which should be legible from a distance of at least 15 m. When the reformer is in operation, the temperature of the fluids inside it may be between 93 and 816°C. This is why the whole system should be housed in a shielded package to prevent operators and technicians from coming into contact with hot surfaces.

Although the quantities of hydrogen onboard are minimal, sensors should be installed on the vehicle, providing an alarm and automatic shutdown when hydrogen or carbon monoxide concentrations in any compartment exceed a preset threshold. Remaining reformate in any system should be vented upon shutdown.

Operation and maintenance – All operators and maintainers of liquid fuel reformer vehicles should receive special training on hydrogen hazards and emergency response. This training should include information on the high temperatures encountered in leaking fluids from the reformer, which can cause serious burns.

The reformers should be designed as "line replaceable units" in a way that the mechanic will not open the hot box to make repairs. Line connections should be periodically checked for leaks according to the manufacturer's service manual. Any air and fuel filters, as well as desulfurization traps, must periodically be replaced accordingly.

13.3.2　Guidelines for Hydrogen Fueling Facilities

Fueling stations are supposed to be designed, constructed, and operated taking into account the hazardous properties of hydrogen and according to current codes, standards, and regulations in every country. The following information addresses mainly vehicle operators.

13.3.2.1　Compressed Hydrogen Fueling

The experience gained from compressed natural gas fuel dispensers is indispensable to the development of compressed hydrogen fuel dispensers because both are very similar. The main difference is the supply of the fuel gas; natural gas fuel stations are usually supplied with gas from pipelines, whereas hydrogen gas fuel stations are usually supplied with tube trailers hauled by trucks (Figure 13.2).

FIGURE 13.2　Hydrogen tube trailer. Photo courtesy of Sunline Transit.

Source: [5].

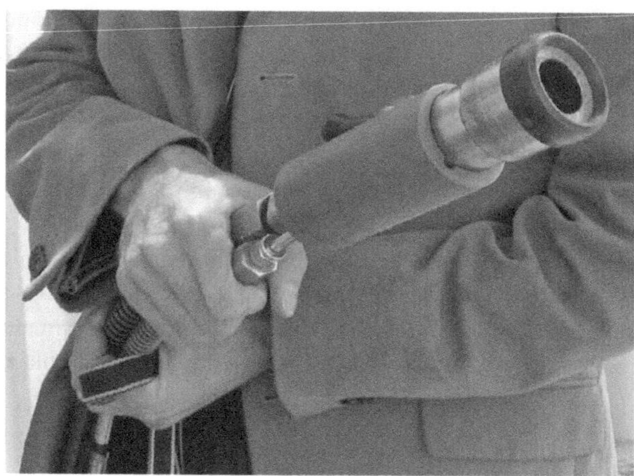

FIGURE 13.3 Compressed hydrogen fuel nozzle. Photo courtesy of Santa Clara Valley Transportation Authority.

Source: [5].

The fueling system for compressed hydrogen is comprised of a coupling on the fueling nozzle that mates with a compatible coupling on the vehicle. These are locked together when the attendant turns a lever (Figure 13.3).

When liquid hydrogen is supplied to the fueling station a vaporizer needs to be installed as well. This is essentially a heat exchanger to vaporize the liquid hydrogen and raise the temperature of the resulting gas to ambient temperature if hydrogen has to be dispersed in a gaseous state (Figure 13.4).

In some cases, hydrogen is produced on-site by reforming natural gas, as occurs with onboard fuel reformers, but in this case the reformer is much larger and a separator has to be installed for the separation of hydrogen from the other ingredients of the reformate.

Design – As is the case with all flammable gases and vapors, hydrogen fuel stations are located outdoors in the open air so that any gases released can be easily and safely dispersed in the air. Any construction should be designed in a way that no collection of escaping gas is allowed. Thus, if, for instance, there is a canopy, this should be constructed with upward slopes (Figure 13.5).

Generally, measures should be taken to protect the station from any damage caused by vehicles, the public, and other nearby activities. According to the NFPA [12], the minimum distance of any component of the fueling station from other buildings or public roadways should be 3 m.

Other recommendations are:

- Measures with special devices that shut off fuel flow should also be taken to prevent accidental or intentional movement of a vehicle with the hose connected.

FIGURE 13.4 Liquid hydrogen storage tank and vaporizer. Photo courtesy of Santa Clara Valley Transportation Authority.

Source: [5].

FIGURE 13.5 Compressed hydrogen fueling station. Photo courtesy of Alameda Contra-Costa Transit District.

Source: [5].

- Explosion-proof electrical equipment must always be installed and all components including the fueling hose must be bonded and grounded. The vehicle must also be bonded and grounded, usually via the fueling hose.
- Open-flame heaters are not allowed at the fueling station; hot air, steam, or hot water should be used for indirect heating instead.
- The fueling system should have provisions for controlling the final pressure so as not to exceed the rated pressure, and for not permitting refilling if the pressure inside the vehicle fuel tanks is less than normal atmospheric pressure. The latter case indicates that air might have entered the tank and so there is a possibility of an explosive mixture having been formed inside the tank.
- All hydrogen storage tanks should be equipped with a PRD (pressure-relief device)/TRD (thermal-relief device). Any hydrogen venting should be routed to an outlet higher than any surrounding structure, which should be equipped with a hydrogen diffuser or flame arrestor.
- If a compressed hydrogen fuel station also has a liquid hydrogen storage tank (cryotank), this should be equipped with a PRD, as well as any line isolated between two closed valves.
- The equipment at the station should include emergency stop systems and hydrogen sensors, including ultraviolet flame sensors to detect flames in the vicinity of the fuel dispenser. Automatic shutdown should occur when the hydrogen concentration in the air is more than 1 vol%.
- Dry powder fire extinguishers should be available at the hydrogen station for firefighting.
- Signs at the fuel station should remind customers of the proper behavior required in the area of the station, as well the appropriate actions in case of an emergency.

Operation and maintenance – All personnel at a hydrogen fueling station should be trained in hydrogen safety and particularly the operational characteristics of these stations. All activities that might result in the creation of sparks, flames, or hot points, such as smoking, use of cell phones, welding, or cutting metals, must be forbidden in the area.

A vehicle should be allowed to be fueled only after the driver has turned off the ignition and immobilized the vehicle. In addition to bonding and grounding via the fuel nozzle, additional grounding by a ground bond strap is recommended. Vehicle maintenance should not be allowed at a hydrogen fuel station and any other maintenance of the station must be performed only with nonsparking tools. At least annual maintenance of emergency equipment, such as hydrogen and flame sensors, should be performed according to the manufacturer's service manual (NFPA [12]).

13.3.2.2 Liquid Hydrogen Fueling

The main difference between compressed gaseous hydrogen fueling stations and liquid hydrogen fueling stations is that compressors and pumps are not used to pump out liquid hydrogen from cryotanks. This is accomplished by the addition of heat to the tank by a suitable safe heater. This energy input increases the pressure inside the tank, thus forcing out liquid hydrogen.

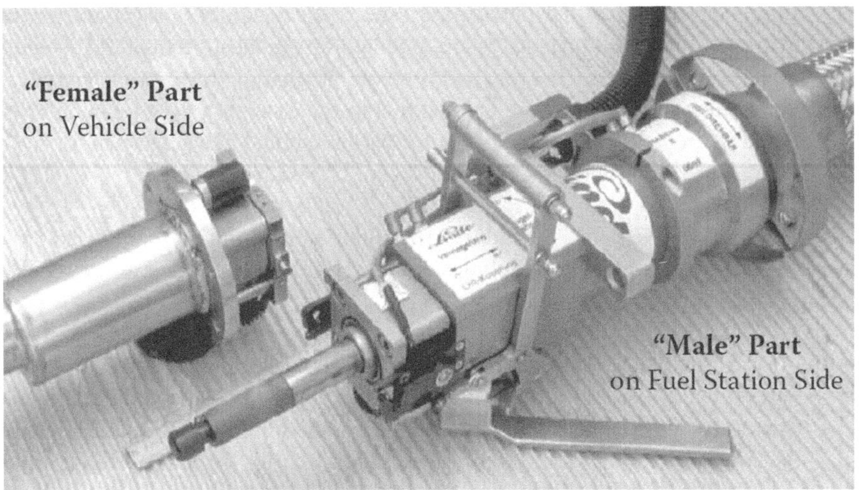

FIGURE 13.6 Liquid hydrogen fueling nozzle and vehicle fuel port. Photo courtesy of Air Products and Chemicals.

Source: [5].

Liquid hydrogen fueling nozzles are more complicated than those used for compressed hydrogen fueling. In this case, the so-called male coupling is found on the fueling nozzle and the so-called female coupling is on the vehicle (Figure 13.6). Both are locked together with the aid of a lever. Only then can fueling begin.

Design – All safety measures applied to compressed gaseous hydrogen fueling stations are also applicable to liquid hydrogen fueling stations. Nevertheless, additional safety measures should be taken owing to cryogenic conditions found in this case. Such additional measures are:

- According to the NFPA [12], hydrogen cryotanks should not be larger than 600 gallons (2.271 m^3) and liquid hydrogen fueling dispensers should be located in the open air.
- The specified minimum set-back distances between liquid hydrogen storage tanks and occupied buildings or public roadways vary between 3 m and 22.8 m, depending on the size of the storage cryotanks. The minimum distance of liquid hydrogen dispensers from occupied buildings should be at least 7.6 m.
- Special note should be taken of the fact that a liquid hydrogen leak could find its way to drains or underground facilities, thus creating a potential explosion hazard after vaporization and formation of an explosive mixture with the air. Consequently, measures such as dikes around the cryotanks should be taken to prohibit this possibility and, in any case, give sufficient time for liquid hydrogen to vaporize before reaching underground facilities.
- Good insulation and an unimpaired outer surface of the insulation are indispensable for safety. Liquid hydrogen leaks are able to liquefy air that is oxygen

enriched. Thus, liquefied air can pose a severe fire hazard if storage tanks, liquid hydrogen lines, and dispensers are mounted on combustible materials such as asphalt. In general, a concrete pad is recommended to prevent this hazard.
- Sometimes ice and frost are created due to the extremely low temperature of liquid hydrogen; such ice and frost formations should be cleaned up for safety reasons using a high-pressure air, nitrogen, or helium supply.

Operation and maintenance – In addition to the safety measures recommended previously for compressed hydrogen fueling stations, the following measures are required for liquid hydrogen fueling stations:

- The extremely low temperature of liquid hydrogen requires special care; otherwise, it will cause severe frostbite to those exposed to this hazard unprotected. Thus, persons connecting and disconnecting the liquid hydrogen nozzle from the vehicle should wear safety glasses and a full-face shield, loose fitting insulated or leather gloves, leather boots at ankle height or higher, a long-sleeved shirt, and long pants without cuffs. Pant legs should be worn outside of the boots.
- Since ice and frost may be formed on the nozzle, the attendant should check the mating surfaces before fueling. If this happens, the attendant may use high-pressure air, nitrogen, or helium to remove this ice and frost to prevent eventual leaking during fueling.
- Maintenance (at least every six months) of emergency equipment such as hydrogen and flame sensors should be performed according to the manufacturer's service manual (NFPA [12]).

13.4 STORAGE, HANDLING, AND DISTRIBUTION OF LIQUID HYDROGEN

In 2002 the European Industrial Gases Association (EIGA) prepared a Code of Practice (COP) on safety in the storage, handling, and distribution of liquid hydrogen [6]. This COP deals with many issues concerning hydrogen safety, including:

- Physical, chemical, and biological properties of hydrogen.
- Customer installations (layout and design features, access to installations, testing and commissioning, decommissioning and removal of tanks, operations and maintenance, customer information).
- Transport and distribution of liquid hydrogen (road transport, tank container, transport by railway, waterways, and sea).
- Training and protection of personnel (gas supplier and customer, work permits).

The first two of these items are generally in accordance with the detailed information and recommendations given in Sections 13.1 and 13.2, which are based on the American National Standard ANSI/AIAA G-095A-2017, *Guide to Safety of Hydrogen and Hydrogen Systems* [4] and on the *Guidelines for Use of Hydrogen*

Fuel in Commercial Vehicles of the US Department of Transportation [5]. However, the last two items give additional information and recommendations addressed to those dealing with the transport of liquid hydrogen by all transport means (with the exception of pipelines and portable containers such as pallet tanks and cylinders). An overview of the EIGA's COP follows.

13.4.1 CUSTOMER INSTALLATIONS

Concerning customer installations, minimum safety distances are recommended depending on the endangered object or item. Safe distances are based on hydrogen cloud dispersion models, heat flux effects of a hydrogen flame, and local overpressure due to flame ignition as measured from the release point. Some of these recommended safety distances are shown in Table 13.1.

These safety distances can be reduced if certain protection measures are taken (such as a water spray curtain) between the liquid hydrogen installation and the exposure area. Nevertheless, such protection measures can be applied only for certain cases, such as items 1, 2, 5, and 8 in Table 13.1.

Provisional measures, in addition to those suggested in the *Guidelines for Use of Hydrogen Fuel in Commercial Vehicles* of the US Department of Transportation [5], such as the construction of hydrogen storage installations in the open air, include barriers or bollards to eliminate vehicular impact. The hydrogen transfer should only take place within the user's premises. In an emergency, at least two separate outward-opening exits, at least 0.8 m wide, should be provided in technical buildings of the hydrogen installation.

The installation and operation of electrical systems in hydrogen installations (within the distance given in Table 13.1) need to be in accordance with national regulations, standards, and codes of practice, and especially with the last amendment

TABLE 13.1
Selected Recommended Minimum Safety Distances for Liquid Hydrogen

	Items	Distance (m)
1	Technical and unoccupied buildings	10
2	Occupied buildings	20
3	Other LH_2 fixed storage	1.5
4	Other LH_2 tanker	3
5	Flammable gas storage	8
6	Open flame, smoking, welding	10
7	Public establishments	60
8	Railroads, roads, property boundaries	10
9	Overhead power lines	10

Source: [6]. With permission.

of Directive 79/196/EEC (electrical equipment for use in potentially explosive atmospheres). Only explosion-proof electrical equipment must be used inside the safety distances given in Table 13.1.

Bonding and grounding of all components is a necessity to preclude development of static electricity, especially when mechanical abrasion occurs or when a fluid containing particles flows past the surface of a solid. Most synthetic materials can very easily generate static electricity charges. The resulting electric sparks are usually sufficiently powerful to ignite the highly sensitive hydrogen.

All hydrogen vents need to be connected to a vent stack arranged to discharge in a safe place in the open air. The height of the vent stack outlet should be either 7 m above ground level or 3 m above the top of the tank, whichever is greater.

The EIGA code of practice also gives some recommendations for the rare occasion that gaseous hydrogen pipelines have to run in the same duct or trench as electrical cables. In this case, all joints of the hydrogen pipelines need to be welded or brazed and these pipelines located higher than other pipelines.

Care should also be taken in the use of certain detection instruments that are not normally compatible with safety precautions required for hydrogen, such as gas chromatographs and flame ionization detectors.

13.4.2 Transport and Distribution of Liquid Hydrogen

13.4.2.1 Road Transport

According to the EIGA, all means of transport of dangerous goods by road should follow the European Agreement Concerning the International Carriage of Dangerous Goods by Road (ADR [Accord Européen Relatif au Transport International des Marchandises Dangereuses par Route]). The ADR, under Council Directive 94/55/EC, harmonizes the law across the European Union. A consolidated "restructured" edition of ADR was published in 2005. The latest ADR rules apply from January 1, 2023.

The COP of the EIGA covers all operations, from the departure of the vehicle from the filling plant until it has completed all deliveries given in the route plan. These operations include route planning, periodic checking, parking, breakdown, product transfer, emergency procedures, and driver training.

Route planning. Planning needs to be very detailed, describing the exact itinerary the tanker or tank container will follow. In general, motorways and trunk roads should be preferred and tunnels should be avoided, as well as densely populated areas. Any diversion from the planned route should be made known to the home base as soon as it is safe to do so.

Periodic checking. The vehicle should be fully checked at the departure site and also checked periodically throughout the duration of the trip. The driver should immediately inform the monitoring base of any abnormalities.

Parking. For parking for meals and the like, the specific public parking areas for heavy goods should be preferred; parking should always be in the open air. The driver should take care and avoid obvious hazards, such as overhead power lines or liquid oxygen tankers when choosing a parking place.

Breakdown. The driver should use all available warning means, such as flashers, reflective triangles, and flashing amber lights, in the case of a breakdown on a highway. No hot work by any nonauthorized persons can be allowed on a liquid hydrogen tanker, unless purged and inerted, and a permit to work has been issued.

Product transfer into customer storage facilities. Transfer operation is only allowed by authorized, trained, and certified personnel of the customer. No transfill operation is allowed during a thunderstorm. The driver should wear all personal protective equipment (gloves, eye protection, helmet, overalls, and protective footwear) during the transfer operation. After product transfer, the delivery hose should be purged of hydrogen before disconnection.

13.4.2.2   Tank Container Transport by Railway

Tank containers transported by railway should be in accordance with RID, *Règlement Concernant le Transport International Ferroviaire des Marchandises Dangereuses* (Regulation Concerning the Transport of Dangerous Goods by International Railway). Only approved tank containers should be used for that purpose, and the nitrogen shield vessel should be full. The hydrogen containers should be positioned away from incompatible substances, such as oxidizing agents, and even away from other trains parked nearby in marshalling yards.

The national railway authority should have agreed on the planned route, and where the route crosses borders relevant information should be passed to the next railway authority. Detailed written instructions should be given to the national railway authority and all other concerned persons (e.g., emergency services) along the journey route. The hydrogen containers should be fully checked at the end of the journey and the results should be recorded on a checklist.

13.4.2.3   Transport by Waterways and Sea

The transportation of liquid hydrogen by waterways and sea should be in accordance with the International Maritime Organization's International Maritime Dangerous Goods Code and the guidelines given in the EIGA's IGC Doc 06/19. The latter should also be applied to road tankers.

In addition to the requirements of the International Maritime Organization's emergency procedures, specific instructions should be provided to the pertinent shipping authorities. A checklist should also be supplied to ensure that performance monitoring of the hydrogen tank is recorded at regular intervals.

13.5   EUROPEAN UNION, INTERNATIONAL, AND US LEGISLATION REGULATING HYDROGEN TECHNOLOGIES

13.5.1   EUROPEAN UNION LEGISLATION REGULATING FUEL CELLS AND HYDROGEN TECHNOLOGIES

European Union (EU) legislation on fuel cells and hydrogen (FCH) refers to the hydrogen applications covered by the HyLaw project [13]. Many legislative acts are relevant to the deployment of hydrogen technologies. Some of these acts have an impact on hydrogen technology deployment indirectly, through their relation to a wider regulatory area (e.g., health and safety, environmental law, labor law, transport law). These EU legislative acts result in obligations for developers and manufacturers. Yet their obligations depend also on additional national limitations, differing from country to country.

Nevertheless, many of the obstacles to hydrogen implementation result from a lack of harmonization of rules between Member States, such as those on green hydrogen and certificates of origin, or by unintended discordances between rules at a national level (such as standards for fuel quality and measurement), rather than significant legal and regulatory barriers imposed at the EU level.

EU legislation affects (directly or indirectly) the production, storage, transportation, and distribution of hydrogen, as well as the use of hydrogen as fuel, hydrogen refueling stations, hydrogen vehicles (cars, buses, trucks, bikes, motorcycles, boats, and ships), electricity grid issues for electrolyzers, the gas grid, stationary power, fuel cells and stationary power (micro-CHPs).

The most significant EU legislative act concerning stationary power (micro-CHPs), such as appliances burning gaseous fuels and their fittings, is Regulation (EU) 2016/426. This act prescribes the obligations of manufacturers, importers, and distributors of such appliances upon distribution on the market (Table 13.2).

13.5.2 US Legislation Regulating Hydrogen Utilization

Based on a DOE-funded report by Sandia National Laboratories [14], Table 13.3 shows examples of specific regulatory activities by various US agencies relevant to hydrogen production, storage, and delivery. (With permission from Sandia National Laboratories).

13.5.3 International Legislation Regulating Hydrogen Utilization

References for the international legislation affecting (directly or indirectly) the production, storage, transportation, and distribution of hydrogen, as well as the use of hydrogen as fuel, hydrogen refueling stations, hydrogen vehicles (cars, buses, trucks, bikes, motorcycles, boats, and ships), electricity grid issues for electrolyzers, the gas grid, stationary power, fuel cells and stationary power (micro-CHPs) are shown in Table 13.4 [13].

TABLE 13.2
Complete List of EU Legislation Affecting (Directly or Indirectly) Hydrogen Utilization (HyLAW [13], public domain)

Legislative Act	Applicable Categories/Applications
Directive 2012/18/EU of the European Parliament and of the Council of 4 July 2012 on the control of major-accident hazards involving dangerous substances (so-called Seveso Directive).	• Production and Storage of Hydrogen • Hydrogen Refueling Stations • Boats/Ships • Approval of Landing/Bunkering • Onboard Transport of H_2
ATEX Directive 2014/34/EU, covering equipment and protective systems intended for use in potentially explosive atmospheres.	• Production, Storage of Hydrogen • Hydrogen Refueling • Boats/Ships • Approval of Landing/Bunkering • Onboard Transport of H_2 • Injection of Hydrogen in the Gas Grid (including methanation) – Safety requirements
Directive 2010/75/EU on industrial emissions (integrated pollution prevention and control).	• Production of Hydrogen • Electricity Grid Issues for Electrolyzers – Legal Status of Power to Gas Plants • Injection of Hydrogen in the Gas Grid (including methanation) – Safety Requirements
Directive 2014/68/EU of the European Parliament and of the Council of 15 May 2014 on the harmonization of the laws of the Member States relating to the making available on the market of pressure equipment.	• Production and Storage of Hydrogen • Transport and Distribution of Hydrogen • Hydrogen Refueling Stations • Boats/Ships • Approval of Landing/Bunkering
Directive 2014/52/EU of the European Parliament and of the Council of 16 April 2014 amending Directive 2011/92/EU on the assessment of the effects of certain public and private projects on the environment.	• Production and Storage of Hydrogen • Hydrogen Refueling Stations – Safety Requirements

(continued)

TABLE 13.2 (Continued)
Complete List of EU Legislation Affecting (Directly or Indirectly) Hydrogen Utilization (HyLAW [13], public domain)

Legislative Act	Applicable Categories/Applications
European Parliament Directive 2011/92/EU and of the Council of 13 December 2011 on the assessment of the effects of certain public and private projects on the environment (EIA Directive).	• Production and Storage of Hydrogen
Directive 2001/42/EC on the assessment of the effects of certain plans and programmes on the environment (SEA Directive).	• Production and Storage of Hydrogen • Injection of Hydrogen in the Gas Grid (including methanation) – Safety Requirements
Council Directive 98/24/EC of 7 April 1998 on the protection of the health and safety of workers from the risks related to chemical agents at work.	• Production of Hydrogen
Directive 2004/35/CE of the European Parliament and of the Council of 21 April 2004 on environmental liability with regard to the prevention and remedying of environmental damage.	• Production of Hydrogen
European Regulation (EC) No. 1272/2008 on classification, labelling and packaging of substances (CLP Regulation).	• Production of Hydrogen • Transport and Distribution of Hydrogen
Directive 2009/104/EC of the European Parliament and of the Council of 16 September 2009 concerning the minimum safety and health requirements for the use of work equipment by workers at work.	• Production of Hydrogen
Directive 1999/92/EC of the European Parliament and of the Council of 16 December 1999 on minimum requirements for improving the safety and health protection of workers potentially at risk from explosive atmospheres.	• Production of Hydrogen • The use of Hydrogen as a Fuel – Hydrogen Refueling stations • Boats/Ships • Approval of Landing/Bunkering
Council Directive 89/654/EEC of 30 November 1989 concerning the minimum safety and health requirements for the workplace.	• Production of Hydrogen
Council Directive 89/391/EEC of 12 June 1989 on the introduction of measures to encourage improvements in the safety and health of workers at work.	• Production of Hydrogen
Council Directive 92/43/EEC of 21 May 1992 on the conservation of natural habitats and of wild fauna and flora.	• Storage of Hydrogen

Directive 2009/147/EC of the European Parliament and of the Council of 30 November 2009 on the conservation of wild birds.	• Storage of Hydrogen
Directive 2008/68/EC of the European Parliament and of the Council of 24 September 2008 on the inland transport of dangerous goods.	• Transport and distribution of Hydrogen – Road Transport • Vehicles (Road Vehicles) – Restrictions
ADR European Agreement concerning the international carriage of dangerous goods by roads.	• Transport and Distribution of Hydrogen – Road Transport • Vehicles (Road Vehicles) – Restrictions
Commission Regulation (EU) No. 453/2010 of 20 May 2010 amending Regulation (EC) No. 1907/2006 of the European Parliament and of the Council of 20 May 2010 on the Registration, Evaluation, Authorisation and Restriction of Chemicals (REACH).	• Transport and Distribution of Hydrogen – Road Transport
Directive 2004/54/EC of 29 April 2004 on minimal safety requirements for tunnels in the trans-European roads.	• Transport and Distribution of Hydrogen – Road Transport • Vehicles (Road Vehicles) – Restrictions
Directive 2010/35/EU, the Transportable Pressure Equipment Directive (TPED).	• Transport and Distribution of Hydrogen – Road Transport
Directive 2014/94/EU of the European Parliament and of the Council of 22 October 2014 on the deployment of alternative fuels infrastructure (AFID).	• The Use of Hydrogen as a Fuel and Hydrogen Refueling Stations
Directive (EU) 2015/1513 of the European Parliament and of the Council of 9 September 2015 amending Directive 98/70/EC relating to the quality of petrol and diesel fuels and amending Directive 2009/28/EC on the promotion of the use of energy from renewable sources (Indirect Land Use Change (ILUC) Directive).	• The Use of Hydrogen as a Fuel – Legal Status and Certificates of Origin
Directive 98/70/EC of the European Parliament and of the Council of 13 October 1998 relating to the quality of petrol and diesel fuels and amending Council Directive 93/12/EEC.	• The Use of Hydrogen as a Fuel – Legal Status and Certificates of Origin • The Use of Hydrogen as a Fuel – Quality Requirements
Directive (EU) 2018/2001 of the European Parliament and of the Council of 11 December 2018 on the promotion of the use of energy from renewable sources (RED II) and replacing Directive 2009/28/EC.	• The Use of Hydrogen as a Fuel – Legal Status and Certificates of Origin • Gas Grid Issues – Guarantees of Origin
Council Directive (EU) 2015/652 of 20 April 2015 laying down calculation methods and reporting requirements pursuant to Directive 98/70/EC of the European Parliament and of the Council of 20 April 2015 relating to the quality of petrol and diesel fuels.	• The Use of Hydrogen as a Fuel – Quality Requirements

(continued)

TABLE 13.2 (Continued)
Complete List of EU Legislation Affecting (Directly or Indirectly) Hydrogen Utilization (HyLAW [13], public domain)

Legislative Act	Applicable Categories/Applications
Directive 2006/42/EC of 17 May 2006 on machinery.	• The Use of Hydrogen as a Fuel – Hydrogen Refueling Stations
Directive 2014/35/EU relating to electrical equipment designed for use within certain voltage limits.	• The Use of Hydrogen as a Fuel – Hydrogen Refueling Stations
Directive 2014/29/EU on simple pressure vessels.	• The Use of Hydrogen as a Fuel – Hydrogen Refueling Stations
Directive 2007/46/EC of the European Parliament and of the Council of 5 September 2007 establishing a framework for the approval of motor vehicles and their trailers, and of systems, components and separate technical units intended for such vehicles.	• Vehicles – Cars, Buses, Trucks – Type Approval and Registration
Regulation (EC) No. 79/2009 of the European Parliament and of the Council of 14 January 2009 on type-approval of hydrogen-powered motor vehicles, and amending Directive 2007/46/EC.	• Vehicles – Cars, Buses, Trucks – Type Approval and Registration
Commission Regulation (EU) No. 406/2010 of 26 April 2010 implementing Regulation (EC) No. 79/2009 of the European Parliament and of the Council of 26 April 2010 on type-approval of hydrogen-powered motor vehicles.	• Vehicles – Cars, Buses, Trucks – Type Approval and Registration
Commission Regulation (EU) No. 630/2012 of 12 July 2012 amending Regulation (EC) No. 692/2008 as regards type-approval requirements for motor vehicles fueled by hydrogen and mixtures of hydrogen and natural gas with respect to emissions, and the inclusion of specific information regarding vehicles fitted with an electric powertrain in the information document for the purpose of EC type-approval.	• Vehicles – Cars, Buses, Trucks – Type Approval and Registration
Commission Regulation (EC) No. 692/2008 of 18 July 2008 implementing and amending Regulation (EC) No. 715/2007 of the European Parliament and of the Council of 18 July 2008 on type-approval of motor vehicles with respect to emissions from light passenger and commercial vehicles (Euro 5 and Euro 6) and on access to vehicle repair and maintenance information.	• Vehicles – Cars, Buses, Trucks – Type Approval and Registration • Vehicles – Cars, Buses, Trucks – Service and Maintenance

Regulation (EU) No. 168/2013 of the European Parliament and of the Council of 15 January 2013 on the approval and market surveillance of two- or three-wheel vehicles and quadricycles.	• Vehicles – Motorcycles, Bikes and Quadricycles – Type Approval and Registration
Commission Delegated Regulation (EU) No. 134/2014 of 16 December 2013 supplementing Regulation (EU) No. 168/2013.	• Vehicles – Motorcycles, Bikes and Quadricycles – Type Approval and Registration
Commission Directive 2003/127/EC of 23 December 2003 amending Council Directive 1999/37/EC on the registration documents for vehicles.	• Vehicles – Cars, Buses, Trucks – Type Approval and Registration
Council Directive 1999/37/EC of 29 April 1999 on the registration documents for vehicles.	• Vehicles – Cars, Buses, Trucks – Type Approval and Registration • Vehicles – Motorcycles, Bikes and Quadricycles – Type Approval and Registration
Commission Directive 2003/127/EC of 23 December 2003 amending Council Directive 1999/37/EC on the registration documents for vehicles.	• Vehicles – Cars, Buses, Trucks – Type Approval and Registration • Vehicles – Motorcycles, Bikes and Quadricycles – Type Approval and Registration
Directive 2014/45/EU of the European Parliament and of the Council of 3 April 2014 on periodic roadworthiness tests for motor vehicles and their trailers and repealing Directive 2009/40/EC Text with EEA relevance (with effect from 20 May 2018).	• Vehicles – Cars, Buses, Trucks – Service and Maintenance • Vehicles – Motorcycles, Bikes and Quadricycles – Service and Maintenance
Directive 2009/33/EC of the European Parliament and of the Council of 23 April 2009 on the promotion of clean and energy-efficient road transport vehicles (Clean Vehicle Directive).	• Vehicles – Incentives and Restrictions
Directive 2014/90/EU of the European Parliament and of the Council of 23 July 2014 on marine equipment.	• Boats/Ships – Design and Type Approval
Directive 2009/45 on safety rules and standards for passenger ships.	• Boats/Ships – Design and Type Approval
Directive 2009/16/EC on port State control.	• Boats/Ships – Approval of Landing/Bunkering Operations
Directive 2013/38/EU of the European Parliament and of the Council of 12 August 2013 amending Directive 2009/16/EC on port State control.	• Boats/Ships – Approval of Landing/Bunkering Operations
Commission Regulation (EU) 2016/1388 of 17 August 2016 establishing a Network Code on demand connection.	• Electricity Grid Issues for Electrolyzers – Connecting an Electrolyzer to the Grid
Directive 96/92/EC concerning common rules for the internal market in electricity.	• Electricity Grid Issues for Electrolyzers – Connecting an Electrolyzer to the Grid

(continued)

TABLE 13.2 (Continued)
Complete List of EU Legislation Affecting (Directly or Indirectly) Hydrogen Utilization (HyLAW [13], public domain)

Legislative Act	Applicable Categories/Applications
Directive 2009/72/EC of the European Parliament and of the Council of 13 July 2009 concerning common rules for the internal market in electricity.	• Electricity Grid Issues for Electrolyzers – Legal Status of Power to Gas Plants • Stationary Power – Residential Stationary FC (Micro-CHP)
Commission Regulation (EU) 2017/2195 establishing a guideline on electricity balancing.	• Electricity Grid Issues for Electrolyzers – Power to Gas and Their Role in Electricity Balancing
Directive 2012/27/EU of the European Parliament and of the Council of 25 October 2012 on energy efficiency.	• Electricity Grid Issues for Electrolyzers – Power to Gas and Their Role in Electricity Balancing • Stationary Power – Residential Stationary FC (Micro-CHP) – Price of Electricity and Support Mechanisms
Directive 2009/73/EC concerning common rules for the internal market in natural gas.	• Electricity Grid Issues for Electrolyzers – Legal Status of Power to Gas Plants • Gas Grid Issues • Stationary Power – Residential Stationary FC (Micro-CHP)
Regulation 715/2009 on conditions for access to the natural gas transmission networks.	• Gas Grid Issues
Regulation (EC) No. 713/2009 of the European Parliament and of the Council of 13 July 2009 establishing an Agency for the Cooperation of Energy Regulators.	• Gas Grid Issues
Commission Regulation establishing a Network Code on interoperability and data exchange rules (703/2015/EU).	• Gas Grid Issues
Commission Regulation (EU) 2017/460 of 16 March 2017 establishing a Network Code on harmonized transmission tariff structures for gas.	• Gas Grid Issues – Injection of Hydrogen in the Gas Grid – Payment Issues
Regulation (EU) 2016/426 of the European Parliament and of the Council of 9 March 2016 on appliances burning gaseous fuels and repealing Directive 2009/142/EC.	• Gas Grid Issues – Injection of Hydrogen – Safety Requirements (compliance with safety regulation/risk control expectations and end user safety requirements)
European Parliament resolution of 12 September 2013 on microgeneration – small-scale electricity and heat generation (2012/2930(RSP)).	• Residential Stationary FC (Micro-CHP) – Price of Electricity and Support Mechanisms

Source: [13]. *With permission.*

TABLE 13.3
Examples of Regulatory Activities by US Agencies Relevant to Hydrogen Production, Storage, and Delivery

System	Oversight	Reference	Summary
Production	EPA	40 CFR Part 98	Defines source categories and emissions thresholds for a hydrogen production facility.
Storage	OSHA	29 CFR Part 1910	Dictates the safety of the structural components and operations of gaseous and liquid hydrogen storage and delivery.
	FAA	14 CFR Part 420	Dictates the separation distance requirements for storage of liquid hydrogen and any incompatible energetic liquids.
Transportation by Pipeline	BSEE	43 USC Part 29	Manage compliance programs governing oil, gas, and mineral operations on the outer continental shelf.
	FERC	18 CFR Part 153	Regulation of the siting, routing, and overall construction of the pipeline system, as well as the distribution and interstate and intrastate sale of natural gas.
		18 CFR Part 284	Filing requirements of the siting, construction, and operation of facilities used for the import or export of natural gas.
	PHMSA	49 CFR Part 192	Prescribes minimum safety requirements for pipeline facilities and the transportation of gas, including pipeline facilities and the transportation of gas within the limits of the outer continental shelf.
		49 CFR Part 193	Prescribes safety standards used for LNG facilities that are used to transport gas via pipeline.
		49 CFR Part 195	Prescribes safety standards for pipeline facilities that transport hazardous liquids.
	USCG	33 CFR Part 154	Regulations for facilities transferring hazardous materials back and forth from a vessel to a facility.
Transportation by Road	PHMSA	49 CFR Part 172	Lists and classifies hazardous materials for transportation, and prescribes requirements for papers, markings, labeling, and vehicle placarding.
		49 CFR Part 173	Provides requirements for preparing hazardous materials for shipment, and inspection, testing, and other requirements for transportation containers.
		49 CFR Part 177	Provides additional requirements when transporting hazardous materials via public highways.

(*continued*)

TABLE 13.3 (Continued)
Examples of Regulatory Activities by US Agencies Relevant to Hydrogen Production, Storage, and Delivery

System	Oversight	Reference	Summary
		49 CFR Part 178	Prescribes specifications for packaging and containers used for transportation of hazardous materials.
		49 CFR Part 180	Provides qualification requirements for inspecting and maintaining packages and containers used to transport hazardous materials.
	FMCSA	49 CFR Part 356	Motor carrier routing requirements.
		49 CFR Part 389	General motor carrier safety regulations.
		49 CFR Part 397	Transportation of hazardous materials.
	FHWA	23 CFR Part 924	Regulates highway safety which includes bridges, tunnels, and other associated elements.
	FTC	16 CFR Part 306	Describes the certification and posting of automotive fuel ratings in commerce.
Transportation by Rail	PHMSA	49 USC 5117	Gives the authority to authorize a variance that is still at the same safety level; a special permit is required to use an alternative fuel that does not have a safety standard.
		49 CFR Part 172	Lists and classifies hazardous materials for transportation, and prescribes the requirements for papers, markings, labeling, and vehicle placarding.
		49 CFR Part 173	Provides requirements for preparing hazardous materials for shipment as well inspection, testing, and other requirements for containers, including usage instructions for DOT-113A60W tank cars.
		49 CFR Part 174	Provides additional requirements for transportation of hazardous materials in or on rail cars.
		49 CFR Part 178	Prescribes specifications for packaging and containers used for transportation of hazardous materials.
		49 CFR Part 179	Provides construction requirements for DOT-113A60W tank cars.
		49 CFR Part 180	Provides qualification requirements for inspecting and maintaining containers used to transport hazardous materials, including DOT-113A60W tank cars.
Transportation by Waterways	PHMSA	49 CFR Part 172	Lists and classifies hazardous materials for transportation, and prescribes the requirements for papers, markings, labeling, and vehicle placarding.
		49 CFR Part 173	Provides requirements for preparing hazardous materials for shipment, as well inspection, testing, and other requirements for containers.

		49 CFR Part 176	Requirements for transportation by vessel.
		49 CFR Part 178	Prescribes specifications for packaging and containers used for transportation of hazardous materials.
		49 CFR Part 180	Provides qualification requirements for inspecting and maintaining containers used to transport hazardous materials.
	USCG	33 CFR Part 154	Regulations for transferring hazardous materials back and forth from a vessel to a facility.
		33 CFR Part 156	Transfer of oil or hazardous materials on the navigable waters or contiguous zone of the US.
		46 CFR Part 38	Requirements for transportation of liquefied or compressed flammable gases.
		46 CFR Part 150	Describes incompatibility of hazardous materials and rules for transporting these materials aboard tanks that are loaded and discharged while on the vessel.
		46 CFR Part 151	Regulations for non-self-propelled ships carrying bulk cargo.
		46 CFR Part 153	Regulations for self-propelled ships carrying bulk cargo.
		46 CFR Part 154	Regulations for self-propelled vessels that contain bulk liquefied gases as cargo, cargo residue, or vapor.
Import/Export Terminals	FERC	18 CFR Part 153	Establishes filing requirements to obtain authorization for the siting, construction, operation, place of entry for imports, or place of exit for exports.
	PHMSA	49 CFR Part 192	Prescribes minimum safety requirements for pipeline facilities and the transportation of gas, including pipeline facilities and the transportation of gas within the limits of the outer continental shelf.
		49 CFR Part 193	Prescribes safety standards used for LNG facilities that are used to transport gas via pipeline.
		49 CFR Part 195	Prescribes safety standards for pipeline facilities that transport hazardous liquids.
	USCG	33 CFR Part 154	Regulations for self-propelled vessels that contain bulk liquefied gases as cargo, cargo residue, or vapor.
		33 CFR Part 156	Transfer of oil or hazardous materials on the navigable waters or contiguous zone of the US.
Electricity Production	FERC	18 CFR Part 292	Sets requirements for a small power production or cogeneration facility.
	FE	10 CFR Part 503	Prohibits any new baseload power plant without the ability to use coal or another alternative fuel as a primary energy source.
		10 CFR Part 504	May prohibit existing power plants from using petroleum or natural gas as a primary energy source.

(continued)

TABLE 13.3   (Continued)
Examples of Regulatory Activities by US Agencies Relevant to Hydrogen Production, Storage, and Delivery

System	Oversight	Reference	Summary
Residential and Commercial Heating	FERC	18 CFR Part 284	Provides regulation of energy sales and distribution of natural gas.
	EERE	10 CFR Part 431	Provides regulation of commercial heaters, hot water boilers, and similar heating appliances.
Chemical and Industrial Use	OSHA	29 CFR Part 1910	Dictates the safety of the structural components and operations of gaseous and liquid hydrogen in terms of storage as well as delivery.
	EPA	40 CFR Part 98	Requires reporting of greenhouse gas emission due to combustion or use of products in a process.
Auxiliary Power and Alternative Power Supply	FHWA	49 CFR Part 390	Regulates additional equipment on commercial vehicles to ensure it does not reduce the overall safety of the vehicle.
	FRA	49 CFR Part 229	Regulations for electrical systems, generators, protection from hazardous gases from exhaust and batteries, and crashworthiness for locomotives.
	USCG	46 CFR Part 111	Regulations for power supply systems on ships.
	FAA	14 CFR Part 23 Subpart E	Requirements for electrical generating systems including auxiliary and backup power for normal category airplanes.
		14 CFR Part 25 Subpart E	Requirements for electrical generating systems including auxiliary and backup power for transport category airplanes.
		14 CFR Part 27 Subpart E	Requirements for electrical generating systems including auxiliary and backup power for normal category rotorcraft.
		14 CFR Part 29 Subpart E	Requirements for electrical generating systems including auxiliary and backup power for transport category rotorcraft.
Use in Consumer/ Commercial Vehicles	NHTSA	49 CFR 571	Provides Federal Motor Vehicle Safety Standards for motor vehicles and motor vehicle equipment.
	FHWA	23 CFR Part 924	Regulates highway safety which includes bridges, tunnels, and other associated elements.
Use in Rail	FRA	49 CFR Part 229	Locomotive safety design and crashworthiness requirements.
		49 CFR Part 238	Safety requirements for passenger locomotives.
	FTA	49 CFR Part 659	Provides guidance for rail fixed guideway systems and the oversight of safety, including hazard management and safety and security plans and review.

		49 CFR Part 674	Mandates state safety oversight of fixed guideway public transportation systems.
Use in Maritime	USCG	46 CFR Parts 24–196	Regulation of vessel construction for both passenger and cargo applications as well as general fuel requirements based on the flash point of the fuel.
	FTA	49 USC Chapter 53	Requirements for a National Public Transportation Safety Plan for public transportation that receives federal funding.
Use in Aviation	FAA	14 CFR Part 23	Provides requirements and airworthiness standards for normal category airplanes.
		14 CFR Part 25	Provides requirements and airworthiness standards for transport category airplanes.
		14 CFR Part 26	Provides requirements and airworthiness standards for transport category airplanes.
		14 CFR Part 27	Provides requirements and airworthiness standards for normal category rotorcraft.
		14 CFR Part 29	Provides requirements and airworthiness standards for transport category rotorcraft.
		14 CFR Part 33	Provides requirements and airworthiness standards for aircraft engines.

Source: [14]. *With permission.*

TABLE 13.4

International Legislation Affecting (Directly or Indirectly) Hydrogen Utilization (HyLAW [13], public domain)

Legislative Act	Relevant Category/Application/ Process
United Nations Economic Commission for Europe (UNECE) 134 Hydrogen fueled vehicles	Vehicles
UN Regulation 110 Compressed and liquefied natural gas system components of motor vehicles, issued by the UNECE	Hydrogen shall be limited to 2 vol% when cylinders are manufactured from a steel with an ultimate tensile strength exceeding 980 MPa. This limitation applies to CNG type 1 cylinders and implies that CNG fueling stations, as well as the supplying gas network, are affected by the 2 vol% hydrogen limitation
IGF Code: International Code of Safety for Ships Using Gases or Other Low-Flashpoint Fuels, 2016 Edition (I109E)	• Boats/Ships – Design and Type Approval
United Nations Convention on Conditions for Registration of Ships, IMO Resolution A.600(15), IMO identifications of ships	• Boats/Ships – Registration for Ships
International Management Code for the Safe Operation of Ships and for Pollution Prevention (ISM Code)	• Boats/Ships – Design and Type Approval • Boats/Ships – Registration for Ships
IMO resolution MSC 391(95): Adoption of the International Code of Safety for Ships Using Gases or Other Low-Flashpoint Fuels (IGF Code)	
Resolution MSC.420(97) (adopted on 25 November 2016) Interim recommendations for carriage of liquefied hydrogen in bulk	• Boats/Ships – Onboard Hydrogen Transport
Resolution MSC.370(93) (adopted on 22 May 2014); Amendments to the International Code of the Construction and Equipment of Ships Carrying Liquefied Gases in Bulk (IGC Code)	
International Maritime Dangerous Good Code (IMDG Code)	
International Code of the Construction and Equipment of Ships Carrying Dangerous Chemicals in Bulk (IBC Code)	
ISO 14687-2:2012 Hydrogen fuel – Product specification – Part 2: Proton exchange membrane (PEM) fuel cell applications for road vehicles	• The Use of Hydrogen as a Fuel – Fuel Quality Requirements and Measurement

SAE J2719_201511 Hydrogen fuel quality for fuel cell vehicles	• The Use of Hydrogen as a Fuel – Fuel Quality Requirements and Measurement
ISO/TS 19880-1 Gaseous hydrogen – Fueling stations	• The Use of Hydrogen as a Fuel – • Fuel Quality Requirements and Measurement • Hydrogen Refueling Stations
ISO/CD 19880-8 Gaseous hydrogen – Fueling stations – Hydrogen quality control – Under development	• The Use of Hydrogen as a Fuel – • Fuel Quality Requirements and Measurement • Hydrogen Refueling Stations
IGC Document 155/09/E Best available techniques for hydrogen production by steam methane reforming	• The Use of Hydrogen as a Fuel – Hydrogen Refueling Stations
IGC Document 122/11 Environmental impacts of hydrogen plants	
IGC Document 15/06 Gaseous hydrogen stations	
IGC Document 100/11 Hydrogen cylinders and transport vessels	
IGC Document 121/14 Hydrogen transportation pipelines	
IGC Document 114/09 Operation of static cryogenic vessels	
IGC Document 6/02 Safety in storage, handling and distribution of liquid hydrogen	
IGC Document 23/08 Safety training of employees	
IGC Document 75/07 Determination of safety distances	
IGC Document 134/12 Potentially explosive atmospheres	
ISO/CD 19881 Gaseous hydrogen – Land vehicle fuel tanks	• Vehicles – Type Approval
ISO/CD 19882 Gaseous hydrogen – Land vehicle fuel tanks – Thermally activated pressure-relief devices	
ISO/AWI 17268 Gaseous hydrogen land vehicle refueling connecting devices	
ISO/TS 15869 Gaseous hydrogen and hydrogen blend – Land vehicle fuel tanks	
ISO 13984:1999 Liquid hydrogen – Land vehicle fueling system interface	
ISO 13985:2006 Liquid hydrogen – Land vehicle fuel tanks	
ISO/TR 15916:2015 Basic considerations for the safety of hydrogen systems	
ISO 23273:2013 Fuel cells road vehicles – Safety specifications – Protection against hydrogen hazards for vehicles fueled with compressed	

(continued)

TABLE 13.4   (Continued)
International Legislation Affecting (Directly or Indirectly) Hydrogen Utilization (HyLAW [13], public domain)

Legislative Act	Relevant Category/Application/ Process
UN Regulation 110 Compressed and liquefied natural gas system components of motor vehicles, issued by the UNECE	• Gas Grid Issues – Injection of Hydrogen in the Gas Grid – Legal Framework, Permissions and Restrictions • Gas Grid Issues – Injection of Hydrogen – Safety Requirements Related to End User Equipment
EN 16726:2015 Gas infrastructure – Quality of Gas Group H EN 16723-1 Natural gas and biomethane for use in transport and biomethane for injection in the natural gas network, issued by the European Committee for Standardization (CEN) ISO 13686 Natural gas – Quality designation, issued by the International Organization for Standardization (ISO)	• Gas Grid Issues – Injection of Hydrogen in the Gas Grid – Legal Framework, Permissions and Restrictions
Common Business Practice (CBP) 2005–1/02 Harmonization of gas qualities, issued by the European Association for the Streamlining of Energy Exchange–Gas (EASEE–gas)	• Gas Grid Issues – Injection of Hydrogen in the Gas Grid – Legal Framework, Permissions and Restrictions
UNI–ENI 16723-2 Natural gas and bio-methane for use in transport and bio-methane for injection in the natural gas network	• Gas Grid Issues – Injection of Hydrogen in the Gas Grid (Transmission Level) – Permission to Connect, Inject
UNI TR 11537 Bio-methane injection in the natural gas network	• Gas Grid Issues – Injection of Hydrogen in the Gas Grid (Transmission Level) – Permission to Connect, Inject
IGC 15/06 Gaseous hydrogen stations (includes security measures for injection) IGC 121/04 Hydrogen transportation pipelines, issued by the European Industrial Gas Association	• Gas Grid Issues – Injection of Hydrogen – Safety Requirements (compliance with safety regulation/risk control expectations)
UNI–ENI 16723-2 Natural gas and bio-methane for use in transport and bio-methane for injection in the natural gas network	• Gas Grid Issues – Injection of Hydrogen in the Gas Grid (Distribution Level) – Permission to Connect, Inject

UNI TR 11537 Bio-methane injection in the natural gas network	• Gas Grid Issues – Injection of Hydrogen in the Gas Grid (Distribution Level) – Permission to Connect, Inject
ISO 69674-6 Natural gas – Determination of composition with defined uncertainty by gas chromatography – Part 6: Determination of hydrogen, helium, oxygen, nitrogen, carbon dioxide and C_1 to C_8 hydrocarbons using three capillary columns	• Gas Grid Issues – Injection of Hydrogen in the Gas Grid (Distribution Level) – H_2 Quality Requirements
ISO 6976 Natural gas – Calculation of calorific values, density, relative density and Wobbe indices from composition	• Gas Grid Issues – Injection of Hydrogen in the Gas Grid (Distribution Level) – H_2 Quality Requirements

Source: [13]. *With permission.*

REFERENCES

1. ATEX Directive 2014/34/EU, European Parliament and of the Council, Harmonization of the laws of the Member States relating to equipment and protective systems intended for use in potentially explosive atmospheres (recast), 26 February, 2014.
2. Biennial Report on Hydrogen Safety, *HySafe (Safety of Hydrogen as an Energy Carrier)*, Chap. 6. www.hysafe.org
3. Kotchourko, A. and Jordan, T. (Eds.), *Hydrogen Safety for Energy Applications,* 1st ed., Elsevier, Amsterdam, Netherlands, 2022.
4. ANSI - American National Standard Institute, *Guide to safety of hydrogen and hydrogen systems,* American Institute of Aeronautics and Astronautics, ANSI/AIAA G-095A-2017, Chap. 4. ANSI, Washington, DC, 2017.
5. US DOT - U.S. Department of Transportation, *Guidelines for use of hydrogen fuel in commercial vehicles*, Final Report, US DOT, Washington, DC, 2007.
6. *Code of Practice on Safety in Storage, Handling and Distribution of Liquid Hydrogen*, European Industrial Gases Association (EIGA), Doc 06/19, Brussels, Belgium, 2019.
7. Rigas, F. and Sklavounos, S., Hydrogen safety, in Ram B. Gupta (ed.) *Hydrogen Fuel: Production, Transport and Storage,* CRC Press, Taylor & Francis, Boca Raton, FL, 2008, 563–565.
8. Sklavounos, S. and Rigas, F., Computer simulation of shock waves transmission in obstructed terrains, *Journal of Loss Prevention in the Process Industries*, 17, 407, 2004.
9. HSE - Health and Safety Executive, *Explosion hazard assessment: A study of the feasibility and benefits of extending current HSE methodology to take account of blast sheltering*, Report HSL/2001/04, Sheffield, UK, 2001.
10. Rigas, F. and Sklavounos, S., Evaluation of hazards associated with hydrogen storage facilities, *International Journal of Hydrogen Energy*, 30, 1501, 2005.
11. IAEA - International Atomic Energy Agency, *Hydrogen as an energy carrier and its production by nuclear power*, IAEA-TECDOC-1085, IAEA, Vienna, Austria, 1999.
12. NFPA - National Fire Protection Association, *Vehicular Fuel Systems Code*, National Fire Protection Association, Quincy, MA, 2006.
13. HyLAW, and Floristean, A., *EU regulations and directives which impact the deployment of FCH technologies*, Joint Undertaking supported by the European Union's Horizon 2020 research and innovation programme, Hydrogen Europe and Hydrogen Europe Research, February 2019.
14. Baird, A.R. et al., *Federal oversight of hydrogen systems*, SAND2021-2955, Sandia National Laboratory, Albuquerque, USA, March 2021. https://energy.sandia.gov/wp-content/uploads/2021/03/H2-Regulatory-Map-Report_SAND2021-2955.pdf

14 Epilogue

The prosperity of the present-day developed world has been based on the abundance of fossil fuels as an energy source and carrier. Yet these reserves of stored solar energy are finite and, in addition, are mainly composed of carbon. Thus, even if fossil fuel deposits were unlimited, they should be abandoned in the coming decades, owing to the massive release of carbon dioxide into the atmosphere and the resulting greenhouse effect, subsequently leading to climate change. Herein lies the necessity for new energy sources and carriers that are based on renewable energy sources such as solar energy. In the opinion of many authors, if the ultimate goal is to develop a sustainable energy economy, the most promising candidate to play the role of storage and transport medium in the near future appears to be hydrogen, provided it is produced from renewable energy sources. In addition, hydrogen is mainly stored on Earth in water of any form; its combustion produces again water and its nuclear fusion the inert gas helium.

The author and visionary Jules Verne was the first to envisage the use of hydrogen as a fuel as long ago as 1874 to "furnish an inexhaustible source of heat and light". More than 50 years ago, NASA's program to send a man to the Moon was based on hydrogen fuel cell technologies, and nowadays we are planning to enter the so-called hydrogen economy. Furthermore, the use of hydrogen as a fuel in fusion nuclear reactors to exploit this limitless source of energy is a dream that originated in the 50s but has not yet been implemented. The enthusiasm with which we welcomed this hopeful new technology seems to be wasting away after waiting for so long for the renewed promise every other decade that within the next 30 years we will finally succeed in commercial fusion. Unfortunately, this promise has not yet been fulfilled (Malik [1]).

The anticipated transition to the *hydrogen economy* has raised many concerns as regards hydrogen's safe production, transport, storage, and use, among them environmental concerns. Although it is true that hydrogen has been used safely for many decades, this has occurred until now in activities mainly in the chemical industry, where skillful and highly trained personnel are engaged. No one is certain what would happen when a layperson handles a potentially hazardous material such as liquid hydrogen to refuel our cars. Such thoughts are sustained by the fact that there is still a significant shortage of knowledge on hazardous properties of hydrogen, for

DOI: 10.1201/9781003313007-14

instance on ignition chemistry, deflagration-to-detonation transition (DDT), storage materials, compatibility with other materials, and under extreme conditions of storage and use. Even accident prevention and mitigation measures have not been definitively determined yet, and suggested active measures, such as forced ventilation or water mists, may not be as efficient as expected, even if they do not aggravate the situation. Skepticism is also emerging on the environmental effects of large hydrogen leaks if hydrogen becomes extensively used worldwide (Rigas and Amyotte [2]).

Despite the great progress achieved in hydrogen safety over recent decades, hydrogen utilization still faces many challenges, and countries, regulatory bodies, industries, and research institutions have been working together to develop the technological means to implement the safe and feasible utilization of hydrogen in everyday life.

A summary of critical safety challenges to be addressed on this matter in view of the upcoming hydrogen economy is as follows (Malik [1], Chatsko [3], Lau [4], Morrison [5]):

- Hydrogen properties: Hydrogen has a low energy density, but also poses unique safety challenges such as low ignition energy, wide flammability range, high diffusivity, and construction material compatibility issues.
- Knowledge gaps: There are still some uncertainties and knowledge gaps regarding hydrogen behavior, detection, mitigation, and risk assessment in various scenarios and applications.
- Safety measures: Several initiatives and programs have been launched to develop codes, standards, best practices, and guidelines for safe hydrogen production, storage, distribution, and utilization.
- Public perception and education: The successful adoption of hydrogen technologies relies on public acceptance and trust. Educating the public about the safety aspects of hydrogen and addressing any misconceptions or concerns is essential to building confidence in the hydrogen economy.

Concerning public perception, the emerging hydrogen economy might be very sensitive to major incidents, as such severe events could be capable of rapidly diverting the direction of technological development. History has taught us that major accidents like the Hindenburg explosion (1937), Three Mile Island (1979), Chernobyl (1986), and Fukushima (2011) can derail the technological progress in a growing market. It would be unrealistic to take for granted that no such major incident will occur with the anticipated widespread use of hydrogen. Thus, beyond improved technological tools such as effective hydrogen detectors and safeguard systems, continual education of the public, local regulators, frontline workers, and first responders would also be crucial to success. For the hydrogen economy to be successful, we need to offer a prolonged record of safe and low-risk operations (Malik [1]).

In general, hydrogen should not be considered more dangerous or safer than other energy carriers. It is simply different, thus requiring specific safety knowledge and skills at all stages, starting from the design of hydrogen systems through the certification, permitting, and commissioning of the new safety culture in the use of these technologies by the public. The progress achieved so far cannot be interpreted as the disappearance of safety issues. New processes of hydrogen production, transportation, storage, and use of hydrogen systems are under development and will require both basic and applied research. The opinion that there are no longer safety issues for hydrogen technologies should be considered unprofessional and dangerously misleading to the public.

Regarding the use of hydrogen as a nuclear fuel, the IAEA [6] states that a fusion reaction can only take place under extreme conditions, so a *runaway chain reaction* is impossible. Furthermore, the IAEA asserts that fusion reactions depend on the continuous input of fuel, and since a fusion reaction will terminate by itself within seconds when input stops, the process is considered *inherently safe*. In any case, it looks incredibly difficult to *put the sun in a bottle*. In addition, environmental concerns arise over the probable accidental dispersion of tritium from fusion plants and the huge water demand of fusion reactors, which many countries would not physically be able to support taking into account climate change and the already-occurring drought conditions in some regions of the world.

The *global environmental consequences* of the anticipated extended practical and effective utilization of hydrogen through the emergence of the hydrogen economy are not fully understood yet. Consequently, much more research is needed to complete our knowledge of global phenomena related to the complex hydrogen interactions in the troposphere and the stratosphere in order to be fully aware of the hidden hazards of this colossal change in energy use. Transitioning from the "fossil fuel economy" to the "hydrogen economy" will be difficult. *There is a rough road ahead!*

This book is a contribution to basic knowledge on hazardous properties of hydrogen and the resulting challenges in view of the anticipated hydrogen economy, and has aimed to offer a comprehensive overview on the safe handling and storage of this new energy carrier.

REFERENCES

1. Malik, D.R., *Addressing Safety Challenges of hydrogen-driven future*, Bic Magazine, February 1, 2021. www.bicmagazine.com/departments/lift-transport/addressing-saf ety-challenges-of-hydrogen-driven-future/
2. Rigas, F. and Amyotte, P., *Hydrogen Safety*. 1st ed., CRC Press, Taylor & Francis Group, Boca Raton, 2013.
3. Chatsko, M., *The hydrogen economy faces big challenges*, TheStreet, September 26, 2022. www.thestreet.com/investing/the-hydrogen-economy-faces-big-challenges
4. Lau, C., *The rise of hydrogen economy and safety fears*, Schneider Electric Blog, August 8, 2023. https://blog.se.com/industry/energies-and-chemicals/2023/08/08/ focus-on-hydrogen-safety-aligns-with-growing-economy/
5. Morrison, G., *Addressing hydrogen safety challenges*, Gexcon, January 26, 2021. www.gexcon.com/blog/addressing-hydrogen-safety-challenges/
6. Willis., C., Liou, J., Safety in fusion, an inherently safe process, *Bulletin of IAEA International Atomic Energy Agency*, 62(2), May 2021. www.iaea.org/bulletin/safety-in-fusion

Index

Note: Page numbers in **bold** refer to tables and those in *italic* refer to figures. Endnotes are indicated by the page number followed by "n" and the note number e.g., 111n5 refers to note 5 on page 111.